Sustainable Economic Development

Sustainable Economic Development

Resources, Environment, and Institutions

Edited by

Arsenio M. Balisacan

Ujjayant Chakravorty

Majah-Leah V. Ravago

AMSTERDAM • BOSTON • HEIDELBERG • LONDON
NEW YORK • OXFORD • PARIS • SAN DIEGO
SAN FRANCISCO • SINGAPORE • SYDNEY • TOKYO

Academic Press is an imprint of Elsevier

Academic Press is an imprint of Elsevier
The Boulevard, Langford Lane, Kidlington, Oxford, OX5 1GB
525 B Street, Suite 1800, San Diego, CA 92101-4495, USA

First published 2015

Notices

Knowledge and best practice in this field are constantly changing. As new research and experience broaden
our understanding, changes in research methods, professional practices, or medical treatment may become
necessary.

Practitioners and researchers must always rely on their own experience and knowledge in evaluating and using any
information, methods, compounds, or experiments described herein. In using such information or methods they
should be mindful of their own safety and the safety of others, including parties for whom they have a professional
responsibility.

To the fullest extent of the law, neither the Publisher nor the authors, contributors, or editors, assume any
liability for any injury and/or damage to persons or property as a matter of products liability, negligence
or otherwise, or from any use or operation of any methods, products, instructions, or ideas contained in the
material herein.

British Library Cataloguing in Publication Data
A catalogue record for this book is available from the British Library

Library of Congress Cataloging-in-Publication Data
A catalog record for this book is available from the Library of Congress

ISBN: 978-0-12-800347-3

For information on all Academic Press publications
visit our website at **store.elsevier.com**

Printed and bound in the United States

15 16 17 18   10 9 8 7 6 5 4 3 2 1

For Jim Roumasset
Outstanding economist, mentor, colleague, and friend

Contents

SECTION 2   RESOURCES, ENVIRONMENT, AND SUSTAINABLE DEVELOPMENT

SECTION 4 THE NATURE, CAUSES, AND CONSEQUENCES OF AGRICULTURAL DEVELOPMENT POLICY

CHAPTER 14 The Role of Agricultural Economists in Sustaining Bad Programs

Brian Wright

CHAPTER 15 Agricultural R&D Policy and Long-Run Food Security

Julian M. Alston, William J. Martin and Philip G. Pardey

CHAPTER 16 Energy and Agriculture: Evolving Dynamics and Future Implications

Siwa Msangi and Mark W. Rosegrant

CHAPTER 17 Trends and Fluctuations in Agricultural Price Distortions 293
Kym Anderson

CHAPTER 18 Getting the Price of Thai Rice Right: Episode II........................ 311
Suthad Setboonsarng

SECTION 5 DEVELOPMENT, VULNERABILITY, AND POVERTY REDUCTION

CHAPTER 21 Deviant Behavior: A Century of Philippine Industrialization....... 371

Emmanuel S. de Dios and Jeffrey G. Williamson

CHAPTER 22 Bundling Drought Tolerance and Index Insurance to Reduce Rural Household Vulnerability to Drought 401

Travis J. Lybbert and Michael R. Carter

About the Editors

Arsenio M. Balisacan is the Socioeconomic Planning Secretary and Director-General of the National Economic and Development Authority of the Philippines. Prior to his appointment in the Cabinet of President Benigno Aquino III, he was Professor and Dean of the School of Economics of the University of the Philippines Diliman and, concurrently, Executive Director of the Philippine Center for Economic Development. He also served as the Director of the Southeast Asian Regional Centre for Graduate Study and Research in Agriculture (SEARCA) and Undersecretary of the Philippine Department of Agriculture. Elected as lifetime Academician of the National Academy of Science and Technology for his outstanding contributions to science and technology, he is Adjunct Professor at the Australian National University. His numerous publications cover wide areas of economic development, including poverty, inequality, human development, regional development, agricultural and rural development, and political economy of policy reforms.

Ujjayant Chakravorty is Professor of Economics at Tufts University and a Fellow at the Toulouse School of Economics and CESifo. He is co-editor of the Journal of Environmental Economics and Management. He works on the economics of fossil fuels and clean energy and on the management of water resources with an application to India and China. He has published in top economics journals such as the American Economic Review, the Journal of Political Economy, and Econometrica. He has a BS from the Indian Institute of Technology and a PhD in Resource and Environmental Economics from the University of Hawaii.

Majah-Leah V. Ravago is Assistant Professor at the University of the Philippines School of Economics. She is also currently the Director for Research since 2013. Her research interests include environmental and resource economics, development economics, and energy economics. She has co-authored papers on resource management and sustainability, agriculture, and experimental games. She is Principal Investigator of research grants from the Philippine Center for Economic Development and Metrobank Foundation. She has also received research grants from the University of the Philippines, University of Hawaii, East-West Center, and the Southeast Asian Regional Center for Graduate Study and Research in Agriculture (SEARCA). She received her PhD in Economics from the University of Hawaii in 2012.

About the Authors

Julian M. Alston is Professor in the Department of Agricultural and Resource Economics of the University of California, Davis, and serves as a Member of the Giannini Foundation of Agricultural Economics, Director of the Robert Mondavi Institute Center for Wine Economics, and Associate Director for Science and Technology Policy at the University of California Agricultural Issues Center. He is a Fellow of the American Agricultural Economics Association, a Distinguished Fellow of the Australian Agricultural and Resource Economics Society, a Distinguished Scholar of the Western Agricultural Economics Association, and a Fellow of the American Association of Wine Economists.

Kym Anderson is the George Gollin Professor of Economics at the University of Adelaide, Australia, and Professor of Economics in the Arndt-Corden Department of Economics, Crawford School of Public Policy, Australian National University, Canberra. He was on extended leave at the Economic Research Division of the GATT (now WTO) Secretariat during 1990–1992 and at the World Bank's Research Group as Lead Economist (Trade Policy) during 2004–2007. He has published more than 300 articles and 30 books, among the more recent being a set of four regional and three global books on *Distortions to Agricultural Incentives*. He earned his PhD from Stanford University in 1977.

Kimberly M. Burnett joined UHERO as a Research Economist in 2008. She has authored over a dozen peer-reviewed articles in scientific journals and books on a wide variety of environmental management issues. Her earlier work focused on the management of invasive species, including Miconia, the Brown Tree Snake, and the Coqui frog. Currently, she is developing system approaches for managing terrestrial, freshwater, and marine ecosystems. Recent extramural projects include work on abatement of nitrogen pollution in submarine groundwater discharge, optimal management of a coastal aquifer with nearshore marine ecological interactions, and integrated watershed and groundwater management.

Michael R. Carter is Professor of Agricultural and Resource Economics at the University of California, Davis. He is the author of numerous journal articles and has published several books on the fields of agriculture, development, and poverty dynamics. In 2003, he was awarded the Outstanding Article in the American Journal of Agricultural Economics. He has served as a Visiting Researcher and Professor in South Africa, Chile, and Peru, and is an Elected Fellow of the NBER (National Bureau of Economic Research) and the American Agricultural Economics Association. He obtained his PhD in Economics at the University of Wisconsin-Madison in 1982.

Ramon L. Clarete is Professor at the University of the Philippines School of Economics, specializing in food policy and international economics since 1989. He managed several trade capacity building projects for USAID in the Philippines. He worked on a technical assistance assignment for ADB on ASEAN rice policy (2010–2013). He co-authored *Trusting Trade and the Private Sector for Food Security in Southeast Asia*, a World Bank publication. He was in the Economics Department Faculty of the University of Western Ontario (1985–1988). He obtained his PhD in Economics at the University of Hawaii in 1984.

Emmanuel S. de Dios is Oscar M. Lopez Sterling Professor of Law and Economics at the University of the Philippines School of Economics, where he previously served as dean. He has published in the fields of institutions and governance, international economics, and the history of economic thought. As a book author and editor, he won the Outstanding Book Award in 2007, 2011, and 2014 from the National Academy of Science and Technology (Philippines). He received the Chancellor's Award for Outstanding Teacher in 2006 and was chosen as one of twelve Centennial Fellows of the university in 2007. He obtained his PhD in Economics from the University of the Philippines in 1987.

Lee H. Endress (PhD, Economics, University of Hawaii at Manoa, 1994) is an independent economist specializing in natural resource, energy, and environmental economics. He is also a member of the affiliate graduate faculty of the Economics Department at the University of Hawaii at Manoa. From September 1996 to October 2009, he served as Educational Program Director and then Academic Dean at the Asia-Pacific Center for Security Studies in Honolulu, Hawaii. Previously, he was Senior Economic Advisor to the Commander, US Pacific Command.

Manabu Fujimura is Professor at the College of Economics of the Aoyama Gakuin University. His research focuses mainly on the fields of development and international economics. He is the co-author of several publications on poverty and transport infrastructure. He also has extensive professional experience, having worked as an economist in the Asian Development Bank and the Japan External Trade Organization. Prior to teaching in Aoyama, he did research at the East West Center in Hawaii and at the University of Shizuoka. He earned his PhD in Economics from the University of Hawaii in 1994.

Raghav Gaiha is (Former) Professor of Public Policy at the University of Delhi Faculty of Management Studies. His book (jointly with Shylashri Shankar) *Battling Corruption: Has NREGA Reached India's Rural Poor?* was published by Oxford University Press in 2013. A second book (jointly with Raghbendra Jha and Vani S. Kulkarni) *Diets, Malnutrition and Disease: The Indian Experience* will be published by Oxford University Press in June 2014. He has also served as a Visiting Fellow and Professor at various institutions, including Harvard University, MIT, Stanford, Penn, Yale University, and the University of Cambridge.

Yazhen Gong is Assistant Professor at the School of Environment and Natural Resources at Renmin University of China. She is currently a post-doc fellow at UNESCO'S IHE (Institute for Higher Education on Water). She received her PhD in Forestry with a major in resource economics from the University of British Columbia (UBC) in Canada in 2010. She has a Master Degree in Forest Resource Management from the University of the Philippines and a Bachelor's Degree from the School of Economics and Management in Beijing Forestry University. Her current research interests focus on irrigation water, climate change and agriculture, and payment for environmental services.

Kenneth Hill is Professor of the Practice of Global Health in the School of Public Health, Adjunct Professor of Public Policy and Associate of the Malcolm Wiener Centre for Social Policy, Harvard. He is a demographer with primary interests in the measurement of mortality in developing countries and interpretation of mortality change. He obtained his PhD from the London School of Hygiene and Tropical Medicine.

Sirilaksana Khoman is Advisor to the Commissioner at the National Anti-Corruption Commission in Thailand and a Member of the World Economic Forum Global Agenda Council for Anti-Corruption and Transparency. She was formerly Dean of the Faculty of Economics of Thammasat University. She has also done work for several international organizations including the World Bank and the Asian Development Bank. Her current areas of research are the economics of corruption. She has taught at the Australian National University, the United Nations University in Tokyo, and the University of Oregon, USA. She obtained her PhD in Economics from the University of Hawaii and a Certificate in International Trade Regulation from the Harvard Law School.

Varsha S. Kulkarni is a doctoral student at Indiana University Bloomington, USA. Her research interests lie in the applications of mathematical models to study complex systems. Her work so far has dealt with financial systems although currently, she is also focusing on analyzing social and other complex networks to study a variety of problems. A recent article on analysis of inventory fluctuations will appear in Journal of Applied Nonlinear Dynamics.

Sumner La Croix is Professor in the Department of Economics of the University of Hawaii-Manoa, an affiliate faculty member of the UH-Manoa Center for Chinese Studies and the UH-Manoa Center for Japanese Studies, and a research fellow with the University of Hawaii Economic Research Organization. His research focuses on the economic history and development of economies in the Asia-Pacific region, with an emphasis on issues pertaining to institutional change, property rights in land, and intellectual property rights. He has been co-organizer of the annual Cliometrics Conferences and Congresses since 2009. He earned his PhD in Economics from the University of Washington in 1981.

Travis J. Lybbert is Associate Professor at the Department of Agricultural and Resource Economics of the University of California, Davis. He has published extensively in applied microeconomics on topics ranging from poverty dynamics and vulnerability to technology adoption and technology policy. He was awarded the Outstanding American Journal of Agricultural Economics Article (2010) and the Quality of Research Discovery Award (2011) for work in experimental economics on risk aversion. He has spent time as a Visiting Researcher in Switzerland, Ghana, Germany, and Sweden. He was a Fulbright Scholar in Morocco before earning his PhD in Applied Economics from Cornell University in 2004.

William J. Martin is Research Manager, Agriculture and Rural Development, in the Development Research Group of the World Bank. Before joining the World Bank, he worked as a Researcher and Manager at the Australian Bureau of Agricultural and Resource Economics and as a Senior Research Fellow at the Australian National University. He has published extensively on agricultural trade policy and developing countries, with a particular focus on the World Trade Organization and economic development. He has a particular interest in using detailed data to build up a complete picture of the effects of policies on welfare impacts at national and household levels.

Siwa Msangi is a Senior Research Fellow in the Environment and Production Technology Division of IFPRI and works across a broad range of topics covering the major socio-economic and bio-physical drivers affecting agricultural production and trade and their impacts on nutrition, poverty, and the environment. His research portfolio includes the economic and environmental dimensions of sustainable agricultural intensification, the bio-economics of aquaculture and livestock, bioenergy, adaptation to climate change, and water resources management. He joined IFPRI in August 2004, after obtaining his doctorate in agricultural and resource economics at the University of California, Davis.

Philip G. Pardey is Professor of Science and Technology Policy in the Department of Applied Economics at the University of Minnesota, where he also directs the University's International Science and Technology Practice and Policy (InSTePP) Center. He is a Fellow of the American Agricultural Economics Association, Distinguished Fellow and Past President of the Australian Agricultural and Resource Economics Society, and winner of the Siehl Prize for Excellence in Agriculture. His research deals with productivity measurement and assessment, the finance and conduct of R&D globally, methods for assessing the economic impacts of research, and the economic and policy (especially intellectual property) aspects of genetic resources and the biosciences.

Sittidaj Pongkijvorasin obtained his PhD in Economics from the University of Hawaii at Manoa in 2007. He is currently Assistant Professor at Faculty of Economics, Chulalongkorn University, Thailand. His research interests are in the environmental and resource economics field, involving both theoretical and empirical contributions. He has many papers published in leading journals, for example, Environmental and Resource Economics, Water Resource Research, and Resource and Energy Economics. He was awarded the Outstanding Teacher from the Association of Southeast Asian Institutions of Higher Learning, Thailand in the field of Social Sciences in 2011, as recognition of his distinguished work with students at the Faculty of Economics.

Mark W. Rosegrant is Director of Environment and Production Technology Division, International Food Policy Research Institute (IFPRI). He has extensive experience in research and policy analysis in agriculture and economic development, with an emphasis on water resources, agricultural policy, and global modeling of food supply and demand. He is the author or editor of 12 books and over 100 refereed papers in agricultural economics, water resources, and food policy analysis. He has won numerous awards, is a Fellow of the American Association for the Advancement of Science, and a Fellow of the Agricultural and Applied Economics Association.

James A. Roumasset is Professor of Economics at the University of Hawai. His research interests include water and energy economics, institutional economics, development microeconomics, sustainable development, and public policy. His recent publications and extramural grants have focused on new methods for resource and environmental policy analysis and promoting connections between and within disciplines. He has held positions at UC Davis, University of the Philippines, Australian National University Yale, University of Maryland, the International Rice Research Institute, and the World Bank. He serves/served on the Board of Directors of Western Economic Association International and the editorial boards of several scholarly journals, including the American Journal of Agricultural Economics and Contemporary Economic Policy.

Suthad Setboonsarng is currently one of the Members of the Board of the International Rice Research Institute. In 2009, he was appointed as one of Thailand's Trade Representatives, following his assignment as an ASEAN partner at PriceWaterhouseCoopers from 2000 to 2008. Before joining the private sector, he was a Deputy Secretary General of ASEAN. While at the ASEAN Secretariat, he pioneered the work on customs cooperation and financial cooperation. He was Professor at the Asian Institute of Technology, Research Fellow at the Thailand Development

Research Institute, Lecturer at Thammasat University, and conducted postdoctoral work at Yale University. He obtained his PhD in Economics at the University of Hawaii in 1983.

Nori Tarui is Associate Professor at the Department of Economics, University of Hawaii, where he has been teaching since 2006. Previously, he was Fellow and Instructor at Columbia University and a Visiting Lecturer at the University of Wisconsin, River Falls. His research interests include environmental and resource economics, applied microeconomics, and applied game theory. He is the author of several publications, book chapters, and working papers. He is also the recipient of a number of grants from both public and private institutions. He earned his PhD in Agricultural and Applied Economics from the University of Minnesota in 2004.

Khemarat Talerngsri Teerasuwannajak is Lecturer in Economics at Faculty of Economics, Chulalongkorn University. Her main interests span from theoretical and applied analysis of competition policy, R&D policy to coordination issues in multiagent environments. Although her expertise and publications are in the area of industrial organization, in the past three years she has been highly engaged in fieldwork focusing on investigation of escalating conflicts between sustainable use of natural resources and poverty reduction policy in Nan, an important watershed area of Thailand.

Ganesh Thapa is the Regional Economist of Asia and the Pacific Division of the International Fund for Agricultural Development (IFAD) based in Rome, Italy. Prior to working for IFAD, he worked as Country Director/Programme Leader for Winrock International in Nepal and as Senior Agricultural Economist for the Ministry of Agriculture, Nepal. He has undertaken research and published journal articles, monographs, and books on a number of topics related to agriculture and rural development, rural poverty, sustainable natural resource management, food security, and agricultural marketing. He has a PhD in Agricultural Economics from Cornell University.

Marilou Uy is Senior Adviser to the Office of the President's Special Envoy on the Millennium Development Goals and Financial Development at the World Bank. Previously, she served as Sector Director for the Africa Region's Financial and Private Sector Development Department from 2007 to 2011 and Director of the Financial Sector Operations and Policy Department in the Financial Sector Vice-Presidency as well as Chair of the Financial Sector Board from 2002 to 2007. She was also part of the World Bank's Development Economics Group's research team that prepared "The East Asian Miracle" in 1991, in which she focused on financial sector issues, together with Joseph Stiglitz. Prior to joining the World Bank, she pursued her graduate studies in economics and finance at the University of California in Los Angeles.

Christopher A. Wada is Research Economist at the University of Hawaii Economic Research Organization (UHERO). He earned a PhD in Economics from the University of Hawaii in 2010 and joined UHERO shortly thereafter. He has co-authored a number of journal articles, book chapters, and technical reports on a variety of topics in the realm of resource and environmental economics. His current research focuses on the economics of groundwater and watershed management.

Peter Warr is John Crawford Professor of Agricultural Economics and Head of the Arndt-Corden Department of Economics, Crawford School of Public Policy, College of Asia and the Pacific, at the Australian National University. He obtained a PhD from Stanford University. His current research interest is on the relationship between economic policy and poverty incidence in Southeast

Asia. Much of this work uses general equilibrium models, specially adapted to measure changes in poverty incidence. He is a Fellow of the Academy of Social Sciences in Australia and President of the Australian Agricultural and Resource Economics Society.

Upali Wickramasinghe is Regional Adviser on Poverty Reduction and Food Security attached to the United Nations Centre for Alleviation of Poverty through Sustainable Agriculture (CAPSA). He has held several positions in the past, including Professor of Economics of the University of Sri Jayewardenepura in Sri Lanka, a Policy and Programme Development Consultant at the FAO Regional Office for Asia and the Pacific in Bangkok, and Economic Adviser of the South Asian Association for Regional Cooperation (SAARC) secretariat in Kathmandu. He holds a PhD in Economics from the University of Hawaii.

Jeffrey G. Williamson served as Harvard University's Chairman of the Economics Department (1997−2000) and as the department's Director of Undergraduate Studies (2001−2002 and 2004−2005). He is now the Laird Bell Professor of Economics, Emeritus, Honorary Fellow in the Department of Economics at the University of Wisconsin (Madison), Research Associate at the National Bureau of Economic Research, and Research Fellow for the Center for Economic Policy Research. He earned his PhD in Economics from Stanford University in 1961.

Brian Wright is Professor at the Department of Agricultural and Resource Economics at the University of California, Berkeley. His research interests include economics of markets for storable commodities, market stabilization, agricultural policy, industrial organization, public finance, invention incentives, intellectual property rights, the economics of research and development, and the economics of conservation and innovation of genetic resources. He is also the author and editor of several books and has published extensively in the leading journals of economics and agricultural economics. He was awarded one of two Frank Knox Fellowships for Australian students by Harvard University, where he received his AM and PhD in Economics.

Foreword

The world has been changing rapidly and, consequently, the issues we are facing are changing dramatically. Moreover, they are far more complex than before. For example, when the fear of future famine was keenly felt in Asia in the 1960s, the solution was to increase crop yields, particularly rice yields, by developing fertilizer-responsive, high-yielding varieties, applying more chemical fertilizer, and using irrigation water. The emission of methane due to the inappropriate use of chemical fertilizer and the massive use of surface and groundwater were not serious issues at that time. At present, however, methane emission from paddy fields is an important factor affecting climate change, and water is becoming a major scarce resource in most economies in Asia. Furthermore, the price of chemical fertilizer is likely to increase in the future due to a rise in production costs resulting from curtailing the use of fossil fuels to reduce greenhouse emissions. Also, agricultural land may be used to produce biofuels instead of food grains to replace fossil fuels. In addition, adaptation to climate change is an emerging issue we have to address to mitigate the adverse impacts of such a change. Finally, we have to tackle the issue of poverty reduction in the face of increasing resource and environmental constraints on economic development.

As the title of this book squarely indicates, the most critical issue is how to achieve *Sustainable Economic Development*. In order to achieve *environmentally* sustainable development, we have to use and allocate increasingly scarce natural resources optimally from the global viewpoint, including not only land, water, and forests but also the atmosphere without the excessive accumulation of greenhouse gases. Appropriate institutions and governance systems must be in place to support the realization of such resource allocation, ranging from property rights institutions to land reform laws and support policies for large-scale commercial agriculture. Since agriculture is the industry most vulnerable to climate change and also the majority of the poor depend on agriculture for their livelihood, special considerations should be given to policies conducive to the sustainable development of agriculture. In order to achieve *socially* sustainable development, poverty must be reduced by creating lucrative jobs for the poor in both farm and nonfarm sectors. Relevant policies include labor-intensive industrialization, fair agricultural trade, weather-based index insurance, and speedy relief against natural disasters. The critical linkage between agriculture and energy use and the critical role of agricultural R&D in long-run food security must also be taken into account in achieving such socially sustainable development.

While it is easy to point out the principles to achieve sustainable economic development, it is quite another task to provide evidence-based appropriate policy recommendations. Indeed, coherent, comprehensive, and systematic policy recommendations for sustainable development are lacking in the development economics literature despite the huge and increasing demand for them. It is time for economists and policymakers to seriously contemplate what should be done. I believe that for this purpose it is best to review, refine, and synthesize the great ideas in the past and modern times that have had some bearing on academic and policy thinking. Also indispensable is a proper understanding of the changing empirical realities.

The purpose of this book is precisely to undertake such endeavors by focusing on the economics of resource use, environments, and institutions with a view to integrating them for the formulation of appropriate policy recommendations. It is fortunate that in this volume so many prominent economists in relevant fields have made concerted efforts to contribute to the common objective of deriving policies to achieve sustainable economic development. I have no doubt that readers can learn immensely from this volume about how resource use and management, climate change, and governing institutions are interwoven to shape the paths of economic development and what policies are likely to be most effective in promoting truly sustainable economic development.

Keijiro Otsuka
National Graduate Institute for Policy Studies, Japan

Preface and Acknowledgments

Development research and policy analysis must continually adapt to changing conditions as nations adjust to new constraints and challenges, including poverty, degradation of the environment and natural resources, globalization, and increasingly complex "rules of the game" in trade, finance, and management of the global commons. These changing circumstances have demonstrated the need for the advancement of policy analysis that explicitly focuses on the design of new institutions and the efficient as well as sustainable management of natural as well as produced capital.

The chief objective of this volume is to refine and synthesize classical and modern ideas in development and policy analysis. The title of this volume has gone through several rounds of discussion. Sustainable economic development (SED) in this volume, as we have articulated in Chapter 1, departs from the conventional "less is more" version. SED expands the conventional view of the economy to the *environomy*—the integrated economic and environmental resource system. Whereas the popular view often confuses sustainability with self-sufficiency and espouses renewable energy at almost any cost, our version of SED promotes the efficient use of resources and the accumulation of total capital for the advancement of human welfare, including poverty reduction.

The chapters include policy analyses, state-of-the-art reviews, reports of new empirical results, and conceptual explorations. The chapter contributors are among some of the most eminent academics and policy practitioners in the world, including professors and economists from Harvard University, the University of California, Berkeley and Davis, the World Bank, the International Food Policy Research Institute, and officials from other research and policy institutions. The volume aims to capture the interest of development practitioners and professionals, policy makers, and development academics.

This volume also serves as a *Festschrift* for James A. Roumasset—professor, mentor, and friend. It pays tribute to his work and honors him as an outstanding economist. The breadth and depth of this volume is reflective of the impetus that Jim provided to his former students and of the common interest he has shared with his colleagues. From the areas of economic development policy, behavior, and organization in agriculture, to environmental resources such as energy and water, this volume contributes to providing guidance to the design of policy and institutional reform aimed at achieving sustained growth and development. The contribution of scientifically sound yet applied research is paramount for good policy making. This is what we have learned from Jim Roumasset.

Jim was, and still is, ahead of his time. Many of the problems he identified in the early days of his academic career are major research challenges today. Back in the day when biotechnology was still unheard of, he was already studying the benefits and costs of the adoption of Azolla for Philippine Rice Production.[1] Jim thought about fossil fuel use and global warming long before the issue became a global concern.[2] Born at the time of "wartime prosperity" in the United States, his

---

[1] See his paper "Technology and Agricultural Policy: An Assessment of Azolla for Philippine Rice Production" (with A. Balisacan and M. Rosegrant), *American Journal of Agricultural Economics*, 67(4), 726–732, November 1985.

[2] He later obtained a National Science Foundation grant on energy, global change, and sustainable development (1992–1994) that led to a string of papers co-authored with students and former students, e.g., "Endogenous substitution among energy resources and global warming" (with Ujjayant Chakravorty and Kinping Tse), *Journal of Political Economy*, 105(6), 1201–1234, 1997 and the 2005 paper on optimal and sustainable development (with Lee Endress and Ting Zhou) referenced in Chapters 1, 3, and 5 of this volume.

predilection toward an academic career was most likely influenced by his parents. Charles headed the Bureau of Labor Statistics office for the Western US, including Hawaii, and Helen was an elementary school teacher. Jim's first role model was Carlmont High School English teacher, John Durham, a brilliant non-conformist who also pointed Jim in the direction of his now wife, Jane Kirton. His next biggest influence was Peter Diamond, who showed how teaching policy economics leads to original thought.

Jim joined the Peace Corps after college and was assigned to the province of Catanduanes in the Philippines, where he and a group of seasoned teachers developed a *New Math* curriculum for grades one to six and implemented a training program for a thousand public school teachers. He extended his Peace Corps stint to serve as Agricultural Information Officer for the Bicol Development Planning Board. These experiences started his romance with the Philippines and with agricultural development. He would later recall: "I learned from my own mistakes about the foolishness of self-sufficiency schemes and puzzled about the importance of extra-market forces such as investment interdependencies. There is nothing like realizing the limitations of your own understanding to motivate further study."

He then moved to Hawaii, receiving an MA from the University of Hawaii in 1969 and then on to Wisconsin for a PhD in 1973 (both in Economics). His dissertation "Risk and Choice of Techniques for Peasant Agriculture and the Case of Philippine Rice Farmers" won First place in the 1973 Omicron Delta Epsilon Society competition for best US PhD dissertation in Economics. This work was the basis of his influential book *Rice and Risk: Decision-Making among Low-Income Farmers in Theory and Practice*, in the North-Holland series Contributions to Economic Analysis, edited by Dale Jorgenson.

From 1972 to 1976, Jim taught Public Economics, Econometrics, and Uncertainty at the University of California, Davis. However, his Hawaii connections and the quasi-Asian culture beckoned. He returned home in 1976 and has been with the University of Hawaii ever since.

Jim took a leave of absence from 1978 to 1979 to direct the Agricultural Development Council program in the Philippines, where he started research programs in the New Institutional Economics and the analysis of agricultural policies with students and young faculty members at the University of the Philippines in Los Baños and Diliman. He continued this work in the early 1980s through Honolulu's East-West Center. This presented him an opportunity to bring in students, mostly from Asia, to work as research interns and at the same time study development economics. He would later refer to this decade as the "golden 80s." His students from this and later eras firmly established themselves in academe, government, and international organizations. Jim had and continues to have a knack for spotting excellent students. To date, he has supervised 27 students at the University of Hawaii (UH) and another nine students from other departments of UH and other universities. A number of these students are contributors to this volume.

This volume is a product of an almost decade-long conceptualization, planning, and coordination that started in 2005. The conference on "Risk, Resources, Governance and Development: Foundations of Public Policy" held at the Imin Conference Center in Honolulu, Hawaii on December 7–8, 2012, sealed the commitment of our colleagues to contribute to this effort. The conference discussions and the subsequent papers have shaped much of the form that this volume has taken. Each chapter focuses on a subject to which Jim has made major contributions and identifies possible directions for future academic and policy research.

We especially thank the chapter authors whose support and enthusiasm made this endeavor a success. Our appreciation also goes to the other conference speakers and session chairs James Wilen, Shankar Sharma, Nancy Lewis, Denise Konan, Inessa Love, Byron Gangnes, and John Lynham, whose focus and stimulating discussion advanced the conference objectives. We thank Keijiro Outsuka for readily accepting our request to write the foreword.

A huge thank you to the people and agencies that believed in this effort and provided financial support to this venture. We thank the US Agency for International Development (USAID), Philippines, through the Climate Change and Clean Energy Project for providing funding support to the conference and the production of this volume. Organizing the conference in 2012 was expensive and would not have been possible if not for the generosity of our sponsors—Southeast Asian Regional Center for Graduate Study and Research in Agriculture (SEARCA), University of Hawaii College of Social Sciences (UH-CSS), East-West Center (EWC), International Food Policy Research Institute (IFPRI), University of Hawaii Economic Research Organization (UHERO), University of Hawaii Economics Department (UH Econ), and the UPecon Foundation, Inc. Many thanks also to the following for facilitating the support of their respective agencies: Gil Saguiguit, Director of SEARCA; Denise Konan, Dean of CSS; Nancy Lewis, Director for Research at EWC; Mark Rosegrant, Director, Environment and Production Technology Division at IFPRI; Carl Bonham and Kimberly Burnett, Executive Director and Associate Specialist, respectively, at UHERO; Byron Gangnes, Chair of the Department of Economics at UH; and Ramon Clarete, Chair of the UPecon Foundation, Inc. Thanks also to chapter authors Marilou Uy and Suthad Sethboonsarng for generous support from their personal funds.

The conference and this volume took form due to the concerted effort of the people at the East-West Center, UH Department of Economics, and the UP School of Economics. Chasuta Anukoolthamchote from the Department of Economics at UH provided coordination and assisted in travel logistics for some of the invited speakers. Administration and conference logistics handled by EWC were very ably managed by Carolyn Eguchi, who was in turn assisted by Laura Moriyama. Christine Ablaza, Ma. Christina Epetia, and Rosemarie San Pascual from UPSE provided administration and research assistance during various stages of the project. Reuben Campos designed the conference website, Lily Tallafer provided preliminary editorial assistance, and Rowena Sancio's organizational skills and attention to detail were invaluable assets in bringing the chapters to Elsevier.

We thank our publisher, Elsevier Inc., and the people behind it who made the publication of this volume possible: J. Scott Bentley, Senior Acquisitions Editor, Melissa Murray and Mckenna Bailey, Editorial Project Managers, Lisa Jones, Senior Project Manager, and the rest of the publications team.

Arsenio M. Balisacan, Ujjayant Chakravorty, and Majah-Leah V. Ravago
Manila and Boston
August 2014

INTRODUCTION AND SYNTHESIS

THE PRINCIPLES AND PRACTICE OF SUSTAINABLE ECONOMIC DEVELOPMENT: OVERVIEW AND SYNTHESIS

Majah-Leah V. Ravago*, **Arsenio M. Balisacan***,†, **and Ujjayant Chakravorty**‡

**University of the Philippines, Diliman, Quezon City, Philippines* †*National Economic and Development Authority, Pasig City, Philippines* ‡*Department of Economics, Tufts University, Medford, MA*

CHAPTER OUTLINE

JEL: P48; Q00; O43; O13; Q18; I30

1.1 INTRODUCTION

The past half-century has seen historically rapid growth and development in many parts of the world. Both the pace of economic growth and poverty reduction have never been faster during this period than in comparable periods of economic history. Yet, the sustainability of economic development has never been more intensely questioned. The questions have gone beyond the threat to environmental resources to include increasing vulnerability to climate change and biodiversity loss.

Good policy-making founded on scientifically sound and practical research can go a long way in sustaining economic development. This volume aims to contribute in this respect. Cognizant of the gap between the demands for sustainable development policy analysis (e.g., as articulated in Rio + 20) and the operational utility of the narrow sustainability concepts, this volume starts with the core model of how to balance the management of produced and natural capital and then adds features of development, policy and institutions, uncertainty, and political economy. We discuss public policy in the context of sustainable development in the broadest sense, incorporating resource management, institutional governance, and poverty reduction.

This chapter provides a summary to this volume, which pays tribute to the work of James Roumasset in the areas of economic development policy, behavior and organization in agriculture, and environmental resources such as energy and water. The authors in this volume are either his longtime collaborators and colleagues or his former students and associates, many of whom are making their own contributions in applying sound economic principles toward understanding problems of development, agriculture, environment, and governance.

Section I sets the tone of the book. Chapter 1 presents the volume overview and synthesis by introducing the principles and practice of sustainable economic development. Chapter 2 by Roumasset focuses on the foundations of development policy analysis. He discusses issues with misplaced exogeneity and incomplete conceptual mechanisms and prescribes fundamental analysis of the behavioral and organizational foundations of agricultural and rural development. This assessment points to promising avenues for future research, including *black hole economics* and the coevolution of specialization and economic development. The focus on agricultural development also provides a preview to Sections V and VII.

1.2 RESOURCE MANAGEMENT AND SUSTAINABLE DEVELOPMENT

Long before the 1987 Brundtland Report "Our Common Future" from the World Commission on Environment and Development (WCED), there was "Scarcity and Growth" (Barnett and Morse, 1963). The latter may be the earliest formal articulation of sustainability and concerns about the environment. The ensuing decades saw the continued escalation of concerns about the sustainability of economic growth in the face of natural resource depletion and environmental pollution. In the 1980s, the World Conservation Strategy (WCS) was formed with the objective of integrating economic and environmental management. However, WCS was unsuccessful in conveying either how poor economic policies would degrade the environment or how conservation requires coordinated economic policies (Pearce et al., 1989). The United Nations responded and formed the Brundtland Commission in 1983, enjoined to investigate the interrelationship between human activity and the environment and their implications for economic and environmental policy.

It was the Brundtland Report (WCED, 1987) that successfully established sustainability as a critical part of economic development policy. The Commission defined sustainability as "... development that meets the needs of the present without compromising the ability of future generations to meet their own needs." This rather vague definition has been the source of considerable contention (see Ravago et al., 2010 for the various references cited therein and Chapter 3, of this text). Nonetheless, the Brundtland report has increased the awareness of the importance of the interlinkages between the economy and its dependence on natural resource systems as well as a sense of stewardship for the future and the environment.

In the field of economics, a consensus regarding *sustainable growth* has also been elusive. Arrow et al. (2004) have espoused an expanded version of the *weak sustainability* criterion, specifically that the wealth of a society, including human capital, knowledge capital, and natural capital (as well as produced capital), does not decline over time. On the other hand, Barbier (2007) and others maintain that *strong sustainability*—nondepletion of essential stocks of natural resources—may be more suitable. This lack of consensus presents a challenge to the emerging transdisciplinary field of *sustainability science*, which began in the early 2000s. Sustainability science arises from the recognition that appropriate science and technology must be utilized in pursuit of sustainable development (Clark and Dickson, 2003; Roumasset et al., 2010). This discipline is being espoused by a number of international scientific organizations, chief among them are the U.S. National Academy of Sciences (NAS), The American Association for the Advancement of Science (AAAS), and the National Science Foundation (NSF).

The weak and strong sustainability criteria have been dubbed as *negative sustainability* (Ravago et al., 2010) inasmuch as they only proscribe limits on natural resource exploitation. They do not offer any guidance on what should be the optimal conservation patterns of natural capital nor the optimal buildup of human and produced capital. *Positive sustainability* (see Chapter 3; Endress et al., 2005; Ravago et al., 2010) fills this lacuna by maximizing intertemporal welfare while incorporating system linkages, dynamic efficiency, and intertemporal equity. The requirements for positive sustainability, reviewed in Chapter 3, provide for a solution that is optimal as well as sustainable.

The view of economic development in this volume includes sustainable development. Accordingly, Section II expands the conventional view of the economy to the *environomy*—the integrated economic and environmental resource system. Chapters in Section II expound and expand the basic principles of natural resource management and the links between environmental resource management and sustainable development. They also provide specific examples of sustainable development. Chapter 3 by Endress not only reviews the principles but suggests how sustainability can be extended to specific sectoral issues such as the management of renewable energy. The rest of the chapters in Section II discuss more specific resource management issues and highlight policies that move markets and resource -use patterns toward sustainable development.

Chakravorty and Gong's Chapter 4 discusses the economics of fossil fuels and pollution. They focus on pollution coming from coal, oil, and gas. These resources are nonrenewable in the sense that their use subtracts from their future availability. Limiting pollution from these low-cost sources of energy requires economic policies that make them more expensive and that induce new energy efficient technologies and the substitution of renewable energy sources such as solar and wind power.

Roumasset and Wada's Chapter 5 reviews the principles of groundwater management, including their extension to the optimal coordination of multiple aquifers, groundwater substitutes such as desalination and wastewater recycling, and investments in watershed conservation. By providing principles of pricing and finance, they show how sustainable development can be extended to the management of resource systems instead of a single resource. Chapter 6 by Burnett et al. applies the principles of coordinated resource management to the specific example of managing an invasive tree species that interferes with groundwater recharge. At the optimum, the marginal cost of increasing recharge via watershed conservation should be equal to the shadow value of groundwater in consumption.

Policies that move markets and resource-use patterns toward optimal and sustainable outcomes are exemplified by Pongkijvorasin and Teerasuwannajak in Chapter 7. Green subsidies and coordinated irrigation development are offered as a win−win solution to the problem of maize farming and rapid deforestation in the province of Nan in northern Thailand.

1.3 INSTITUTIONS, GOVERNANCE, AND POLITICAL ECONOMY

Sustainable economic development incorporates sustainable growth and dynamically efficient development patterns. In addition, sustainable development must take into account the lessons from development theory, including how optimal patterns of production, consumption, and trade change with standards of living. While science and technology are commonly employed in the pursuit of sustainability, other factors such as institutions, governance, and political economy are also critical dimensions of sustainable development. This is the focus of Section III: the political, institutional, and policy dimensions of sustainable development.

Deficient institutions, governance, and political economy are typical pathways to unsustainable development. Section III considers the coevolution of institutional arrangements and the environomy. Tarui's Chapter 8 reviews the literature on managing natural resources under different institutional arrangements and provides a model of resource governance that allows for the optimal transition between open access, common property management, and private property, given the transaction cost properties of those institutions. The rest of the chapters in Section II take on specific country experiences in Africa, Asia, and the Pacific.

Chapter 9 by Ravago and Roumasset continues the theme of institutions and resource management by examining the resource curse but generalizing the phenomenon to include other foreign exchange booms that may lead to distortionary policies, using the Philippines as an illustrative example. The focus is on two specific curses of abundance—contraction of the sector with the most growth externalities and the increase in unproductive rent-seeking. The lessons from this chapter are revisited in Chapter 21.

Governance and institutions evolve together. In Chapter 10, Uy considers the opportunity presented by vast amounts of fallow land in Africa and the difficulties of governing foreign investment to promote economic development and food security. Uy discusses mechanisms to coordinate interdependent private and public investments with local interests and knowledge to provide a win−win outcome for investors and communities. Uy also explores ideas for using private investors as sources of improved infrastructure and "governance beyond government."

Relatedly, land policy has been argued to be a critical ingredient of strong economic growth, e.g., land reforms in East Asia. La Croix's Chapter 11 reviews the history of large-scale land confiscations in early modern Europe, the United States, and Hawai'i to provide a foundation for understanding the nature of modern land reform policies. The takeaway from this chapter is the recognition that East Asian states after World War II were *limited access orders* (as opposed to the more democratic *open access orders*) in which the strategy of the new government is to confiscate and redistribute property to strengthen their coalition's position.

Economies that have undergone civil wars and conflicts are also likely to be limited access orders. These economies require a particular set of policy and institutional reforms that contribute to nation-rebuilding. Another strategy is the integration of these economies within a region to help facilitate nation-building. Integrating fragile nations, however, may promote illicit sectors of the economy. Fujimura's Chapter 12 investigates the political economy of narco-nations in the context of Afghanistan and Myanmar, the two largest sources of illicit drugs in Asia. Chapter 12 sets out requirements for benefit−cost analysis of imperfect prohibition and enforcement policies versus regulated legalization.

Khoman's Chapter 13 examines the issue of corruption and governance with particular emphasis on the experience of Thailand. This chapter traces the underlying structural shifts of the country as a result of the 1997 economic crisis to the present time, pointing out the irony of the promising governance trends and the unintended effects of the reformist constitution. This chapter focuses on network relationships that underlie governance failures and suggests possible ways to combat the new forms of corruption.

1.4 THE NATURE, CAUSES, AND CONSEQUENCES OF AGRICULTURAL DEVELOPMENT POLICY

The paradigm of *nature, causes, and consequences* has been successfully developed and applied to agricultural development policy. While this paradigm remains relevant, a whole set of problems have emerged in recent decades, ranging from climatic uncertainties to the scarcity of environmental resources to the survival and advancement of millions of poor people in developing economies. These changing circumstances have demonstrated the need to advance policy analysis for agricultural and rural development, including the design of new institutions and the sustainable management of energy and environmental resources.

Section IV applies and elaborates concepts from Chapter 2 to specific aspects of agricultural policy, thus continuing the theme of the importance of agriculture to food security, poverty reduction, and overall economic development. In Chapter 14, Wright turns to the role of agricultural economists in sustaining bad programs. While a good number of economists have staved off many bad policy proposals, initiated beneficial modifications of others, and seized propitious moments to initiate wholesale agricultural policy transformation, a few others have provided research that places a veneer of respectability over indefensible policies. Wright cites crop and disaster programs as discouraging examples. The financial clout of private insurers and the farm lobby has sometimes diverted economists from the pursuit of the public interest. In particular, institutions that incentivize research on the regulation of the rampant moral hazard and adverse selection of these programs are needed.

Chapter 15, by Alston et al., explores the role of agricultural R&D and related policies in contributing to or mitigating the consequences of variability in agricultural production. This chapter places particular emphasis on the slowdown in agricultural productivity growth in the face of a growing global population combined with the rise of biofuels, the changing climate, and the potential for agricultural innovations to address the attendant challenges for global food security.

Msangi and Rosegrant's Chapter 16 discusses the critical linkages between agriculture and energy and explores the implications for resource management and pricing. Drawing from the key examples of interlinkages in groundwater, fertilizer use, and biofuels, it shows how the future dynamics of agriculture and energy markets are likely to evolve, indicating relevant entry points for policy intervention. This chapter recommends improving use and conversion efficiencies, removing distortions to resource prices, and allowing for freer movement of goods through liberalized trade.

Chapters 17–19 examine the nature and consequences of agricultural price policies. Anderson's Chapter 17 assesses the distortions to agricultural prices within and between countries and the continued propensity for governments to insulate their domestic food market from fluctuations in international prices. Chapters 18 and 19 provide further examples, focusing on Thailand and the

Philippines, respectively. Thailand established a Paddy Mortgage Scheme (PMS) in 2011, meant to stockpile rice exports in order to increase world prices. Setboonsarng's Chapter 18 demonstrates that the PMS has been successful in transferring income to rice growers, temporarily increasing world prices, and stabilizing farm gate prices. However, the negative impacts of PMS include a huge budgetary loss incurred by the government, the near collapse of market mechanisms in guiding efficient resource allocation, and the decline of rice quality and diversity. On the other hand, the Philippines is implementing its Food Self-Sufficiency Program, which, among other objectives, aims to make the country self-sufficient in rice. Having missed its first deadline of 2013, the country is now aiming for rice self-sufficiency by 2016. As the 2016 deadline approaches, the Philippines has restricted rice imports, supported *palay* prices, and has invested billions of pesos largely in irrigation facilities. In Chapter 19, Clarete describes and suggests possible rationales for the program, discusses its pitfalls, and suggests appropriate reforms.

Wickramasinghe's Chapter 20 explores the link between production specialization and market participation by small-farm households. The key premise of this chapter is that transaction costs play a fundamental role in specialization, which in turn affects household decisions regarding the extent of market participation. Based on the evidence shown for the different market environments, the chapter recommends adapting policy tools to specific environments in order to promote market participation.

1.5 DEVELOPMENT, VULNERABILITY, AND POVERTY REDUCTION

Section V addresses poverty, development, risks, and vulnerability from the perspective of specific countries. Chapter 21 by de Dios and Williamson zeros in on development lessons from the Philippine experience. Examining a century of the country's deviant behavior in industrialization sheds light on the various antidevelopment forces, including institutions, liberalization policies, labor emigration, and Dutch disease that brought down the growth trajectory of the country. Understanding the past avenues of unsustainable development helps prevent history from repeating itself and promotes sustainable development and poverty reduction. Disaster management has emerged as a major issue in recent years, both in developed and developing countries. Given the increased damage from natural disasters, economists have increasingly started to focus their attention in that direction. Farmers' vulnerability to weather shocks such as drought will remain central to global poverty concerns. In Chapter 22, Lybbert and Carter investigate the proper bundling of two innovations, drought tolerant crops and drought index insurance, and how to leverage the complementarities between them. They calibrate such a package for Ecuador to illustrate alternative policy options.

Chapter 23 by Gaiha et al. identifies factors associated with the frequency of natural disasters and the resulting impacts on mortality. This chapter concludes that the payoff from learning from the experience of natural disasters is high; even moderate learning from responding to it can save a large number of lives. More rapid economic growth can also help avert deaths by providing resources for disaster prevention and mitigation. A challenge for development assistance is to combine accelerated growth with better disaster forecasting, rapid response, and speedy relief in order to reduce vulnerability to natural disasters.

In Chapter 24, Balisacan brings the reader back to the Philippines. While the Philippine experience illustrates the various sources of unsustainable development (see Chapters 9, 19, and 21), history also shows that the country has had episodes of substantial growth (Balisacan and Hill, 2003; Canlas et al., 2011). The subsequent question is whether these spurts of growth helped in lifting the poor out of poverty. Using the Alkire−Foster aggregation methodology, which preserves the "dashboard" of poverty dimensions, to systematically assess the magnitude, intensity, and sources of multidimensional poverty over the past two decades and across subpopulation groups, Balisacan finds that poverty did actually decline. While income-based poverty remained largely unaffected by economic growth during the past decade, the multidimensional metric of poverty has decreased. From a policy perspective, this result reinforces the view that economic growth, even in the short term, is required to reduce poverty.

The last but certainly not the least chapter in this volume, by Warr, has taken the discussion of structural transformation (see Chapter 21) and measures of poverty reduction (see Chapter 24) to the Mekong economies consisting of Cambodia, Laos, Myanmar (Burma), Thailand, and Vietnam and including two provinces of China: Guangxi and Yunnan. Lessons from this study strengthen the view highlighted in Chapters 8 and 9 that poverty reduction is strongly related to growth of real gross domestic product per person but that the sectoral composition of this growth also matters.

1.6 CONCLUSION

As demonstrated in this volume, recent research has enriched our understanding of the nature, causes, and consequences of policy responses to the development challenges of our time. The examples provided for certain agricultural, resource, and environmental policy concerns have shown the richness of incorporating transaction costs, risks, institutions, and political economy into traditional models of economy−society−environment interactions. Yet, the efforts to unravel fundamental explanations to many recent and emergent development patterns have only just begun. For students of development, the field is fertile for research, both at the theoretical and empirical levels. For example, the link between security and sustainability through the concept of risk is yet to be analytically clarified. Clarifying the concepts of and interrelationship among disaster management, risk, hazard, vulnerability, resilience, and sustainability would contribute greatly in our understanding of the economics of disaster.

The role of institutions in natural resource use can be further explored to include transitions across different forms of institutions, the role of government and its interaction with resource users, and economic development. It is useful, for example, to explore how certain political economy models of rent-seeking and citizen voting can affect resource use outcomes and the sustainability of economic development. Development of theoretical foundations of "learning-by-lobbying" would be likewise useful in shedding light on the transmission effects of the use of resources and the potential for a resource curse.

From a policy perspective, advancing the search for fundamental explanations will go a long way toward informing what works and what does not in efforts to achieve sustainable economic development. For one, we hope to mitigate the tyranny of fads, fancies, and myths over concrete and logically thought-out proposals to achieve sustainable development. For another, a clear understanding of the nature, causes, and consequences of policy helps inform the deployment of appropriate metrics for monitoring and evaluating progress—or the lack of it—in achieving sustainable development.

There is much room for improving the methods for measuring various capital stocks, shadow prices, rates of depreciation and depletion, and rates of investment as the different economies move forward to green accounting—inclusion of environment and natural resource considerations in, say, the National Income Accounts.

The framework of nonrenewable resource and a ceiling on the stock of pollution can also be used in the evaluation of regulatory policies, such as in modeling the impacts of the Keystone Pipeline System in Canada and the United States as well as the discovery of shale gas reserves in China and the United States. Complementing the formal models of institutional development and natural resource management with empirical studies will further increase the value of these tools toward a more fully comprehensive approach to sustainable development.

REFERENCES

Arrow, K.J., Dasgupta, P., Goulder, L.H., Daily, G.C., Ehrlich, P., Heal, G., et al., 2004. Are we consuming too much?. J. Econ. Perspect. 18 (3), 147–172.

Balisacan, A.M., Hill, H. (Eds.), 2003. The Philippine Economy: Development, Policies, and Challenges. Oxford University Press, New York, NY.

Barbier, E., 2007. Natural Resources and Economic Development. Cambridge University Press, New York, NY.

Barnett, H., Morse, C., 1963. Scarcity and Growth. Resources for the Future, Washington, DC.

Canlas, D.B., Khan, M.E., Zhuang, J. (Eds.), 2011. Diagnosing the Philippine Economy: Toward Inclusive Growth. Asian Development Bank, Manila.

Clark, W.C., Dickson, N.M., 2003. Sustainability science: the emerging research program. Proc. Natl. Acad. Sci. USA 100 (14), 8059–8061.

Endress, L., Roumasset, J., Zhou, T., 2005. Sustainable growth with environmental spillovers. J. Econ. Behav. Organ. 58 (4), 527–547.

Pearce, D., Markandya, A., Barbier, E.B., 1989. Blueprint for a Green Economy. Earthscan, London.

Ravago, M., Roumasset, J., Balisacan, A., 2010. Economic policy for sustainable development vs. Greedy growth and preservationism. In: Roumasset, J.A., Burnett, K.M., Balisacan, A.M. (Eds.), Sustainability Science for Watershed Landscapes. Institute of Southeast Asian Studies/Southeast Asian Regional Center for Graduate Study and Research in Agriculture, Singapore/Los Baños, Philippines, pp. 3–45.

Roumasset, J., Burnett, K.M., Balisacan, A.M. (Eds.), 2010. Sustainability Science for Watershed Landscapes. Institute of Southeast Asian Studies/Southeast Asian Regional Center for Graduate Study and Research in Agriculture, Singapore/Los Baños, Philippines.

World Commission on Environment and Development (WCED), 1987. Our Common Future. Oxford University Press, <http://www.un-documents.net/wced-ocf.htm> (accessed May 6, 2009).

REFLECTIONS ON THE FOUNDATIONS OF DEVELOPMENT POLICY ANALYSIS

2

James A. Roumasset

Department of Economics, University of Hawai'i at Mānoa, Honolulu, HI, USA

CHAPTER OUTLINE

JEL: O12; O13; B41; B52

2.1 INTRODUCTION

Apart from an outstanding group of mentors, the main influence on my life as an economist has been collaboration with a diverse group of independent and courageous scholars, many of whom are contributors to this volume. Our overriding vision has been that designing institutions, including public policies, requires fundamental foundations for understanding behavior and organization. Yet development economics is replete with policy recommendations based on *ad hoc* models and empirical findings that leave room for alternative explanations. By understanding the fads and fancies of development economics, we are more likely to avoid repeated pitfalls and chart a productive research future.

In the sections that follow, I attempt to explain how nonfundamental development microeconomics has gone through stages of modernism and postmodernism. Modernism is associated with the social engineering paradigm of A.C. Pigou, Paul Samuelson, and Jan Tinbergen (McCloskey, 1994, 1996, 1998). Postmodernism in development economics derives from the nihilism of nonconvexities, multiple equilibria, and the inability of competitive equilibrium to reach even a constrained Pareto equilibrium (Greenwald and Stiglitz, 1986). Nonfundamental models, characterized by misplaced exogeneity (Nozick, 1974), proliferate. Subsidized social insurance and smart fertilizer subsidies, for example, are justified by exogenous risk preferences, arbitrarily assumed market failure, and presumed behavioral anomalies. Modern empirical applications including randomized design, natural experiments, and instrumentation for quasi-randomization often suffer from the similar problem of missing mechanisms. By focusing on market and behavioral failures, postmodern development economics often fails to deal with the central problems of economic development—bad institutions, including bad economic policies.

There has been substantial progress in building more fundamental theories. Both New Classical Microeconomics and the New Institutional Economics are notable in this regard but are still incomplete. What is needed in both cases is to marry the theories of social optimization with appropriate notions of equilibrium that will permit the consequences of alternative policy reforms to be assessed.

Prescribing policy reforms and other changes in institutional design requires understanding the nature, causes, and consequences of the institutions in place. An alternative paradigm—*postmodern structuralism*, rooted in Coasean fundamentalism—is described and illustrated in this chapter. Particular attention is devoted to agricultural development policy, e.g., by contrasting the misplaced exogeneity of the small-is-beautiful perspective with the fundamental theory of the agricultural firm. Suggestions are offered for developing the fundamentalist paradigm further, particularly regarding institutional change, specialization, and sustainable development. The challenge of assessing development policy analysis also affords me the welcome opportunity to reflect on and synthesize a few of my own writings on the subject.

2.1.1 THE FOUR STAGES OF RESEARCH IN DEVELOPMENT ECONOMICS

Mookherjee (2005, p. 4330) has identified four stages of research and related them to the evolution of development economics. *Stage 1* consists of the identification of "empirical regularities that need to be explained by a suitable theory." *Stage 2* involves "the formulation of a relevant theory, including derivation of potentially observable (hence falsifiable) implications." *Stage 3* deals with "the testing and estimation of theories, a stage which may lead back to modification or replacement of the previous theories, in an iterative back and forth with *stage 2*." *Stage 4* "employs the least unsuccessful theory from the standpoint of empirical verification for purposes of prediction and policy evaluation."

Mookherjee describes classical development economics as "dealing with stage 1 empirics (e.g., Kuznets, Myrdal, and Rostow) and *stage 2* theories, especially those of Rosenstein-Rodan, Hirschman, Leibenstein, Lewis, Nurkse, Scitovsky, and Sen." He describes development economics in the 1970s and 1980s as most notable for departing from the Walrasian paradigm (e.g., using agency and game theories) to explain the stylized facts of agrarian institutions.[1] While Walrasian economics provided a unifying foundation, the new theories, based on arbitrary game theoretic and transaction cost specifications, provided "an embarrassment of riches."

These *stage 2* theories were then said to prepare the way for *stage 3* empirics. As the stage of testing and measurement progressed, the field became increasingly focused on removing econometric biases, leading in turn to the obsession with randomized experiments, both natural and controlled.

> Research papers tend to get evaluated almost exclusively in terms of their success in combating the econometric problems, often to the exclusion of the importance of the context or issues addressed by the analysis, the imaginativeness of the underlying hypotheses formulated or tested, or the importance of the findings from a wider standpoint...The research is consequently increasingly microscopic in character.
>
> **Mookherjee, 2005, p. 11.**

A similar critique by Deaton (2010) concludes:

> As with IV methods, RCT-based evaluation of projects, without guidance from an understanding of underlying mechanisms, is unlikely to lead to scientific progress in the understanding of economic development.

Rather than rejecting experimental methods, Deaton prescribes returning to the four stages and applying modern empirical methods more to the evaluation of theories than to projects.

Nor is reaching stage 4 the end of the game. As Ben-Porath (1980) describes, the stage of testing and measurement gives rise to new theories to explain statistical patterns not consistent with existing theory (think of Krugman's, 1979, new trade theory). A natural dialectic between theory and evidence should continue that produces new and better theories and an improved mapping between the domain of underlying conditions and appropriate theories. Instead, the current trend in development economics appears to have become somewhat fixated in stage 3.

Through the examples that follow, I argue that empirical testing should not be the only criterion for navigating what Keynes called the "slippery problem of passing from statistical description to inductive generalization" (Keynes, as cited in Mookherjee, 2005, p. 566),[2] especially when the theories are used as the basis for policy evaluation. Inasmuch as the dialectic interaction between

---

[1]My own work in this genre includes Roumasset (1978), Roumasset and Uy (1980, 1987), and Roumasset (1994).
[2]See also Wolpin (2013).

stages 2 and 3 is an ongoing and never ending process, we must be suitably cautious when moving to stage 4 policy prescription. As noted by Coase (1994), economics as objective, logical positivism is what economists pretend to do, not what they actually do.[3] Additional criteria for evaluating theories may be appropriate, even if they are subjective. In development policy analysis, the researcher must decide whether the underlying model abstracts from reality in a useful and meaningful way or whether the policy prescription derives from an *ad hoc* assumption.

2.1.2 THE NATURE-CAUSES-CONSEQUENCES PARADIGM FOR DEVELOPMENT POLICY ANALYSIS

The *nature, causes, and consequences* paradigm for policy analysis involves first describing policy according to its *economic nature* via summary statistics such as nominal protection rates or implicit tariffs.[4] Next the analyst assesses the consequences of a particular policy change according to its *consequences* in a particular modeling framework, e.g., using the distorted (but uniformly priced) competitive equilibrium as the benchmark (Clarete and Roumasset, 1987; Roumasset and Setboonsarng, 1988; Gardner, 1988; Krueger et al., 1991). This paradigm abstracts from transaction costs and implicitly assumes well-functioning markets.

Surely a characteristic of developing countries is that markets are not well developed, and the absence of well-functioning credit and insurance markets means that risk aversion may play a more important role in farm-household decisions than otherwise (Chetty and Looney, 2005, 2006). Policy evaluation in developing agriculture therefore requires behavioral and organizational foundations that go beyond the conventional paradigm.

Inasmuch as policy analysis needs to be based on a positive theory that goes beyond the Walrasian paradigm, Sections 2.2 and 2.3 are devoted to the behavioral and organizational foundations of agricultural development. In particular, Section 2.2 explores progress and research needs regarding behavioral theories of the farm household. It begins with a comparison of traditional and *ad hoc* theories of farmer decision making under uncertainty with the approach of *strong structuralism* (as in Section 2.2.1). The section continues with a discussion of how the modern culture of economic empiricism has become an end in itself, instead of contributing to a dialectic improvement of quantified conceptual models needed for policy evaluation. Section 2.3 also compares *ad hoc* and fundamental theories, this time about agricultural organization. I attempt to demonstrate how misplaced exogeneity has led to a regression of economic development thinking back to the defunct structure—conduct—performance paradigm. In contrast, the fundamental approach promotes a healthy respect for the diversity of institutions. Section 2.3 also includes suggestions for how lessons from agricultural institutions can contribute to a unified version of *the new institutional economics* that applies to industrial as well as agricultural organization. Inasmuch as policy recommendations need to be also based on an understanding of the existing political equilibrium, the unified perspective is extended to include the *causes* of public policy. Section 2.4 provides a brief critique of modern theories of institutional and market failure and the need to return to fundamental explanations. Section 2.5 is an inquiry into the evolution of specialization in the organization of

---

[3]See also McCloskey (1994) and Medema (1995).
[4]An improved measure is the nominal rate of assistance (Anderson et al., 2008; Chapter 18 of this volume).

agricultural labor and how the nature of the agricultural firm evolves to incentivize the specialized tasks of labor and management. Section 2.6 discusses *black-hole* policy failures and the role of the economist in opposing them.

2.2 BEHAVIORAL FOUNDATIONS FOR AGRICULTURAL DEVELOPMENT POLICY

For the last half century, the study of agricultural development has been overwhelmingly devoted to the problem of how, in Arthur Mosher's (1966) parlance, to *get agriculture moving*. As such, agricultural development is best seen as a problem of public economics. Public economics is an explicitly normative (prescriptive) discipline.[5] Just as surgery requires an understanding of anatomy, agricultural policy analysis requires a positive theory in order to evaluate the potential *consequences* of policy reform. A positive theory suitable for evaluating the consequences of rural development policy needs to include two central components—a theory of farm-household behavior and a theory of agricultural organization. The current section is devoted to the first of these; part 3 is concerned with the second.

2.2.1 TOWARD FUNDAMENTAL EXPLANATIONS OF FARM-HOUSEHOLD BEHAVIOR

There is a long tradition in the economics of agricultural development that behavior and organization in developing countries are inefficient. Low-income farmers were once thought not to be rational profit maximizers but tradition-bound, uninformed, and possibly indolent. Even when Theodore Schultz's *Transforming Traditional Agriculture* effectively showed that low-income farmers are privately efficient, numerous theories remained about the social inefficiency of traditional agriculture.

On the behavioral front, farmers are often assumed to be risk averse, and many economists have concluded that, in the absence of government provided credit and social insurance programs, the actions of even privately efficient farmers are socially inefficient. Hazell et al. (1986, p. 293) sum up the conventional wisdom:

> Yield and price risks induce farmers to allocate their resources conservatively. Farmers pursue more diversified cropping patterns than is socially optimal, and they are sometimes reluctant to adopt improved technologies because of the increased risks associated with their use.

Similarly, Binswanger and Rosenzweig (1993) conclude that weather risk is a significant cause of inefficiency. This view has continued to the present time (e.g., World Bank, 2008; Chetty and Looney, 2006; Karlan et al., 2014).

The conventional wisdom has been questioned at several levels, however. The hypothesis that *risk aversion implies underinvestment* (RAUI) is based on the proposition that farmers have

---

[5]Indeed a standard graduate text (by Richard Tresch) is entitled, *Public Finance: A Normative Theory*.

markedly concave utility functions of current period income, especially in the neighborhood of subsistence levels of living (e.g., Chetty and Looney, 2006) and that modern techniques display greater variance than traditional ones. As reviewed in Roumasset (1978b, 1979a), however, logical problems with the RAUI hypothesis abound, including

1. Rationality does not imply the existence of utility functions defined in current income (Spence and Zeckhauser, 1971).
2. Even if utility functions in current income exist, it is unlikely that they should be uniformly concave in the presence of transaction costs.
3. Aversion to risk should be based on a definition of risk, not on the second derivative of a contrived utility function, as in the rather backward definition that risk is what risk averters will pay to avoid (e.g., Rothschild and Stiglitz, 1970).
4. Closeness to "subsistence" (which is empirically elusive) may imply desperation rather than conservatism (see also Banerjee, 2000).

One way of dealing with these issues is to directly model behavior as aversion to the chance of loss, where the loss threshold depends on the farm-household's asset liability position and then estimating the frequency distribution of farm-operator profits for different techniques. This provides the information to test the RAUI hypothesis versus the null hypothesis that farmers maximize expected profits without additional regard to risk. When this comparison was done, the expected profit model outperformed the RAUI hypothesis (Roumasset, 1976). This model also illustrates Mookherjee's four stages. Stage I is the discovery of the stylized fact that most farmers did not follow fertilizer recommendations. Stage II is the development of a suitable behavioral theory to test RAUI (appropriate behavioral model and methodology for measuring frequency distributions). Stage III was the test (hypothesis rejected), and Stage IV (policy implications) concluded that the *prima facie* case for crop insurance and fertilizer subsidies was likewise rejected and that more attention was needed on the diversity of agroclimatic, economic, and institutional settings of various farm households. Despite the lack of well-founded arguments to the contrary, however, the RAUI case for subsidies continues to be made (World Development Report [WDR] 2008).[6]

The "Rice and Risk" procedure just described also illustrates what might be called "strong structuralism." Instead of using first-order conditions to derive an appropriate statistical model and letting the data determine coefficients, strong structuralism requires the model to directly provide a numerical estimate of farmer choice, say the predicted amount of nitrogenous fertilizer per hectare. All that is left for regression analysis is to compare the goodness of fit of the alternative and null hypotheses. In order to capture the economic diversity of farmer-households, the structuralist approach can also be extended to capture individualized shadow prices, as in Roumasset (1981), Roumasset and Smith (1981), Stiglitz (1984), and de Janvry et al. (1991). If the household is a net seller of grain, for example, the household shadow price is the market price minus the transportation and seller's transaction costs. Similar computations can be made according to whether the household hires out or hires labor, borrows or lends, etc.[7] Incorporating idiosyncratic shadow prices into a strong structural explanation,

---

[6]Notably, Duflo et al. (2008) reject risk aversion as an explanation for underuse of fertilizer among Kenyan farmers.

[7]See Evenson and Roumasset (1986) for an empirical application of the transaction cost model to the explanation of rural fertility decline.

we found that the expected profit model explains 91% of the variation in levels of mechanization among a sample of Nepalese farmers, even without allowing for an intercept different than zero or a slope coefficient different than one (Roumasset and Thapa, 1983).

Even in cases where risk aversion does apparently impact decision making, one cannot conclude that market intervention, e.g., through mandates and subsidies, is warranted. Firstly, much of what appears to be risk aversion (farmers hate to lose more than they love to gain) is the result of transaction costs, especially in credit and output markets. Since transaction-cost-induced risk aversion is idiosyncratic, efficiency cannot necessarily be enhanced by even costless risk-sharing mechanisms. Indeed government policies, such as subsidized insurance that cause farmers to ignore transaction costs, are likely to decrease, not increase, efficiency (Roumasset, 1979a, 2010).

Secondly, the mere absence of markets (e.g., for products such as multiperil crop insurance) does not imply inefficiency (Arrow, 1969), notwithstanding the common characterization of missing and imperfect markets as "market failure" (e.g., World Bank, 2008; de Janvry et al., 1991). If the private sector fails to offer some insurance products to some customers, it is largely because the residual moral hazard and adverse selection in the face of available private governance outweighs the corresponding benefits (see Chapter 23). Intervention requires a specific rationale for why government has a comparative advantage in promulgating alternative institutions. Fundamental models are needed that derive behavioral implications from dynamic optimization in the presence of inevitable transaction costs. Only then can the consequences of policies to relax credit and risk "constraints" (as termed by Karlan et al., 2014) be properly weighed against the costs.

2.2.2 MODERN TRENDS IN EMPIRICAL ANALYSIS

Recent trends in the fads and fancies of farm-household behavior have tended away from fundamental explanations. In particular, the search for data that allow for natural experiments or quasi-experimental methods via "clever instrumental variables" (Bardhan, 2005) leads many analysts away from fundamental questions of dynamic decision making. While instruments may be external, they are typically not exogenous in the sense required by consistent statistical estimation (orthogonality).[8] Consider Deaton's example of the effect of railroad construction on poverty reduction. Poverty and railroad construction clearly do not affect the presence of earthquakes (the instrument), and hence the instrument is external to the model. However, the propensity of earthquakes may well have an effect on poverty, in which case earthquakes would not be orthogonal to poverty (in the sense that earthquakes may affect poverty through channels other than railroad construction/destruction). Even randomized-control trials (RCTs) do not entirely escape selection problems and often suffer from problems of external validity.[9] It is easy to think that selection in RCTs involves only randomly dividing the sample into a treatment and control group, but in fact there are actually two stages of selection. In the first stage, researchers start with the whole population and choose a group in this population, e.g., one village out of all villages. The selected village is often chosen in consideration of other factors, such as convenience or politics, which changes the representativeness of the village. While RCTs are lauded for their internal validity, i.e., the ability to tease out causality, the causal connection in one situation may not be appropriate in another situation such

---

[8]See also Wolpin (2013).
[9]For a full discussion of these problems, see especially Deaton (2010), Ravallion (2011), and Rosenzweig (2012).

that replication may not necessarily lead to the same results. In the economics profession at least, Rodrik (2008) points out the lack of incentives for researchers to undertake these validation exercises. With all these attendant problems in mind, Deaton (2010) prescribes using RCT methods to better understand the causes ("mechanisms") of economic development in order to better guide policy prescription.

As if in answer to Deaton, Duflo, Kremer, and Robinson (DKR) (Duflo, Kremer, and Robinson, 2011) use RCTs to explore the causes of Kenyan farmers' failure to use fertilizer in economically optimal doses. DKR posit a model of stochastic impatience: Boundedly rational farmers "naively" put off the inconvenience of buying fertilizer after harvest, when they are patient. Then, if they turn out to be impatient in the last period that fertilizer investment is still profitable, they will fail to invest. The model predicts, and a RCT confirms, that many farmers will buy fertilizer vouchers at the end of the previous harvest, given a small subsidy to offset the convenience costs of buying fertilizer. The vouchers are said to serve as an effective commitment device for farmers who know that they may be impatient in the relevant period but underestimate the extent of their impatience.[10] This is an important example of Banerjee and Duflo's (2011) mission to establish the importance of *nudge economics* (from Thaler and Sunstein, 2008) in development policy.

However, DKR do not consider more fundamental explanations of why small Kenyan maize farmers are normally reluctant to use fertilizer but why many respond positively to free-delivery vouchers. DKR's estimate that applying half a teaspoon of nitrogenous fertilizer per corn plant yields a return of 15−27.2% may be somewhat optimistic. Farmers may have believed that the experimental years were better than average or (implicitly) that using a point estimate overestimates the optimal rate of fertilization—not because of risk aversion (which the authors rejected in DKR, 2008) but because of fundamental nonlinearities in gains and losses due, for example, to transaction costs.

Full rationality is an abstraction that, as Friedman (1953) explained, is meant to be useful but false. Once we enter the world of bounded rationality, there are too many possible departures to test. Given that the absolute gains of the voucher program were small (in the neighborhood of $10, as noted by Rosenzweig, 2012), small departures from full rationality could be decisive. Among these is the placebo effect (entertainment value, social benefit, or "warm glow") associated with going along with what the celebrity experimenters offered. These psychic phenomena provide an alternative explanation of why most farmers who purchased the vouchers did not use fertilizer in subsequent seasons. The policy implications of alternative formulations of bounded rationality are different. Procrastination may imply the kind of small subsidies that DKR recommend. If the novelty of the experiment was the inducement, one cannot expect that this would carry over to a government voucher program. (Even if it did, the administrative costs of the small-subsidy programs may be larger than the welfare benefits to farmers.) In other words, the policy implications of one particular formulation of bounded rationality are not robust with respect to other possible formulations. Moreover, bounded rationality explanations are subject to *Lucas critique*—a change in policy may change the behavioral mechanism.[11]

---

[10]Almost all subject farmers said that they planned to use fertilizer in future seasons, but few followed through. DKR conclude that this implies minimal learning by farmers about their own departures from full rationality.

[11]One resolution of this dilemma is to provide a full optimality theory with behavioral foundations such as lexicographic safety first (Roumasset, 1976, Chapter 2).

A more promising approach is to develop the dynamic foundations of full rationality. Taking a one-period utility function as a characterization of farmer risk preferences is clearly *ad hoc*. One-period risk preferences, if they exist at all, may have more to do with transaction costs than they do with lifetime risk aversion. Indeed the one condition under which indirect utility functions can be derived as a function of current income is that the lifetime utility function is separable in utilities of different years (Roumasset,1979a). However, if the lifetime utility function is separable, then one cannot distinguish between the elasticity of substitution between consumption in different years and the degree of relative risk aversion (Epstein and Zin, 1989). So one loses no generality by assuming that contemporaneous risk aversion is created by the degree of intertemporal substitution and transaction costs. Under plausible assumptions then, risk aversion is not a distinct characteristic of preferences but derivable from more fundamental preferences.

Nonetheless the view of risk aversion as a *primitive* persists in modern development economics, as does the presumption that risk aversion constrains development. Theory and evidence that insurance programs for low-income households have high costs relative to benefits (e.g., Cole et al., 2013) have simply resulted in calls for modifying insurance products to lower costs, including bundling them with microfinance (e.g., ILO, 2013). Ironically, empirical findings that bundling actually detracts from insurance demand (Banerjee et al., 2014) and thereby decreases the present value of such programs have apparently not swayed the modern school from its priors. Instead, the *nudge school* of development economics (Banerjee and Duflo, 2011) rationalizes lack of insurance demand as a departure from full rationality, even in the face of their own finding that households' unwillingness to accept health insurance was "correct *ex post*" because of implementation failures (Banerjee et al., 2014). Instead of calling for an inquiry into the fundamental causes of risk aversion and whether that undermines the case for subsidized insurance, the authors seem to feel that natural experiments and RCTs are an end in themselves.

2.3 ORGANIZATIONAL FOUNDATIONS FOR DEVELOPMENT POLICY ANALYSIS: THE NEW INSTITUTIONAL ECONOMICS

2.3.1 EXAMPLES OF NONFUNDAMENTAL EXPLANATIONS

Traditional development economics abounds with theories of market failure and the need for government intervention. Possibly the most pervasive of these is the theory of dualism between the agricultural sector and the "modern" sector. The theory was spawned by the (*stage 1*) stylized fact that the relative size of the industrial to the agricultural labor force is correlated with per capita income. The (*stage 2*) theory of dualistic economic growth that Fei and Ranis (1964) developed to explain the correlation presumed that wages in the industrial sector were set exogenously by institutional factors and that the marginal product of labor in agriculture was zero. This exemplifies the problem of misplaced exogeneity: The very thing that matters for policy is assumed, not derived. The Fei-Ranis model was discredited by Jorgenson (1969), who not only provided a fundamental model to explain the stylized fact in question but showed that his model provided a superior explanation of the empirics regarding the changing labor-force composition between the agricultural and industrial sectors. In spite of this, the theory of dualistic development remains popular to the present day, possibly because of its ideological appeal.

A variant of the dualism model has been used to explain the stylized fact that there is an inverse relationship between farm size and productivity (IRSP). Again, the wage in the modern (commercial) agricultural sector was said to be exogenously set by "institutional" factors and the wage in the agricultural sector was assumed to be substantially lower, implying a lower marginal product and a higher intensity of cultivation (Sen, 1962, 1966; Mazumdar, 1965; Berry and Cline, 1979). This *small-is-beautiful* theory has often been used to advocate land-to-the-tiller reforms (World Bank, 2008; Lipton, 2009). Again we have misplaced exogeneity. The policy implications are in effect assumed, not derived.

Recent empirical tests of IRSP have several flaws. For example, World Bank (2009) concludes that such a relationship exists for Philippine rice and corn farms, following similar methodology from Feder (1985), Binswanger et al. (1995), and Benjamin (1995).[12] These studies regress farm profits on farm size in the presence of several "controls," e.g., for land quality and tenure status. The problem is that farm size is endogenous (Roumasset and James, 1979). The correct conceptual definition of land quality is potential rent per hectare. So if farm size were exogenous, one had an exact measure of land quality, and if farmers maximized profits, P, the relationship between P and land quality is an identity. When one attempts to estimate an identity, the estimated coefficients are simply a reflection of measurement error.

The same dualism is said to underpin the advantage of a *unimodal* development strategy (Johnston and Kilby, 1975; World Bank, 2008). Land distribution and farm size in many Latin American countries has been described as "bimodal," i.e., the mode of the farm size distribution is small but the mode of the land distribution is large.[13] Concentrated landownership, the argument goes, leads to agricultural development being focused on capital intensive technology, which is antipoor. In unimodal economies, however, technical change is more likely to be capital saving and labor using, i.e., pro-poor. Again the idea is that large landowners face different factor prices than small farmers. The key is explaining why this is the case. If it is transaction costs, pursuing an agricultural strategy that ignored those costs is unlikely to be efficient. If transaction costs can be reduced, e.g., through land reform, then the benefits of reducing those costs should be compared with the costs of the reform program. Moreover, the reform program may inadvertently increase other transaction costs, especially in credit, marketing, and downstream coordination.

Another example of a nonfundamental explanation is provided by Stiglitz's (1974, 1993) model of share tenancy.[14] This time the stylized fact to be explained is the mere existence of share contracts. Stiglitz's canonical explanation holds that share tenancy is privately efficient but socially inefficient since, by transferring the land to the tenants, the government could remove the perverse labor incentives that share tenancy confers. As discussed in more detail below, this model mischaracterizes farmers as workers and share tenancy as a labor contract instead of as a partnership. Before turning to the task of providing a fundamental theory of the agricultural firm, and how share tenancy fits in, we need to consider the nature of a fundamental explanation of institutional change due to Ronald Coase, including his colleagues and descendents.

---

[12]See Ali and Deininger (2014) for a more recent example.

[13]This is not optimal semantics inasmuch as each of the two separate distributions is thought to have a unimodal distribution.

[14]See Hayami and Otsuka (1993) for an intuitive exposition of the Stiglitz model.

2.3.2 FROM THE COASE THEOREM TO FUNDAMENTAL EXPLANATIONS OF AGRARIAN CONTRACTS

The "UCLA equivalence version" of the Coase theorem is that absent contracting costs, competitive contracting is equivalent to universal, competitive markets which are equivalent in turn to markets with Pigouvian corrections (Roumasset, 1979b). The upshot of this equivalence is that institutional evaluation must rest on a comparison of transaction costs. In Demsetz's (1967) famous example, European contact with North America raised the implicit price of beaver pelts to the point where the benefits of private property rights became greater than the transaction costs of enforcement. In the case of bison hides however, the price increase was insufficient to induce institutional change because the potential enforcement costs of private property would have been too high in the case of roaming herds.

The Demsetz proposition is that: "Property rights develop to internalize externalities when the gains of that internalization become larger than the costs of internalization" (Demsetz, 1967, p. 350). This was further clarified by Anderson and Hill (1975, 1990) who argued that the advent of barbed wire in the American West sufficiently lowered the cost of property enforcement to render the costs of private property less than the benefits.[15] The thesis can also be seen as underlying "the idea that the common law evolves toward efficient rules" (Merrill, 2002) and the *New Economic History*, wherein the English Enclosures were said to evolve in response to the benefits versus the costs of private property (North and Thomas, 1973).

The efficiency view has been used successfully in explaining share tenancy "as an understandable market response" (Reid, 1973, 1976). With the advent of agency theory (Jensen and Meckling, 1976), the efficiency view was further developed to explain pervasive patterns of agricultural contracts in the Philippines using the methodology of *The New Institutional Economics of Agricultural Organization* (Roumasset, 1978a). The version of the New Institutional Economics (NIE) that we used in the Philippines was inspired by Coase and Demsetz but was more explicit in developing first- and second-best theories.

The first-best theory stems from the Cheung—Coase proposition that absent transaction costs, a contracting equilibrium provides a perfect substitute for a universal market solution (including a market for spillover effects) and a Pigouvian corrective mechanism. The competitive contracting equilibrium of a share tenancy economy is equivalent to the solution obtained by landlords hiring workers from a competitive labor market and to lessees renting from a competitive land market. The competitive contracting equilibrium in an apple—honey economy is likewise equivalent to the competitive market solution, including a market for pollination services and to a solution with Pigouvian pollination subsidies. The Cheung—Coase proposition can be proved by showing that the core of a bilateral contracting economy (whether landlords and tenants or interdependent apple and honey producers) shrinks to a competitive equilibrium (Roumasset, 1979b; Johansson and Roumasset, 2002).

The first-best theory can be used to explain statistical patterns regarding the terms of agricultural contracts. For example, higher quality land is organized in smaller family farms with higher landlord shares than medium quality land (Roumasset and James, 1979). Landlord shares also vary according to crop: high for capital intensive crops, such as coconut, and low for labor intensive crops such as abaca (Roumasset, 1986). By endogenizing the terms of contracts, the first-best

---

[15]As discussed by Hornbeck (2010), the property law was already in place. Barbed wire provided feasible enforceability that completed "the bundle of property rights."

theory reveals that the inverse relationship between farm size and yield per hectare and the mere existence of share tenancy do not constitute *prima facie* evidence of inefficiency.

As Dixit (1998) explains, second-best theories examine the efficiency of economic organization in the face of transaction costs. In the Philippine version of the NIE, we use positive agency theory (Jensen, 1983) to explain for example why piece-rate contracts are favored over wages when the intermediate output of a specific task is readily observable (Roumasset and Uy, 1980). Agency theory is also used to explain the relative advantages of share tenancy to owner managed and fixed lease arrangements due to difficulties of mitigating the labor shirking of wage workers and "land shirking" of renters (Roumasset and Uy, 1987; Roumasset, 1995).

These stage 2 theories were followed by explicit comparisons of the efficiency model with the Stiglitz inefficiency model. Using data on agricultural contracts in the United States, Allen and Lueck (2003) reject the Stiglitz risk-aversion model in favor of an efficiency model with double moral hazard regarding both land and labor shirking. (The Coasean influence in both Allen and Lueck's "The Nature of the Farm" and my "The Nature of the Agricultural Firm," 1995, is unmistakable.) Deweaver and Roumasset (2002) reject Stiglitz's canonical model on both conceptual and empirical grounds. For parameters obtained from Philippine data, the Stiglitz model predicts that tenant shares should decrease from 100% to 80% and then increase back to 100% (fixed lease) as the tenant's risk aversion increases, contradicting Stiglitz's central theoretical finding that optimal tenant shares are declining in risk aversion. Moreover, since actual tenant shares are predominantly clustered around one-half and two-thirds, there must be one or more disadvantages of fixed lease contracts that the canonical model neglects.

Stiglitz implicitly assumes that the more tenants work, the more is at risk. However, much of labor is precautionary, e.g., pest control. By incentivizing labor, rent contracts also incentivize precaution. Moreover, hard work under rent contracts reduces the chance of loss, simply by moving the distribution of profits to the right. Accordingly, rent contracts afford two advantages—mitigation of labor shirking and partial mitigation of risk. If share and fixed lease contracts were really alternatives to wage labor, as Stiglitz assumes, then they would be almost nonexistent. In contrast, the theory of the agricultural firm—wherein rent contracts have the additional disadvantage of land shirking and share tenancy incentivizes tenant management as well as labor—is consistent with evidence from the Philippines, Nepal, and the United States (Roumasset, 1995; Allen and Lueck, 2003).

2.3.3 ASSUMPTIONS, LEVELS OF ANALYSIS, AND CATEGORICAL VERSUS NONCATEGORICAL THEORIES

The Philippine version of the New Institutional Economics proffers three levels of analysis. As just discussed, both of the first two levels use efficiency models—first-best efficiency in the absence of transaction costs and second-best efficiency in the presence of transaction costs, particularly agency costs. The third level of analysis admits inefficiency via rent-seeking. Dixit (1998) identifies third-best with "endogenous coalition formation." In *positive* third-best analysis, the goal is to explain the *causes* of public policy. For example, Balisacan and Roumasset (1987) explain the pervasive statistical correlation between agricultural protection and per capita income by modeling investment in political influence as a public good within each coalition. Coalitions endogenously form for investing for and against agricultural protection according to the group benefits versus the coalition costs (affected by the number of potential beneficiaries, individual stakes, group homogeneity, and other public choice variables). These considerations determine reaction functions, and agricultural protection is determined as a Nash equilibrium (see also Gardner, 1988, Chapter 12). In *normative* third-best economics, the understanding of the causes of rent-seeking is focused in turn on issues of constitutional design (Section 2.3.6).

Levels of analysis are then chosen according to the problem at hand. For each level of analysis, there is a noncategorical theory (theoretical framework). In the literature discussed above, the first-best theory is used to explain the terms of agricultural contracts. The second-best theory (minimum agency cost) is used to explain the forms of contracts and firms. The third-best theory is used to explain the *causes* of economic policies. Noncategorical theories are irrefutable. In application, restrictive assumptions convert the noncategorical theories into categorical theories. For example, the results that landlord shares are increasing in land quality and rent/wage factor prices require the restrictive assumptions of Ricardian land quality and land/labor elasticity of substitution being less than one (Roumasset and James, 1979). When categorical theories are rejected, the implication is that the theory does not fit the situation described by the data. The theory may still apply in a different situation.[16]

As a further example, consider the theory of induced technical change. Hayami and Ruttan (H-R) (Hayami and Ruttan, 1985) assert that agricultural innovations in the United States were predominantly labor saving in face of the high and rising relative price of labor to land but predominantly land saving in Japan when the opposite was true. By providing examples of labor-saving innovations that occurred when agricultural land prices were rising faster than wages, Olmstead and Rhode (O-R) (Olmstead and Rhode, 1993) claimed to have refuted H-R's theory of induced technical change. However, since the theory of induced institutional change is an irrefutable, noncategorical theory, what they have implicitly done is to refute a categorical theory. The problem is that neither H-R nor O-R have formally provided conditions under which changing factor prices will induce factor-biased technical change and when they won't.

Here is a sketch of a categorical theory that fills the Hayami–Ruttan requirements. Let agricultural output be a nested production function of augmented land and labor, where capital can be used to augment land, labor, or both. Assume that existing knowledge of production technology is represented by a "meta-isoquant" in the augmented land and labor plane that displays a unitary elasticity of substitution between augmented labor and capital. Now consider a potential meta-isoquant obtained by shifting all points of the original one a fixed distance toward the origin, where the distance corresponds to a fixed level of R&D expenditures. However, each point on the new isoquant corresponds to a particular composition of R&D expenditures, i.e., only one point on the new isoquant can be attained by directing research expenditures in a particular factor-saving manner. We can now show that R&D will be optimally deployed along the ray to the origin determined by the new relative prices of land and labor. The noncategorical theory here is the framework of optimal R&D expenditures.[17] All of the other assumptions about elasticity and uniformity of innovation ease render the theory a categorical one. Those restrictive assumptions may be false; in particular it may be less costly to invent a labor-saving technique than a land-saving one and the predictions of the theory may be found not to accord with observation, which explains the O-R rejection of the noncategorical theory, but only for a particular domain.

---

[16]For a further discussion of categorical and noncategorical theories in economics, see Walsh (1987).

[17]For a version of this theory with exogenous R&D, as well as a categorical application, see de Janvry et al. (1995).

The noncategorical theory remains intact. Indeed, Acemoglu's (2002) theory of directed technical change can be seen as a much less restrictive, but still categorical, application of induced technical change that possibly could survive the O-R critique.

While categorical theories rest on restrictive assumptions that are meant to be true for a particular domain, noncategorical theories rest on abstracting assumptions that are meant to be useful but false. When we abstract from transaction costs, for example, it is not because they don't exist but because of our (subjective) belief that we can explain phenomena, such as contractual terms, without them and that the potential improvement from a more complicated explanation is not worth the cost. Thus the agenda following the Olmstead and Rhode findings (see also Olmstead and Rhode, 2008, on biological innovations) is to go back to the drawing board and combine induced-innovation theory with different restrictive assumptions, such as those of Acemoglu's theory of directed technical change, that have some grounding in reality and which provide explanations of the elusive phenomena at hand.

The distinction between categorical and noncategorical theories also helps make sense out of Coase's (1994) perspective on economics as rhetoric and his proposition that methodologies are at least partly chosen on the grounds of what is likely to convince other economists. Indeed, the wise economist practices both the art of deciding what to abstract from, by subjectively choosing a non-categorical theory according to the problem at hand, as well as the science of deriving, testing, and parameterizing categorical theories.

2.3.4 TOWARD A UNIFIED VERSION OF THE NEW INSTITUTIONAL ECONOMICS

Williamson's New Institutional Economics (e.g., 1985) is founded on the pillars of bounded rationality and opportunism. Because of bounded rationality, contracts are assumed to be incomplete in the face of uncertainty. Because of opportunism and the knowledge that agents may behave strategically, parties to an agreement will seek mechanisms of governance (Williamson, 1998, 1996b). However, Baumol (1986) and Samuelson (1983) have criticized the Williamsonian NIE as being nonoperational, by which they meant that hypotheses were only suggested, not derived.

In contrast, the efficiency version of NIE (at the first- and second-best levels) has provided theoretically sound explanations of the terms and forms of agricultural contracts and firms without the contrivances of bounded rationality and opportunism. In agriculture, terms tend to be set according to the competitive shadow prices of factors and organizational forms are determined in accordance with their comparative abilities to minimize agency costs.

The efficiency version has also been usefully employed to advance comparative institutional analysis of resource management. This literature has been unfortunately concentrated on the study of "common property resources," but this label obscures the fundamental research agenda of comparing resource management by private property, common property, no property, government control, or some combination of these. The four resources of central attention by the International Association for the Study of Common Property, the group that launched Elinor Ostrom to fame, were all renewables—forests, fish, water, and land—yet until fairly recently resource economics was rarely part of the discussion. Ostrom (1990) set the stage for institutional analysis of resource management by showing that private property and governmental control were not the "only ways" to manage resources and that common property (not to be confused with open access) was often a highly effective institution. She then identified supply-side factors, such as group homogeneity, that helped explain the success of common property institutions.

Copeland and Taylor (2009) and Roumasset and Tarui (2010) explicitly combined the efficiency version of institutional analysis with renewable resource economics to obtain theories of which resources will be better served by which institutions and how institutions will optimally change in response to increasing resource scarcity. While the efficiency theories explicitly included governance costs, the primary emphasis was on demand-side factors such as changes in resource price.

Why should we have separate NIEs, one for agriculture and natural resources and another for industrial organization? I believe that the starting point for a reconciliation is being explicit about different levels of analysis. While agricultural production in small-farm Asian economies is competitive, the same cannot be said for industry. Thus the first-best level can be appropriately extended to allow for market power. At the second-best level, the Williamson and efficiency versions of the NIE are closer than appearances indicate. Williamson (1985) notes that the economic function of institutions is to "economize on transaction costs." Inasmuch as "economize" and "transaction costs" are somewhat vague, the efficiency version stipulates that economic institutions evolve to minimize contracting and agency costs, much as firm behavior evolves to approximate profit maximization (Alchian, 1950).[18] This proposition is in line with Coase's (1937) theory of when the producer of an intermediate product, such as Fisher Body, should be left as an independent contractor or be integrated into the firm as a division (of General Motors). Coase's idea was that the profit-maximizing firm would choose to integrate if and when the costs of internal organization (what we now call "agency costs") became less than the costs of dealing with the independent contractor (which we might call "contracting costs"). In other words, the profit-maximizing firm chooses the organizational architecture that minimizes the sum of agency and contracting costs.[19]

The Coasean theory of the firm can serve as a starting point for a synthesized version of the NIE. Instead of starting with behavioral departures from full rationality and strategic interaction of agents due to opportunism, we posit that efficient organization forms are those that minimize the sum of contracting and agency costs. As discussed above, this second-best theory goes a long way in explaining statistical patterns about agricultural organization in competitive settings. Departures from the efficiency benchmark may then be explained by either departures from full rationality due to information and decision costs or forces that block cooperation such as opportunism. These require, in turn, mechanisms of governance. At the second-best level of analysis, we take the relationship between expenditures on governance (monitoring, bonding, etc.) and shirking (departures from first-best performance) as given. The constitutional-design aspect of determining those relationships takes place at the third-best level.

---

[18]Oliver Williamson and I were on the same panel at the January 1975 meetings of the Public Choice Society. He presented a version of what became Chapter 1 of Williamson (1975), wherein he coined the term *The New Institutional Economics*. I presented "Induced Institutional Change, Welfare Economics, and the Science of Public Policy," originally written for Vernon Ruttan's 1974 conference on Induced Institutional Change. In my 1974 paper, I stated that institutions evolve to "economize on transaction costs." At the time, I was aware of the ambiguity of this expression, but (as Jensen and Meckling had not yet invented positive agency theory) I was content to leave clarification to future research (Roumasset, 1974). When Williamson later used the same expression (e.g., Williamson, 1985), it is not clear whether he understood that the ambiguity was a barrier to formalization. In Roumasset (1995), I show that minimizing agency costs, in the Jensen−Meckling sense, is equivalent to maximizing profits, inclusive of monitoring and bonding costs.

[19]This idea has been extended and formalized by Oliver Hart and coauthors (e.g., Grossman and Hart, 1986; Hart and Moore, 1990).

2.3.5 MORE ON BIG VERSUS SMALL FARMS

On the other side of the spectrum from the small-is-beautiful perspective reviewed in Section 2.3.1, we have the proponents of large-farm commercial agriculture. The latter contend that large farms have advantages in management, technology, specialization, mechanization, marketing, and credit (e.g., Collier, 2012). For purposes of their empirical investigation, Ali and Deininger (2014) have summarized this debate by saying that *either* small farms are efficient and that policy should focus on value chains *or* that they are inefficient and policy should focus on consolidation.

Why should it be either/or? Ay, there's the rub. If farm size were exogenous, it might make sense to investigate whether small or large farms are more efficient. However, as reviewed in Section 2.3.1, farm size is decidedly *endogenous*. The size of family farms depends largely on the size of the family and land quality (Roumasset and James, 1979) as well as shadow prices of credit as a determinant of the degree of mechanization (Roumasset and Thapa, 1984). The determinants of commercial farm size may also depend on land quality and shadow prices but in a very different way. For example, Uy (1979) finds an inverse relation between land quality and farm size for family farms but a direct relationship for commercial farms!

Indeed, in an efficient contracting environment, but with diverse land and water resources and idiosyncratic shadow prices, there is no conceptual reason to expect the evolution of contracts to favor *either* small or large farms. Rather we should expect that diverse conditions would lead to diverse farm sizes, the coexistence of which provides no evidence of inefficiency whatsoever.

If diversity is the natural order of things, agricultural policies should be formulated to increase productivity on large as well as small farms. Given the huge amount of fallow land in Africa, Uy (Chapter 10 of this volume) reviews the potential for private investment in large-farm agriculture. Inasmuch as many early attempts were beset by land grabs, speculation instead of investment, and displacement of smallholders, she lays out a strategy for avoiding those mistakes: Temporally space sales in phases, prequalify investors, agree on business plans and monitoring procedures, and involve the local community in both the planning and the enterprises themselves. For example, Ghana has had notable success in nuclear farming, an arrangement whereby a commercial enterprise leases and manages a large tract of land and provides planting materials and marketing for surrounding smallholders. This allows the large commercial enterprise to pursue its comparative advantage in enterprise design, finance, and marketing and the smallholders to specialize in labor intensive aspects of cultivation and in helping to provide security against outside interference. Senegal and Mozambique provide additional examples of creative engagement of local communities in contract design.

2.3.6 THE ECONOMICS OF THE THIRD-BEST: A CONSTITUTIONAL APPROACH TO GOVERNING RENT-SEEKING

As in the perspective of Vincent and Elinor Ostrom, the constitution is not necessarily a written document but the set of rules establishing membership, rights and responsibilities, decision-making procedures, and penalties/rewards for compliance and performance. The constitutional perspective is suggestive of an expanded view of the third-best level of analysis. From a positive perspective, third-best models seek to explain how individuals seek to appropriate extra-efficiency rents and the mechanisms that arise to combat this rent-seeking. For the case discussed above about coalitions for and against agricultural protection, one can imagine constitutional limits on levels of protection that would decrease rent-seeking. As it happens, these limits are effected largely at the international level, through institutions such as the WTO. The rent-seeking perspective also applies to contracting parties and individuals within firms as the firm entrepreneur seeks a constitutional architecture that provides the largest value added. The nature of the firm literature (e.g., Coase, 1937; Grossman and Hart, 1986; Hart and Moore, 1990; Roumasset, 1995) can be usefully viewed from this perspective.

The efficiency gains that can be garnered by institutions such as private property can be considered as demand-side factors and analyzed at the second-best level. However, institutional change does not happen spontaneously; there are political costs to be paid. North and Thomas (1973) and Davis and North (1970) implicitly recognized these in their reference to a "political action group" (Dixit's "endogenous coalition") that must supply the effort and resources to change an institution, e.g., to lobby the English Parliament to pass an act of enclosure on a specific tract of land. There is likely to be some opposition, and passage can be viewed as the outcome of a noncooperative game (*ala* Becker, 1983). In Demsetz's beaver example, demand-side factors include the high price of beaver pelts induced by European contact as well as the enforcement costs. Acceptance of what Demsetz calls private property was presumably facilitated by the implicit tribal constitution that provided a mechanism for changing the rules. Similarly, private property in the American West that Anderson and Hill associate with the advent of cheap fencing (barbed wire) was facilitated by the supply-side fact that legal property rights were already in place and that only enforcement was needed to complete "the bundle of property rights" (Hornbeck, 2010). This helps to reconcile the Demsetz view that institutional change will evolve when the benefits exceed the costs of enforcement with the view of "the new economic history" (North and Thomas, 1973) that change will happen when the net benefits exceed the political costs of change.

The third-best perspective also provides a promising approach to explaining the resource curse. Discovery of exportables, e.g., minerals, tourism, plantation agriculture, and other sources of foreign exchange, confer two potentially negative effects on growth. First, the booming sector draws resources away from other sectors, which may have greater growth externalities. Second, the potential boom stimulates the formation of a coalition to promote a particular exportable. In Hawaii, this took the form of a primary action group, led by prospective pineapple and sugar planters, to first lobby for private property and then for annexation as a U.S. territory. The export boom in turn stimulates import competition, and the dominant coalition joins with domestic commercial interests to seek import protection (Ravago and Roumasset, this volume). The "resource curse" is that the loss of growth externalities and rent-seeking shrink the economy relative to its potential. To the extent that rent-seeking coalitions persist, it is possible that lower growth rates persist into the future. What is needed is to develop a dynamic theory of the resource curse and appropriately confront it with data, ideally regarding the size of excess burden over time. Timing of the resource boom is also critical. European discoverers of Latin America found mineral abundance and designed extractive political institutions (Acemoglu and Robinson, 2012). These were not possible in North America, and by the time that gold was discovered in California, political institutions were already firmly in place.

At the normative (policy) level, governments should seek constitutional arrangements that limit rent-seeking. That is, given that individuals and groups are seeking to enlarge their share of income, even at the expense of shrinking the economic pie, government can seek arrangements that reduce such efficiency losses. I believe this is what the late Armen Alchian was getting at when he proclaimed that economics is the study of efficiency but that there are different levels of efficiency to be understood.[20] The problem of infrastructure, specifically irrigation projects, is instructive. Irrigation, especially in developing countries, is notoriously inefficient. Project location, design, operation, and maintenance are susceptible to the iron triangle (Lowi, 1969) of rent-seeking comprised of politicians, bureaucrats, and special interests. Projects end up being too capital intensive with a mismatch between design and operation such that farmers in the tail of the system are underserved. Projects are undermaintained and delivered water declines over time. Irrigation fees are low and underenforced (Repetto, 1986; Roumasset, 1987). The same paradigm can be used to analyze policy failures in general. For example, renewable energy subsidies may be the result of a strong coalition of environmentalists and special interests in the energy sector versus many ill-informed taxpayer/consumers facing relatively small individual stakes and high coalition costs (Olson, 1965).

In their theories of "just taxation," Wicksell and Lindahl argued that the key to reducing rent-seeking is benefit taxation. This idea can be extended to the fiscal constitution of irrigation design, operation, and finance (Roumasset, 1989). The centerpiece of the fiscal architecture is the water-user association, which chooses the irrigation design and operates the irrigation system. Direct beneficiaries (farmers) are charged according to their individual benefits, which are only a proportion of total benefits, rendering full-cost recovery inappropriate. To the extent that benefit taxation of indirect beneficiaries (consumers, downstream suppliers, and commercial interests that benefit from increased demand) is administratively infeasible, their share of benefits can be financed out of general taxation under the auspices of the ministry of finance or treasury. The national irrigation association provides menus of irrigation design and consulting services of engineers. Federations of water-user associations may also be appropriate. The incentives for rent-seeking are greatly reduced. Contractors are not overpaid because the water-user association economizes on costs. The political-patronage benefits to politicians are reduced by both the reduced rents to contractors and the fact that farmers are paying their fair share. Also, decentralization of design authority limits the power of central bureaucracies. This exemplifies the normative side of third-best analysis.

2.4 MODERN THEORIES OF MARKET AND INSTITUTIONAL FAILURE: SHOCKS, TRAPS, NETS, AND LADDERS

As represented in the WDR 2008, theories abound regarding reasons that markets and institutions in developing countries are inadequate for efficiency and that government intervention is therefore required. Possibly the most far reaching argument is the Greenwald–Stiglitz theorem, according to which government can "almost always" improve on the market allocation (Stiglitz, 1991). This formulation commits a sophisticated version of the *Nirvana Fallacy* (Demsetz, 1969) by portraying market equilibrium as a straw man and implicitly assuming that only *the government* can do governance. Private governance structures and multilateral contracting are not allowed.

Chetty and Looney (C-L) (Chetty and Looney, 2005, 2006) have taken this line of argument to a new level. They begin by reviewing arguments by Townsend (1994) and Morduch (1995, 1999) who conclude that, inasmuch as the coefficients of variation of consumption are similar in

---

[20]Remarks by Armen Alchian during a 1977 training program for "promising" young professors.

developed and developing countries, poor households are equally good at smoothing consumption as higher income households. C-L correctly point out that this conclusion does not follow. It is logically possible that poor households are highly risk-averse (have low elasticities of intertemporal substitution) and make extreme sacrifices in order to smooth consumption. The similar coefficients of variation could logically result from higher risk aversion in developing countries combined with higher cost of smoothing or from developing countries having the same levels of risk aversion and smoothing costs as developed countries. Since the latter is unlikely, C-L conclude that this provides a *prima facie* case for government investment in social insurance.

However, this is another case of misplaced exogeneity. While C-L's argument establishes that it is *plausible* that there are substantial benefits to social insurance, many steps remain before concluding that particular forms of social insurance are warranted. In particular, the ease of smoothing consumption depends on transaction costs. Without transaction costs, borrowing and lending costs would be equal, credit constraints would be equal to total household wealth, and the only impediment to perfect smoothing would be shocks larger than wealth. Shocks would only matter according to the amount by which they decrease wealth. Because of moral hazard, the benefits of social insurance in this transaction-costless world would be negative. Meanwhile, the costs of crop and other insurance programs are likely to be high because of countermeasures against moral hazard and adverse selection as well as means-testing and other administrative procedures (Roumasset, 1978a, 1979a). Even in the United States, the costs of crop insurance can be as much as 3.5 times premia.[21]

To the extent that social insurance programs are likely to have benefits less than costs, a direct assault on transaction costs may be a more suitable approach. Inasmuch as transaction costs are the source of the problem, programs that reduce them (e.g., through physical and legal infrastructure) will facilitate greater private smoothing. Poverty can also be addressed through subsidies of basic needs, e.g., through conditional cash transfers. The conceptual justification of basic-need supports has little to do with shocks and theoretically sophisticated "poverty traps" but rather is founded on the principle of consumption externalities, i.e., that nonpoor taxpayers are made better off by increased food, shelter, health, and education of the poor (Harberger, 1978, 1984). We do not have to know whether the benevolence of the taxpayer comes from his altruism or self-interest (e.g., less slums to breed crime). Either case is appropriately represented as a consumption externality. Accordingly, basic needs consumed by poor households are public goods and can be efficiently provided according to the Lindahl equilibrium, which also satisfies the equity criterion of benefit taxation.[22]

The rural poor are often characterized as trapped in poverty by nonconvexities (e.g., Banerjee and Duflo, 2011).[23] These arguments often overlook the availability of institutional ladders for climbing out. For example, James and Roumasset (1984) document nonconvexities in the return-to-investment function facing migrants to newly settled areas of the Philippines. The highest rate of return accrued to those with adequate capital to fund the establishment of their own homesteads. However, institutions were also available for low-asset migrants. Those without savings could begin as share tenants. After typically five years, these migrants had established homesteads. Despite earning a lower rate of return, they became middle-class households. An intermediate

---

[21]Conference presentation by Brian Wright.
[22]See the discussion of "the Lindahl Voucher" in Roumasset (1993).
[23]This is a relatively new expression of an old idea. As Billie Holiday put it, *"Them that's got shall get, them that don't shall lose, so the Bible says, and it still is news."*

rate-of-return arrangement called "land borrowing" was available for migrants with an intermediate amount of savings. After typically three years farming and improving borrowed land, these migrants were able to return the use rights to the owner and farm their own homesteads. That is, despite nonconvexities, each group had a path to a higher socioeconomic position.

The ability of contracts to provide an agricultural ladder has also been documented for alternative tenancy contracts in Philippine rice production (Roumasset and Smith, 1981; Hayami and Otsuka, 1993). Future farmers may start as agricultural workers, become share tenants, advance to leaseholders or part owners, and may eventually become owners. When these institutions are available, there is little reason to believe the nonconvexity imperative—that nonconvexities automatically create a trap. Ironically, well-intentioned government actions, such as outlawing share tenancy and land-to-the-tiller reforms, may have the unintended effect of removing institutions from the agricultural ladder, possible trapping individuals on a low rung of the ladder.

Another possible source of agricultural stagnation is the lack of market outlets for inputs such as fertilizer. This has the appearance of a classic chicken–egg problem. Dealers don't sell fertilizer because of low farmer demand, but farmers must first learn the benefits of fertilizer. This coordination problem has led to the Pigouvian prescription of temporary subsidies to boost demand and/or supply. An alternative prescribed by the facilitation approach is for government to enter into public–private partnerships to supply fertilizer combined with extension efforts to teach the best composition, amount, and timing of fertilizer application for farmers in diverse agroclimatic and economic environments. The "trap" may also be illusory. Fertilizer is commonly supplied by multipurpose stores who cater to consumers as well as farmers and any indivisibilities in fertilizer delivery may be small. As reviewed in Section 2.2.2, the claim that low-income farmers irrationally underuse fertilizer remains controversial.

2.5 THE ANATOMY OF SPECIALIZATION

While specialization is the engine of growth (Smith, 1976; Young, 1928; Lucas, 2002), it remains largely confined to a black box. Agricultural specialization is a good laboratory because there are so many firms.

Consider the evolution of specialization in the agricultural labor markets of developing countries. In the induced-innovation paradigm, the Asian green revolution in rice and wheat was motivated by land scarcity caused by population pressure. The new varieties were developed to warrant greater amounts of labor and especially chemicals per unit of land, which increased yields per hectare. The same population pressure and indivisibilities regarding smaller farm sizes led to a social bifurcation between the landed and landless and the rapid expansion of hired labor (Roumasset and Smith, 1981). This initial specialization between management and labor was followed by horizontal specialization among different tasks and to further vertical specialization (e.g., teams of transplanting men headed by a team leader acting as both recruiter and supervisor; Roumasset and Uy, 1980). The organization pyramid of the agricultural firm became both wider (increased horizontal specialization) and higher, with workers taking direction from supervisors taking direction from farm operators. The evolution of horizontal specialization over time is illustrated by the "specialization pyramid" (e.g., Roumasset and Lee, 2007), where bars increase in width, reflecting more labor per hectare as we descend the pyramid, and display ever new forms of specialization, going from family to exchange to wage labor and then to piece-rate and piece-rate-with-teams.[24]

The coevolution of intensification and specialization stands in stark contrast to the neo-Marxian critique, whereby the new rice and wheat technologies spawned increased commercialization and immiserization of the peasantry. This view, embraced by a much publicized Asian Development Report in 1979, seemed to explain the negative correlation between adoption of the new high-yielding varieties and agricultural wages. This conclusion commits the fallacy of *post hoc ergo propter hoc*, a form of misplaced exogeneity. In the more fundamental induced-innovation paradigm (Binswanger and Ruttan, 1978), the modern varieties were the endogenous response to increased land scarcity, which they helped to ameliorate. Since agricultural production with new varieties represented only a small part of Asian economies, their positive effect on wages was swamped by the negative force of population pressure. Once we understand that the new varieties were induced by land scarcity and relatively falling wages, we can understand their positive effect.

Agricultural specialization has also been prominent in output markets. The *supermarket revolution* (Reardon and Timmer, 2007) describes the advent of vertical coordination between farmers and "big box" retailers, largely by means of specialized wholesalers who are dedicated to specific supermarket chains. This vertical coordination allows retailers to cater to specific product niches and to tailor farmer orders accordingly. This exemplifies how the expansion of Adam Smith's extent of the market begets vertical specialization and coordination, which begets more horizontal specialization, which increases the value-added from agriculture via Dixit–Stiglitz preferences for variety and improved marketing efficiency (product transformation over space, time, and form).

Perhaps the most revealing inconsistency of the small-farm bias (Sections 2.3.1 and 2.3.5) is ignoring the implications of the supermarket revolution for efficient farm size. For example, the WDR 2008 uses the alleged transaction cost advantage of small farms (with their lower fractions of hired labor) to advocate policies that would further shrink average farm size. However, the supermarket revolution places small farms at a transaction cost disadvantage in marketing (e.g., Lipton, 2009; Roumasset, 2010), further augmenting the disadvantage they already face in credit markets. Proponents of the small-is-beautiful view seem to overlook this advantage, believing instead that any advantage of small farms is "natural," but any advantage of large farms results from a "policy bias." In effect, the alleged political economy of Latin America (e.g., de Janvry, 1981; IFAD, 2001) has been presumed to be accurate for the world as a whole. This led the authors of WDR 2008 to advocate small-farm subsidies, including public support of marketing and credit cooperatives

Again, such policy implications are premature absent a fundamental explanation of behavior and organization in developing agriculture. Despite an abundant literature on the supermarket revolution, the nature of specialization inherent in new agriculture has not, to my knowledge, been adequately explained. In the old marketing system, inevitable differences in products across space, time, and form were dealt with by wholesalers and others in the marketing chain, who sorted and processed products after production. A farmer would not know the destination of his product in advance and could not tailor it to the particular tastes of that niche market. In the new agriculture,

---

[24]In the 1980s, wages increased sufficiently to motivate labor-saving mechanization in Laguna, Philippines such that intensification continued, as evidenced by higher yields per hectare, but labor per hectare began to decline.

the varieties, quality, and timing of farm products can be prearranged through a contracting process with particular farmers. In this way, vertical coordination begets horizontal specialization and produces external economies of the sort lauded by Lucas (2009).

Even leaving aside the transaction cost advantages of large farms regarding credit, marketing, and self-insurance, the alleged transaction cost disadvantages of hiring labor are countered by their advantages in facilitating specialization. As shown by the labor market pyramid (Roumasset and Lee, 2007), more farm labor facilitates horizontal specialization by task within the farm and vertical specialization whereby worker teams or firms (especially for land preparation, planting, weeding, and harvesting) provide intermediate inputs. Because of these internal and external economies of specialization through which hired labor can increase productivity, it may well be that promoting small-farm agriculture through regulations, confiscations, and subsidies will exacerbate preexisting distortions that have rendered many farms smaller than efficiency would dictate.

A related omission is the extent to which misguided land reform policies have impeded the natural evolution of economic organization, e.g., economies of vertical integration and horizontal specialization of food markets. Surely some policies result from rent-seeking large farmers, but the potential inefficiencies of large-farm dissolution and the unseen prevention of consolidation are not typically considered.

The transaction cost wedge allegedly facing large farms is the centerpiece of both the case for land reform ("small-farm entry") and small-farm "competitiveness" subsidies for fertilizer, credit, insurance, and producer cooperatives. The support for these market interventions rests on the claim that existing equilibria are dualistic. Dualism, a resilient cockroach[25] from the 1960s and 1970s, rests in turn on the notorious claim that the inverse relationship between farm size and yield per hectare is evidence of labor market duality, which has had a long life in agricultural development circles (see, for example, Sen, 1962, 1966; Berry and Cline, 1979; World Bank, 2008).

Ignoring all explanations for the inverse relationship except the one most suggestive of market failure is fallacious. There are at least three efficiency explanations. The first is that larger farm size on poorer quality land is a device for equalizing the marginal product of labor on family farms (Roumasset, 1976; Roumasset and James, 1979). The second is the transaction cost wedge whereby larger farms have a greater incidence of hired labor and therefore hire less labor per hectare (e.g., Sah, 1986). The third is endogenous occupation choice given heterogenous farm skills (Assuncao and Ghatak, 2003). Thus on logical grounds alone, one cannot jump to the conclusion that the inverse relationship is *prima facie* evidence of inefficiency nor that corrective intervention is required. Assuncao and Braido (2007) find empirical support for the land quality hypothesis but not for the other two. This does not imply that the second and third explanations are invalid. Rather, the measurement and specification problems, especially regarding transaction costs, may have rendered those relationships insignificant.

There are other logical problems with jumping to the conclusion that small farms are more efficient. First, small, productive farms in low-wage Asian economies often do hire labor at the margin (e.g., Roumasset and Smith, 1981; Hayami and Kikuchi, 1982).[26] Labor turnover costs (Stiglitz, 1984),

---

[25]As Krugman (1994) has noted, the role of economists is to flush bad ideas, but like New York cockroaches, he notes, the bad ideas just keep coming back.

[26]For example, the rapid emergence of the labor market for rice production in Laguna, Philippines, during the 1970s was a small farm phenomenon.

including recruiting, training, and negotiating costs, are subject to economies of scale. This may put small farms that hire labor at a transaction cost disadvantage. Small farms that don't hire labor refrain not because family labor is cheaper but because they cannot capture the economies of scale, including those of specialization, that larger farms afford. Also, as the relative cost of farm equipment declines relative to labor, the scale economies of farm mechanization drive the optimal farm size up. This does not mean that government should promote reverse land reform to encourage consolidation but that biasing land policy to promote small farms may increase inefficiency and stagnation.

The economics of specialization, partly developed from the stylized facts of agricultural organization, can potentially provide more fundamental foundations for understanding the industrial revolution. For millennia, population and agriculture moved together such that per capita incomes increased over the long run at 0.1% or less and median incomes remained almost stagnant (Lucas, 1993). The few long swings of increasing per capita incomes were eventually overcome by Malthusian pressures until the eighteenth century, when wages began their secular increase until the present day. Why was this associated with industrialization? Agriculture also has numerous opportunities for specialization that can raise (at least temporarily) total factor productivity, but these "Boserup effects" are few and far between in agriculture (Boserup, 1965; Roumasset, 2008). In contrast, the opportunities for specialization in industry are almost endless. While vertical specialization in agriculture is largely limited to specialized management and labor, manufacturing affords ever continuing layers of intermediate products and product differentiation. This greater density of specialization possibilities was a key driver of the growth acceleration that accompanied the rural to urban transition.

Economics is still struggling with the problem of embedding the coevolution of institutions and specialization into dynamic, general equilibrium models. A promising path forward is suggested by the New Classical Economics (see especially Yang, 2003), which models the evolution of specialization as a response to falling unit transaction costs. Incorporating endogenous transaction costs into such models and combining them with an appropriate equilibrium concept is a major challenge that will require advanced mathematics, possibly borrowed from electrical engineering. See, for example, Kratz et al. (2001).

2.6 BLACK-HOLE ECONOMICS

There is a family resemblance between policies to prohibit alcohol, drugs, prostitution abortion, illegal immigration, and guns. Enforcement in all cases is, was, or would be ineffective—leading to a black hole of spiraling enforcement costs without necessarily improving results. This common problem may have the common solution of *high-fence, wide-gate.*

2.6.1 PROHIBITION OF ALCOHOL AND DRUGS

Black-hole economics is perhaps best illustrated by the U.S. experience with alcohol prohibition. Prohibition initially cut alcohol consumption by two-thirds, but by the end of prohibition, consumption had more than doubled, even as enforcement expenditures multiplied more than sixfold. The greater the enforcement expenditures, the greater the difference between the street price of alcohol and its cost of production. These potential rents stimulated technical and transactional (bribery) innovations and changes in market structure that lowered the risk of being punished for illegal activities, thus lowering the effective price and increasing consumption. These in turn stimulated new rounds of innovations and so on—a black hole of expenditures that just made the problem worse, ending only with the repeal of prohibition (Roumasset and Thaw, 1999).

The same inexorable dynamic renders the U.S. "war on drugs" counterproductive. Welfare economics suggests that the policy goal toward currently illicit drugs should be harm reduction, especially the externality of harm to others. Distinguishing between high-externality drugs (e.g., crystal meth, crack, and date rape drugs) and low-externality drugs, such as marijuana and heroin,[27] would allow a *high-fence, wide-gate* policy to be employed in drug, enforcement. The set of legal drugs would be expanded, and the enforcement effort against illegal drugs would be increased. Legalization combined with revenue-maximizing taxation has been shown to be welfare increasing. By spending only 11% of revenues (in the case of cocaine) on treatment programs, drug use can be held constant, leaving 89% for welfare improvement, not even counting the benefits to treated addicts. By spending even more on demand reduction, legalization can be a pathway to reducing consumption (Johansson and Roumasset, 1997).

2.6.2 ILLEGAL IMMIGRATION

Immigration policy is another black-hole phenomenon in the sense that the laws are routinely violated. About half a million unauthorized immigrants were entering the United States prior to the Great Recession, down from about 850,000 per year in the early 2000s. Legal immigration has been about double illegal immigration, although that ratio may be falling. As suggested by Senator Chuck Schumer, "Americans are for legal immigration but against illegal immigration." There is some indication that this is good economics. Immigration does not need to be decreased, but its composition needs to be changed. Legal immigration may promote economic and wage growth because of the complementarity of the imported skills and entrepreneurship while illegal immigrants are more likely to retard wage growth, expand entitlement spending, and increase negative externalities (crime).[28] To the extent that current enforcement efforts are not as effective as desired, a *high-fence, wide-gate* policy is promising. At the same time that illegal enforcement is enhanced by actual or metaphorical fences,[29] legal immigration controls could be relaxed, such that the high fence and the wide gate combine to discourage illegal immigration by both increasing its expected penalty and elevating the prospects of the alternative. As illegal immigration is decreased, legal immigration is increased. As in the drug case, intermediate forms between legal and illegal immigration are under consideration, including faster pathways to citizenship for workers in high demand and guest worker programs. *High-fence, wide-gate* helps draw attention away from framing

---

[27]Heroin is a low-externality (addicts are nonviolent) but high- "internality" (highly addictive) drug. Intervention should be based on merit-good grounds, such as behavior modification via methadone (see also Gruber, 2002, 2003).

[28]See, for example, the mechanisms discussed by Williamson (1996a). Perhaps because of data limitations, the growth effects of legal and illegal immigration have not been well developed. One report attributes three out of every four patents filed to first-generation immigrants ("Patent Pending: How Immigrants Are Remaking the American Economy").

[29]For example, stiffer penalties for employers who hire illegals.

issues in a polarizing way—drugs and immigrants must be good or bad—and focuses instead on developing a screening mechanism to increase the benefits and decrease the harm.

Immigration has long been an important source of U.S. growth. Immigration has provided a supply of hard-working and creative labor that supports the growth engine of specialization. From the point of view of resource economics, we can further analyze this as an optimal control problem. Two control variables are required to regulate the additions to the two stocks, legal and illegal immigrants.

To the extent that border enforcement has been too lax (the metaphorical fence too low) and the barriers to legal immigration too stiff (gate too narrow), the current stock of legal immigrants is too small and that of illegals is too large. This means that the control on legal immigrants should be temporarily even lower than its long run optimal level and the control on illegal immigrants higher (i.e., wider fence and higher gate in the interim). Even with these controls in place, it may take decades for levels of the two types of immigrants to reach their optimal levels. Accordingly, additional instruments are needed to deal with the stock of illegals that has accumulated, including a path to citizenship for those who fulfill appropriate conditions and a path to deportation for those who don't. Policies are also needed to allow greater market guidance in the selection of legal immigrants as opposed to selection committees. Guest worker programs may be a useful complement to immigration policies. These presume an adequate registration program for appropriate enforcement.

2.6.3 ABORTION AND PROSTITUTION

Applying the same principle to abortion policy suggests that the goal of abortion policy should be harm reduction. The high fence in this case might be the ban on late term abortion as applied in many countries and most states of the United States.[30] The gate can include not only legalization of less-harmful abortions but alternatives including birth control and adoption. The abortion policy analogy to the drug and immigration cases would be to legalize abortion but to tax it, not subsidize it through the health care system. Instead, it is the alternatives that should be subsidized (just as drug treatment programs).

Similarly, prostitution policy should begin with the recognition that all prostitution is not equal. Legalizing some forms of prostitution would also help focus enforcement expenditures on human trafficking for sexual exploitation, child prostitution, and other harmful practices that naiveté prevents me from describing. Licensing and taxation provide both revenue and a means of regulation, e.g., regarding sexually transmitted diseases.

2.6.4 BANS AND SUBSIDIES: PARASTATALS, RENEWABLE ENERGY, AND SUSTAINABILITY

Prohibition sometimes goes hand in hand with subsidization. Consider grain-marketing parastatals that are justified on the grounds that price fluctuations ("shocks") are bad for consumers and producers, though it has not been established why. If the source of domestic price instability is international price variability, even costless stabilization is likely to be welfare reducing inasmuch as consumers gain more consumer surplus from low prices than they lose from high prices, and

---

[30]Generally these bans allow exceptions when there is a threat to the life or health of the mother. See, for example, Guttmacher Institute (2007).

producers gain more from high prices than they gain from low ones. The analysis is different if the source of unstable prices is fluctuations in domestic supply. In this case, costless stabilization via stable international prices or an omniscient operator of an unlimited domestic buffer-stocking program would be welfare increasing, but feasible acquisition and release strategies are likely to be welfare reducing due to the limited potential stabilization and high costs. Moreover, empirical evidence suggests that attempts to stabilize grain prices succeed only in raising prices (Roumasset, 2000, 2003b), and theoretical analysis shows that stabilization strategies involving buffer stocks tend to be destabilizing in the long run due to the probability that stocks, storage capacity, or available budgets will eventually be exhausted (Williams and Wright, 2005).

Nonetheless, misguided critics of globalization justify price-stabilization programs on the grounds that consumers and producers should be insulated from "the vagaries" of international markets. This view legitimizes the creation of grain parastatals and the banning of international trade in grain staples such as rice. These interventions have often led to large losses in consumer welfare by inducing consumer prices much higher than what would prevail under free trade. In the Philippine case, producer prices are also higher than what free trade would deliver but by much less than the consumer markup. Excess burden results from lost consumer surplus, excess production costs, tax friction, and foregone revenue that could have been obtained with a system of tariffs and taxes that would have delivered the same prices. Remarkably, while the Philippine parastatal was buying low and selling high, they actually lost money in the process necessitating substantial government subsidies. In 1999, these static losses amounted to more than a billion dollars for rice alone (Roumasset, 2000). Black-hole subsidies increased in subsequent years as the National Food Authority program expanded to other crops and agricultural produce. Moreover, the static losses hide a source of dynamic efficiency. When governments outlaw international trade, and engage in major transport and storage operations, investment in private marketing is curtailed and the patterns of production, trade, storage, and processing are disrupted. The very things that led to increasing specialization and rising value added in the supermarket revolution, which occurred primarily outside of government regulations, are curtailed.

In a recent review of "Why Government Fails so Often," Yuval Levin (2014) notes that the twin temptations of social science are nostalgia and utopianism. Nothing is more nostalgic and utopian than the quest for self-sufficiency. Due to public ignorance of the *gains from trade*, politicians are able to convince voters that letting money flow out of the domestic economy for imports will shrink the economy. Another bogeyman is the "exploitative middleman." Marcos and his cronies were able to monopolize coconut processing and trading under the pretense that ethnic Chinese traders were gouging both producers and consumers. When the government took over, the wedge between producer and consumer prices increased dramatically (Clarete and Roumasset, 1982).

For most economies, such ideological imperatives (including stabilization of prices at pleasing levels for both consumers and producers and the elimination of drugs, guns, abortion, and prostitution) are missions impossible. When such missions fail, the bureaucracies that have grown up to promote them inevitably clamor for more funding (Schuck, 2014). In most cases, however, society will not tolerate the black-hole economic costs that would be required. Instead the actual policies are illusory "rope-a-dope" programs meant to both curry favor with the public and transfer rents to special interests.[31]

---

[31]Thanks to University of the Philippines Professor Noel de Dios for this insightful term.

The quest for self-sufficiency is confounded with popular (noneconomic) conceptions of sustainable development. In Hawaii, for example, the mantra of sustainability is used to support a plethora of prohibitions, mandates, and subsidies in the name of renewable energy. In the push for the goal of 40% renewable production of electricity by 2030, for example, Hawaii has embarked on a series of extremely costly policies. Nonrenewable ventures, such as the use of natural gas, have been tacitly discouraged while a wide variety of renewable ventures have been promoted by tax credits, capital supports, and purchasing guarantees. The state has adopted a regulatory policy of decoupling utility revenues from sales, allegedly to promote renewables, that has resulted in electricity prices three times the national average (Endress, 2013; Chapter 3 of this volume). Now the state is about to launch a project of unknown feasibility and expense to connect the "outer islands" to the population center of Oahu. It's looking more and more like a black hole.

As Joseph Stiglitz pointed out in reference to Hawaii's high electricity prices, (unnecessarily) "high prices shrink the economy."[32] This example of mandates and subsidies leading to high electricity prices in Hawaii illustrates another dimension of the black hole. Both the commercial sector and the general public forego large benefits due to prices far in excess of necessary costs. Remarkably, the mandates and subsidies in question are justified by the quest for sustainability.

Sustainability need not be a separate goal from optimality. Optimality is usually sustainable but not the other way around (Heal, 2003). Heal concludes,

> Instead of proselytizing about sustainability as a social goal, perhaps environmental economists should work to refine the concept of optimality [such that it] includes an understanding of human dependence on environmental systems.

In other words, sustainability can be pursued through optimizing models. In order to incorporate the concerns of sustainable development, we need to include not only the interdependence between environmental-resource systems and the economy but also the social concern for intergenerational equity (Chapters 1 and Chapter 3 of this volume). Far from these apparent burdens leading to a dismal outcome, the pursuit of optimality can provide a win-win path to abundance (Roumasset and Endress, 1996).

In his review of Schuck (2014), Levin notes that the reason why government fails so often is that it is essentially impossible for centralized managers to consolidate information to the degree necessary to manage complex social systems, and bureaucracies respond to failure by demanding even more power. In contrast, government succeeds more often when it focuses on establishing the circumstances for success (as with the GI bill and the interstate highway system in the United States) instead of on prescribing specific institutions and behaviors. Thus, Levin (2014) writes, "Government's best practice is to set goals and arrange incentives so society's knowledge can be put to use by its dispersed possessors."

Moreover, it is not just bad policies that threaten economic resources from being pulled into a black hole but the deadweight loss of the arm-race-like lobbying efforts (Magee et al., 1989). In Chapter 9 of this volume, Majah Ravago and I discuss how a resource or other foreign-exchange boom may stimulate a rent-seeking coalition to garner a share of the new largesse. Rent-seeking begets more seeking through learning-by-doing and by stimulating retaliatory action by losers.

---

[32]Remarks in his Stephen and Marylyn Pauley Seminar in Sustainability, University of Hawaii, Manoa, 2012.

2.6.5 **THE ROLE OF THE ECONOMIST**

The role of the economist in mitigating policy failures was famously anticipated by Adam Smith:

> People of the same trade seldom meet together, even for merriment and diversion, but the conversation ends in a conspiracy against the public, or in some contrivance to raise prices.
> **Wealth of Nations, Book I, Chapter 10**

As noted by Voltaire, government officials may be involved in the conspiracy: "In general, the art of government consists of taking as much money as possible from one class of citizens to give to another." Smith himself realized the potential for government involvement in the passage subsequent to the above: "But though the law cannot hinder people of the same trade from sometimes assembling together, it ought to do nothing to facilitate such assemblies; much less to render them necessary."

Later economists suggested the more general proposition that economic policies will tend to be biased toward relatively small coalitions with high potential per-member benefits relative to any potential blocking coalition. As Arrow (1969, p. 51) puts it: "It is not the presence of bargaining costs per se but their bias that is relevant." More specifically, Mancur Olson famously postulated: ". . . the larger the group, the less it will further its common interests" (Olson, 1965, p. 36).

The first step to formalizing what has been called "Olson's Law" is to distinguish two forces at work. The first, called the "Hume Theorem," is that the Nash equilibrium extent of public-good underprovision increases with the number of potential beneficiaries (Roumasset, 1991). The second, more in the realm of cooperative games, is that the costs of collective action per member eventually rise as the group expands to include members with lower individual gains (Olson, 1965, p. 50). Thus for heterogeneous groups, the marginal costs of cooperation as a function of group size increase for two reasons. First, the required individual contributions compared to the noncooperative solution increase with the number of group members. Second, additional members tend to have lower individual gains compared to the high-benefit members (who presumably are among the earlier members) and will thus will be harder to entice into membership. On the benefit side, marginal benefits decrease with group size so long as members are ordered as indicated. This implies an optimal group size for collective action, which in turn provides a basis for reaction functions providing a Nash equilibrium determination of public policy (Becker, 1983; Balisacan and Roumasset, 1987; Gardner, 1988).

If policies are determined by factors such as optimal coalition sizes and benefits and costs of investing in political influence by members, this would seem to suggest a rather nihilistic result regarding the role of the economist. Perhaps as Blinder (1988) suggests, policy-makers (or coalitions) will simply select the economist who best defends their predetermined position.[33] Nonetheless, many economists, including Blinder himself, strive mightily to correct what they see as policy mistakes. As Kenneth Arrow once told me, "it is natural in academic life to lean against the wind." Milton Friedman observed that "only a crisis—actual or perceived—produces real

---

[33]To his great credit, Blinder continues to ardently explain the organization of economic affairs and to draw out policy implications (e.g., Blinder, 2013).

change. When that crisis occurs, the actions that are taken depend on the ideas that are lying around. That, I believe, is our basic function: to develop alternatives to existing policies, to keep them alive and available until the politically impossible becomes the politically inevitable."[34]

In terms of the political economy of public policy, by making the machinations of the iron triangle more transparent, the economist can change the rent-seeking calculus of politicians, bureaucrats, and special interests and of taxpayer/consumers in opposing them. A cadre of courageous economists, willing to speak to and for the general public and expose the nature, causes, and consequences of policies and policy proposals, can act as an informal transparency agency that militates against efforts to plunder the general public.

REFERENCES

Acemoglu, D., 2002. Directed technical change. Rev. Econ. Stud. 69 (4), 781−809.

Acemoglu, D., Robinson, J., 2012. Why Nations Fail: The Origins of Power, Prosperity, and Poverty. Crown Business, New York.

Alchian, A., 1950. Uncertainty, evolution and economic theory. J. Polit. Econ. 58 (3), 211−221.

Ali, D., Deininger, K., 2014. Is there a Farm-Size Productivity Relationship in African Agriculture? Evidence from Rawanda. World Bank, Washington, DC (World Bank Policy Research Working Paper 6770).

Allen, D., Lueck, D., 2003. The Nature of the Farm: Contracts, Risk, and Organization in Agriculture. MIT Press, Cambridge, MA.

Anderson, K., Kurzweil, M., Martin, W., Sandri, D., Valenzuela, E., 2008. Methodology for Measuring Distortions to Agricultural Incentives. World Bank, Washington DC (Agricultural Distortions Working Paper 02).

Anderson, T., Hill, P., 1975. The evolution of property rights: a study of the American West. J. Law Econ. 18 (1), 163−179.

Anderson, T., Hill, P., 1990. The race for property rights. J. Law Econ. 33 (1), 177−197.

Arrow, K., 1969. The organization of economic activity: issues pertinent to the choice of market versus non-market allocations. In: Analysis and Evaluation of Public Expenditures: The PPB-System, Joint Economic Committee, 91st Cong., 1st session.

Assuncao, J., Braido, L., 2007. Testing household-specific explanations for the inverse productivity relationship. Am. J. Agric. Econ. 89 (4), 980−990.

Assuncao, J., Ghatak, M., 2003. Can unobserved heterogeneity in farmer ability explain the inverse relationship between farm size and productivity. Econ. Lett. 80 (2), 189−194.

Bahrampour, T., 2010. Number of Illegal Immigrants in US Drops, Report says. The Washington Post, September 1. Retrieved July 30, 2011.

Balisacan, A., Roumasset, J., 1987. Public choice of economic policy: the growth of agricultural protection. Weltwirtsch. Arch. 123 (2), 232−248 (Review of World Economics).

Banerjee, A., 2000. The two poverties. Nord. J. Polit. Econ. 26 (2), 129−141.

Banerjee, A., Duflo, E., 2011. Poor Economics: A Radical Rethinking of the Way to Fight Global Poverty. Public Affairs, New York, NY.

Banerjee, A., Duflo, E., Hornbeck, R., 2014. Bundling health insurance and microfinance in India: there cannot be adverse selection if there is no demand. Am. Econ. Rev. 104 (5), 291−297.

Bardhan, P., 2005. Theory or empirics in development economics? Econ. Polit. Wkly XL (40), 4333−4335.

Baumol, W., 1986. Review: Williamson's the economic institutions of capitalism. Rand J. Econ. 17 (2), 279−286.

---

[34]From the 1982 preface (Friedman, 2002).

Becker, G., 1983. A theory of competition among pressure groups for political influence. Q. J. Econ. 98 (3), 371−400.

Ben-Porath, Y., 1980. The F-connection: families, friends, and firms and the organization of exchange. Popul. Dev. Rev. 6 (1), 1−30.

Benjamin, D., 1995. Can unobserved land quality explain the inverse productivity relationship? J. Dev. Econ. 46 (1), 51−84.

Berry, A., Cline, W., 1979. Agrarian Structure and Productivity in Developing Countries. Johns Hopkins University, Baltimore, MD.

Binswanger, H., Rosenzweig, M., 1993. Wealth, weather risk, and the composition and profitability of agricultural investments. Econ. J. 103 (416), 56−78.

Binswanger, H., Ruttan, V., 1978. Induced Innovation: Technology, Institutions, and Development. Johns Hopkins University Press, Baltimore, MD.

Binswanger, H., Deininger, K., Feder, G., 1995. Power distortions revolt and reform in agricultural land relations. In: Behrman, J., Srinivasan, T.N. (Eds.), Handbook of Development Economics, vol. III. Elsevier, Amsterdam, The Netherlands, 2659−2772.

Blinder, A., 1988. Hard Heads, Soft Heart: Tough-Minded Economics for a Just Society. Basic Books, New York, NY.

Blinder, A., 2013. After the Music Stopped: The Financial Crisis, the Response, and the Work Ahead. Penguin Press, New York, NY.

Boserup, E., 1965. The Conditions of Agricultural Growth: The Economics of Agrarian Change Under Population Pressure. Aldine, Chicago, IL.

Chetty, R., Looney, A., 2005. Income Risk and the Benefits of Social Insurance: Evidence from Indonesia and the United States. National Bureau of Economic Research, Inc. (NBER Working Paper 11708).

Chetty, R., Looney, A., 2006. Consumption smoothing and the welfare consequences of social insurance in developing economies. J. Public Econ. 90 (12), 2351−2356.

Clarete, R., Roumasset, J., 1982. Economic Policy and the Philippine Coconut Industry. Philippine Institute for Development Studies, Makati (PIDS Working Paper 83-08).

Clarete, R., Roumasset, J., 1987. A Shoven−Whalley model of a small-open economy: an illustration with Philippine Tariffs. J. Public Econ. 32 (2), 247−261.

Coase, R., 1937. The nature of the firm. Economica 4 (16), 386−405.

Coase, R., 1994. Essays on Economics and Economists. University of Chicago Press, Chicago, IL.

Cole, S., Gine, X., Tobacman, J., Topalova, P., Townsend, R., Vickery, J., 2013. Barriers to household risk management: evidence from India. Am. Econ. J. Appl. Econ. 5 (1), 104−135.

Collier, P., 2012. An overview of African development prospects. In: Aryeetey, E., Deverajan, S., Kanbur, R., Kasekende, L. (Eds.), The Oxford Companion to the Economics of Africa. Oxford University Press, New York, NY, 26−32.

Copeland, B., Taylor, M., 2009. Trade, tragedy, and the commons. Am. Econ. Rev. 99 (3), 725−749.

Davis, L., North, D., 1970. Institutional change and American economic growth: a first step towards a theory of institutional innovation. J. Econ. Hist. 30 (1), 131−149.

de Janvry, A., 1981. The Agrarian Question and Reformism in Latin America. Johns Hopkins University Press, Baltimore, MD.

de Janvry, A., Fafchamps, M., Sadoulet, E., 1991. Peasant household behavior with missing markets: some paradoxes explained. Econ. J. 101 (409), 1400−1417.

de Janvry, A., Fafchamps, M., Sadoulet, E., 1995. Transaction costs, public choice, and induced technological innovations. In: Koppel, B. (Ed.), Induced Innovation Theory and International Agricultural Development: A Reassessment. Johns Hopkins University Press, Baltimore, MD, 151−168.

Deaton, A., 2010. Instruments, randomization, and learning about development. J. Econ. Lit. 48 (2), 424−455.

Demsetz, H., 1967. Toward a theory of property rights. Am. Econ. Rev. 57 (2), 347−359.

Demsetz, H., 1969. Information and efficiency: another viewpoint. J. Law Econ. 12 (1), 1−22.

Deweaver, M., Roumasset, J., 2002. A correction and prima-facie test of the canonical theory of share tenancy. Econ. Bull. 15 (4), 1−16.

Dixit, A., 1998. The Making of Economic Policy: A Transaction Cost Politics Perspective. MIT Press, Cambridge, MA.

Duflo, E., Kremer, M., Robinson, J., 2008. How high are rates of return to fertilizer? evidence from field experiments in Kenya. Am. Econ. Rev. 98 (2), 482−488.

Duflo, E., Kremer, M., Robinson, J., 2011. Nudging farmers to use fertilizer: theory and experimental evidence from Kenya. Am. Econ. Rev. 101 (6), 2350−2390.

Endress, L., 2013. Sustainable Development and the Hawaii Clean Energy Initiative. UHERO (Working Paper 2013-4).

Epstein, L., Zin, S., 1989. Substitution, risk aversion, and the temporal behavior of consumption and asset returns: a theoretical framework. Econometrica 57 (4), 937−969.

Evenson, R., Roumasset, J., 1986. Markets, institutions and family size in rural Philippine households. J. Philipp. Dev. 23 (18), 141−162.

Feder, G., 1985. The relation between farm size and farm productivity: the role of family labor, supervision and credit constraints. J. Dev. Econ. 18 (2−3), 297−313.

Fei, J.C.H., Ranis, G., 1964. Development of the Labor Surplus Economy: Theory and Policy. Richard A. Irwin, Inc., Homewood, IL.

Friedman, M., 1953. Essays in Positive Economics. University of Chicago Press, Chicago, IL.

Friedman, M., 2002. Capitalism and Freedom, Fortieth Anniversary Edition. University of Chicago Press, Chicago, IL.

Gardner, B., 1988. The Economics of Agricultural Polices. McGraw-Hill, New York.

Greenwald, B., Stiglitz, J., 1986. Externalities in economies with imperfect information and incomplete markets. Q. J. Econ. 101 (2), 229−264.

Grossman, S., Hart, O., 1986. The costs and benefits of ownership: a theory of vertical and lateral integration. J. Polit. Econ. 94 (4), 691−719.

Gruber, J., 2002. Taxes and health insurance. Tax Policy Econ. 16, 37−66.

Gruber, J., 2003. Smoking internalities. Regulation Winter, 52−57.

Guttmacher Institute, 2007. State policies on later-term abortions. State Policies in Brief. Guttmacher Institute, Washington, DC.

Harberger, A.C., 1978. On the use of distributional weights in social cost−benefit analysis. J. Polit. Econ. 86 (2), S87−S120.

Harberger, A.C., 1984. Basic needs versus distributional weights in cost−benefit analysis. Econ. Dev. Cult. Change 32 (3), 455−474.

Hart, O., Moore, J., 1990. Property rights and the nature of the firm. J. Polit. Econ. 98 (6), 1119−1158.

Hayami, Y., Kikuchi, M., 1982. New rice technology and income distribution—a perspective from villages in Java. Econ. Rev. 33 (1), 1−11.

Hayami, Y., Otsuka, K., 1993. The Economics of Contract Choice: An Agrarian Perspective. Clarendon Press, Oxford.

Hayami, Y., Ruttan, V., 1985. Agricultural Development: An International Perspective. Johns Hopkins University Press, Baltimore, MD.

Hazell, P., Pomareda, C., Valdes, A., 1986. Crop Insurance for Agricultural Development: Issues and Experience. Johns Hopkins University Press, Baltimore, MD.

Heal, G., 2003. Optimality or sustainability? In: Arnott, R., Greenwald, B., Kanbur, R., Nalebuff, B. (Eds.), Economics for an Imperfect World: Essays in Honor of Joseph E. Stiglitz. MIT Press, Cambridge, MA, 331−348.

Hornbeck, R., 2010. Barbed wire: property rights and agricultural development. Q. J. Econ. 125 (2), 767−810.

IFAD, 2001. International Fund for Agricultural Development. Oxford University Press, New York, NY, Rural Poverty Report.

ILO, 2013. MicroInsurance Innovation Facility, as cited in Banerjee et al. (2014).

James, W., Roumasset, J., 1984. Migration and the evolution of tenure contracts in newly settled regions. J. Dev. Econ. 14 (1), 147−162.

Jensen, M., 1983. Organization theory and methodology. Account. Rev. 58 (2), 319−339.

Jensen, M., Meckling, W., 1976. Theory of the firm: managerial behavior, agency costs and ownership structure. J. Finan. Econ. 3 (4), 305−360.

Johansson, P., Roumasset, J., 1997. Prohibition vs. Taxification: Drug Control Policy in the USA. University of Hawaii, Department of Economics (Working Paper 96-8).

Johansson, P., Roumasset, J., 2002. Apples, bees and contracts: a Coase−Cheung theorem for positive spillover effects. World Econ. Forum 6, 1−11.

Johnston, B., Kilby, P., 1975. Agriculture and Structural Transformation: Economic Strategies in Late Developing Countries. Oxford University Press, New York, NY.

Jorgenson, D., 1969. The role of agriculture in economic development: Classical vs. Neoclassical Models of growth. In: Wharton, C. (Ed.), Subsistence Agriculture and Economic Development. Aldine Publishing Co., Chicago, IL, 320−347.

Karlan, D., Osei, R., Osei-Akoto, I., Udry, C., 2014. Agricultural decisions after relaxing credit and risk constraints. Q. J. Econ. 129 (2), 597−652.

Kratz, F., Roumasset, J., Syrmos, V., 2001. Optimal Storage in Time and Space: Primal, Dual, and Comparative Statics. Unpublished manuscript, University of Hawaii.

Krueger, A., Schiff, M., Valdes, A., 1991. The Political Economy of Agricultural Pricing Policy, vol. 1: Latin America. Johns Hopkins University Press, Asia, Baltimore, MD.

Krugman, P., 1979. Increasing returns, monopolistic competition, and international trade. J. Int. Econ. 9 (4), 469−479.

Krugman, P., 1994. Peddling Prosperity: Economic Sense and Nonsense in the Age of Diminished Expectations. Norton, New York, NY.

Levin, Y., 2014. Book Review: "Why Government Fails So Often" by P. Schuck (Princeton). New York Times, June 9.

Lipton, M., 2009. Land Reform in Developing Countries: Property Rights and Property Wrongs. Routledge, New York, NY.

Lowi, T., 1969. The End of Liberalism: The Second Republic of the United States. W.W. Norton, New York, NY.

Lucas, R., 1993. Making a miracle. Econometrica 61 (2), 251−272.

Lucas, R., 2002. Lectures on Economic Growth. Harvard University Press, Cambridge, MA.

Lucas, R., 2009. Ideas and growth. Economica 76 (301), 1−19.

Magee, S., Brock, W., Young, L., 1989. Black Hole Tariffs and Endogenous Policy Theory: Political Economy in General Equilibrium. Cambridge University Press, Cambridge.

Mazumdar, D., 1965. Size of farm and productivity: a problem of Indian peasant agriculture. Economica 32 (126), 161−173.

McCloskey, D., 1994. Knowledge and Persuasion in Economics. Cambridge University Press, Cambridge, MA.

McCloskey, D., 1996. Ask what the boys in the Sandbox Will Have, Times Higher Education Supplement (London), reprinted in the introduction to The Vices of Economists—The Virtues of the Bourgeoisie.

McCloskey, D., 1998. The Rhetoric of Economics. University of Wisconsin Press, Madison, WI.

Medema, S., 1995. Ronald Coase on economics and economic method. Hist. Econ. Rev. 24, 1−22.

Merrill, T.W., 2002. Introduction: the Demsetz thesis and the evolution of property rights. J. Legal Stud. 31 (2), 331−338.

Mookherjee, D., 2005. Is there too little theory in economic development today? Econ. Polit. Wkly XL (40), 4328−4333.

Morduch, J., 1995. Income smoothing and consumption smoothing. J. Econ. Perspect. 9 (3), 103−114.

Morduch, J., 1999. Between the state and the market: can informal insurance patch the safety net? World Bank Res. Obs. 14 (2), 187−207.

Mosher, A.T., 1966. Getting Agriculture Moving: Essentials for Development and Modernization. Praeger, New York, NY.

North, D., Thomas, R., 1973. The Rise of the Western World: A New Economic History. Cambridge University Press, Cambridge.

Nozick, R., 1974. Anarchy, State, and Utopia. Basic Books, Oxford.

Olmstead, A., Rhode, P., 1993. Induced innovation in American agriculture: a reconsideration. J. Polit. Econ. 101 (1), 100−118.

Olmstead, A., Rhode, P., 2008. Creating Abundance: Biological Innovation and American Agricultural Development. Cambridge University Press, Cambridge.

Olson, M., 1965. The Logic of Collective Action: Public Goods and the Theory of Groups. Harvard University Press, Cambridge, MA.

Ostrom, E., 1990. Governing the Commons: The Evolution of Institutions for Collective Action. Cambridge University Press, Cambridge.

Ravallion, M., 2011. On the Implications of Essential Heterogeneity for Estimating Causal Impacts Using Social Experiments. World Bank, Washington, DC (Policy Research Working Paper Series 5804).

Reardon, T., Timmer, P., 2007. Transformation of markets for agricultural output in developing countries since 1950: How has thinking changed? Handbook of Agricultural Economics. Elsevier, Amsterdam.

Reid, J., 1973. Sharecropping as an understandable market response: the post-bellum south. J. Econ. Hist. 33, 106−130.

Reid, J., 1976. Sharecropping and agricultural uncertainty. Econ. Dev. Cult. Change 24 (3), 549−576.

Repetto, R., 1986. Economic Policy Reforms for Natural Resource Conservation. World Resources Institute, Washington, DC.

Rodrik, D., 2008. The New Development Economics: We Shall Experiment, but How Shall We Learn? Harvard University, John F. Kennedy School of Government (Working Paper Series rwp08-055).

Rosenzweig, M., 2012. Thinking small: poor economics: a radical rethinking of the way to fight global poverty: review essay. J. Econ. Lit. 50 (1), 115−127.

Rothschild, M., Stiglitz, J., 1970. Increasing risk: I. A definition. J. Econ. Theory 2 (3), 225−243.

Roumasset, J., 1974. Institutional Change, Welfare Economics, and the Science of Public Policy (Working Paper 46, UC Davis).

Roumasset, J., 1976. Rice and Risk: Decision-Making among Low-Income Farmers in Theory and Practice. North-Holland Publishing Co., Amsterdam.

Roumasset, J., 1978a. The new institutional economics and agricultural organization. Philipp. Econ. J. 17 (3), <http://ideas.repec.org/p/pra/mprapa/13175.html>.

Roumasset, J., 1978b. The case against crop insurance for developing countries. Philipp. Rev. Bus. Econ. March 1978 (also circulated as an Agricultural Development Council reprint).

Roumasset, J., 1979a. Risk aversion, agricultural development and the indirect utility function. In: Roumasset, J., Boussard, J.M., Singh, I.J. (Eds.), Risk, Uncertainty and Agricultural Development. SEARCA/ADC, Philippines, 93−113.

Roumasset, J., 1979b. Sharecropping, production externalities, and the theory of contracts. Am. J. Agric. Econ. 61 (4), 640−647.

Roumasset, J., 1981. Positive Methods of Agricultural Decision Analysis. Australian National University Development Studies Center Occasional Paper No. 27. ANU Development Studies Center, Canberra.

Roumasset, J., 1986. The Welfare Economics of Rural Credit and the Benefits of Land Titles. World Bank manuscript.

Roumasset, J., 1987. The Public Economics of Irrigation Management and Cost recovery. Mimeo, World Bank, Washington, DC.

Roumasset, J., 1989. Decentralization and local public goods: getting the incentives right. Philipp. Rev. Bus. Econ. 26 (1), 1−13.

Roumasset, J., 1991. Constitutional management of free riders and common property resources. Osaka Econ. Pap. 40 (3−4), 321−333.

Roumasset, J., 1993. Development Is Letting Go of Fear. WP series, UH Manoa, Program on Conflict Resolution.

Roumasset, J., 1995. The nature of the agricultural firm. J. Econ. Behav. Organ. 26 (2), 161−177.

Roumasset, J., 2000. Blackhole Security. University of Hawaii (Working Paper 00-5).

Roumasset, J., 2003. Agricultural parastatals and pro-poor economic growth. International Workshop on Agribusiness: From Parastatals to Private Trade—Why, When and How? 15−16 December. New Delhi, India.

Roumasset, J., 2008. Population and agricultural growth. In: Durlauf, S., Blume, L. (Eds.), The New Palgrave Dictionary of Economics, second ed. Palgrave Macmillan, Basingstoke. (The New Palgrave Dictionary of Economics Online).

Roumasset, J., 2010. Wither the economics of agricultural development? Asian J. Agric. Dev. 7, 1.

Roumasset, J., Endress, L., 1996. The Yin and Yang of sustainable development: A case for win-win environmentalism. J. Asia Pac. Econ. 1 (2), 185−194.

Roumasset, J., James, W., 1979. Explaining variations in share contracts: land quality, population pressure and technological change. Aust. J. Agric. Econ. 23 (2), 116−127.

Roumasset, J., Lee, S., 2007. Labor: decisions, contracts and organization. In: Handbook of Agricultural Economics, vol. III: Agricultural Development: Farmers, Farm Production and Farm Markets. Elsevier, New York, NY, 2705−2740.

Roumasset, J., Setboonsarng, S., 1988. Second-best agricultural policy: getting the price of Thai rice right. J. Dev. Econ. 28 (3), 323−340.

Roumasset, J., Smith, J., 1981. Population, specialization, and landless workers: explaining transitions in agricultural organization. Popul. Dev. Rev. 7 (3), 401−419.

Roumasset J., Tarui, N., 2010. Governing the Resource: Scarcity-Induced Institutional Change. University of Hawaii at Manoa Economics Department (Working Paper 10-15).

Roumasset, J., Thapa, G., 1983. Explaining the mechanization decision: a policy oriented approach. J. Dev. Econ. 12 (3), 377−395.

Roumasset, J., Thaw, M., 1999. The Economics of Prohibition: Price, Consumption and Enforcement Expenditures during Alcohol Prohibition. Hawaii Reporter, Honolulu.

Roumasset, J., Uy, M., 1980. Piece rates, time rates and teams: explaining patterns in the employment relation. J. Econ. Behav. Organ. 1 (4), 343−360.

Roumasset, J., Uy, M., 1987. Agency costs and the agricultural firm. Land Econ. 63 (3), 290−302.

Sah, R., 1986. Size, supervision and patterns of labor transactions. J. Philipp. Dev. 13 (23), 163−175.

Samuelson, P., 1983. My life philosophy. Am. Econ. 27 (2), 5−12.

Schuck, P., 2014. Why Government Fails so Often: And How it Can Do Better. Princeton University Press, Princeton, NJ.

Sen, A., 1962. An aspect of Indian agriculture. Econ. Wkly Annual Number 14 (1962), 243−246.

Sen, A., 1966. Peasants and dualism with or without surplus labor. J. Polit. Econ. 74 (5), 425−450.

Smith, A., 1976. An Inquiry into the Nature and Causes of the Wealth of Nations. University of Chicago Press, Chicago, IL.

Spence, A., Zeckhauser, R., 1971. Insurance, information, and individual action. Am. Econ. Rev. 61 (2), 380−387.

Stiglitz, J., 1974. Incentives and risk sharing in sharecropping. Rev. Econ. Stud. 41 (2), 219−255.

Stiglitz, J., 1984. Price rigidities and market structure. Am. Econ. Rev. 74 (2), 350−355.

Stiglitz, J., 1991. Symposium on organizations and economics. J. Econ. Perspect. 5 (2), 15−24.

Stiglitz, J., 1993. Post Walrasian and Post Marxian economics. J. Econ. Perspect. 7 (1), 109−114.

Thaler, R., Sunstein, C., 2008. Nudge: Improving Decisions about Health, Wealth, and Happiness. Yale University Press, New Haven, CT.

Townsend, R., 1994. Risk and insurance in village India. Econometrica 62 (3), 539−591.

Uy, M., 1979. Contractual Choice and Internal Organization in Philippine Sugarcane Production. School of Economics, University of the Philippines, Diliman.

Walsh, V., 1987. Models and theory. In: Eatwell, J., Milgate, M., Newman, P. (Eds.), The New Palgrave: A Dictionary of Economics, first ed. Palgrave Macmillan, Basingstoke, 482−483.

Williams, J., Wright, B., 2005. Storage and Commodity Markets. Cambridge University Press, Cambridge, MA.

Williamsón, J., 1996a. Globalization, convergence, and history. JEH 56 (2), 277−306.

Williamson, O., 1975. Markets and Hierarchies: Analysis and Antitrust Implications. Free Press, New York, NY.

Williamson, O., 1985. The Economic Institutions of Capitalism. Free Press, New York, NY.

Williamson, O., 1996b. The Mechanisms of Governance. Oxford University Press, New York, NY.

Williamson, O., 1998. The institutions of governance. Am. Econ. Rev. 88 (2), 75−79.

Wolpin, K., 2013. The Limits of Inference Without Theory. The MIT Press, Cambridge, MA.

World Bank, 2008. World Development Report. World Bank, Washington, DC.

World Bank, 2009. Land Reform, Rural Development, and Poverty in the Philippines: Revisiting the Agenda. World Bank, Washington, DC.

Yang, X., 2003. Economic Development and the Division of Labor. Blackwell, New York, NY.

Young, A., 1928. Increasing returns and economic progress. Econ. J. 38, 527−542.

RESOURCES, ENVIRONMENT, AND SUSTAINABLE DEVELOPMENT

SCARCITY, SECURITY, AND SUSTAINABLE DEVELOPMENT

3

Lee H. Endress

Economics Department, University of Hawaii at Manoa, Honolulu, HI

CHAPTER OUTLINE

JEL: Q01; Q56

3.1 INTRODUCTION

Various groups bring differing views to the issues of scarcity, security, and sustainability, and their interconnections. Economists share a disciplined way of thinking about these issues, although this has not led to a consensus perspective. While the profession has made tremendous contributions to relevant literature, it has perhaps been less successful than it would like to be in influencing public policy through effective economic diplomacy.

Intertemporal welfare (i.e., welfare over time and across generations) is at the heart of economic policy analysis and design, and should be what sustainable development is all about, but often is not. This chapter reviews the notion of sustainable development and the three pillars of sustainability and comments on energy policy and its emphasis on current technology renewable energy. The final section outlines possible research opportunities and offers brief thoughts on economic diplomacy, education, and the role economists can play in helping build human capital and critical thinking for the future.

3.2 SCARCITY AND SECURITY

Scarcity is a prominent topic within strategy and security communities, relating traditionally to non-renewable resources (e.g., minerals, oil) deemed to be economically critical and militarily strategic. Typically, the primary concern is physical scarcity rather than economic scarcity, so notions like "peak oil" and "supply vulnerability" get a lot of attention, often with presumed links to potential violent conflict. (Titles like "Resource Wars" and "Blood and Oil" [Klare, 2001, 2004] are representative.) Seriously missing in this line of thinking is the role of prices in signaling economic scarcity and influencing the incentives that evolve to address it.

In contrast, energy and natural resource economists take the view that price and marginal user cost, or scarcity value, are the best indicators of scarcity. Krautkraemer (2005) provides an excellent primer on the economics of scarcity in the latest volume (Simpson et al., 2005) of the trilogy on scarcity published by Resources For the Future (RFF). Of course, when prices are not directly observable, the challenge is to impute shadow prices. Even when market prices are observable, shadow prices are often required to reflect scarcity in the second-best environment of policy distortion.

An accompanying paper by Pearce (2005) in the same RFF volume argues that concern for resource scarcity has moved beyond nonrenewable resources, which are usually managed by the private sector, to renewable and environmental resources, which often exhibit public good characteristics (e.g., ecosystem services). Here, with missing markets, pricing is even more of a challenge.

In the realm of scarcity, food and water, as well as energy, have notably become issues of concern among security practitioners in Asia, not only because of the potential for conflict within and between nations, but because these issues open new possibilities for regional cooperation. Nonetheless, there is little understanding of, or trust in, the role of prices and market mechanisms. Strategy and policy are usually viewed as matters of command and control. What is often overlooked is that fear of scarcity can lead to lose—lose policies that actually create artificial scarcity. The 2008 rice crisis in Asia is a good example (see Charles, 2011 for a summary.)

In many corners of strategic thinking and planning, security is a matter of all or nothing, rather than a matter of more or less; if security is associated with protection against loss, how much is enough? Economists are more inclined to see security as the opposite of risk, which could be expressed analytically as security (1—risk), where risk is a probability—in particular, the probability of a specific bad outcome (e.g., a natural disaster or terrorist attack made specific by intervals of timing, location, magnitude, and impact).

Realist schools of international relations and strategic studies view security primarily in terms of competing national interests in an international environment of anarchy. The realist worldview assigns a low probability to the prospects for international cooperation. In contrast, schools of liberal institutionalism see the emergence of multilateral organizations and international cooperation as almost inevitable.

Area studies and international relations specialists of both realist and liberal persuasion, avoiding the quantitative rigor of the rational choice approach in political science, have paid little attention to game theoretic formulations of the collection action problem within and among nations.

This author's view has long been that cooperation (joint action for mutual gain) is neither impossible nor inevitable. It is not *kumbaya*. Cooperation is tough stuff, as compellingly mapped out by Schelling (1980), Axelrod (1984), and a series of essays by Roumasset and Barr (1992). The pursuit of cooperation strategies is the rationale for organizations like the Asia-Pacific Center for Security Studies, the East West Center, and Pacific Forum, all located in Honolulu.

Security thinking at the U.S. Pacific Command and in some quarters of the U.S. Department of Defense has gone beyond the narrow realm of national defense. There appears to be an expanded willingness to consider strategic implications of security beyond the nation state to international security in one direction and to human security in the other.

At the same time, concepts of security have expanded outside the military domain to economic, energy, natural resource, environmental, and public health concerns. (This has engendered skepticism among some traditionalists who fault the security establishment for embracing "new age security.") One think tank in Washington, DC, is marketing the idea of "natural security." Rising concern at the Pacific Command and the Asia-Pacific Center on the energy−water−food security complex in Asia is another expression. At the international level, cooperation on these issues appears to be an ideal mechanism for building trust and habits of security cooperation. Unfortunately, this thinking has not always led to win−win strategies. Rather, it has sometimes reinforced notions of economic and natural resource nationalism. The false allure of self-sufficiency and economic independence arises from such notions.

3.3 SUSTAINABLE DEVELOPMENT: WHAT IS IT ANYWAY?

It's hard to be against sustainability. In fact, the less you know about it, the better it sounds.

Robert Solow, 1991

RFF published the first volume on "Scarcity and Growth" in 1963 (Barnett and Morse, 1963; Smith, 1979; Simpson et al., 2005). As worldwide concern continued to mount in the 1960s, 1970s, and 1980s over natural resource scarcity, limits to growth, and environmental security, it seems inevitable, from the present vantage point, that these issues would coalesce around a unifying theme, a global bumper sticker. So arose the notion of sustainable development, which the Brundtland Commission introduced in its famous 1987 report. While the well-known Brundtland definition of sustainable development is vague and not particularly operational, it does convey a sense of stewardship for the future. The specific means and mechanisms of sustainable development have been debated for some 25 years now, and still no universal consensus has been reached.

The interpretation of sustainability most favored by environmentalists, ecologists, and the no-growth, anti-capitalism left has become known as "strong sustainability." The mandate of "strong sustainability" is to retard, if not inhibit, global economic growth and to prohibit the depletion of natural or ecological capital and, in its extreme form, the use of particular natural resources. (For some hard-core advocates, "sustainable development" itself is an oxymoron.)

Economists generally have a different perspective, effectively represented by Robert Solow in a 1991 lecture delivered at the Woods Hole Institution in Massachusetts. In discussing what the present owes the future, Solow remarked "...*what we are obligated to leave behind is a generalized capacity to create well-being, not any particular thing or any particular natural resource.*" This perspective is behind the alternative formulation of "weak sustainability," which requires the summed value of produced capital, natural capital, and even human capital—that is, the value of total capital in the economy—to remain constant or increase over time.

The distinction between strong and weak sustainability can be expressed mathematically. Let K represents the stock of physical capital, N the stock of natural capital, and H the stock of human capital in the economy. Each type of capital is associated with a shadow price, P_X, $X = K, N, H$, representing the contribution of a unit of that type of capital to the overall value of total capital, or total wealth, W. Strong sustainability then requires that the time rate of change of the resource stock, dN/dt, be nonnegative. In contrast, weak sustainability requires that the time rate of change of the value of total capital in the economy remains nonnegative, which can be expressed as $dW(t)/dt = P_K (dK/dt) + P_N (dN/dt) + P_H (dH/dt) \geq 0$.

As articulated by Solow (1991, p. 179 (in Stavins); 1992), it is the economy's productive base, represented by the totality of capital in its various forms, that provides the capacity for creating well-being now and in the future. The prominent idea is intertemporal well-being or welfare, giving rise to a "sustainability criterion" that intertemporal social welfare does not decrease over time. $W(t)$, the value of total capital at time t, is a key determinant of intertemporal social welfare $V(t)$, looking forward from time t. Moreover, shadow prices equivalently represent the contribution of a unit of each type of capital to intertemporal social welfare. The sustainability criterion can be specified as $dV(t)/dt \geq 0$.

Arrow et al. (2004) present a mathematical proof that weak sustainability is a necessary condition to achieving the sustainability criterion. That is, if the sustainability criterion, $dV(t)/dt \geq 0$, is satisfied, then it necessarily follows that $dW(t)/dt \geq 0$. *Ad hoc* sustainability constraints, inhibitions, and prohibitions are not required. The proof also mathematically establishes sufficiency, but for policy implications, a caveat is in order. Weak sustainability guarantees that future generations will have the opportunity to keep intertemporal welfare from declining; it does not guarantee that policymakers will not squander that opportunity. (See discussion of public policy and political economy in Section 3.4.2.)

The degree of substitutability between different forms of capital, especially between produced and natural capital, remains a contentious issue, dividing proponents of strong sustainability from advocates of weak sustainability. This controversy presents a challenge to the emerging, multidisciplinary field of "sustainability science," as discussed below in connection with the three pillars of sustainability.

In the context of weak sustainability, two main approaches have been constructed to link an economy's wealth to sustainability and intertemporal welfare. The first approach is associated with the World Bank and the work of Kirk Hamilton and Michael Clemens, as surveyed in "Where is the Wealth of Nations?" (World Bank, 2006). The theoretical foundation goes back to the classic paper by Weitzman (1976) titled "On the Welfare Foundations of National Product in a Dynamic Economy."

So-called "comprehensive wealth" is defined as the discounted flow of consumption over an infinite time horizon, where, as in Weitzman (1976), consumption is assumed to follow an optimum trajectory. Hamilton and Clemens showed that "genuine saving," or net adjusted saving, defined as the change in real asset values, is equal to the change in social welfare, or comprehensive wealth, in an optimal economy.

An earlier and related result by Hartwick (1977) showed that if genuine saving is equal to zero over all time (all natural resource rents are invested in capital accumulation to offset resource depletion), then utility of consumption remains constant. This result, popularly known as Hartwick's Rule, is valid only under stringent conditions: elasticity of substitution between natural and produced capital must be greater than or equal to one; output elasticity of produced capital in production must exceed that of natural capital; and population growth must be zero in the absence of technological change.

The second approach to weak sustainability is that developed by Arrow et al. (2004), based on the idea of "inclusive wealth," as extensively discussed in the "Inclusive Wealth Report 2012" of the United Nations (UN, 2012). Inclusive wealth in the UN framework differs from the World Bank's comprehensive wealth in that wealth is now defined as the "shadow value" of all capital assets a country owns. (When observable market prices are not available, shadow prices or shadow values must be imputed.) No assumption is made on the economy following an optimal path of consumption. The virtue of this approach is its applicability in imperfect, distorted economies, though estimating shadow prices in the presence of distortions is an added challenge.

While the theoretical foundations of sustainable development are still being debated even within the realm of weak sustainability, the larger challenge is empirical—the measurement of the various capital stocks, rates of depreciation and depletion, and rates of investment as basis of a comprehensive or inclusive wealth accounting. Compounding this challenge is the lack of uniform accounting standards, methods, and data across countries, which makes cross-country analysis difficult. Both the World Bank 2006 publication and the UN 2012 report discuss measurement techniques and associated issues in great detail.

3.4 TRILOGIES, TRIADS, AND TRIANGLES

The RFF trilogy on scarcity was published over a period of 42 years (1963, 1979, and 2005). Only the third volume addresses the concept of sustainability and its link to scarcity and growth; none of the three volumes touches on the field of sustainability science, which gained momentum and visibility in the early 2000s. The U.S. National Academy of Sciences (NAS), the American Association for the Advancement of Science (AAAS), and the National Science Foundation (NSF) represent a triad of strong advocates of sustainability science.

On its website (www.pnas.org), the Proceedings of the National Academy of Sciences describes sustainability science as "an emerging field of research dealing with the interaction between natural and social systems and how these interactions affect the challenge of sustainability..." There is now an impressive sustainability literature, summarized in a "reading list" published by Harvard University's Center for International Development (Kates, 2010).

3.4.1 POSITIVE SUSTAINABILITY AND THE THREE PILLARS

The connection between sustainability science and the economics perspective on sustainable development is more recent. Roumasset et al. (2010) applied sustainability science and economics to the case of Pacific watersheds. Burnett et al. (2014) consider sustainability across space and time in the management of renewable resources. This line of research has been inspired by a formulation of "positive sustainability" founded on three principles or pillars. The first was influenced by sustainability science and the other two were put forward by Stavins et al. (2003): (i) adopting a complex systems approach to modeling and analysis, integrating natural resource systems, the environment, and the economy; (ii) pursuing dynamic efficiency, i.e., efficiency of both time and space in the management of the resource–environment–economy complex to maximize intertemporal well-being; and (iii) enhancing stewardship for the future through intertemporal equity, which is increasingly represented as intergenerational neutrality or impartiality.

3.4.1.1 A systems approach

Although influenced by the field of sustainability science, the concept of integrating natural resource, environmental, and economic systems is not new. Resource and environmental economists have long included natural resources as inputs to production and environmental amenity as a component of utility in economic models, including models of growth.

Feasibility constraints in these models are typically expressed in terms of the factors affecting the rate of change of pertinent quantities or stock levels. For example, the time rate of change of a stock level is equal to the difference between the rate of its growth and the rate of depletion. (Think fill and drain, birth and death, investment and depreciation.) This type of equation has wide application and is used to describe the dynamics of produced capital, natural capital, and population. In the case of natural capital with stock level N, the mathematical representation would be $dN/dt = G(N) - R$, where $G(N)$ is a biological growth function for a renewable resource and R is the rate of harvest or extraction. In the case of a nonrenewable resource, $G(N) = 0$.

Clark (1990) considers the dynamics of economic and biological systems (fisheries, forests), including predator–prey and source–sink models, that exhibit coupling. In such models, the rates of growth and depletion of one form of capital depend on the stock levels of other forms of capital in complex, possibly nonlinear, modes of interaction. These models readily become computationally challenging and, like many multidimensional dynamical systems, often defy closed-form analytical solutions. Numerical analysis is then required.

Much interdisciplinary research, both theoretical and empirical, remains to be done, and this is where sustainability science has the potential to make a substantial contribution. For example, more sophisticated analysis of substitutability among different varieties of capital, especially in coupled systems, appears possible. Notions of network effects, emergence, spontaneous order, nonlinearities, critical transitions, bifurcations, regime shifts, tipping points, and chaos in adaptive, complex systems, though regarded as "faddish" in some circles, are now being incorporated in advanced models. "Critical Transitions in Nature and Society" by Scheffer (2009) is a good example.

The temptation here is to regard everything as connected to everything else in an incredible web of interactions and feedbacks, both positive and negative. The challenge is to identify those connections that really matter for sustainability as intertemporal welfare. Nonetheless, in pursuing sustainable development, caution must be exercised to avoid single-issue agendas, initiatives, and programs that neglect or undermine the integrity of other dimensions of the economic–ecological complex. That is the essence of the systems approach.

3.4.1.2 Dynamic efficiency

This principle or pillar has been a mainstay of neoclassical natural resource and environmental economics, going all the way back to 1931 and Harold Hotelling's classic paper on the economics of exhaustible resources (Hotelling, 1931). (Mathematical details of welfare maximization and dynamic efficiency are presented in the technical appendix.) What has become known as Hotelling's Rule says that an optimal program of exhaustible resource extraction exhibits a special trajectory of resource prices: at each time t, price is equated to the marginal cost (i.e., unit cost) of extraction plus marginal user cost, where the latter reflects the scarcity value of the unit of resource just extracted (or, equivalently, the opportunity cost of future use foregone).

As the field of resource and environmental economics has matured, Hotelling's Rule has been extended to renewable resources; David Pearce, a British economist, extended it to environmental externalities, such as pollution. (Pearce and Turner, 1990 is an excellent reference.) The dynamic efficiency condition for optimal resource management, with possible environmental externalities, has evolved into what has become known as the Pearce equation.

Along the optimal trajectory in the Pearce formulation, price P, at every time t, is equal to the sum of three factors: marginal extraction or harvesting cost (MC), marginal user cost (MUC), and marginal externality cost (MEC, e.g., pollution damage cost) attributed to the unit of resource just harvested and used. The Pearce equation can be written as $P = MC + MUC + MEC$. The sum of the three terms on the right-hand side is often called marginal opportunity cost (MOC) of resource harvesting and use.

A second dynamic efficiency condition, complementing the Pearce equation, is attributed to the mathematical philosopher, Frank Ramsey (1928, p. 543), and is known as the Ramsey condition: $r = \rho + \eta g$. At each point in time along the optimal path of consumption, the market rate of interest, r, is equal to the consumption rate of interest (so designated by Little and Mirrlees, 1974), where the latter is determined by three factors: the rate of time preference, ρ; the elasticity of intertemporal substitution ($1/\eta$); and the rate of consumption growth, g. (Details are given in the technical appendix.)

Parameters ρ and η are keys in the theory of optimal growth. The rate of time preference, ρ, represents the degree of impatience on the part of economic agents, reflecting a preference for consumption sooner rather than later; the higher the value of ρ, the greater the degree of impatience. The other parameter, η, registers a preference for smoothing out or evening out the consumption level over time. A higher η reflects a preference for more consumption smoothing. (At the extreme, an infinite η would represent a preference for a constant level of consumption.)

Endress et al. (2005) present a mathematical representation of the planner's maximization problem in the case of a nonrenewable resource with environmental spillovers; in the case of a renewable resource with possible amenity value, see Endress et al. (2014). The Pearce equation and the Ramsey condition are special cases of the efficiency condition that pervades neoclassical economics: to achieve optimality with respect to some economic objective (e.g., consumer welfare, producer profit) and continue the policy or program until the marginal benefit and marginal cost of so doing are equated. In the Pearce equation, the marginal benefit to the resource manager is price P. The marginal cost is MOC. This is a nontrivial formulation. When the systems approach (first sustainability pillar) is rigorously adopted, the construction and computation of MOC can be a challenging endeavor.

In most cases, the efficiency condition implies an interior solution to the problem: the optimal policy or program is not a matter of all or nothing. So, with rare exception (plutonium comes to mind), zero level of pollution would entail a huge net cost and would not be optimal; neither would a 100% recycling program. Nonetheless, there are strong advocates for such all-or-nothing approaches to policy as moral imperatives.

Economists think, write, and talk a lot about efficiency, so it is not surprising that policy-makers and the general public have the common, but erroneous, view that economists see efficiency as an end in and of itself. On the contrary, as economists see it, efficiency relates to eliminating economic waste and is a means—in fact, a necessary condition—for enhancing well-being in society, which should be the objective or end of economic policy, including that for sustainable development.

3.4.1.3 Intertemporal equity

The third pillar concerning how we weight the welfare of future generations relative to our own is perhaps the most controversial. It has everything to do with intergenerational justice and goes back to the sentiment of British philosopher Frank Ramsey, expressed in his classic 1928 paper on the mathematical theory of saving: "...*it is assumed that we do not discount later enjoyments in comparison with earlier ones, a practice which is ethically indefensible and arises merely from weakness of the imagination.*" The neutral or impartial weighting of generations across time implies that the rate of time preference be equal to zero. The debate among economists, ecologists, and political philosophers pertaining to time preference and intergenerational weighting is far from settled, especially in the context of sustainability and climate change.

Zero time preference was first rejected on technical grounds. Attempting to maximize welfare over an infinite time horizon without discounting led to sums (in discrete time models) or integrals (in continuous time models) that were infinite: they didn't converge, and so alternative time paths of consumption and utility could not be ranked to yield the optimum path that maximized intertemporal welfare.

Koopmans and Diamond demonstrated this technical difficulty in the early 1960s, but in a 1965 paper, "On the Concept of Economic Growth," Koopmans developed a solution to his own problem by constructing a clever integral transformation using Solow's "golden rule" steady state result (Koopmans, 1965). The transformed integral was shown to converge for a nonempty set of so-called eligible paths compatible with the dynamic constraints of the economy. Endress et al. (2005) extended this result to a production economy with a natural resource as an approach to modeling sustainability (see also Section A.2 of the technical appendix).

Current objections to zero time preference have been put forward on ethical rather than technical grounds. A common complaint is that zero time preference, as might be imposed by a planner, violates consumer sovereignty and a natural, positive rate of impatience. In the recent book, "Carbon Crunch," Dieter Helm (2012), a British energy economist, goes so far as to characterize intergenerational impartiality as a form of radical socialism that has actually impeded significant progress on global reduction of carbon emissions. (Helm is by no means a climate change denier.)

In response to these ethical objections, Endress et al. (2014) adapted and extended to a full production economy a model of overlapping generations with natural resources (Burton, 1993) that distinguishes intratemporal impatience from generational weighting. In the adaptation, the distinction permits neutral weighting across generations without violating consumer sovereignty. Public policy approximating neutral weighting could include a carbon tax, a national consumption tax (in lieu of an income tax) to encourage saving and investment, and meaningful reduction of the US national debt so as not to pass on a growing debt burden to future generations.

3.4.2 PUBLIC POLICY: PROSUSTAINABILITY OR NOT?

The project of sustainable development comes with problems standard in natural resource and environmental economics: externalities (spillover effects like pollution, with missing markets); underprovision of public goods (goods that exhibit nonrivalry in consumption, e.g., protection of ecosystems from invasive species); and open access natural resources (open ocean fisheries, global climate), which are vulnerable to ruinous depletion (Hardin, 1968).

Appropriate policies for addressing these problems will have to rest on the three pillars outlined above. They will also be shaped by systematic consideration of the policy environment at three levels of analysis, or the three-tiered, analytical hierarchy of economic policymaking: first-best, second-best, and third-best. Dixit (1996) offers a transaction cost perspective of economic policy along these same lines.

The **first-best** world features an idealized, frictionless economy without information costs, contracting costs, or agency costs. Information is complete; contracts cover all contingencies and can be fully monitored and enforced. Crime, corruption, and rent-seeking (pursuit of special privilege) can be costlessly nullified. At this level, government intervention and private contracting can be equivalent solutions to the problem of missing markets. This result is sometimes labeled the "Coase equivalency theorem." Prudent policy analysis and design start by getting it right at the first-best level.

Second-best brings information, contracting, and agency costs to the forefront. Because information gathering is no longer costless, asymmetric information is a common occurrence: one party to a transaction has much more accurate and timely information than the other. Information, legal, and administrative costs result in incomplete contracts. Agency costs prevent full monitoring and enforcement. The optimal levels of crime, corruption, and rent-seeking are no longer likely to be zero when the net costs of combating them are considered. The default view, often associated with Nobel laureate economist, Joseph Stiglitz, is that second-best costs are pervasive and government intervention (via taxes, subsidies, and regulation) is necessary to improve social welfare. Since government faces the same types of cost and information constraints as the private sector, it cannot be a foregone conclusion that government can always do it better; private sector solutions with market competition may in some cases be superior. (This is the essence of the healthcare debate in the United States.)

Third-best is the world of political economy, wherein costs and benefits directly influence the formation of coalitions that compete for political and economic advantage in society. The pursuit of such advantage is called "rent-seeking" in economics and typically involves activities such as lobbying, public relations campaigns, political contributions, and, sometimes, outright bribery. Unfortunately, the expansion of government that accompanies intervention on second-best grounds can facilitate rent-seeking at the third-best level. *Corollary*: The antidote to rent-seeking is not more government and regulation, but competition.

This was a key insight of Adam Smith (1776, Book I, Chapter X, p. 148), who observed: *"People of the same trade seldom meet together, even for merriment and diversion, but the conversation ends in a conspiracy against the public, or in some contrivance to raise prices."*

A particularly powerful type of rent-seeking coalition, long studied in political science, is termed "the iron triangle," because of the strength of the collaborative relationships among a triad of actors: politicians who seek campaign contributions, votes, and reelection; government bureaucrats who aspire to expand fiefdoms and budgets; and private sector interest groups who seek special privileges in the form of political access, favorable legislation, subsidies, protection of monopoly positions, and lucrative government contracts. The iron triangle is durable and impenetrable because it functions as a highly efficient, three-cornered, rent-seeking machine.

Nowhere (except perhaps in health care) do third-best politics sink first-best and second-best economic considerations as deeply as in the realm of energy policy. In assessing energy policy in Europe and the United States, Helm (2012) is especially critical of policymakers' obsession with current technology renewable energy, which is not yet commercially viable without major government subsidies and mandates. Deficiencies of current technology include intermittency (wind and solar), lack of low-cost energy storage capability, and the need for costly system redundancy to maintain base load capacity. Consequently, renewables have remained ineffective in lowering energy prices, creating green jobs, and reducing carbon emissions worldwide. The result is high costs for little gain. In a review of Helm's book, "The Carbon Crunch," *The Economist* (October 20, 2012 issue) highlights Helm's observation that the entire renewable sector has become an "orgy of rent-seeking." This outcome is not compatible with the sustainability criterion.

Mandates and subsidies (e.g., solar tax credits) are notoriously inefficient because they reduce consumer and taxpayer welfare. The loss of government revenue is a direct burden, but the overall loss is even worse. "Excess burden" is the additional welfare loss to society because subsidies distort prices and incentives in the economy, inefficiently drawing resources from other production sectors into the renewable energy sector. (The renewable sector gains at the expense of jobs and income in the rest of the economy.) On top of that is the added excess burden of tax friction: every dollar of tax revenue raised to finance subsidies costs the economy about another 25 cents. Economists refer to this friction as the social cost of public funds.

So, what are the alternatives? Helm (2012) offers some constructive recommendations for rational energy policies in Europe and the United States: (i) institute carbon taxes, (ii) increase investment in research and development for advanced renewable technologies, and (iii) adopt natural gas as a transition fuel until advanced technology renewables are economically viable without subsidy.

Other economists favor carbon taxes, a carbon dioxide emission trading scheme, or some combination of the two policy instruments. Nordhaus (2013) is a strong advocate of a carbon tax: "A carbon tax is the closest thing to an ideal tax... (it) will increase economic efficiency because it reduces the output of an undesirable economic activity (emitting CO_2)... A carbon tax can buttress or replace many inefficient regulatory initiatives and thereby provide yet further improvements in economic efficiency." He cites a third advantage of a carbon tax: the possibility of helping to reduce growing fiscal deficits in the United States.

3.5 **RESEARCH OPPORTUNITIES**

The foregoing discussion of scarcity, security, and sustainable development points to several research opportunities for sustainability scholars and graduate students alike. Five suggested areas are outlined below:

1. linking security and sustainability through the concept of risk, tidying up definitions and vocabulary: disaster management, risk, hazard, vulnerability, impact, capacity, and resilience;
2. designing strategies to induce and sustain multilateral cooperation in the absence of a central coordinating authority;
3. drawing on the field of sustainability science to incorporate notions of complexity, network effects, emergence, nonlinearities, critical transitions, bifurcations, regime shifts, tipping points, and chaos in models of sustainable development;
4. improving methods for measuring various capital stocks, shadow prices, rates of depreciation and depletion, and rates of investment as the basis of a comprehensive or inclusive wealth accounting in imperfect economies; and
5. adapting the Pearce and Ramsey efficiency conditions to models with complex, nonlinear, multidimensional, coupled dynamics.

3.6 **THOUGHTS ON ECONOMIC DIPLOMACY AND EDUCATION**

Can economists make a difference, or are they just hopelessly leaning against gale-force political winds? In the shorter run, economic diplomacy may not be able to constrain rent-seeking activity and its distortion of public policy. But at the least, economists should not facilitate rent-seeking by slipping, even inadvertently, from analysis to advocacy. U.S. President Harry Truman has been quoted to have said *"Give me a one-handed economist. All my economists say, on the one hand...And then on the other."* I say, beware the one-handed economist; the invisible hand may be holding the hidden agenda.

So, what should economists do? This question, posed by James Buchanan back in 1964, has prompted a variety of perspectives from prominent economists (all Nobel laureates), including Buchanan himself (Buchanan, 1964). In Buchanan's view, economists' excessive focus on scarcity and optimal resource allocation is misplaced. Instead, economists should concern themselves primarily with the institutions of voluntary exchange, which have the potential to displace coercive, rent-based, political relationships in society.

The notion that *"there ain't no such thing as a free lunch"* is commonly attributed to Milton Friedman, although he may not have said it first. Schelling (1995, p. 22) prefers an alternative truth: *"there are free lunches all over, waiting to be discovered or created."* What Schelling has in mind is that free lunches, and even banquets, are available to economies that can remove excess burden, i.e., eliminate economic waste. So, another thing that economists should do is to find those free lunches. A good place to start looking would be where the light does not always shine because of lack of transparency: at wasteful government policies that perpetuate mandates, subsidies, handouts, and monopoly power in the economy.

Krugman (1994, p. 292) seems to have taken on the role of a fearless cockroach warrior, fighting bad ideas that never die, implying that all courageous economists should suit up. From his brief time in government, Krugman warns that *"fighting bad ideas is like flushing cockroaches down the toilet; they just come right back."* The daddy of all cockroaches could be "self-sufficiency," followed closely by "keeping the money at home" (Endress, 2012), and the politically convenient contrivance of the "magic multiplier" as justification for government expenditure, if only for digging holes and filling them in again. Protectionism is another candidate, especially when justified on grounds of security (e.g., the Jones Act) or patriotism. English writer Samuel Johnson is credited by his biographer, James Boswell, for the pronouncement (1775), *"Patriotism is the last refuge of the scoundrel."*

Then there is Milton Friedman, who died with his boots on at age 94, defending capitalism and freedom (Friedman, 2002, p. xiv). From his 1982 book of the same title, Friedman gives this charge to economists: *"Only a crisis—actual or perceived—produces real change. When that crisis occurs, the actions that are taken depend on the ideas that are lying around. That, I believe, is our (i.e., economists) basic function: to develop alternatives to existing policies, to keep them alive and available until the politically impossible becomes the politically inevitable."*

It is hard to improve on all of this collective advice, other than perhaps to recommend that economic diplomacy be seasoned with a sharper sense of humor. A little more Marx might help—Groucho Marx, that is, and the twisted insights of American humorist and moviemaker, Woody Allen, can provide inspiration: *"Money is better than poverty if only for financial reasons."*

In the long run, education may be the realm where economists can have the strongest influence. Building human capital for the future is certainly a vital contribution to inclusive wealth and sustainability. In this regard, economists have a comparative advantage in imparting critical thinking skills, quantitative reasoning, "guesstimation" (making order-of-magnitude approximations), and the art of public diplomacy (leaving the equations and graphs behind). Three areas (another triad!) can be seen as foundational to the theme of this chapter, where these skills and arts can be directly and profitably applied: basic welfare economics and cost–benefit analysis; discounting and the economics of time; and the collective action problem along with strategies of cooperation.

Finally, as we contemplate the prospects for sustainable development in the face of climate change, natural disasters, terrorist attacks, conflict in the Middle East, fiscal cliffs, suffocating excess burden, and cockroach plagues, here are some encouraging words from Woody Allen's "My Speech to the Graduates" (Allen, 1981, p. 81): *"More than any other time in history, mankind faces a crossroads. One path leads to despair and utter hopelessness. The other, to total extinction. Let us pray we have the wisdom to choose correctly."* As a proud member of the dismal profession, this author feels a moral obligation to be optimistic.

REFERENCES

Allen, W., 1981. My speech to the graduates. In: Allen, W. (Ed.), Side Effects. Ballantine Books, New York, NY.

Arrow, K., Dasgupta, P., Goulder, L., Daily, G., Ehrlich, P., Heal, G., et al., 2004. Are we consuming too much? J. Econ. Perspect. 18 (3), 147–172.

Axelrod, R., 1984. The Evolution of Cooperation. Basic Books, New York, NY.

Barnett, H., Morse, C., 1963. Scarcity and Growth: The Economics of Natural Resource Availability. Johns Hopkins University Press for Resources for the Future, Baltimore, MD.

Buchanan, J., 1964. What should economists do? South Econ. J. 30 (3), 213–222.

Burnett, K., Endress, L., Ravago, M., Roumasset, J., Wada, C., 2014. Islands of sustainability in time and space. Int. J. Sustain. Soc. 6 (1/2), 9–27.

Burton, P., 1993. Intertemporal preferences and intergenerational equity considerations in optimal resource harvesting. J. Environ. Econ. Manage. 24 (2), 119–132.

Charles, D., 2011. How fear drove world rice markets insane. The Salt, November 2011. Available at: http://www.npr.org/blogs/thesalt/2011/11/02/141771712/how-fear-drove-world-rice-markets-insane.

Clark, C.W., 1990. Biomathematical Economics. second ed. Wiley, New York, NY.

Dixit, A., 1996. The making of economic policy: a transaction-cost policy perspective. Munich Lectures in Economics. MIT Press, Cambridge, MA.

Endress, L., 2012. Keeping the money at home! Economic Currents. University of Hawaii Economic Research Organization (UHERO).

Endress, L., Roumasset, J., Zhou, T., 2005. Sustainable growth with environmental spillovers. J. Econ. Behav. Organ. 58, 527–547.

Endress, L., Pongkijvorasin, S., Roumasset, J., Wada, C., 2014. Intergenerational equity with individual impatience in a model of optimal and sustainable growth. Resour. Energy Econ. 36 (2), 620–635.

Friedman, M., 2002. Capitalism and Freedom. University of Chicago Press, Chicago, IL.

Hardin, G., 1968. The tragedy of the commons. Science 162, 1243–1247.

Hartwick, J., 1977. Intergenerational equity and the investing of rents from exhaustible resources. Am. Econ. Rev. 67 (5), 972–974.

Helm, D., 2012. The Carbon Crunch. Yale University Press, New Haven, CT.

Hotelling, H., 1931. The economics of exhaustible resources. J. Polit. Econ. 39, 137–175.

Kates, R., 2010. Readings in sustainability science and technology. Working Paper No. 213, Center for International Development at Harvard University, Cambridge, MA.

Klare, M., 2001. Resource Wars: The New Landscape of Global Conflict. Metropolitan Books, New York, NY.

Klare, M., 2004. Blood and Oil: The Dangers and Consequences of America's Growing Dependency on Imported Petroleum. Metropolitan Books, New York, NY.

Koopmans, T., 1965. On the concept of economic growth. The Economic Approach to Development Planning. Rand McNally, Chicago, IL.

Krautkraemer, J., 2005. The economics of scarcity: the state of the debate. In: Simpson, R.D., Toman, M., Ayres, R. (Eds.), Scarcity and Growth Revisited: Natural Resources and the Environment in the New Millennium. Resources for the Future, Washington, DC, pp. 54–77.

Krugman, P., 1994. Peddling Prosperity. W.W. Norton & Company, Inc., New York, NY.

Little, I.M.D., Mirrlees, J.A., 1974. Project Appraisal and Planning for Developing Countries. Heinemann, London.

Nordhaus, W., 2013. The Climate Casino: Risk, Uncertainty, and Economics for a Warming World. Yale University Press, New Haven, CT, p. 232.

Pearce, D., 2005. Environmental policy as a tool for sustainability. In: Simpson, R.D., Toman, M., Ayres, R. (Eds.), Scarcity and Growth Revisited: Natural Resources and the Environment in the New Millennium. Resources for the Future, Washington, DC, pp. 198–224.

Pearce, D., Turner, R., 1990. Economics of Natural Resources and the Environment. Johns Hopkins University Press, Baltimore, MD.

Ramsey, F., 1928. A mathematical theory of saving. Econ. J. 138, 543–559.

Roumasset, J., Barr, S., 1992. The Economics of Cooperation: East Asian Development and the Case for Pro-Market Intervention. Westview Press, Boulder, CO.

Roumasset, J., Burnett, K., Balisacan, A. (Eds.), 2010. Sustainability Science for Watershed Landscapes. Institute for Southeast Asian Studies, Singapore.

Scheffer, M., 2009. Critical Transitions in Nature and Society. Princeton Studies in Complexity. Princeton University Press, Princeton, NJ.

Schelling, T., 1980. The Strategy of Conflict. Harvard University Press, Cambridge, MA.

Schelling, T., 1995. What do economists know? Am. Econ. 39 (1), 20−22.

Simpson, R.D., Toman, M., Ayres, R. (Eds.), 2005. Scarcity and Growth Revisited: Natural Resources and the Environment in the New Millennium. Resources for the Future, Washington, DC.

Smith, A., 1776. The Wealth of Nations. 1994 Modern Library Edition. Modern Library, New York, NY.

Smith, K. (Ed.), 1979. Scarcity and Growth Reconsidered. Johns Hopkins University Press for Resources for the Future, Baltimore, MD and London.

Solow, R., 1991. Sustainability: an economist's perspective. In: Stavins, R. (Ed.), Economics of the Environment, fifth ed. W.W. Norton, New York, NY. (Presented as the Eighteenth J. Seward Johnson Lecture to the Marine Policy Center, Woods Hole Oceanographic Institution, Woods Hole, MA.)

Solow, R., 1992. An almost practical step toward sustainability. (A lecture on the occasion of the 40th anniversary of Resources for the Future. Published 1993.) Resour. Policy 19 (3), 162−172.

Stavins, R., Wagner, A., Wagner, G., 2003. Interpreting sustainability in economic terms: dynamic efficiency plus intergenerational equity. Econ. Lett. 79 (3), 339−343.

The Economist, 2012. Climate Change: How to Fix It. (October 20, 2012).

UN (United Nations), 2012. Inclusive Wealth Report. United Nations University International Human Dimension Programme and United Nations Environment Programme. Cambridge University Press, New York, NY.

Weitzman, M., 1976. On the welfare significance of national product in a dynamic economy. Q. J. Econ. 90 (1), 156−162.

World Bank, 2006. Where is the Wealth of Nations? Measuring Capital for the 21st Century. World Bank, Washington, DC.

TECHNICAL APPENDIX

MAXIMIZING INTERTEMPORAL WELFARE

A.1. The case with discounting

We start with utility of consumption, $U[C(t)]$, as a measure of well-being or welfare at a particular time, t. A typical functional form for utility of consumption is given by $U(C) = -C^{-(\eta-1)}$ for $\eta > 1$. For this functional form, $U(C)$ is negative, but increasing in C, i.e., $U'(C) > 0$.

Summing up this instantaneous well-being over time, starting at time t, yields a measure of intertemporal welfare, denoted $V(t)$. In the general case, future utilities are discounted according to the rate of time preference, or impatience, designated by the parameter ρ. In continuous time, the "summing up" is achieved through the process of integration. With discounting, intertemporal social welfare at time t can be expressed as

$$Z(t) = \int_t^\infty U[C(s)]e^{-\rho(s-t)}ds \qquad (A.1)$$

where s is the variable of integration.

A goal of economic policy should be to maximize the intertemporal welfare of society starting from initial time $t = 0$, subject to the feasibility constraints of the economy. These constraints relate to two key sectors in the economy: one, to the management of natural capital, and the other, to the accumulation of physical capital over time. Because harvesting or extracting a natural resource entails cost, the constraints are linked.

The dynamic equation for the stock level of a general natural resource is written as

$$\dot{N} = G(N) - R$$

(Dotted variables represent time derivatives, e.g., $\dot{N} = dN/dt$.) Note that all terms in this equation are flow variables. $G(N)$ is the resource growth function, often modeled as a logistic equation for renewable resources

$$G(N) = rN(1 - N/N_{max})$$

where r is called the intrinsic growth rate and N_{max} represents the environmental carrying capacity or saturation level of the resource system. The rate of harvest or extraction is R. Unit harvesting cost, $\Phi(N)$, is a function of the stock level and is assumed to be nondecreasing as the stock N is drawn down: $\Phi'(N) \le 0$.

Output of the economy, $F(K,R)$, at each time t, is a function of physical capital K and the flow of natural resources R into the production process. Replenishment of capital, which depreciates at a continuous rate δ, the cost of natural resource harvesting or extraction, and aggregate consumption in the economy are all flow variables drawn from the flow of output. What is left over is allocated to the buildup of new physical capital

$$\dot{K} = F(K, R) - \delta K - \Phi(N)R - C$$

Letting $Z = Z(0)$, the problem can now be stated as

$$\text{Max } Z = \text{Max} \int_0^\infty U[C(s)]e^{-\rho s}ds$$

subject to

$$\dot{K} = F[K(t), R(t)] - \delta K(t) - \Phi[N(t)]R(t) - C(t)$$
$$\dot{N} = G[N(t)] - R(t)$$

(A.2)

Using the mathematical technology of optimal control, the solution of this problem can be expressed in terms of two necessary first-order conditions.

The first condition is known in the theory of economic growth as the Ramsey rule, or Ramsey condition, in recognition of contributions presented in the classic paper by Ramsey (1928).

$$F_K - \delta = \rho + \eta\{\dot{C}\}[1/C(t)]$$

(A.3)

The parameter $\eta = -[C(t)]\{U''[C(t)]/U'[C(t)]\}$ is the (absolute value of the) consumption elasticity of marginal utility, which is typically assumed to be constant over all levels of consumption. It reflects preference for the smoothing of consumption over time; a higher value of η represents greater preference of consumption smoothing. $F_K = \partial[F(K,R)]/\partial K$ is the marginal product of capital. The notation in the Ramsey rule can be simplified by setting $F_K - \delta = r$, the real rate of interest, and $\dot{C}[1/C(t)] = g$, the growth rate of consumption. Then $r = \rho + \eta g$, as in Section 3.4.1.2.

The second condition is an extension of the classic rule of Hotelling (1931) to the case of a renewable resource with biological growth function, $G(N)$. $F_R = \partial[F(K,R)]/\partial R$ is the marginal product of the natural resource.

$$[F_R - \Phi(N)] = [1/(F_K - \delta - \rho)]\{\dot{F}_R + [F_R - \Phi(N)][G'(N) - \rho] - \Phi'(N)G(N)\}$$

(A.4)

As an aside, for the case of a nonrenewable resource, $G(N) = 0$, and the Hotelling rule reduces to $[F_R - \Phi(N)] = [1/(F_K - \delta - \rho)][\dot{F}_R]$. For a competitive natural resource market, the resource price P would be equal to the marginal product of the resource, F_R. Setting the unit extraction cost equal to the constant c allows the Hotelling rule to be written in the classic form $r[P - c] = \dot{P}$. The term $[P - c]$ is often called the marginal user cost or the scarcity value of the resource.

The Ramsey (A.3) and Hotelling (A.4) conditions guide the optimal, welfare maximizing paths of consumption, physical capital accumulation, and natural resource management toward the steady state, where $\dot{K} = \dot{C} = 0$ and $\dot{N} = \dot{F}_R = 0$. The steady state versions of (A.3) and (A.4) become

$$F_K(K^*, R^*) = \rho + \delta$$

(A.3)

and

$$F_R(K^*, R^*) = \Phi(N^*) + \{\Phi'(N^*)G(N^*)\}/[G'(N^*) - \rho]$$

(A.4)

Conditions (A.3′) and (A.4′), for the case $\rho > 0$ are usually characterized as the Modified Golden Rule. Starred quantities, K^*, N^*, and R^*, are the steady state values satisfying (A.3′) and (A.4′). Steady state consumption is then easily solved as

$$C^* = F(K^*, R^*) - \delta K^* - \Phi(N^*)R^*$$

(A.5)

A.2. The case with no discounting: $\rho = 0$

If we take $\rho = 0$, neutral weighting of utility across time as a pillar of sustainability (Endress et al., 2005, 2014), then the steady state results are

$$F_K(\hat{K}, \hat{R}) = \rho$$

(A.6)

and

$$F_R(\hat{K}, \hat{R}) = \Phi(\hat{N}) + \{\Phi'(\hat{N})G(\hat{N})\}/[G'(\hat{N})] \tag{A.7}$$

$$\hat{C} = F(\hat{K}, \hat{R}) - \delta\hat{K} - \Phi(\hat{N})\hat{R} \tag{A.8}$$

These steady state conditions, with hatted steady state values $\hat{K}$, $\hat{N}$, $\hat{R}$, and $\hat{C}$, comprise the classic "Golden Rule" of capital accumulation and resource management in the context of long-run sustainability.

Golden Rule utility, $U(\hat{C})$, is an important component in the formulation of social well-being across time with zero discounting. One measure of well-being at a particular point in time t (instantaneous well-being) is the gap between $U[C(t)]$ and $U[\hat{C}]$. Observe that for $t < \infty$,

$$\{U[C(t)] - U[\hat{C}]\} < 0$$

As the economy grows and both $C(t)$ and $U[C(t)]$ increase, the gap narrows, becoming less negative.

The summing up of instantaneous well-being of society over time through the process of integration, starting at time t, yields intertemporal welfare, $V(t)$.

$$V(t) = \int_t^\infty \{U[C(s)] - U[\hat{C}]\} ds \tag{A.9}$$

where s is the variable of integration.

Observe that $V(t) < 0$.

This representation is an integral transformation, due to Koopmans (1965), that yields a finite value of the integral over an infinite time horizon without discounting (i.e., $\rho = 0$). Endress et al. (2005) further discuss this technique and its application to sustainable growth.

Again, the planner's problem is to maximize intertemporal welfare of society starting from initial time, $t = 0$, subject to the feasibility constraints of the economy. Setting $V = V(0)$, the problem can now be stated as

$$\text{Max } V = \text{Max} \int_0^\infty \{U[C(s)] - U[\hat{C}]\} ds \tag{A.10}$$

subject to

$$\dot{K} = F[K(t), R(t)] - \delta K(t) - \Phi[N(t)]R(t) - C(t)$$

$$\dot{N} = G[N(t)] - R(t)$$

Solution of this problem leads directly to the following first-order efficiency conditions

$$\text{Ramsey: } F_K - \delta = \eta\{\dot{C}\}[1/C(t)] \tag{A.11}$$

$$\text{Hotelling: } [F_R - \Phi(N)] = [1/(F_K - \delta)]\{\dot{F}_R + [F_R - \Phi(N)]G'(N) - \Phi'(N)G(N)\} \tag{A.12}$$

Note that in maximizing V, we are choosing a path for $U[C(s)]$ that makes the integral $\int_0^\infty \{U[C(s)] - U[\hat{C}]\} ds$ the least negative. A key consequence of following the optimal path, as guided by the Ramsey and Hotelling rules, is that intertemporal welfare is nondeclining over time, i.e., $0 \le \dot{V}$. Thus optimal and intergenerationally neutral growth satisfies the Arrow et al. (2004) sustainability criterion without the contrivance of a sustainability constraint. Figure A.1 depicts the

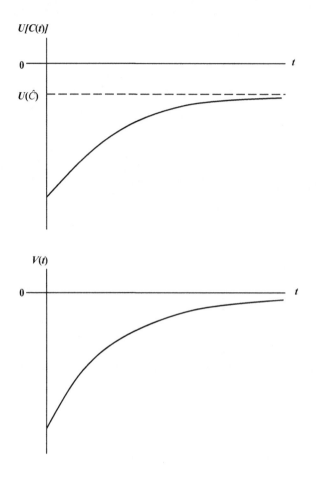

FIGURE A.1

Optimal trajectories for $U[C(t)]$ and $V(t)$.

optimal paths of $U(C)$ and V as functions of time. The economy approaches the steady state as t approaches infinity; $U[C(t)]$ approaches $U[\hat{C}]$ and $V(t)$ approaches zero from below.

Several extensions to this model are possible. Endress et al. (2005) considered a nonrenewable resource with a backstop technology along with fund and then stock pollutants. The Hotelling condition is thereby modified along the lines of the Pearce formulation discussed in Section 3.4.1.2 to account for marginal externality costs. Another extension is to attribute amenity value to the renewable resource by including the resource stock as well as consumption in the utility function: $U(C,N)$. This extension is developed in Section 5 of Endress et al. (2014).

THE ECONOMICS OF FOSSIL FUELS AND POLLUTION

4

Ujjayant Chakravorty* and Yazhen Gong[†]

*Department of Economics, Tufts University, Medford, MA
[†]School of Environment and Natural Resources, Renmin University of China, Beijing, China

CHAPTER OUTLINE

JEL: Q30; Q48; Q54

4.1 INTRODUCTION

Most carbon emissions in the atmosphere come from the burning of fossil fuels, such as coal, oil, and natural gas, which together supply an overwhelming share (more than 80%) of the world's commercial energy (International Energy Agency, 2013). Thus, stabilizing the climate implies reducing the emissions from fossil fuel combustion and switching to less carbon-intensive fuels, such as a substitution away from coal to natural gas in electricity generation, or a larger share of renewable fuels, such as wind and solar energy in the energy mix.

From an economics perspective, it is important to think about policy instruments that may facilitate this transition away from fossil fuels to cleaner energy sources. However, this issue is somewhat complicated because all of the major fossil fuels are nonrenewable in nature. Of course, we can debate as to whether the stocks of oil, coal, and gas can be assumed to be fixed, as in standard economic models of nonrenewable resources, or if the significant rate of new discoveries and technological change suggests that we will never encounter a situation in which the supply of these resources is physically constrained.

From the point of view of economic theory, it is important to think about how the exhaustibility of resources may affect pollution taxes and resource use, as well as the time of transition from a typical fossil fuel, such as oil, to an alternate clean resource, such as solar energy. In this chapter, we consider a specific form of a damage function, which is a ceiling on the stock of carbon emissions. We posit a scenario in which an extension to the Kyoto Protocol or another international agreement imposes a binding target for atmospheric carbon.[1] One may think of this as the 2°C limit imposed by the IPCC beyond which scientists suggest that the earth may see catastrophic changes to its ecological balance. From an economic modeling point of view, this assumption is convenient because it imposes a limit on the stock of emissions from the use of a fossil fuel, and we avoid considering an explicit damage function. As previous studies have suggested, the path of taxes and energy prices may be highly sensitive to the derivatives of a damage function (Farzin and Tahvonen, 1996).

In Section 4.2, we outline a simple model of environmental regulation with only one fossil fuel. In Section 4.3, we extend the analysis to two resources with different pollution characteristics. Section 4.4 concludes by discussing the limitations of the framework and suggesting possible analytical and empirical applications.

4.2 THE FRAMEWORK WITH NONRENEWABLE RESOURCE AND A CEILING ON THE STOCK OF POLLUTION

Here we introduce the basic ideas. Imagine a stock of oil which is fixed, and a social planner extracting it with some discount rate $r > 0$ and a utility function given by $U(q)$, where q is the rate of extraction of oil in each time period t. Suppose the cost of extraction of oil is constant and given by $b > 0$. Then this is the well-known Hotelling (1931) problem that is written as

$$\text{Max} \int_0^\infty [U(q) - bq]e^{-rt}dt \tag{4.1}$$

subject to the constraint

$$\dot{X}(t) = -q, \ X(0) = \overline{X} > 0 \tag{4.2}$$

where r is the discount rate of the social planner. The solution to this problem is quite familiar— resource prices go up over time, there is a markup between the price of the resource and the extraction cost, and consumption goes down over time. Although we do not do this here, one can

---

[1]Since fossil fuels account for 75% of global emissions (the rest is deforestation) the effect of an international environmental agreement (e.g., the Kyoto Protocol) can be assumed to be a direct restriction on carbon emissions from the production of fuels such as coal. Many empirical studies on the effect of environmental regulation on energy use have concluded that the energy sector that is likely to be most impacted by climate regulation is electricity because it mainly uses coal, the dirtiest of all fuels. Clean substitutes for oil in the transportation sector or natural gas in residential heating (the cleanest of all fossil fuels) are much more expensive than the substitutes available in electricity generation such as hydro and nuclear power. That is, relative to coal, oil and natural gas have strong comparative advantage in their respective uses.

reformulate this model by specifying some price, say $\bar{p}$, at which some backstop resource (say bio-fuels or solar) is available in abundant quantities, in which case the terminal price at which there is a complete transition occurs at this price $\bar{p}$.

Now consider a ceiling on the stock of pollution. For convenience, suppose we assume that one unit of oil used emits one unit of pollution. Then we can define the stock of pollution as z with some initial level of stock at $z(0) = z_0$. For the model to make sense, the ceiling has to be above the initial stock of pollution. One can imagine an initial carbon concentration of say 400 ppm and a ceiling imposed by the IPCC of 450 ppm. By imposing the ceiling, we can rewrite the above problem by introducing another constraint, which can be written as

$$\dot{z}(t) = q - \alpha z, \quad z_0 \leq \bar{z} \tag{4.3}$$

where the parameter $\alpha > 0$ represents the natural dilution of the carbon in the atmosphere. In such a simple model, note that we do not include an explicit damage function in which utility may be a decreasing function of the stock of carbon emissions. However, implicitly by imposing a constraint on the stock, we have assumed that damages are zero until the stock hits $\bar{z}$ but infinite afterward. The model will not allow the stock to rise above $\bar{z}$. However, utility comes from burning fossil fuels, and there is no benefit to stay below the ceiling, thus the solution will lead the stock to rise and hit the ceiling at some point in time. Because at the ceiling the amount of energy that can be consumed is limited—we must only use what we can dilute—we have the relationship $\dot{z}(t) = 0$ which implies that $\bar{x} = \alpha \bar{z}$, where $\bar{x}$ is the amount of fossil energy consumed while at the ceiling. The solution is shown in Figure 4.1. The oil consumed rises until the stock hits the ceiling, stays

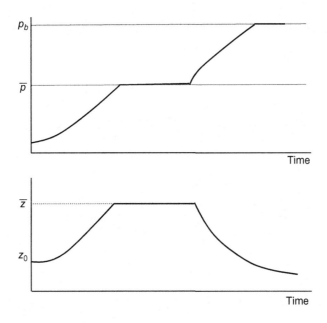

FIGURE 4.1

With a ceiling, the price of the fossil fuel increases, stays constant for a while, and finally rises to hit the backstop fuel. The stock of pollution rises, stays at the ceiling, then declines to zero.

there for some time, and finally declines when the price of oil has risen sufficiently that consumption is no longer constrained by the imposed ceiling. In the figure, we have assumed that at a sufficiently high price defined by $\bar{p}$, an alternative clean source of energy (say solar) is economical.

At the time when the price of energy is equal to the backstop price, the oil is completely exhausted (see top panel of Figure 4.1). Note that compared to the standard Hotelling model without any stock constraint, the price of energy will include a tax that is equal to the multiplier on condition (3). Burning a unit of fossil fuel will create pollution which will bring the stock closer to the ceiling, and hence there will be a higher price for the fossil fuel. The consumption of the fossil fuel will be the converse of what is shown in Figure 4.1. It will go down from the beginning, keep steady at the ceiling, and finally decline further until exhaustion. The bottom panel shows the stock of pollution, which starts at a low level, rises to the mandated ceiling, stays at the ceiling for some time, and finally slides below the ceiling when the endowment of oil is no longer a problem for the regulation, and a new unregulated Hotelling price path starts exactly from the end point of the ceiling. When the renewable takes over completely, there are no more emissions in the atmosphere, and the stock of pollution declines asymptotically to zero because of the natural dilution.

It is clear from the above picture that if the price of the backstop was below the price of energy at the ceiling given by $\bar{p}$, then at the ceiling there would be joint use of the two resources, until all oil is exhausted and the backstop takes over. This has been shown by Chakravorty et al. (2006). Along the same lines, one could think of more complicated scenarios such as the deployment of the backstop making it more cost effective. For example, if in order to limit greenhouse gas emissions to 450 ppm an economy was to use a large volume of solar panels, then there may be significant learning effects from this process, which will in turn increase solar use and reduce fossil fuel use. In that case, a deployment of solar may be justified even before the ceiling is attained.

4.2.1 ABATEMENT OF POLLUTION

The model presented above does not consider the option of reducing pollution through abatement options such as carbon capture and sequestration (CCS). Moreover, the natural dilution parameter α may be small, as some scientists point out, so that abatement may involve costly investment. With carbon capture, the adoption depends on the cost structure of the abatement technology. If the unit cost of CCS technology is assumed constant, then there is no benefit to its early adoption, in which case, as we may expect, the adoption will only happen at the ceiling, where energy use is constrained. Abatement of pollution will then allow for burning more fossil fuels, yet keeping the aggregate net emissions to zero. Adoption is not beneficial before the ceiling is attained or after the ceiling is crossed. Once the ceiling no longer binds, the shadow price of pollution is zero. However, if abatement through technologies (such as CCS) generates benefits in terms of learning by doing and the unit cost declines with use, then there may be use of this technology strictly before the ceiling, so that the benefits of use can accrue.

4.2.2 NONSTATIONARY DEMAND

Things get more complicated when we consider nonstationary demand. This may happen if there is an increase in global population or an increase in per capital consumption of energy driven by income growth. With increasing demand, there may not be constant energy consumption at the ceiling. Energy consumption may increase even before the ceiling in spite of a growth in resource prices because of rising scarcity rents and pollution taxes. If demand was shifting exogenously, the pattern of resource use could change. For example, if demand was high during the ceiling and the clean substitute (solar energy) was relatively cheap, both resources may be used at the ceiling. However, after the ceiling period is over, we revert back to the exclusive use of coal, and again when the coal is exhausted completely, there is a complete switch to solar energy. So we may have two disjoint periods when the clean substitute is used.

4.3 CEILING WITH FOSSIL FUELS WITH DIFFERENT POLLUTION INTENSITIES

An important issue to consider is the issue of substitution of a fossil fuel not only by a clean fuel, as we discuss above, but by other fuels, with potentially different emission characteristics. For example, almost all energy resources have substitutes—coal and natural gas are substitutes in power generation, where gas is much cleaner than coal in terms of carbon emissions. In transportation, biofuels are a substitute for gasoline, the former being somewhat cleaner in terms of life cycle emissions than gasoline, although it is an issue that is actively debated by scientists. The question arises, how would a ceiling on carbon affect resources that are perfect substitutes, yet have different pollution intensities?

Chakravorty et al. (2008) study an extension of the above model in which they consider two resources, say coal and natural gas, each with different pollution intensities. In their model, coal is dirty while gas is a clean resource. Both resources come with known reserves, and emissions face an aggregate stock constraint as discussed earlier. Now given two resources, say 1 for gas and 2 for coal, condition (2) becomes

$$\dot{X}_i = -q_i, \quad i = 1, 2 \tag{4.4}$$

If each of the resources has pollution emissions per unit given by $\theta_i > 0$, then we get the stock constraint as

$$\dot{z} = \sum_i \theta_i q_i - \alpha z, \quad z^0 \leq \bar{z} \tag{4.5}$$

with $\bar{z}$ as the imposed limit on pollution as before. We write z^0 as the initial stock to differentiate from resources which are denoted by subscripts. Assuming away any extraction cost for the two fossil fuels, and incorporating a backstop resource such as solar energy whose consumption is given by y and unit costs written as c_r, we can write the new optimization problem as

$$\text{Max} \int_0^\infty \left[U\left(\sum_i q_i + y \right) - c_r y \right] e^{-\rho t} dt \tag{4.6}$$

with the same set of constraints (4) and (5). Note that with two resources, the model gets quite complicated, even though we have simplified and assumed away any extraction costs for fossil fuels. The best way to understand the solution is to think of starting from the ceiling and checking how the two resources will be used. If the starting pollution stock is at the ceiling, then note that

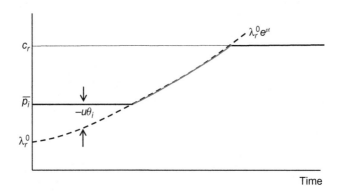

FIGURE 4.2

Resource price with one abundant resource.

the higher the initial stock of a resource, say coal, the lower its initial price (since costs are zero, price equals scarcity rent). Thus if we start from the ceiling, there is a critical stock at which the ceiling does not bind. We can call this X_i^H for resource i where the superscript H denotes Hotelling, meaning the maximal stock at which Hotelling rules and there is no constraint on extraction. However, beyond this stock, the extraction will be bounded for a time, otherwise the stock of pollution will be exceeded. This is shown in Figure 4.2.

If the starting stock of pollution is below the ceiling, the stock of pollution increases until it hits the ceiling, then decreases, as we have discussed before. However, initial stock of resources implies a longer stay at the ceiling. This defines some maximal level of stock of resource $X_i^H(Z^0)$ which is the resource stock beyond which Hotelling no longer holds. The higher the initial stock of pollution Z^0, the smaller this resource stock X_i^H.

With two resources things get complicated because they are perfect substitutes and all that matters is the aggregate stock of pollution. Thus there may be opportunities to substitute depending on the abundance of the two resources. For example, if natural gas is abundant, then one may burn a lot of gas at the beginning of the time horizon, then move to a joint use of the two resources at the ceiling. The reason the social planner would like to burn a lot of gas initially is because it is cleaner, and hence more energy can be consumed in earlier periods, rather than later, generating profits at an earlier time. On the other hand, if coal was abundant, and gas was scarce, one would burn coal and get to the ceiling where energy consumption is constrained and thus preserve the stock of limited gas to be used when it is needed the most—when the ceiling binds. Figure 4.3 shows the resource prices starting from the stock at the ceiling.

When both resources are abundant, the planner consumes the maximum amount of gas first because it is clean and generates the highest current utility. The price is fixed at $\bar{p}_1$. Recall that the stock of emissions is at the ceiling and does not change. However, once some gas is used up, we need to transit to a regime with only coal use. Since coal is dirtier, we can use less of coal and hence generate less utility. The transition from gas price $\bar{p}_1$ to coal price $\bar{p}_2$, shown in Figure 4.3, occurs by using some combination of the two such that we stay at the ceiling, and both prices are equal. The transition is gradual with a small amount of coal substituting for gas at the beginning

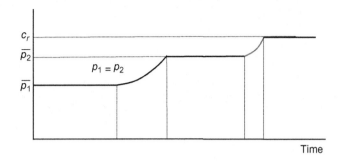

FIGURE 4.3

Resource prices when both resources are abundant and the stock of pollution starts at the ceiling.

and finally a small amount of gas which gives way to exclusive coal use. Once price $\bar{p}_2$ is reached all gas is exhausted. Then we use coal until we arrive at a stock that allows for an unconstrained transition to the backstop (e.g., solar energy) at price c_r.

If the initial stock is lower than the regulated stock of emissions, then we must find a path to the ceiling through the use of the two resources. If there is a lot of coal, then one can show that it may be beneficial to burn more coal initially and "race" to the ceiling from below the ceiling. This is because the "free" dilution of the carbon given by αz increases the stock of carbon z. Because of this proportionality, the planner benefits from rushing to increase the stock, if the current stock lies below the ceiling. In some sense, this is a perverse result: without regulation, the standard Hotelling theory tells us, we will be indifferent between using the two resources since they are perfect substitutes and equally costly (or costless). However, with regulation there may be a preference to burn the dirty resource first in order to benefit from free dilution.

The discussion of pollution intensities suggests that the effect of environmental regulation may be difficult to determine *ex ante*. The profile of resource use is a function of the pollution characteristics of the two resources, the discount rate, and their relative abundance. If one was to remove the assumption of zero cost of the two resources, then the solution may be even more complicated. If natural gas was cheaper to extract, then it would have a double advantage of being both cleaner and cheaper than coal. Since we care about current benefits will push the extraction of gas relative to coal toward the present and move some coal extraction to the future because it is higher cost.

4.4 CONCLUSION

In this chapter, we have considered the economics of regulating fossil fuel use by using a very simple framework. The regulation we consider is somewhat specific—we limit a catastrophic buildup of pollution but do not worry about contemporaneous pollution. In the case of global warming and the accumulation of greenhouse gases, our assumption of an imposed ceiling may be quite realistic. Essentially, we propose a marginal damage function that exhibits low damages until we hit some limit and extremely high damages beyond this value. However, if we consider pollution from coal and oil use in terms of sulfur emissions that cause acid rain, this assumption may be unrealistic.

In that case, the damage from emissions even when we are below some maximum limit of pollution stock may be significant. This may lead to more joint use of resources and less of the sharp transitions we see in our model (for a general equilibrium model with two resources which are imperfect substitutes, see Smulders and van der Werf, 2008).

In our model, the taxes on pollution are given by the shadow price of the stock constraint. In the model with one resource, the tax increases over time and finally reaches a maximum when the ceiling is attained. It declines throughout the ceiling period and equals zero exactly when the ceiling no longer binds. This result of an inverted U-shaped tax path is very similar to what has been obtained for more general damage functions, as in Tahvonen (1997).

There is a large literature on the Herfindahl (1967) conjecture which suggests that resources with the least cost must be used first. Many authors have found conditions under which this principle is violated.[2] The framework we have discussed above can be thought of as an extension to this literature where we use resource abundance as a proxy for costs—the higher the stock of the resource, the lower is its scarcity rent and hence its true marginal cost. The basic idea in this literature is that relative costs matter, and we represent this by looking at relative stocks of resources.

Finally, one may wonder how such a framework can be applied to examine resource and energy use under environmental regulation. There is significant work in this area starting with the pioneering contribution of Nordhaus (1973) who developed a calibration model to study the effect of OPEC oil price shocks on the US economy and modified in the context of learning effects and cost reductions in solar photovoltaics (Chakravorty et al., 1997) and land allocation induced by biofuel mandates (Chakravorty et al., 2013). These empirical models take many other factors into account that make the analytical model complicated—for example, they assume exogenous rates of technological change, new resource discoveries, and energy demand shifts driven by population and income growth. One could use this framework to examine contemporary issues in regulation, such as the effect of the Keystone pipeline or the discovery of shale gas reserves in China and the United States.

ACKNOWLEDGMENTS

The authors would like to thank David Joseph U. Anabo for a careful read of the paper and helpful comments.

REFERENCES

Chakravorty, U., Krulce, D.L., 1994. Heterogeneous demand and order of resource extraction. Econometrica 62 (6), 1445–1452.

Chakravorty, U., Roumasset, J., Tse, K., 1997. Endogenous substitution of energy resources and global warming. J. Polit. Econ. 105 (6), 1201–1234.

Chakravorty, U., Magne, B., Moreaux, M., 2006. A Hotelling model with a ceiling on the stock of pollution. J. Econ. Dyn. Control 30 (12), 2875–2904.

---

[2]See, for example, Kemp and Van Long (1980), Lewis (1982), Chakravorty and Krulce (1994), and a more recent application of the same idea to landfill sites (Gaudet et al., 2001).

Chakravorty, U., Moreaux, M., Tidball, M., 2008. Ordering the extraction of polluting nonrenewable resources. Am. Econ. Rev. 98 (3), 1128–1144.

Chakravorty, U., Hubert, M.-H., Moreaux, M., Nostbakken, L., 2013. The Long-Run Impact of Biofuels on Food Prices. SSRN Working Paper 2104053.

Farzin, Y.H., Tahvonen, O., 1996. Global carbon cycle and the optimal time path of a carbon tax. Oxf. Econ. Pap. 48 (4), 518–536.

Gaudet, G., Moreaux, M., Salant, S.W., 2001. Intertemporal depletion of resource sites by spatially distributed users. Am. Econ. Rev. 91 (4), 1149–1159.

Herfindahl, O.C., 1967. Depletion and economic theory. In: Gaffney, M. (Ed.), Extractive Resources and Taxation. University of Wisconsin Press, Madison, WI, pp. 63–90.

Hotelling, H., 1931. The economics of exhaustible resources. J. Polit. Econ. 39, 137–175.

International Energy Agency, 2013. Key world energy statistics. Retrieved from < http://www.iea.org/publications/freepublications/publication/KeyWorld2013.pdf >.

Kemp, M.C., Van Long, N., 1980. On two folk theorems concerning the extraction of exhaustible resources. Econometrica 48 (3), 663–673.

Lewis, T.R., 1982. Sufficient conditions for extracting least cost resource first. Econometrica 50 (4), 1081–1083.

Nordhaus, W.D., 1973. The allocation of energy resources. Brookings Pap. Econ. Act. 4 (3), 529–576, Economic Studies Program, The Brookings Institution.

Smulders, S., van der Werf, E., 2008. Climate policy and the optimal extraction of high- and low-carbon fossil fuels. Can. J. Econ. 41 (4), 1421–1444.

Tahvonen, O., 1997. Fossil fuels, stock externalities, and backstop technology. Can. J. Econ. 30 (4), 855–874.

INTEGRATED GROUNDWATER RESOURCE MANAGEMENT

James A. Roumasset*[†] and Christopher A. Wada[†]

*Department of Economics, University of Hawai'i at Mānoa, Honolulu, HI
[†]University of Hawai'i Economic Research Organization, Honolulu, HI

CHAPTER OUTLINE

JEL: Q20; Q25

5.1 GROUNDWATER MANAGEMENT: FROM SUSTAINABLE YIELD TO DYNAMIC OPTIMIZATION

In many parts of the world, irrigation and other freshwater uses are largely dependent on groundwater. Figure 5.1 portrays the cross-section of a typical *aquifer*, or subsurface layer of water-bearing, porous materials. Over time, an aquifer is recharged naturally from precipitation that infiltrates below ground. It can also be recharged via irrigation return flow, due either to canal leakage or excess applied water not consumed by crops. The cost of withdrawing water is a direct function of lift, which is the distance between the water table and the surface. In some cases, water can also naturally discharge from the aquifer to adjacent water bodies, or in the case of a coastal aquifer, into the ocean. One aquifer can also be recharged from an adjacent (and more elevated) aquifer.

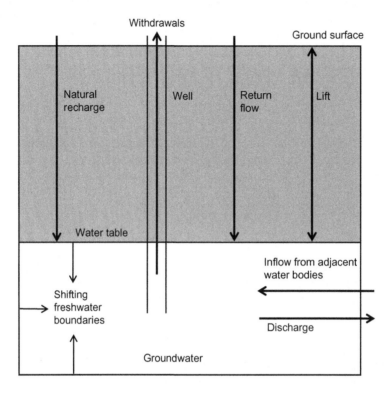

FIGURE 5.1

Single-cell aquifer model.

Thus groundwater satisfies both characteristics of the canonical renewable resource: left unharvested, the stock grows, and the rate of growth depends on the stock. The management problem is to determine how much groundwater to withdraw over time.

Analogous to biological recommendations for fisheries and forests, a common recommendation by hydrologists is to limit extraction of a renewable resource (e.g., groundwater) to the maximum sustainable yield (MSY)—the amount of resource regeneration that would occur at the stock level that maximizes resource growth. In the single-cell aquifer case, this may be given by the minimum freshwater stock before further withdrawals become salty or which would otherwise damage the integrity of the aquifer. Economists often criticize the MSY criterion for harvesting from a renewable resource, noting that the steady-state resource stock is likely to be above the maximum yield level in order to conserve on harvesting (extraction) costs. In the case of groundwater, however, the change in lift from its initial condition to the point of MSY is sufficiently small that the MSY may well be the optimal steady state (Roumasset and Wada, 2012). This still leaves open the question of what sequence of groundwater withdrawals over time maximizes the present value (PV) of a single groundwater aquifer or system of aquifers. The next section begins with a framework for the optimal management of a single groundwater resource. The model is then extended to allow for multiple water resources, watershed conservation, and endogenous resource governance. The chapter concludes with a discussion of key policy implications and directions for further research. To a large extent, the lessons below apply to other renewable resources as well.

5.2 OPTIMAL MANAGEMENT OF A SINGLE GROUNDWATER AQUIFER

Figure 5.1 shows that groundwater has all of the features of a generic renewable resource and can be modeled the same way. The standard groundwater management model is extended to allow for desalination as a backstop resource.[1] Keeping return flow constant for now and including it in recharge, the extraction problem is formally represented as:

$$\max_{q_t,b_t} \int_0^\infty e^{-rt}[B(q_t + b_t) - c_q(h_t)q_t - c_b b_t]\, dt \tag{5.1}$$

subject to:

$$\gamma \dot{h}_t = R - D(h_t) - q_t \tag{5.2}$$

where r is a positive rate of discount; B is the benefit of water consumption, measured for example as consumer surplus or farm profits; q is the amount of water extracted for consumption; b is the quantity of a backstop (such as desalinated water) that may be used to supplement groundwater; $c_q(h)$ and c_b are the unit costs of groundwater extraction and backstop production, respectively, with the former a convex function of the groundwater stock; the head level (h), or the distance from some reference point to the top of the water table, is an index for groundwater volume; R is the amount of exogenous recharge to the aquifer; D is the amount of groundwater that discharges from the aquifer naturally (e.g., to the sea), net of inflows from adjacent water bodies; and γ converts head level to water volume.[2] The optimal steady-state head level, where extraction equals net recharge, will depend on a variety of factors, including the aquifer's physical characteristics and the demand for water. When water demand is rising, it may be optimal to gradually draw down the groundwater stock to the MSY level, and thereafter supplement with an alternative water source such as desalinated brackish water.

5.2.1 TRANSITIONAL DYNAMICS

Calculating the optimal steady-state head level is generally straightforward, but that level will rarely coincide with the initial state of the system. Optimal extraction in each period is determined by withdrawing groundwater until the marginal benefits (MBs) of water fall to equal the full marginal cost (FMC) of withdrawal. The history of extraction determines, in turn, the path of the head level as it transitions from its initial state to the optimal long-run target. Since the FMC is determined only after the solution to the dynamic optimization problem is known, one cannot characterize ex ante the extraction and stock paths. A few general results have been established, however, with respect to time dependence. For a single resource, if the demand and cost functions are stationary over time, the paths of extraction and head will be monotonic (Kamien and Schwartz, 1991).

---

[1]If the aquifer has already been depleted below its steady-state level, desalination may be employed as a "frontstop" (Roumasset and Wada, 2012).
[2]For example, in Gisser and Sánchez (1980) the aquifer is modeled as a rectangular "bathtub" such that γ is the area of the rectangle.

That is, if the initial head level is above (below) the optimal steady-state level, it will fall (rise) smoothly over time until it reaches the target level. If demand is growing over time, however, it may be appropriate to accumulate groundwater initially before drawing it down and finally stabilizing groundwater stock at the optimal steady-state level (Krulce et al., 1997).

5.2.2 THE PEARCE EQUATION AND PRICING FOR OPTIMAL EXTRACTION

A measure of social welfare ideally includes not only the consumption benefits and physical extraction costs of the resource but also nonuse benefits and environmental damage costs. Thus, the FMC of resource consumption should include any externality cost (e.g., irrigation-induced salinization of underlying aquifers) and user cost, which is defined as the cost of using the resource today in terms of forgone future benefits. In the case of groundwater, extracting a unit of water today lowers the water table—thus increasing stock-dependent extraction costs in all future periods—and forgoes capital gains that could be obtained by leaving the resource *in situ* to be harvested at a later date. David Pearce (Pearce and Markandya, 1989; Pearce et al., 1989; Pearce and Turner, 1989) suggested that efficient resource extraction satisfies:

$$MB_t = c_t + MUC_t + MEC_t \equiv FMC_t \tag{5.3}$$

where FMC includes marginal extraction cost (c); marginal user cost (MUC); and marginal externality cost (MEC). Setting the price of the resource equal to the MB along the trajectory described by Eq. (5.3) ensures optimal resource management.[3]

Inasmuch as the "Pearce equation" integrates microeconomics, resource economics, and environmental economics, it is important to provide a rigorous definition and to explore under what conditions the equation holds.[4] The equation is quite standard for the case of fund pollution, wherein there is no MUC and the corrective tax is set equal to MEC, which is simply the contemporaneous marginal damage cost. For the case of carbon pollution from burning coal, MB is the value of marginal product of coal, c and MUC are the extraction and marginal user costs of coal, and MEC is the incremental PV of damages from the carbon emissions of the marginal unit of coal (Nordhaus, 1991; Farzin, 1996; Perman et al., 2003; Endress et al., 2005). Again, the corrective tax is set equal to MEC.

However, if the externality arises indirectly from the impact that depletion of one resource has on another resource stock, optimality requires $p = c + MUC$, where externalities are accounted for within the MUC (Pongkijvorasin et al., 2011; Roumasset and Wada, 2013a). For example, depletion of a groundwater aquifer reduces submarine groundwater discharge, which supports brackish ecosystems in estuaries and bays. In this case, the full MUC is increased beyond the value of water for human extraction, and because these external effects are included in the MUC, a separate MEC term is not necessary (Pongkijvorasin et al., 2010).

Since the FMC exceeds the physical costs of extraction and distribution, a public utility may not be legally allowed to charge the optimal price for all levels of consumption. Another complication arises from the fact that a price increase across the board may decrease welfare disproportionately for lower income users. One potential solution that addresses both efficiency and equity is an increasing block pricing (IBP) structure. If consumers respond to prices at the *margin*, the only

---

[3]For a nontechnical exposition of market-based instruments, see Pearce (2005).
[4]This was suggested by Ed Barbier (personal communication), who worked closely with David Pearce.

requirement for efficiency is that the price for the last unit of water is equal to FMC in every period, i.e., the price can be lower for inframarginal units of water. In the simple case of two price blocks, the first-block price can even be set to zero to ensure that all users can afford water for basic living needs. Any units of water beyond the first block would be priced at FMC. If designed properly, the IBP would induce efficient consumption, while returning would-be surplus revenue to consumers via the free block.

5.3 EXTENSIONS AND EXCEPTIONS TO THE PEARCE EQUATION

The Pearce equation corresponding to a standard resource management problem includes three terms on the cost side: marginal extraction cost, marginal user cost, and marginal externality cost (Eq. (5.3)). However, resource management problems may involve one or more constraints in addition to the resource's equation of motion. This section discusses the possibility of corner solutions wherein the standard Pearce equation should be modified and/or should account for one or more shadow price terms.

5.3.1 PEARCE EQUATION FOR MULTIPLE WATER RESOURCES

When multiple water resources are available, optimality requires that the MB of water consumption be equal to the FMC of extraction, as in the standard case. However, if at least one resource stock is below its long-run equilibrium level, there will be periods in transition to the steady state during which one or more of the resources are not being used, i.e., a corner solution. For the resources not in use, the Pearce equation does not apply (Roumasset and Wada, 2012). That is, the necessary conditions for the maximization problem require only that MB = FMC for the resource(s) with positive extraction. Zero harvest is optimal for some resources precisely because MB < FMC during certain stages of extraction. For an arbitrary demand sector i and resources $j = 1, \ldots, J$, the following modified version of the Pearce equation ensures optimal extraction from the resource system:

$$\text{MB}_t^i = \min\left\{\text{FMC}_t^{i1}, \text{FMC}_t^{i2}, \ldots, \text{FMC}_t^{iJ}\right\} \tag{5.4}$$

The *min* operator in Eq. (5.4) requires that the MB for sector i be equal to the least FMC. For all other resources, MB < FMC and extraction is optimally zero.

For the case of multiple aquifers on Oahu, for example, the Pearl Harbor Aquifer (PHA) is initially a lower cost resource compared to the Honolulu aquifer because of its greater leakage. Optimal management calls for drawing down the "leakier" aquifer first until it reaches the minimum head level (defined by the EPA limit of 2% of ocean salinity). Thereafter, PHA is no longer governed by the Pearce equation, but maintained at its minimum level by setting extraction at the MSY. Instead the Pearce equation governs the Honolulu aquifer once PHA is at its minimum and accordingly drawn down until it too reaches the minimum head level. Once both aquifers are being maintained at their MSY levels, additional increases in quantity demanded at backstop cost are satisfied by desalination. This joint management reduces the waste of independent management by $4.7 billion (Roumasset and Wada, 2012).

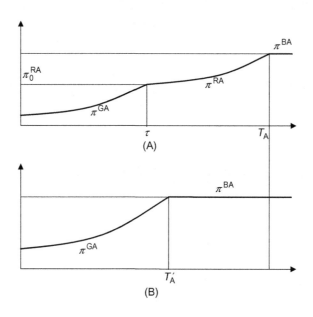

FIGURE 5.2

Hypothetical time paths of FMCs (π): (A) agricultural sector with water recycling and (B) agricultural sector without water recycling.

Source: *Roumasset and Wada, 2011.*

As another example, consider the case where groundwater can be supplemented by recycled wastewater and/or desalinated seawater.[5] For simplicity, water is consumed in either the agricultural (A) or household sector (H), and recycled water (R) is a perfect substitute for groundwater (G) in the agricultural sector only—that is, treated water does not meet the minimum quality standard for household consumption. Desalination (B) is a perfect substitute for groundwater in both sectors. If groundwater is relatively abundant, then the price path is likely to follow a kinked, upward sloping path (Figure 5.2A). In the first stage (until year τ), groundwater is used exclusively in the agricultural sector. As water scarcity rises, groundwater is eventually supplemented by recycled water ($p = \pi^{GA} = \pi^{RA} < \pi^{BA}$). At year T_A, all FMCs are equal, and groundwater is supplemented by both alternatives in the steady state. The price as determined by the modified Pearce equation (Eq. (5.4)) is graphically a lower envelope of the FMC curves for the various water resources. Figure 5.2B illustrates how the need for costly groundwater supplementation can be pushed much closer to the present when all alternatives are not optimally included in a resource management plan.

5.3.2 **PRICING AND FINANCE OF WATERSHED SERVICES**

Now assume a privately owned watershed whose quality (stock of suitably measured natural capital) affects the quantity of recharge. While often mentioned as a supply side groundwater

---

[5]For a detailed discussion of results and derivations, see Roumasset and Wada (2011).

management instrument, watershed conservation is typically undertaken independently of ground-water extraction decisions. Thus, sizeable potential welfare gains generated from joint optimization of groundwater aquifers and their recharging watersheds go to waste under current water management programs. This section builds on the basic theoretical framework introduced in Section 5.2 to illustrate management principles that are capable of capturing those potential gains.

The objective of the problem is still to maximize the PV of groundwater, but Eq. (5.1) must be modified to incorporate the cost of watershed conservation measures. In the simplest case, a unit of investment in conservation (I_t) has a constant cost c_I such that the total investment cost paid in period t is $c_I I_t$. The equation of motion for the aquifer head level (Eq. (5.2)) must account for the fact that investment affects recharge via its contribution to the conservation capital stock (N). Thus, recharge is transformed from a scalar to a function of conservation capital: $R(N_t)$. Although conservation capital is modeled as a single stock, there are in reality a variety of instruments capable of enhancing groundwater recharge, e.g., fencing for feral animals, reforestation, and man-made structures such as settlement ponds. For the purpose of illustrating the joint optimization problem, it is sufficient to assume a generic capital stock, such that recharge is an increasing and concave function of N.[6] This presumes that investment expenditures are allocated optimally among available instruments. The first units of capital are most effective at enhancing recharge, and the marginal contribution of N tapers off. Assuming no natural growth of the capital stock but an exogenous rate of depreciation δ (e.g., a fence), conservation capital changes over time according to the following:

$$\dot{N}_t = I_t - \delta N_t \tag{5.5}$$

Given proper boundary conditions, the problem can be solved using optimal control, and the necessary conditions can be used to derive a Pearce equation, albeit with the constant recharge term replaced by $R(N_t)$. Since the conservation capital stock enters the MUC of groundwater through the recharge function, independent management of the aquifer and watershed would clearly yield different results. An analogous efficiency condition can be derived for the conservation of natural capital (Roumasset and Wada, 2013b). At the margin, the resource manager should be indifferent between conserving water via watershed investment and demand-side conservation:

$$\frac{c_I(r + \delta)}{R'(N_t)} = \lambda_t \tag{5.6}$$

The right-hand side of Eq. (5.6) is the MUC of groundwater, or the marginal future benefits obtained from *not* consuming a unit of groundwater in the current period. The left-hand side of Eq. (5.6) can be interpreted as a supply curve for recharge. Given that the marginal productivity of capital in recharge is diminishing, the marginal cost of producing an extra unit of groundwater recharge is upward sloping. If the marginal cost of recharge were less than the MUC of groundwater, welfare could be increased by investing more in conservation because the value of the gained recharge would more than offset the investment costs. Thus, the "system shadow price" of groundwater, λ, governs both optimal groundwater extraction and optimal watershed investment decisions.

---

[6]One could also specify a direct relationship between recharge and investment expenditures if parameterization of such a recharge function is feasible for the application of interest.

In many cases, the optimal management program can be implemented with a system of ecosystem (recharge) payments to private watershed owners. One option is to pay landowners for all service units, starting from zero. That seems excessive, however, inasmuch as providing zero units (e.g., for recharge) is not a feasible option. Another approach is to integrate conservation financing into a block pricing scheme for water. To properly serve the public interest, the public utility must not only be constrained by the zero-profit condition, but it should charge the FMC in order to incentivize efficient extraction so that the value of a set of aquifers is maximized. Pricing each unit of water at FMC would generate a surplus. Part of the surplus can be returned to consumers through lower priced inframarginal units of water, and the remainder can be used to finance ecosystem payments. Landowners are paid the shadow price for marginal units of recharge but not for inframarginal units below the level of conservation required by zoning.

5.3.3 MEASURING NATURAL CAPITAL

The Pearce equation can also be used to measure changes in natural capital. Suppose for example that a resource manager wants to evaluate the potential benefits of a watershed conservation project that will prevent the loss of groundwater recharge services. Satisfying the Pearce equation for groundwater extraction in each period, with and without conservation, provides the PVs for the two scenarios. The difference in the PVs gives the groundwater benefits of watershed conservation, to which can be added other conservation values (Kaiser and Roumasset, 2002).

From this perspective, the value of natural watershed capital is always relative to some alternative land cover. Even if there were no flora whatsoever, there would still be some recharge. So if this method were to be used to estimate the total value of watershed capital, it would have to be relative to the hypothetical scenario wherein all ecological services are zero. In this sense, natural capital is that which provides ecosystem services. Caution is needed, however, in attributing the value of natural capital to something specific, such as trees.

Another difficulty relating to the use of the Pearce equation in resource valuation has to do with the question of which side of the Pearce equation should be used. Some economists have recommended using net price as a proxy for shadow price on the grounds that it is observable. But from the above discussion, and as others have shown (Arrow et al., 2003; UNU-IHDP and UNEP, 2012), net price is only equal to the resource shadow price (its MUC) when the resource is being optimally extracted. In the more typical case of overextraction, the net price undervalues the MUC. In fact, a resource harvested at the open access equilibrium has a net price of zero.

5.3.4 PEARCE EQUATION WITH ENDOGENOUS GOVERNANCE

Because common pool resources may face overuse by multiple consumers with unrestricted extraction rights, additional governance may be warranted if the gains from governance exceed the costs (Demsetz, 1967; Ostrom, 1990). The optimal solution may be unattainable when enforcement and information costs are considered. Which of several institutions (e.g., privatization, centralized ownership, user associations) maximizes the net PV of the groundwater resource depends on the relative benefits generated from each option net of the governance costs involved in establishing the candidate institution. For example, if the initial demand for water is small and the aquifer is large, the gains from management are likely to be small, and open access might be preferred (NMB_0 in

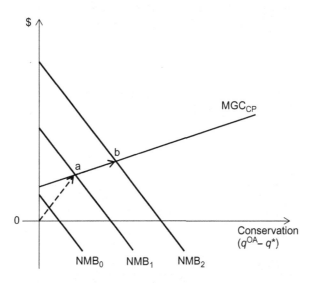

FIGURE 5.3 Governance increases with resource scarcity.

The net marginal benefit of water (NMB), defined as the difference between MUC and the MB of consumption, shifts outward over time as water scarcity increases. The marginal governance cost (MGC) is an increasing function of conservation. In period 0, the marginal cost of common property (CP) governance exceeds the NMB_0 curve for all levels of consumption, i.e., open access (OA) is optimal. In period 1, the MGC curve is less than NMB_1 up to some positive quantity, meaning a CP arrangement like a user association becomes optimal (point a). In future periods, costly governance increases as water becomes scarcer (point b). Note that we are considering the long run, i.e., initial costs are treated as capital and the implicit rental cost of capital is included in the marginal cost.

Source: *Roumasset and Tarui, 2010.*

Figure 5.3). As demand grows over time and water becomes scarcer (net benefits shift out to NMB_1 and eventually NMB_2), however, a CP arrangement such as a user association may become efficient. Eventually, another institution such as a water market, with lower initial MC but lower slope, may become optimal. It is also possible that an intermediate institution such as CP may never be optimal, in the case wherein the lower slope/higher initial MC institution dominates for all levels of positive governance.

Figure 5.4 illustrates the dynamics of the full marginal net benefit (FMB) and shadow price of a resource in transition to governance using a relatively simple constant price model. The simulation assumes a constant price $p = 2$, an extraction cost function $c(S) = 1/S$, an initial resource stock $S_0 = 9.9$, a discount rate $\rho = 0.03$, a resource carrying capacity $K = 10$, an intrinsic resource growth rate $r = 0.5$, and a marginal governance cost $g = 1$. The FMB of harvesting is defined by $p - c(S_t) + g$ and the shadow price (λ) is the costate variable associated with the resource stock under the optimal solution. The noninstantaneous convergence of the FMB and λ appears to contradict the tendency to associate scarcity with net price or MUC, the justification being that they are all equal along the optimal path. However, this puzzle is resolved by examining the necessary conditions for the maximization problem. Solving a modified version of the standard groundwater problem

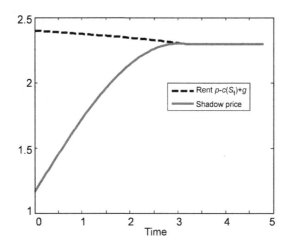

FIGURE 5.4

Marginal net benefit and shadow price under transition to governance ($c = 0$).

(Eq. (5.1)) with MGC in the objective functional and a nonnegativity constraint on governance results in the following optimality condition:

$$p - c(S_t) + g = \lambda + \theta \tag{5.7}$$

where θ is the Lagrangian multiplier associated with the governance constraint. Without g and θ, the condition can be described as net price = MUC or equivalently that price = FMC (Eq. (5.3)). If the FMB is interpreted as the net price + g, then optimality requires that FMB = MUC + θ or FMB = FMC, where the FMC includes the shadow value of meeting the nonnegative governance constraint. Even though the FMB is declining over time, the shadow price (the true scarcity value) is in fact rising as the resource becomes scarcer. In this particular example, the FMB declines by roughly 0.1 units in the three periods required for the system to reach a steady state, while the shadow price rises by almost 1 unit. The declining value of θ reflects the fact that harvest is moving (from above) toward open access. Thus, the standard Pearce equation needs to be modified to allow for θ, the difference between FMB and MUC, when governance is endogenous.

5.4 OPEN ACCESS AND THE GISSER—SÁNCHEZ EFFECT

In many parts of the world, groundwater is characterized as a *common pool resource*, i.e., without appropriate governance, it can be accessed by multiple users who may ignore the social costs of resource depletion. In the limit, it is individually rational for competitive users to deplete the groundwater until MB equals unit extraction cost. In this *open access* equilibrium, each user ignores the effect of individual extraction on future value. Gisser and Sánchez (1980) found that under certain circumstances, the PV generated by the competitive solution is almost identical to that generated by the optimal solution. The surprising result that the potential welfare gain from groundwater management is negligible has come to be known as the Gisser—Sánchez (GS) effect.

Welfare gains may be larger, however, when one or more of the original model's simplifying assumptions are relaxed, e.g., when extraction costs are nonlinear, demand is nonstationary, the discount rate is low, and the aquifer is severely depleted at the outset. From the perspective of Section 5.3.4, the GS effect, under conditions when it is operative, can be recast as prima facie evidence that open access is at least nearly optimal.

The welfare effects of open access also tend to increase dramatically when spatial groundwater pumping externalities are a concern. Pumping groundwater to the surface generates an effect known as a cone of depression, wherein the water table within a certain radius is pulled down toward the well. As a result, nearby users face an increase in lift and consequently extraction costs. Thus, the pumping externality varies over space and depends on the relative locations of the wells. Recent work in this area (Brozović et al., 2006, 2010) has integrated spatial dynamic flow equations into the equation of motion for an aquifer (Eq. (5.2) in the basic nonspatial case). Although this increases the complexity of the optimization procedure and has more stringent data requirements (e.g., the spatial locations of all wells in the aquifer), welfare gains can be potentially large under certain circumstances. For example, if wells are clustered, gains from optimal spatial pumping management are likely to be substantial.

5.5 **POLICY IMPLICATIONS AND DIRECTIONS FOR FURTHER RESEARCH**

Standard renewable resource economics techniques can be applied to the management of a single groundwater resource. In particular, FMC pricing—which takes into account the scarcity value of water—incentivizes optimal consumption. Even when spatial pumping externalities are not considered, FMC pricing creates substantial welfare gains, except in certain specific circumstances, e.g., when the aquifer is particularly large relative to the quantity demanded at extraction cost. Although FMC prices are typically much higher than the marginal extraction cost, zero excess-revenue restrictions that may be imposed on the water manager can be maintained through appropriate block pricing. Because only the marginal price block needs to be equal to the FMC to incentivize optimal consumption, lower blocks can be reduced to offset revenue gains. Relatedly, first-block price reductions can be used to ensure that price reform does not disadvantage the poor or (as in the case of watershed management) the present generation.

When the standard problem is extended to include joint management of additional groundwater resources, simultaneous watershed management, or endogenous governance, the solution method becomes more complicated but the same basic principles apply. For example, when multiple resources are considered, pricing at the *minimum* FMC of all available resources incentivizes optimal consumption, and there are likely to be transitional stages of extraction wherein some resources are not used at all. Whether or not the additional benefit of joint management warrants incurring the additional computational costs will depend on the particular situation, but anecdotal evidence suggests that welfare gains can be large. For example, Roumasset and Wada (2012) found that jointly optimizing the two aquifers underlying South Oahu (Hawaii) would generate a PV welfare gain of over USD 4 billion relative to independent management.

Whether the management problem involves a single resource or multiple water resources, efficiency pricing promises substantial welfare gains. Although across the board price increases would likely be viewed as undesirable to current taxpayers, pricing policies that effect efficiency gains can be win—win if designed correctly. In the case of payments for watershed services, for example, sizeable investments in conservation during earlier periods may be optimal, even though the biggest gains are realized far into the future when water prices are much higher. In other words, the cost of investment in earlier periods is likely to outweigh the contemporaneous benefits. Nonetheless, financing the investment can be based on the principle of benefit taxation. Because the beneficiaries are water users, the price blocks for water can be adjusted to incorporate a lump sum conservation charge that is proportional to the water benefits received. Provided that the PV sum of conservation fees collected from water users is equal to the PV cost of investment—much of which is incurred in earlier periods—a bond could be issued to finance the project and the fees used to pay off the bond.

Groundwater economics can also be generalized to incorporate the spatial dimension. For a single integrated demand network, this simply involves adding transportation costs to the right-hand side of the Pearce equation for each location in the system. The resulting solution is a matrix of efficiency prices over time and space. There is a minimum cost, wholesale shadow price for each time, and the location-specific shadow prices are given by this *system* shadow price plus the distribution cost (Pitafi and Roumasset, 2009; Roumasset and Wada, 2012). A natural extension of this system would be to allow for endogenous demand networks and multiple water resources (e.g., separate aquifers, watershed capital, conjunctive use of surface and groundwater, and freshwater substitutes such as desalinated water and treated wastewater). Separate networks would then be characterized by the condition that the shadow prices of any two locations in the separate networks would differ by less than the transportation costs, such that no interdistrict transport is economic (Jandoc et al., in press). Note, however, that this condition need only hold for a particular point in time. As scarcities (or transport costs) change, so may the system boundaries. This theoretical development could provide an important tool for planning water transportation infrastructure.

The above is still partial equilibrium in nature, however, inasmuch as demand functions at various locations are taken as given, as are interest rates. A further step would be to endogenize the accumulation/depreciation of produced capital for the rest of the economy, investment in water infrastructure, and watershed capital. Presumably, the necessary conditions would involve a generalized version of the extended Hotelling conditions discussed above plus a Ramsey condition for the accumulation of produced capital (Endress et al., 2005). This formulation would allow for the exploration of water (or resource)-related limitations to sustainable development.

REFERENCES

Arrow, K.J., Dasgupta, P., Mäler, K.-G., 2003. Evaluating projects and assessing sustainable development in imperfect economies. Environ. Resour. Econ. 26 (4), 647–685.

Brozović, N., Sunding, D.L., Zilberman, D., 2006. Optimal management of groundwater over space and time. In: Berga, D., Goetz, R. (Eds.), Frontiers in Water Resource Economics, Natural Resource Management and Policy Series, vol. 29. Springer, New York, NY, p. 275.

Brozović, N., Sunding, D.L., Zilberman, D., 2010. On the spatial nature of the groundwater pumping externality. Resour. Energy Econ. 32, 154–164.

Demsetz, H., 1967. Toward a theory of property rights. Am. Econ. Rev. 57 (2), 347–359.

Endress, L., Roumasset, J., Zhou, T., 2005. Sustainable growth with environmental spillovers. J. Econ. Behav. Organ. 58 (4), 527–547.

Farzin, Y.H., 1996. Optimal pricing of environmental and natural resource use with stock externalities. J. Public Econ. 62, 31–57.

Gisser, M., Sánchez, D.A., 1980. Competition versus optimal control in groundwater pumping. Water Resour. Res. 31, 638−642.

Jandoc, K., Juarez, R., Roumasset, J., in press. Towards an Economics of Irrigation Networks. In: Burnett, K., Howitt, R., Roumasset, J., Wada, C.A. (Eds.), Routledge Handbook of Water Economics and Institutions, Routledge.

Kaiser, B., Roumasset, J., 2002. Valuing indirect ecosystem services: the case of tropical watersheds. Environ. Econ. Dev. 7 (4), 701−714.

Kamien, M.I., Schwartz, N.L., 1991. Dynamic Optimization: The Calculus of Variations and Optimal Control in Economics and Management. North-Holland, New York, NY.

Krulce, D.L., Roumasset, J.A., Wilson, T., 1997. Optimal management of a renewable and replaceable resource: the case of coastal groundwater. Am. J. Agric. Econ. 79, 1218−1228.

Nordhaus, W.D., 1991. To slow or not to slow: the economics of the greenhouse effect. Econ. J. 101 (407), 920−937.

Ostrom, E., 1990. Governing the Commons: The Evolution of Institutions for Collective Action. Cambridge University Press, Cambridge.

Pearce, D., 2005. Environmental policy as a tool for sustainability. In: Simpson, R.D., Toman, M., Ayres, R. (Eds.), Scarcity and Growth Revisited: Natural Resources and the Environment in the New Millennium. Resources for the Future, Washington, DC, pp. 198−224.

Pearce, D.W., Markandya, A., 1989. Marginal opportunity cost as a planning concept. In: Schramm, G., Warford, J.J. (Eds.), Environmental Management and Economic Development. The Johns Hopkins University Press, Baltimore, MD, pp. 39−55.

Pearce, D.W., Turner, R.K., 1989. Economics of Natural Resources and the Environment. The Johns Hopkins University Press, Baltimore, MD.

Pearce, D.W., Markandya, A., Barbier, E.B., 1989. Blueprint for a Green Economy. Earthscan Publications, Ltd., London.

Perman, R., Ma, Y., McGilvray, J., Common, M., 2003. Natural Resource and Environmental Economics. Pearson Addison Wesley, New York, NY.

Pitafi, B., Roumasset, J., 2009. Pareto-improving water management over space and time. Am. J. Agric. Econ. 91 (1), 138−153.

Pongkijvorasin, S., Roumasset, J., Duarte, T.K., Burnett, K., 2010. Renewable resource management with stock externalities: coastal aquifers and submarine groundwater discharge. Resour. Energy Econ. 32, 277−291.

Pongkijvorasin, S., Roumasset, J., Pitafi, B., 2011. Pricing resource extraction with externalities: the case of indirect stock-to-stock effects (unpublished manuscript).

Roumasset, J., Tarui, N., 2010. Governing the resource: scarcity-induced institutional change. Working Paper Series #10−15, Department of Economics, University of Hawai'i.

Roumasset, J., Wada, C.A., 2011. Ordering renewable resources: groundwater, recycling, and desalination. B. E. J. Econ. Anal. Policy—Contributions 11 (1). Available from: http://www.degruyter.com/view/j/bejeap.2011.11.issue-1/bejeap.2011.11.1.2810/bejeap.2011.11.1.2810.xml?format=INT, (Article 28).

Roumasset, J.A., Wada, C.A., 2012. Ordering the extraction of renewable resources: the case of multiple aquifers. Resour. Energy Econ. 34, 112−128.

Roumasset, J.A., Wada, C.A., 2013a. Economics of groundwater. In: Shogren, J.F. (Ed.), Encyclopedia of Energy, Natural Resources and Environmental Economics, vol. 2. Elsevier, Amsterdam, pp. 10−21.

Roumasset, J.A., Wada, C.A., 2013b. A dynamic approach to PES pricing and finance of interlinked ecosystem services: watershed conservation and groundwater management. Ecol. Econ. 87, 24−33.

UNU-IHDP and UNEP, 2012. Inclusive Wealth Report 2012. Measuring Progress toward Sustainability. Cambridge University Press, Cambridge.

OPTIMAL JOINT MANAGEMENT OF INTERDEPENDENT RESOURCES: GROUNDWATER VERSUS KIAWE (*PROSOPIS PALLIDA*)

Kimberly M. Burnett[*], **James A. Roumasset**[†], **and Christopher A. Wada**[*]

[*]*University of Hawai'i Economic Research Organization, Honolulu, HI*
[†]*Department of Economics, University of Hawai'i at Mānoa, Honolulu, HI*

CHAPTER OUTLINE

JEL: Q00; Q20; Q25

6.1 INTRODUCTION

It is common in resource economics to solve for optimal harvest rates of an implicitly independent resource (e.g., a forest stand, fishery, groundwater aquifer, or oil reserve). Yet, the premise of ecological economics is that resources are interdependent. The objective of this chapter is to help extend the principles of resource economics to deal with the joint management of interdependent resources. It particularly considers the case where the groundwater uptake of an invasive species detracts from the aquifer stock.

The standard economics approach of maximizing the present value (PV) of net benefits generated by a natural resource specifies the optimal steady state stock level and characterizes the path of optimal resource extraction leading up to that steady state. Decision rules for the dynamically efficient (PV-maximizing) allocation of groundwater were first developed almost half a century ago (Burt, 1967; Brown and Deacon, 1972). More recent efforts have refined the hydrogeological aspects of the management framework, developed instruments for implementing optimal extraction, and considered the welfare implications of various management strategies (Gisser and Sánchez, 1980; Feinerman and Knapp, 1983; Moncur and Pollock, 1988; Tsur and Zemel, 1995; Krulce et al., 1997; Brozović et al., 2010). Few, however, have considered the simultaneous management of natural resources that are interconnected with the aquifer of interest. Those that have modeled resource interdependency (both within and outside the groundwater literature) typically focused on management of a single resource, taking harvest from the adjacent resource as exogenous, e.g., shrimp farms and offshore fisheries (Barbier et al., 2002) and groundwater and nearshore species such as seaweed (Duarte et al., 2010). In the model presented herein, management decisions consider tradeoffs both between resources (groundwater and kiawe) and over time.

Throughout Hawai'i, kiawe (*Prosopis pallida*), a nonnative tree introduced to the islands in the early nineteenth century, can be found in both coastal wetlands and upland ecosystems, covering 58,766 ha or 3.55% of the state's total land area. Kiawe, a nitrogen-fixing legume, can potentially reduce groundwater quality by providing nitrogen-rich organic material for leaching, as well as reduce regional groundwater levels via deep taproots (Richmond and Mueller-Dombois, 1972). In an application to the Kona Coast on the island of Hawai'i, a basic groundwater management model was modified to include water uptake by kiawe. When kiawe is removed, groundwater extraction is higher in every period, corresponding to a lower water scarcity value. In addition, the need for an alternative backstop resource such as desalinated brackish water to meet growing demand is delayed. Both factors contribute to higher welfare in PV terms. PV gains from kiawe management were compared with PV costs of removing and maintaining kiawe using several different methods. The net present value (NPV) is positive for each method ranging from USD 17.0 million to 31.8 million.

6.2 GROUNDWATER–KIAWE MANAGEMENT FRAMEWORK

Although kiawe can affect nearshore ecosystems via increased nutrient loads, the study focused only on its ability to reduce regional groundwater levels. A single-cell coastal aquifer model was modified to include groundwater uptake by kiawe and was integrated into a management framework, the objective of which is to maximize the PV of net benefits from water consumption.

6.2.1 **GROUNDWATER DYNAMICS**

Given that the study is interested in the long-run aquifer-level implications of management decisions (i.e., it abstracts from spatial externalities associated with short-term pumping decisions, such as cones of depression), a single-cell aquifer model is used to determine changes in groundwater stock over time. Under certain conditions (detailed in Section 6.3), the stored volume of water in a coastal aquifer is approximately related to the head level (h)—the distance between mean sea level and the water table—by a constant factor of proportionality (γ). Recharge from precipitation or adjacent water bodies (R) is assumed constant and exogenous. Stock-dependent natural leakage along the freshwater–saltwater interface (L) is an increasing and convex function of the head level; a high head level implies a larger groundwater lens, which exerts greater pressure along a larger surface area. The quantity of groundwater extracted (q) is determined by the resource manager in every period, and uptake (U) is an increasing function of the kiawe stock (K). In what follows, a dot over a variable indicates its derivative with respect to time. The head level evolves over time according to the following relationship:

$$\gamma \dot{h}_t = R - L(h_t) - q_t - U(K_t) \tag{6.1}$$

6.2.2 **KIAWE DYNAMICS**

Kiawe provides some stock (e.g., pollen for the honey production industry) and extraction (e.g., charcoal) benefits to users in the region. However, the study views such benefits as small enough relative to the potential value of water salvage that they can be abstracted from. In the more general case where the benefits provided by both resources are relatively large, the model can be easily adjusted to include those benefits in the objective functional. The stock of kiawe increases according to its natural net growth function (F) and decreases with anthropogenic removal (r):

$$\dot{K}_t = F(K_t) - r_t \tag{6.2}$$

The general framework is amenable to other invasive plant species, provided that one can parameterize the net growth and damage (in this case groundwater uptake) functions. The integrated terrestrial–hydrological system is depicted in Figure 6.1.

6.2.3 **PV MAXIMIZATION**

The resource manager's problem is to choose the rates of groundwater extraction (q), desalination (b), and kiawe removal (r) in every period to maximize the NPV, i.e.:

$$\max_{q_t, b_t, r_t} \int_{t=0}^{\infty} e^{-\rho t} [B(q_t + b_t, t) - c_q(h_t)q_t - c_b b_t - c_r(K_t)r_t] dt \tag{6.3}$$

subject to Eqs. (6.1) and (6.2), given a positive discount rate ρ. Benefits (B) are a function of water consumption (e.g., the area under the inverse demand curve). The unit extraction cost of groundwater, $c_q(h)$, is a function of the head level because it is determined primarily by the energy required to lift groundwater; when the water table is lower, groundwater must be lifted further to reach the surface. The cost of kiawe removal $c_r(K)$ is also stock dependent because management would entail targeting the lowest cost (e.g., the most accessible) areas first. Desalinated water serves as a costly backstop resource, which can be used to supplement groundwater at a constant unit cost c_b.

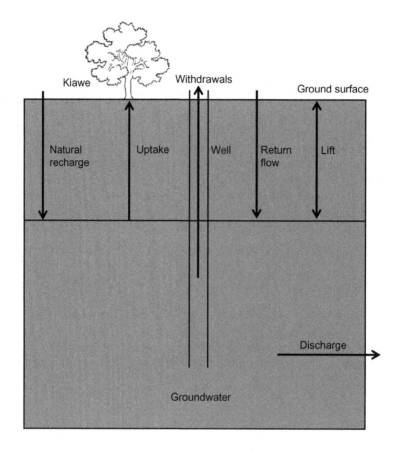

FIGURE 6.1

Coastal aquifer cross section.

If the price at which the marginal benefit and marginal cost of water extraction are equal is defined as $p_t \equiv B'(q_t + b_t, t)$, then it is straightforward to construct the following efficiency price equation for water (see the Appendix for a detailed derivation):

$$p_t = c_q(h_t) + \frac{\dot{p}_t - \gamma^{-1}c_q'(h_t)[R - L(h_t) - U(K_t)]}{\rho + \gamma^{-1}L'(h_t)} \tag{6.4}$$

The second term on the right-hand side of Eq. (6.4) is the marginal user cost (MUC) or the loss in PV resulting from an incremental reduction of the groundwater stock in period t. It is identical to the usual MUC associated with groundwater extraction, except that the net recharge term is adjusted by natural leakage and natural groundwater uptake from kiawe. All else equal, the larger the uptake term, the larger the MUC. Intuitively, this is because uptake adds to anthropogenic extraction in drawing down the head level, thus creating higher future extraction costs and hence larger PV losses than would be realized in the absence of kiawe.

An optimal management rule can also be constructed for the stock of kiawe (see the Appendix for a detailed derivation):

$$\gamma^{-1}\lambda_t = \frac{c_r(K_t)[\rho - F'(K_t)] - F(K_t)c_r'(K_t)}{U'(K_t)} \tag{6.5}$$

Kiawe should be removed until its marginal benefit in terms of avoided uptake, i.e., the shadow value of water $(\gamma^{-1}\lambda)$, is equal to its marginal cost. The numerator of the cost term accounts for the forgone interest that would have accrued had the income not been spent on tree removal, as well as the effect on future kiawe growth and removal costs. Removing a tree today means that future removal of the remaining trees is more expensive on a per-unit basis. It also means that the rate of future kiawe growth is higher or lower, depending on where the stock resides on the growth curve (F). The denominator converts the units of the numerator from dollars per tree to dollars per unit of water. At the optimum, the manager should be indifferent between conserving water via tree removal and via consumption reduction.

6.2.4 THE OPTIMAL STEADY STATE

The optimality conditions (Eqs. (6.4) and (6.5)) must hold in every period, even when the system is in a long-run equilibrium or steady state. By definition, the costate and state variables remain constant in a steady state, i.e., $\dot{h}_t = \dot{K}_t = \dot{\lambda}_t = \dot{\mu}_t = 0$, which implies that $\dot{p}_t = 0$. If demand for water grows over time as a result of rising per capita income and population expansion, desalination will eventually be required at a finite time T to supplement groundwater withdrawals. For $p_T = c_b$, Eqs. (6.4) and (6.5) can be solved for unique values of head and kiawe stock, h^{ss} and K^{ss}, respectively. If the solution yields a negative value for K^{ss} and/or $h^{ss} < h_{min}$, however, one must instead conclude that $K^{ss} = 0$ (i.e., eradication) and $h^{ss} = h_{min}$, where h_{min} is a minimum allowable head level beyond which further pumping yields water of unacceptable quality.

The terminal conditions for head and kiawe stock can then be used in conjunction with initial field measurements for h_0 and K_0 to numerically solve the system of equations (Eqs. (6.1), (6.2), (6.4), (6.5)). Intuitively, any set of paths that satisfies Eqs. (6.1) and (6.2) is feasible, but optimality requires that the state variables also satisfy Eqs. (6.4) and (6.5) in every period. Solving the problem thus involves selecting the endogenous terminal time T such that the resulting paths are feasible, optimal, and consistent with the initial conditions.

6.3 AN APPLICATION TO THE KONA COAST OF HAWAI'I

A simplified version of the model presented in Section 6.2 was applied to data from the Kiholo aquifer and its surrounding watershed on the Kona Coast of Hawai'i Island. The main departure from the general framework is the absence of kiawe stock dynamics (Eq. (6.2)). Although the simplification rules out the possibility of dynamic tree management, the results still illustrate the trade-off between the recharge benefits and costs of kiawe removal.

6.3.1 HYDROLOGY

The Kiholo aquifer is a thin basal or Ghyben−Herzberg lens of freshwater floating on underlying denser seawater. Given the high porosity of the aquifer and hence its relatively thin brackish transition zone (Duarte, 2002), the freshwater−saltwater mixing region was modeled as a sharp interface. Although not amenable to characterizing spatial disequilibrium relationships in the short run, a one-dimensional aquifer model is still useful for identifying the long-run optimal extraction path. The equation of motion for the head level of a one-dimensional, sharp-interface, coastal aquifer can be expressed as $\dot{h}_t = (2000/41\theta WE)[R - L(h_t) - q_t]$, where θ is porosity, W is the aquifer width, and E is the aquifer length (Mink, 1980). For the values $\theta = 0.3$, $W = 6000$ m, and $E = 6850$ m, the volume (thousand gallons) to head (feet) conversion factor (γ^{-1}) for the Kiholo aquifer is 0.0000000492.

Following Pongkijvorasin et al. (2010), the aquifer's natural recharge is assumed constant and equal to 3,992,700 thousand gallons per year (tg/yr). However, leakage or discharge from the aquifer, as discussed previously, is not constant. Mink (1980) derived a structural expression for discharge as a function of head: $l(h) = kh^2$, where k is an aquifer-specific coefficient. Since the leakage function needs to satisfy current conditions, a discharge rate of 3,883,330 tg/yr and head level (h_0) of 5.74 ft. imply that k is equal to 117,864.

6.3.2 GROUNDWATER EXTRACTION AND DESALINATION COSTS

The cost of extracting groundwater is primarily determined by the energy required to lift water from the subsurface aquifer to the ground level (e). Duarte (2002) estimated the energy cost of lifting groundwater from the Kiholo aquifer to be USD (2001) 0.00083/m^3 per meter. In 2012 dollars, the cost is USD 0.00108/m^3 per meter or equivalently USD 0.00125/tg/ft. Given that the average ground elevation relative to mean sea level is 1322.5 ft., the unit cost of groundwater extraction as a function of the head level can be expressed as $c_q(h) = 0.00125 (1322.5 - h)$.

Pitafi and Roumasset (2009) used a straightforward amortization procedure for capital costs (e.g., treatment facility construction) in combination with cost projections for annual operation and maintenance (e.g., wages, materials, energy) of a reverse osmosis desalination plant to estimate the unit cost of desalination: USD (2001) 7/tg. After adjusting for inflation, c_b was estimated to be USD (2012) 9.07/tg.

6.3.3 DEMAND FOR WATER

The County of Hawai'i Department of Water Supply charges a fixed "standby charge," a volumetric "power cost charge," and a volumetric "general use" rate that varies discretely by water quantity blocks. Assuming that the average family falls into the second price block—which is consistent with the average household use of roughly 13,000 gallons per month on O'ahu—the retail price for water in the region was USD (2008) 4.80/tg.

At the price of USD 4.80/tg, 1074.4 m^3 of groundwater were extracted for consumption in 2008. Based on Griffin (2006) and Dalhuisen et al. (2003), the price elasticity of demand for water (η) is assumed equal to -0.7, which corresponds to a constant elasticity demand function of the form $q_t = 850,983 p_t^{-0.7}$, measured in tg/yr. With the development of projects in the area, extraction is expected to increase to 3809 m^3/yr (Pongkijvorasin, 2007). A 5% growth rate of demand is consistent with a 25-year period to project completion and similar growth thereafter. However, a more reasonable assumption may be that in the years following completion of the projects, population and per capita income growth would converge to a lower level. Therefore, it is assumed that demand grows at an average rate of 3% per annum, such that period t demand is determined by $q_t = 850,983p_t^{-0.7}e^{0.03t}$.

6.3.4 GROUNDWATER UPTAKE BY KIAWE

Ideally, a relationship between kiawe and water uptake could be constructed using a time series of relevant data. In the absence of the requisite data, however, potential water salvage of kiawe removal can be roughly estimated using such a relationship for a similar type of tree. Saltcedar (*Tamarix* spp.), for example, is also known to lower water tables via deep taproots, particularly in the southwestern United States. Barz et al. (2009) estimated that removal of 8954 acres of saltcedar from the Texas Pecos River Basin would release 7.41 million m^3 of water per year. This translates to an annual recharge gain of 218.62 tg of water per acre of trees removed.

Assuming that kiawe is roughly distributed in proportion to land area for each of the islands across the state and that one-fourth of the kiawe habitat on Hawai'i Island lies on the Kona Coast in close proximity to the Kiholo aquifer, the total potential water salvage associated with removing all of the kiawe in Kiholo is 2,936,570 tg/yr. These assumptions are summarized in Table 6.1, along with the other functions and parameters discussed in Sections 6.3.1−6.3.3.

6.3.5 KIAWE REMOVAL COSTS

Several previous studies had estimated the cost of removing kiawe using a variety of methods, ranging from bulldozing to aerial broadcast of herbicides to controlled burning. The initial per acre costs ranged from a low of USD (2012) 7 for burning to as much as USD (2012) 295 for bulldozing. Follow-up treatment for each method tended to also vary, suggesting that a PV approach to calculating costs is necessary to ensure that streams of costs accruing in different time periods are converted to comparable units. Thus for an initial treatment cost of X followed by maintenance treatment every Y years at a cost of Z, the PV cost of removal is calculated as $X + \sum_{t=1}^{\infty}[\$Z(1+\rho)^{-Yt}]$ per acre. Per acre costs and PV costs for removing all existing acres of kiawe in the Kiholo region are presented in Table 6.2.

Table 6.1 Equations and Parameters

Description	Unit	Equation or Value
State equation for water	tg/yr	$\dot{h}_t = 0.0000000492[R - L(h_t) - q_t]$
Recharge	tg/yr	$R = 3,992,700$
Leakage	tg/yr	$L(h_t) = 117,864\,h_t$
Extraction cost	USD/tg	$c(h_t) = 0.00125(1322.5 - h_t)$
Desalination cost	USD/tg	$c_b = 9.07$
Water demand	tg/yr	$q_t = 850,983p_t^{-0.7}e^{0.03t}$
Kiawe uptake	tg/yr	$U = 2,936,570$

Table 6.2 Kiawe (*P. pallida*) Removal Costs in 2012 Dollars

Author	Year	Location	Method	Cost (USD/acre)	Follow-up	PV (Million USD)[a]
Campbell et al.	1996	Australia	Single pull	30		1.57
			Double pull	64		3.36
			Bulldoze	295		15.5
March et al.	1996	Australia	Aerial spray	133		6.98
			Blade plough	61		3.20
Teague et al.	1997	Texas	Hand spray	35	Retreat every 10–12 years	1.84
			Spray + chain	56	Chain again after 2 years; spray every 10–12 years	2.83
			Roller chopping	92	Retreat every 6–8 years	7.60
			Root plowing + reseed	127	Grub every 12 years	3.34
			Fire	7	Burn every 5–7 years	0.68
			Grub	106	Retreat every 10–15 years	5.56

[a]*If the study does not provide recommendations for follow-up treatment, it is assumed that the initial treatment is repeated every 10 years in perpetuity.*

6.4 RESULTS

The maximization problem (Eq. (6.3)) is solved for $U = 0$ (all kiawe removed) and $U = 2,936,570$ (no kiawe removed). The removal of kiawe significantly affects water price, head level, and consumption trajectories in a manner that increases benefits to society. Specifically, it allows the aquifer to build for a period before being drawn down to its steady state level. Because water is relatively abundant at the outset and demand is growing, the time path of the head level is nonmonotonic; the aquifer is allowed to replenish initially in anticipation of future scarcity. This is not to say that groundwater consumption is lower under kiawe management. On the contrary, the water salvaged from kiawe ensures a lower price and higher consumption in every period, in addition to delaying the need for a costly alternative such as desalination by nearly 40 years. The price, head, water extraction, and consumption paths are presented in Figure 6.2.

Quantitatively, the benefits of kiawe management are calculated as the difference between the PVs of the aquifer with and without kiawe removed. While the PV benefits are net of the costs associated with extracting groundwater, they do not yet account for the cost of controlling kiawe. The NPV was calculated by subtracting the PV cost of kiawe treatment for each of the methods outlined in Table 6.3. The NPV is positive for each method, ranging from a low of USD 17.0 million for bulldozing to a high of USD 31.8 million for fire (Table 6.3).

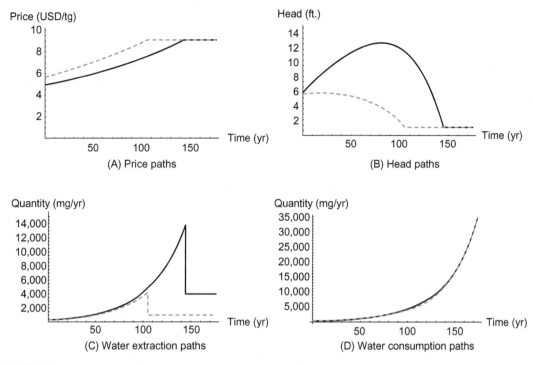

FIGURE 6.2

Price, head, extraction, and consumption trajectories for no kiawe management (dashed lines) and the case where all kiawe is removed (solid lines): (a) price paths, (b) head paths, (c) water extraction paths, and (d) water consumption paths.

Table 6.3 NPV Calculations for Various Kiawe Management Instruments			
Method	**PV Benefit (Million USD)**	**PV Cost (Million USD)**	**NPV (Million USD)**
Single pull	32.5	1.57	30.9
Double pull	32.5	3.36	29.1
Bulldoze	32.5	15.5	17.0
Aerial spray	32.5	6.98	25.5
Blade plough	32.5	3.20	29.3
Hand spray	32.5	1.84	30.6
Spray + chain	32.5	2.83	29.6
Roller chopping	32.5	7.60	24.9
Root plowing + reseed	32.5	3.34	29.1
Fire	32.5	0.68	31.8
Grub	32.5	5.56	26.9

6.5 CONCLUSION

This study derived welfare-maximizing decision rules for the dynamic management of two interacting resources: groundwater and kiawe (*P. pallida*). The optimal quantity of groundwater extraction satisfies the condition that the marginal benefit of water consumption is equal to the sum of extraction and marginal user cost, where the latter is a function not only of the groundwater stock but also of the kiawe stock via its ability to uptake groundwater. Analogously, the optimal decision rule for kiawe control is dependent on the stock of groundwater; kiawe should be removed until the marginal benefit in terms of water salvage is equal to the marginal cost of removal. At the optimum, which can be achieved only through joint management of the resources, the costs of conserving water via tree removal and consumption reduction are equal. One way of implementing the optimal solution is to set the marginal water price equal to the cost of providing water through both mechanisms.

An application of the model to the Kona Coast of Hawai'i showed that the PV cost of removing existing kiawe trees is outweighed by the benefits, measured as the difference in PV welfare to water consumers with and without the trees removed. Among the 11 removal methods considered, management by fire yields the lowest PV cost (USD 0.68 million) and hence the highest NPV (USD 31.8 million), while management by bulldozing yields the highest PV cost (USD 15.5 million) and the lowest NPV (USD 17.0 million). However, the NPV estimates considered only the direct costs of management (e.g., wages, materials, equipment rental). Each removal instrument may generate additional costs that must be accounted for when devising a management strategy. Fire, for example, does not require much labor or rental of expensive machinery, but the potential for spread to nontargeted areas may not be trivial, especially in dry leeward areas conducive to kiawe growth. The smoke generated might also cause discomfort to surrounding residents. Herbicide application is effective, but has the potential to affect nontarget native species and to compromise the quality of underlying groundwater sources. Obtaining permits for aerial broadcast of herbicides may be prohibitively costly or difficult. Thus, Table 6.3 should be viewed primarily as a starting point for the development of kiawe management policy.

Regardless of method, kiawe removal may disrupt other activities that generate benefits. For instance, although it is an invasive species in Hawai'i, kiawe is valued for its role in honey production and as firewood. These industries are small relative to the state's economy, but the potential welfare loss should be incorporated into a comprehensive benefit−cost analysis of various management instruments under consideration. While a detailed analysis of the impact on the local economy is beyond the scope of this chapter, any losses suffered by those industries are believed to likely be outweighed by the potential recharge benefits of management, inasmuch as the kiawe under consideration in Kiholo composes only a fraction of the forest stands on the island and in the state.

The framework developed herein can be applied to a variety of settings around the world, where the presence of one natural resource stock affects the quantity, quality, or availability of another. Other appropriate applications include jointly managing upstream forests and downstream waterways, invasive pest control in agriculture, and groundwater management and linked nearshore marine ecosystems. To the extent that the relationships between natural stocks, as well as the implications of changing one or the other, can be characterized, optimal management trajectories for maximizing their joint benefit can be obtained.

From a policy perspective, one can draw several lessons from the framework developed. First, resource scarcity can be largely affected by interlinkages between different types of ecosystems or natural resources, so management decisions such as price reform should consider those interlinkages. Relatedly, managing resources independently—e.g., a groundwater aquifer and an invasive species such as kiawe that affects the aquifer—overlooks potentially large welfare gains that may be obtained from joint management. Lastly, even when currently available data are not sufficient to jointly optimize both resources, management scenarios, such as removing all of the invasive species in the current period, may serve as a useful approximation (or lower bound) of NPV benefits to justify financing a joint management approach.

The analysis presented can be extended in a variety of ways. The NPV calculations assume that kiawe reduction would occur immediately, when in fact it may make sense to delay the removal of the trees. If the discount rate is large (future benefits and costs are not valued highly from today's standpoint) and groundwater is initially relatively abundant, consumers may prefer to delay the cost of kiawe management. In that case, the problem becomes one of optimal timing: at what point in the future should kiawe trees be removed to maximize PV? An even more ambitious extension would involve determining the optimal dynamic path of kiawe reduction, provided that data are available to parameterize detailed uptake and growth functions.

ACKNOWLEDGMENTS

This research was funded in part by NSF EPSCoR Grant No. EPS-0903833.

REFERENCES

Barbier, E.B., Strand, I., Sathirathai, S., 2002. Do open access conditions affect the valuation of an externality? Estimating the welfare effects of mangrove-fishery linkages in Thailand. Environ. Resour. Econ. 21, 343–367.

Barz, D., Watson, R.P., Kanney, J.F., Roberts, J.D., Groeneveld, D.P., 2009. Cost/benefit considerations for recent saltcedar control, Middle Pecos River, New Mexico. Environ. Manage. 43, 282–298.

Brown, G., Deacon, R., 1972. Economic optimization of a single cell aquifer. Water Resour. Res. 8, 552–564.

Brozović, N., Sunding, D.L., Zilberman, D., 2010. On the spatial nature of the groundwater pumping externality. Resour. Energy Econ. 32, 154–164.

Burt, O.R., 1967. Temporal allocation of groundwater. Water Resour. Res. 3, 45–56.

Campbell, S.D., Setter, C.L., Jeffrey, P.L., Vitelli, J., 1996. Controlling dense infestations of *Prosopis pallida*. Proceedings of the 11th Australian Weeds Conference, Melbourne, Australia, September 30–October 3.

Chiang, A.C., 2000. Elements of Dynamic Optimization. Waveland Press, Inc., Long Grove, Illinois, IL.

Dalhuisen, J.M., Florax, R.J.G.M., de Groot, H.L.F., Nijkamp, P., 2003. Price and income elasticities of residential water demand: a meta-analysis. Land Econ. 79 (2), 292–308.

Duarte, T.K., 2002. Long-term management and discounting of groundwater resources with a case study of Kuki'o, Hawaii (Ph.D. dissertation). Department of Civil and Environmental Engineering, Massachusetts Institute of Technology.

Duarte, T.K., Pongkijvorasin, S., Roumasset, J., Amato, D., Burnett, K., 2010. Optimal management of a Hawaiian coastal aquifer with nearshore marine ecological interactions. Water Resour. Res. 46, W11545; http://onlinelibrary.wiley.com/doi/10.1029/2010WR009094/abstract.

Feinerman, E., Knapp, K.C., 1983. Benefits from groundwater management: magnitude, sensitivity, and distribution. Am. J. Agric. Econ. 65, 703—710.

Gisser, M., Sánchez, D.A., 1980. Competition versus optimal control in groundwater pumping. Water Resour. Res. 31, 638—642.

Griffin, R.C., 2006. Water Resource Economics: The Analysis of Scarcity, Policies and Projects. The MIT Press, Cambridge, MA.

Krulce, D.L., Roumasset, J.A., Wilson, T., 1997. Optimal management of a renewable and replaceable resource: the case of coastal groundwater. Am. J. Agric. Econ. 79, 1218—1228.

March, N., Akers, D., Jeffrey, P., Vietlli, J., Mitchell, T., James, P., 1996. "Mesquite (*Prosopis* spp.) in Queensland". Pest Status Review Series—Land Protection Branch. Department of Natural Resources and Mines, Queensland, Australia.

Mink, J.F., 1980. State of the Groundwater Resources of Southern Oahu. Board of Water Supply, Honolulu, HI.

Moncur, J.E.T., Pollock, R.L., 1988. Scarcity rents for water: a valuation and pricing model. Land Econ. 64 (1), 62—72.

Pitafi, B.A., Roumasset, J.A., 2009. Pareto-improving water management over space and time: the Honolulu case. Am. J. Agric. Econ. 91 (1), 138—153.

Pongkijvorasin, S., 2007. Stock-to-stock externalities and multiple resources in renewable resource economics: watersheds, conjunctive water use, reefs and mud (Ph.D. dissertation). Department of Economics, University of Hawai'i at Mānoa, Hawai'i.

Pongkijvorasin, S., Roumasset, J., Duarte, T.K., Burnett, K., 2010. Renewable resource management with stock externalities: coastal aquifers and submarine groundwater discharge. Resour. Energy Econ. 32, 277—291.

Richmond, T.d.A., Mueller-Dombois, D., 1972. Coastline ecosystems on Oahu, Hawaii. Plant Ecol. 25, 367—400.

Teague, R., Borchardt, R., Ansley, J., Pinchak, B., Cox, J., Foy, J.K., 1997. Sustainable management strategies for mesquite rangeland: the Waggoner Kit Project. Rangelands 19 (5), 4—8.

Tsur, Y., Zemel, A., 1995. Uncertainty and irreversibility in groundwater resource management. J. Environ. Econ. Manage. 29, 149—161.

APPENDIX

Recall that the objective is to maximize Eq. (6.3) subject to state Eqs. (6.1) and (6.2). Optimal control is implemented to characterize the necessary conditions for the maximization problem. The corresponding current value Hamiltonian is:

$$H = B(\bullet) - c_q(h_t)q_t - c_b b_t - c_r(K_t)r_t + \lambda_t \gamma^{-1}[R - L(h_t) - q_t - U(K_t)] + \mu_t[F(K_t) - r_t] \qquad (A.1)$$

and the Maximum Principle requires that the following conditions are satisfied (Chiang, 2000):

$$\frac{\partial H}{\partial q_t} = B'(q_t + b_t, t) - c_q(h_t) - \gamma^{-1}\lambda_t \le 0 \qquad \text{if} < \text{then } q_t = 0 \qquad (A.2)$$

$$\frac{\partial H}{\partial b_t} = B'(q_t + b_t, t) - c_b \le 0 \qquad \text{if} < \text{then } b_t = 0 \qquad (A.3)$$

$$\frac{\partial H}{\partial r_t} = -c_r(K_t) - \mu_t \le 0 \qquad \text{if} < \text{then } r_t = 0 \qquad (A.4)$$

$$\dot{\lambda}_t - \rho\lambda_t = -\frac{\partial H}{\partial h_t} = c_q'(h_t)q_t + \gamma^{-1}\lambda_t L'(h_t) \qquad (A.5)$$

$$\dot{\mu}_t - \rho\mu_t = -\frac{\partial H}{\partial K_t} = c_r'(K_t)r_t + \gamma^{-1}\lambda_t U'(K_t) - \mu_t F'(K_t) \qquad (A.6)$$

$$\dot{h}_t = \frac{\partial H}{\partial \lambda_t} = \gamma^{-1}[R - L(h_t) - q_t - U(K_t)] \qquad (A.7)$$

$$\dot{K}_t = \frac{\partial H}{\partial \mu_t} = F(K_t) - r_t \qquad (A.8)$$

An efficiency price equation for water that is dependent only on constant parameters and the two state variables can be derived using the above conditions. First, define the price for which the marginal benefit and marginal cost of water extraction are equal as $p_t \equiv B'(q_t + b_t, t)$. Then assuming groundwater extraction is positive, Eq. (A.2) becomes:

$$p_t = c_q(h_t) + \gamma^{-1}\lambda_t \Leftrightarrow \lambda_t = \gamma[p_t - c_q(h_t)] \qquad (A.9)$$

Taking the time derivative of Eq. (A.9) yields:

$$\dot{\lambda}_t = \gamma[\dot{p}_t - c_q'(h_t)\dot{h}_t] \qquad (A.10)$$

Next, replace $\dot{h}_t$ in Eq. (A.10) with the right-hand side of the equation of motion (Eq. (A.7)). Finally, substitute all λ_t and $\dot{\lambda}_t$ terms in Eq. (A.5) with Eqs. (A.9) and (A.10):

$$p_t = c_q(h_t) + \frac{\dot{p}_t - \gamma^{-1}c_q'(h_t)[R - L(h_t) - U(K_t)]}{\rho + \gamma^{-1}L'(h_t)} \qquad (A.11)$$

Similarly, a condition can be derived to describe the optimal removal of kiawe over time. When removal is positive, $\mu_t = -c_r(K_t)$. Taking the time derivative of the costate variable yields:

$$\dot{\mu}_t = -c_r'(K_t)\dot{K}_t \tag{A.12}$$

Eq. (A.12) can be further simplified by replacing $\dot{K}_t$ with the right-hand side of the equation of motion (Eq. (A.8)). Substituting all μ_t and $\dot{\mu}_t$ terms in Eq. (A.6) results in the following equimarginality condition:

$$\gamma^{-1}\lambda_t = \frac{c_r(K_t)[\rho - F'(K_t)] - F(K_t)c_r'(K_t)}{U'(K_t)} \tag{A.13}$$

WIN–WIN SOLUTIONS FOR REFORESTATION AND MAIZE FARMING: A CASE STUDY OF NAN, THAILAND

7

Sittidaj Pongkijvorasin and Khemarat Talerngsri Teerasuwannajak

Faculty of Economics, Chulalongkorn University, Bangkok, Thailand

CHAPTER OUTLINE

JEL: Q150; Q230; Q280; Q560

7.1 INTRODUCTION

In 1973, Thailand had 22.17 million hectares (ha) of forestland, which accounted for approximately 43.21% of the country's total land area of 51.31 million ha. Over the past 40 years, forestlands have been disappearing: in 2013, only 16.3 million ha (31.57% of total land area) remains, of which 55% can be found in Northern Thailand (Department of Forestry, 2014). In recent years, forestlands have been shrinking drastically due to massive conversion to maize farming in northern Thailand.

A notorious example of deforestation is found in Nan, a province in northern Thailand. Of its land area of 1,147,200 ha, 74% (849,600 ha) was categorized as forestland in 2004. By 2009, this figure had decreased to 70.4% (806,400 ha). One of the main reasons for deforestation in Nan is the rapid expansion of agricultural areas, especially maize farming (e.g., see UNDP, 2012; Meechoui and Surapornpiboon, 2008). In 2005, maize farms in the province covered 46,934 ha only. The number had almost tripled to 136,355 ha by 2009. Nan has become the second largest maize producing province in the country (Nan Provincial Agricultural Extension Office, 2011).

Such rapid increase in maize farm area meant massive forest clearing, creating a significant environmental impact. For instance, farmers usually burn their fields after harvest to prepare them for the next planting. This causes severe air pollution. In February—March 2012, the highest 24-h average particulate matter (PM_{10}) level in Nan reached 216.4 $\mu g/m^3$, far above the 120 $\mu g/m^3$ standard level. The high PM_{10} level increases the rate of respiratory diseases (Nan Provincial Office of Natural Resources and Environment, 2012).

Moreover, maize farming tends to overuse pesticides, which, in turn, degrade the soil and contaminate water. For instance, water from the river is no longer fit for human consumption. A large degree of forest conversion to maize farming in the highland has also aggravated soil erosion. These adverse effects on the environment will inevitably decrease people's income and worsen their quality of life.

Perhaps the most important factor of the rapid maize farming expansion has been the triple increase in the domestic maize price in 2005—2011, from just around USD 0.13/kg to almost USD 0.37/kg. This supports the argument that there is a positive relationship between agriculture payoff and deforestation (Deacon, 1995; Ehui et al., 1990). Evidence of such trade-off between economic development and environment degradation can be found in many developing countries where monitoring and law enforcement are not effective.

Looking at this issue from a longer horizon perspective, some believe that even with no intervention, forest deforestation will stabilize over time (Berck, 1979; Vincent, 1992; Hyde et al., 1996). The cost of deforestation in a more distant or marginal land, which includes transportation cost and risk of being pressed for criminal charges, tends to increase significantly. Continuous increase in the crop supply would also result in a fall in average price over time. The higher costs and lower returns of farm expansion would stabilize deforestation over time. This trend has been well observed in North America, Europe, and East Asia. However, what is left of the forest area in the case of a no-state intervention cannot be considered an optimal level from a social viewpoint, because important nonmarket value of forests, such as biodiversity and effects of forest destruction on climate change and society as a whole, are not taken into account (Vincent and Gillis, 1998).

The Environmental Kuznets Curve (EKC) relationship suggests that once society has evolved and reached a certain state of economic development, people would place high importance on environmental issues and forests. This would exert pressure for reforestation and more effective protection of natural forests. In this view, a win—win situation, where society prospers economic-wise and environment is well protected, can be attained when people have gained decent income levels (Shafik and Bandhyopadhyay, 1992; Cropper and Griffiths, 1994; Bhattarai and Hammig, 2001) or realized the environment's true value, which results in supportive political institutions and good governance (Bhattarai and Hammig, 2001).

In a similar vein, Gunatileke and Chakravorty (2003) show that, given effective control of a protected forestland, a higher return from farming or agricultural activities can lead to a better forest condition. The forest regains its richness as farmers rely less on gathering or selling forest products (nontimber products) and focus more on making the highest possible return from their farms. Gunatileke and Chakravorty (2003) emphasize the important roles of monitoring and affirm that a win−win solution between economic development and environmental protection is possible given an adequate level of law enforcement. This finding contrasts with Deacon (1995) who argues that a higher return in farming will increase deforestation.

This chapter analyzes farmers' incentives and possible win−win solutions to maize farming and deforestation in Nan, Thailand. Interesting features and drastic consequences of maize farming in the highlands of Nan deserve urgent attention. Nan is one of Thailand's most important watershed areas, thus holding a high ecological value. As such, to wait for a natural adjustment to a long-term equilibrium where forests become well preserved again may come at an unbearable environmental cost. Moreover, to expect farmers to become aware of the overall benefits of forests anytime soon seems impractical: the majority of highland maize farmers still receive low incomes and live in poverty in spite of maize prices continuously increasing over the past 10 years (Talerngsri and Pongkijvorasin, 2012). Yet, amid the critical deforestation problem, the Thai government has continuously given signals that encourage maize farm expansion, such as the maize pledging scheme.[1] More importantly, notwithstanding evidence of a continuous and massive degree of forest encroachment, the government is still in denial of its ineffective enforcement of forest laws and regulations and is reluctant to find more effective policies that would create mutual benefits for farmers and the environment. Its inaccurate presumption of the effectiveness of government control over forestlands has led to ill-designed policy that has worsened the problem, both economically and environmentally.

Although the critical state of deforestation in Nan warrants studies from academics in various fields, no study has yet analytically examined how highland maize farmers trade benefits from the forest against income from maize farming. Using survey data from Nan, this study provides a simple analytical tool to examine such a trade-off. It shows that, as income from maize farming exceeds benefits from the forest and soil quality is not too low, the farmer perceives an incentive to convert a forestland into a maize farm. Current crop subsidy policies (e.g., pledging schemes), help raise farmers' incomes, but also further encourage the farmer to expand his/her farming area (as had happened in Mexico, according to Deininger and Minten, 1999). The scheme has resulted in continuing deforestation activities. Using the proposed tool, this chapter shows that a win−win solution to the problem is possible, whereby farmers' incomes increase along with the forest area. The farmers win in the sense of having a higher income, and environment improves due to less incentive for deforestation. A green subsidy, irrigation, and awareness raising are examples of such win−win measures. The following section presents a summary of current practices and maize farming in Nan and estimates of income streams derived from maize farming. Section 7.3 reviews previous studies on the value of community forests and estimates the changes in the value of forest products over time. Section 7.4 examines the farmers' incentive to convert forest to maize farm using a simple analytical tool and analyzes the current situation. Section 7.5 discusses the impacts of the crop subsidy policy. Section 7.6 puts forward possible win−win solutions. Section 7.7 concludes and provides policy recommendations.

---

[1]The price support scheme was in the form of a price guarantee under the previous government led by Mr. Abhisit Vejjajiva. It had a similar impact on expansion of maize farming areas.

7.2 MAIZE FARMING IN NAN PROVINCE

Nan, a province to the east of northern Thailand, bordering with Laos, is considered one of the most important watershed areas in the country. It is surrounded by a mountain range. Nan River, one of the principal tributaries of Chao Praya River (the major river in Thailand), originates here. Given these geographical characteristics, the forests in Nan carry a very important ecological role (Figure 7.1).

Although both maize price and farm area have increased in recent years, most farmers in Nan still live in poverty. While farmers could sell their maize at relatively high prices, farm yields have declined, mainly due to soil degradation as a result of intensive cultivation and increased cost of production factors (e.g., fertilizer, seed, and wage). In 2004, only 12.59% of Nan's population lived

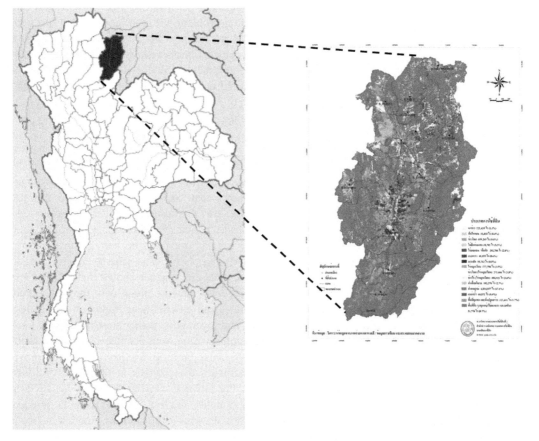

FIGURE 7.1

Map of Nan, Thailand.

Retrieved from Wikipedia, n.d. (left) and Land Development Department, 2011 (right).

in poverty, making Nan the 26th poorest province in the nation. By 2010, the poverty level had increased to 14.97%, moving Nan to the 11th rank among the poorest provinces (Office of the National Economic and Social Development Board, 2011).

Maize farming in Nan are of two types: lowland and highland. Lowland farmers grow maize on land drained by the Nan River and irrigation systems. They start soil preparation early in the year and begin to seed around the start of the rainy season (May). Seeding to harvesting normally takes three months. By August, farmers start to harvest and sell their produce. The short production cycle is suitable for lowland cultivation, as harvest must be completed before the area becomes flooded around the end of the rainy season. Lowland farmers sell their "fresh seed" crop to middlemen or merchants immediately after harvest. Fresh seed receives a lower price in the market because of its moisture content, which, on the other hand, makes it weigh more than dry seed. Thus, land grown to fresh seed type of maize appears to have a higher yield (total weight/area) than the dry seed maize. After harvesting maize, the same piece of land can still be used to cultivate other crops (e.g., rice, chili) or another round of maize. The environmental impacts of lowland maize farming are not obvious and are, to a lesser degree, comparable with those of highland maize farming since lowland farming does not particularly involve deforestation. Most farmers cultivate land that they legally own.

On the other hand, faced with limited access to water resources, highland farmers can cultivate only once a year. They produce "dry seed" maize, which takes longer to harvest, usually 4−5 months from seeding. The harvesting and selling period is from from November to February. Dry seed has a lower moisture content, thus, can be sold at a higher price than fresh seed. The highland fields are left idle after harvest due to lack of water, unlike the lowland fields where other crops are cultivated after maize. Thus, highland maize farmers rely heavily on income from once-a-year maize cultivation.

Talerngsri and Pongkijvorasin (2012) show that highland maize farmers are trapped in a vicious cycle of poverty and environmental degradation. Faced with many constraints, such as water scarcity, financial constraint, and high transportation cost, highland farmers have a limited choice on the crop to grow, resorting to maize cultivation, which can be done once a year only. This has pressured farmers to expand their farmlands and increase maize output as much as possible. Moreover, major government agricultural policies (i.e., the price insurance scheme and pledging scheme) incentivize farmers to increase their maize output and expand their cultivation areas to benefit more from the policies. Altogether, these physical and institutional limitations and misplaced government policies have drastically increased farmed areas and encouraged more intensive farming practices, which require farmers to invest more in fertilizers, seeds, and pesticides. This has resulted in environmental degradation due to severe deforestation and contamination. The increasing cost of farming inputs adds to trapping these farmers deeper in the cycle of debt and poverty.

Table 7.1 summarizes the data on highland maize cultivation gathered from in-depth interviews with stakeholders and a survey of 64 maize farm households (2010−2012). Highland maize cultivation yield averaged 4.7 tons/ha in 2011, with a total production cost (comprising material cost, labor cost, and transportation cost) of USD 705/ha.[2] Dry seed maize had an average moisture level of 18%, which farmers sold at USD 0.27/kg. This means that 1 ha of maize could generate USD 1,244 in revenue. Given the production cost, the net revenue was USD 539/ha. With an average

---

[2]The production cost shown here omits noncash cost items, such as health cost and environmental cost from heavy use of pesticides.

Table 7.1 Summary of Survey Data on Highland Maize Cultivation	
Details of highland maize cultivation (1 crop/year)	
Crop length (month)	5
Output per area (ton/ha)	4.7
Cost per area (USD/ha)	705
– Material cost	
– Fertilizers	191
– Pesticides	27
– Seed	75
– Fuel and others	7
– Labor cost	318
– Milling and transportation cost	86
Average moisture (%)	18
Average price (USD/kg)	0.27
Total revenue per area (USD/ha)	1,244
Net revenue per area (USD/ha/year)	**539.20**
Average cultivation area per household (ha/household/year)	5.76
Average net annual revenue per household (USD/household/year)	**3,106**

cultivation area of 5.76 ha/household, the net revenue per year for a highland farm household amounted to USD 3,106.

However, when yield changes across time were considered, the adverse effect of maize farming on soil quality became evident. Yields have gradually decreased, which also means declining farmers' incomes, giving farmers justification to heavily use fertilizers to sustain yields. To capture this nature of yield reduction, data were obtained from Ban Had-rai of the Sanna-Nongmai subdistrict, which still has areas that can be cleared for more maize farming. The yield value from such fertile land is as high as 7.5 tons/ha. On the contrary, data from Ban Fang-min, Ainalai subdistrict can be used to guage maize yield from bad soil due to repetitive and heavy use of land over a long period of time. Here, the yield value is as low as 2.8 tons/ha.[3] Given that maize farming in most areas of Nan started around 1987, it is reasonable to assume that it takes about 20–25 years for the yield value to fall from 7.5 to 2.8 tons/ha. Using an inverted S curve to approximate gradual reduction in yield over time, the changes in yield value and corresponding farmer income per hectare across time[4] can then be represented by Figure 7.2.

Figure 7.2 illustrates that growing maize on fertile land would give farmers a maximum income of USD 1,250/ha. However, after a certain period of time (25 years in most cases in Nan), soil quality would drastically decline, resulting in low yields (as low as 2.8 tons/ha) and reductions in farmers' incomes to as low as USD 41.7/ha.

---

[3]Survey data suggest that the production process and amount of fertilizers and pesticides used by highland farmers from various areas in Nan are not significantly different. Hence, these farmers are assumed to share the same production functions; any difference in yield is presumed to be attributable to the difference in soil quality or severity of soil erosion.

[4]Price of maize and cost of production are held constant across time at USD 0.27/kg and USD 705/ha, respectively.

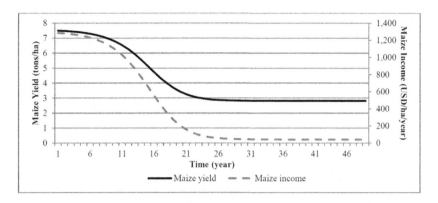

FIGURE 7.2

Changes in yield and associated farmers' incomes across time.

Table 7.2 Estimated Value of Forest Products from Community Forests

Author	Site	Year of study	Area (ha)	Value (USD)	Value at 2011 price (USD)	Value/area at 2011 price (USD/ha)
Chaitup (2003)	Lampoon	1998	400	33,283	45,500	114
Roongtawanreongsri (2006)	Pattalung	2004	320	80,678	101,163	316
Bookaew (2009)	Mahasarakham	2007	683	104,125	116,724	171

7.3 VALUE OF COMMUNITY FOREST PRODUCTS

A number of valuation studies on the benefits of community forests in Thailand have been done. The most relevant ones are Chaitup (2003), Roongtawanreongsri (2006), and Bookaew (2009). Focusing on different regions of Thailand, these studies assessed how villagers benefit from community forests by collecting forest products and estimating the value of these products.[5] Table 7.2 summarizes the results from these studies, with the last column showing estimations of forest product values per area (hectare) at the 2011 price. For example, Chaitup (2003) showed that the value of forest products collected from a 400-ha community forest in Lampoon, Thailand, by 300 households living around this forestland is USD 33,283/year, or USD 45,500/year (at the 2011 price) or USD 114/ha (at the 2011 price).

Note that the value of the community forest generated purely from forest product collection ranges from about USD 114/ha to 316/ha. As such, the average value of USD 208/ha will be used henceforth to facilitate calculations and estimations.

---

[5]The focus was on the cash value of collecting forest products as these values are tangible from the farmers' point of view and exert direct impact on farmers' decision making. The forest's noncash value, such as biodiversity protection and reduction of flood/drought frequency, which would help reduce welfare loss of farmers, is not included here.

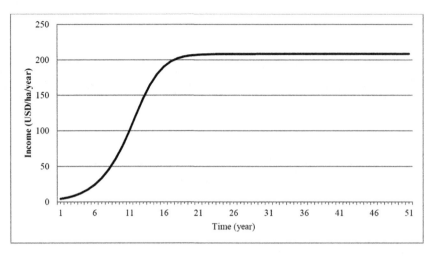

FIGURE 7.3

Income per year derived from selling forest products (USD/ha).

After degraded land is reclaimed for reforestation, forests need time to be fully restored. Hence, farmers may receive very small or almost no income from forest products during the initial phase of reforestation. According to in-depth interviews with former maize growers and villagers at Praputhabat subdistrict in Chiang-Grang district, northern Nan, who were able to reclaim degraded maize farmlands for reforestation, the rehabilitation period will take almost 20 years before villagers can obtain benefits again from forest products. Again, it is reasonable to assume that changes in income derived from selling forest products over the years can be approximated by an S curve, as shown in Figure 7.3. The maximum income level is attained 20 years after reforestation is begun.

Figures 7.2 and 7.3 indicate that although the annual income per hectare from maize farming far exceeds that from forest products during the first 15 years, it will eventually fall to around USD 41.7/ha only, which is relatively low compared with possible gains from forest products. It would take about 19 years for income from maize farming to fall to about USD 208/ha, equal to the maximum annual value of forest products.

7.4 FARMERS' INCENTIVE TO CONVERT FOREST TO MAIZE FARM

7.4.1 PERFECT FORESIGHT VIEW

The net present value (NPV) of incomes from both options (i.e., to grow maize or reforest) was computed at an 8% discount rate[6] using data illustrated by Figures 7.2 and 7.3. Figure 7.4 shows the results obtained.

---

[6]Results obtained under 5% and 10% discount rates were not different substantially.

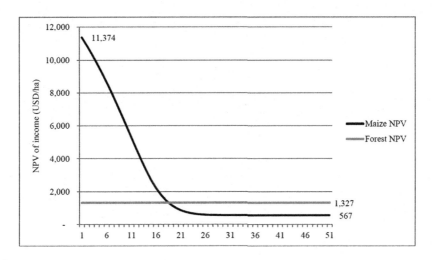

FIGURE 7.4

NPV of income per hectare from reforestation and maize cultivation.

    In deciding which option to take, farmers with perfect foresight compare possible future gain from maize farming with that of reforestation. Under constant maize price and production cost assumptions, a relatively high NPV of income from maize compared with that of forest products during the first 17 years gives farmers an incentive to convert forests to maize farms.[7] It is only when the NPV of forest income outperforms that of maize income that farmers will find the other option more appealing. This implies that if farmers have perfect foresight, they will naturally switch to reforestation after some period of time (i.e., at the start of year 18 in this case) without any government intervention (Table 7.3).

7.4.2 SHORTSIGHTED VIEW

However, in reality, maize is continuously grown year after year. Even when soils are already badly degraded and farmers feel the need to find better options, they are very reluctant to change. There are two main explanations behind this unwillingness to change.

7.4.2.1 Myopia

Farmers lack foresight. The shortsighted farmers tend to put the highest importance on income received from selling their maize, virtually giving no weight to a stream of maize income that can be derived in the future. Switching to reforestation could mean having empty stomachs for some

---

[7]This NPV value considers only the cash value from collecting forest products. It is constant at USD 1,327/ha because regardless of starting point, the forest is assumed to take about 20 years to be fully rehabilitated, at which point annual income from collecting forest products reaches its maximum of about USD 208/ha/year and continues at that level.

Table 7.3 Comparison Between NPVs of Maize and Forest Product, Various Years

Year	Yield (tons per ha)	NPV Maize (USD per ha)	NPV Forest (USD per ha)	Difference (USD per ha)
0	7.5	11,374	1,327	10,046
5	7.3	8,659	1,327	7,331
10	6.5	5,280	1,327	3,953
15	4.7	2,237	1,327	910
16	4.4	1,817	1,327	489
17	4.0	1,477	1,327	149
18	3.7	1,212	1,327	−115
19	3.4	1,015	1,327	−313
20	3.3	872	1,327	−455
25	2.9	603	1,327	−725
30	2.8	567	1,327	−760

period of time. Even when the land is terribly degraded and the annual income is only USD 41.7/ha, farmers still prefer maize farming to reforestation.

7.4.2.2 Financial constraints

Due to the relatively low income derived from forest products during the initial phase of reforestation, farmers would have financial difficulty supporting daily household expenses and debt repayment. Highland maize farmers mostly rely on informal credits from middlemen (or maize collectors), who lend both money and raw materials for growing maize. Switching to reforestation would certainly deprive these farmers of an informal but very important channel to obtain loans. Furthermore, as previously mentioned, highland maize farmers are trapped in a vicious cycle of debt and poverty from which they have difficulty breaking free.

Given that farmers are shortsighted and that the monetary benefits of maize growing exceed those of reforestation in most years, farmers perceive an incentive to increase their farm areas. If government fails to control forest conversion to maize farms, forestlands will soon disappear. The PF_F line in Figure 7.5 illustrates this trade-off relationship between income from maize and private benefits from the forest for a representative farmer (economic value vis-a-vis environmental value).[8] The vertical axis represents the farmer's income from maize farming, while the horizontal axis represents monetary benefits from a community forest. Thus, the PF_F line illustrates the maximum possible level of income the representative farmer can derive from all accessible and available land, including the community forest in the area where he or she lives. The horizontal intercept, B_1, represents the case where the farmer wants to rely purely on income from forest product and thus would be willing to dedicate all of his or her land to reforestation. On the other hand, the

---

[8]Benefits from the forest are assumed to be monotonically related to the forest area.

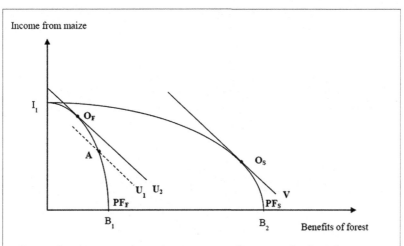

From a farmer's perspective, point *A* represents the current situation where return from maize is larger than private benefit of forest. A farmer has incentive to convert land to maize farm so as to reach O_F. A horizontal difference between *A* and O_F reflects current government efforts in protecting forest land. A social optimal outcome is illustrated by O_S, implying that current forest area is socially suboptimal.

FIGURE 7.5

Investigation of trade-off between maize farming and benefits from forests.

vertical intercept, I_1, presents the scenario where the farmer wants to obtain his or her total income from maize farming and thus would convert most accessible land to maize farm.

From a farmer's perspective, however, the larger the scale of forest conversion, the higher the risk of getting caught for causing deforestation, in addition to the problem of rising transportation cost. The rising marginal costs of forest conversion and maize cultivation are captured by the concavity of benefit possibility frontier curve (PF_F). The curvature of this line reflects a ratio of revenue gained from maize farming and that of collecting forest products. Moreover, as the farmer derives utility from income received from maize and forest products, in monetary terms, incomes from these two sources perfectly substitute each other, suggesting a straight line indifference curve with the slope of −1. The optimal point for a utility-maximizing farmer is represented by the tangency point, O_F, whereby the marginal return from maize farming is equal to that of collecting forest products. However, with current government control, the present ratio of revenue per area of maize farm to forest is approximately 2.5 (according to data in Table 7.1). The present trade-off point faced by the farmer (denoted by A in Figure 7.5) can be located where the slope of the PF_F line is equal to −2.5. This means that a utility-maximizing farmer would have the incentive to further convert forestland to maize farm, as shown by the movement from point A to O_F. The horizontal difference between point A and O_F implies government efforts in preventing forest degradation in the current situation.

In a broader perspective, however, forests provide social benefits (ecological functions, such as carbon sequestration, soil protection, and water sources, and also income from forest products) that far outweigh the farmers' private benefits (i.e., income generated purely from gathering forest products). The social benefit frontier is represented by the PF_S curve. The horizontal intercept is now at B_2, showing the social benefit if all land is dedicated to reforestation. The social optimal point is represented by point O_S. Since the social benefits of forest exceed private benefits to the farmer, the level of socially optimal forest area is higher than that determined by the farmer, implying a situation of excessive forest conversion from society's point of view.

7.5 LIMITATIONS OF CURRENT GOVERNMENT POLICIES

Crop subsidy is a popular government policy aimed at stabilizing farmers' incomes and alleviating rural poverty; it can come in the form of price insurance or pledging. Either way, crop subsidy causes a shift in possibility frontier from PF_F to PF_F', extending the frontier of income derived from maize farming (Figure 7.6). The optimal point under crop subsidy is depicted by O_F'. Theoretically, it is possible that when maize price increases, farmers can cultivate in a smaller plot of land while receiving higher maize income. More specifically, income from maize can increase together with reforestation (i.e., win—win outcome). However, the crop subsidy tampers with the ratio of return from maize farming and collecting forest products such that the utility-maximizing

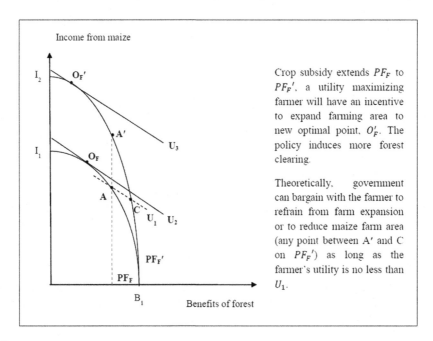

Crop subsidy extends PF_F to PF_F', a utility maximizing farmer will have an incentive to expand farming area to new optimal point, O_F'. The policy induces more forest clearing.

Theoretically, government can bargain with the farmer to refrain from farm expansion or to reduce maize farm area (any point between A' and C on PF_F') as long as the farmer's utility is no less than U_1.

FIGURE 7.6

Effects of crop subsidy policy.

farmer would increase his or her utility by moving from point A to O_F', where O_F' is to the left of O_F. Hence, the crop subsidy incentivizes farmers to expand their farm areas, resulting in more forest conversion and deforestation.

Theoretically, the government can use crop subsidy as a tool to negotiate with farmers so that they would refrain from expanding their farm areas (A′), or even to the extent that farmers may agree to give back part of their land for reforestation (anywhere between A′ to C) and still get a higher than or equal level of utility. For example, the government can specifically subsidize only maize grown in existing farmlands. However, implementing such restriction, as well as its monitoring and control, is very costly in practice. The government would have difficulty distinguishing maize cultivated in existing farmlands and maize cultivated in newly converted areas. The high administrative and transaction costs prevent the possibility of a win–win outcome to be reached.

7.6 ALTERNATIVE WIN–WIN POLICIES

7.6.1 GREEN SUBSIDY

A policy option to deal with poverty and conservation problems is green subsidy, whereby government subsidizes reforestation. Green subsidy extends the frontier of income derived from the forest. It can be used to both internalize external benefits of forests and provide incentive for farmers to conserve the forest.[9] Although costly, a well-designed green subsidy can be used to restore the social optimum and is considered as a win–win policy because it can increase both forest area and the farmer's income. Analytically, subsidy will shift the PF line in Figure 7.5 to the right by increasing returns from the forest. If subsidy is set equal to the marginal external benefit of forest, PF line will be at PF_S. This shift will increase the incentive of farmers to switch from maize cultivation to reforestation.

Here, subsidy is viewed as a tool to incentivize farmers to be more willing to leave maize farming for reforestation; this section does not aim to estimate the size of subsidy that would restore the social optimal level of forestland. For example, at the current year, a farmer gets USD 539/ha from maize cultivation. If government decides to intervene to facilitate the shift to reforestation, the farmer has to be subsidized for not growing maize by at least USD 539/ha in the first year (assuming that no forest products can be collected in the first year of reforestation). The size of subsidy decreases over time as the value of forest products increases and the yield of maize farming decreases. From Table 7.3, if the government wants to reclaim 400 ha from highland farmers who have grown maize for 15 years, it would need to spend at least the NPV of USD 363,833 per village to bring about the expected change.

The estimated value of subsidy reflects only the minimum sum of money that can help farmers see the stream of private benefits derived from collecting forest products. It does not, by any means, shed light on the much larger sum of money needed to bring about the social optimal level of forest benefits.

---

[9]See Bull et al. (2006) for a review of various forms of green subsidy for industrial plantation.

7.6.2 IRRIGATION

Developing irrigation systems is another win—win policy. Lack of water is one major constraint in highland maize farming in Nan. Without adequate water in the dry season, farmers can cultivate only one crop a year. This limitation incentivizes farmers to expand their farm areas as much as possible (Talerngsri and Pongkijvorasin, 2012). Irrigation can help solve farmers' poverty problems. Irrigation in the highlands can be small scale only (e.g., check dam or water storage basin) to pose no excessive adverse effect on the environment. However, some people (including the government) fear that there is a trade-off between development (poverty alleviation) and environment—that is, development comes at the cost of environmental degradation. Consequently, irrigation, once viewed as a tool for development, has become a factor of environmental degradation. In particular, the perspective is that providing farmers with more water will lead to more extensive agriculture and consequently more environmental degradation. This belief has hindered the development of irrigation systems in highland areas.

With adequate water supply, on the other hand, highland farmers can have at least two crops a year. Farmers can grow other crops that are suitable for highland cultivation in order to bring in higher and more stable income. However, there is a possibility that farmers have become so used to maize cultivation, they would find it difficult to adopt alternative crops. A survey of highland farmers in Nan showed that a farming household cultivates maize on 5.76 ha on the average and obtains a net revenue of USD 3,106/year for one crop only. With irrigation, farmers can grow two crops on the same plot of land and double their annual income to USD 6,212.

Theoretically, providing irrigation would work in the same manner as crop subsidy does on farmers' decisions to grow maize because it increases farmers' incomes from cultivating maize on a piece of land.[10] Figure 7.7 shows the effects of irrigation on the farmer's decision making. Irrigation will shift the PF_F line up to PF_F''. Assuming that all farmers have access to this irrigation system, it is expected that, like crop subsidy, they will expand their farm areas, resulting in a fall in forest area (i.e., a shift from O_F' to O_F'').

With irrigation, a farmer's utility can improve from U_1 and reach the maximum of U_4. All points along the portion I_3F of the PF'' curve are welfare improving from the farmer's point of view. For example, point D represents the case where a farmer does not expand his or her cultivation area but just restricts to 5.76 ha of maize farm. The farmer gets a total income of USD 6,212/year, twice the amount received before irrigation is provided, while the forest area is unchanged.

Point E represents the scenario where a farmer reduces by half (i.e., 2.88 ha) his or her maize farming area. With irrigation, he or she can cultivate twice a year, thus maintaining the same level of income derived from maize farming (USD 3,106/year). In this particular case, the forest area will increase by 432 ha (2.88 ha/household × 150 households). Once the forest is fully restored (i.e., after 20 years), a farming household can get an additional income of about USD

---

[10]Irrigation also has nonmonetary benefits. For example, a risk averse farmer will have higher utility, as he or she can have a more continuous flow of income (i.e., at least two crops a year), mitigating the risk incurred from relying on one crop a year. However, these nonmonetary benefits are not captured in this chapter.

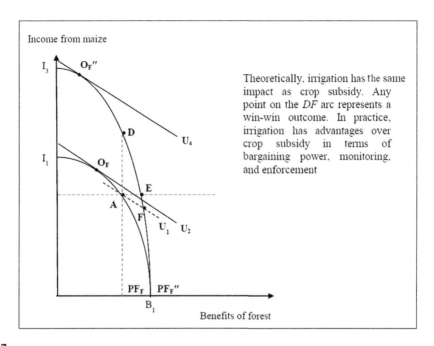

Theoretically, irrigation has the same impact as crop subsidy. Any point on the *DF* arc represents a win-win outcome. In practice, irrigation has advantages over crop subsidy in terms of bargaining power, monitoring, and enforcement

FIGURE 7.7

Effects of irrigation.

600/year (2.88 ha/household × USD 208/ha/year) from forest products. Hence, point E can be considered as one of the win—win outcomes in which the farmer gets higher income and the forest area increases.

Point F represents the case where the government reclaims the largest land possible from the farmers. In this case, the additional income from the forest will just compensate for the lower income from maize farming. Therefore, the farmer's income level at point F is the same as that in the case of no irrigation. In other words, the farmer's utility level is unchanged.

To achieve a win—win outcome whereby both reforested area and farmers' incomes increase, the government can negotiate or offer irrigation to farmers on certain conditions so as to increase both farmers' incomes and forest area. All points along the portion DF of the PF″ curve are considered win—win outcomes.

The crucial advantage of using an irrigation system as a bargaining chip is that providing the system and subsequently monitoring it are not too difficult to undertake. Under crop subsidy, it is difficult (almost impossible) for the government to distinguish maize grown by farmers who agree to give back part of their land for reforestation from maize cultivated by farmers who continue to convert forest areas to maize farms. On the other hand, government can use irrigation as an incentive for farmers to refrain from expanding their farmlands through scoping of irrigation areas. For example, point D in Figure 7.7 can be used to capture the situation where the government offers an irrigation system that can irrigate only 2.88 ha per household on the average. Instead of causing environmental degradation, irrigation, when used wisely, can be viewed as a win—win environmental and economic development instrument.[11]

Raising farmers' awareness of the stream of benefits from reforestation and the future trend and drawbacks of continued maize cultivation is important in the implementation of the two policies discussed above. As discussed in Section 7.4, a farmer with perfect foresight is expected to automatically switch from maize cultivation to reforestation after a certain period (i.e., year 18), even with no government intervention, due to reduction in yield and returns from maize farming over time as a result of soil degradation. However, if farmers lack foresight and are concerned only with immediate monetary returns, the switch cannot occur without government intervention. Raising awareness via multiple means of communication also helps transform a farmer with a myopic view into a more foresighted one, as well as promotes more involvement from other stakeholders (e.g., public and private sectors, local and national players). Awareness raising enhances better understanding and realization of the problem at hand, hence it is a key driver in the effective implementation of green subsidy and irrigation development policies.

7.7 CONCLUSION

Thailand's upland forests are in a critical stage. The forest area in Nan, one of the most ecologically important areas in the country, has decreased continuously and rapidly over the past decade. This has been mainly due to the expansion of maize farm areas, which, in turn, was triggered by the rapid increase in maize price, the crop subsidy policy, and failure of the government to control illegal forest clearing. This situation is a classic example of a case where government attempts to increase the farmer's returns but fails to control the expansion of farming areas, consequently resulting in continuous forest clearing (Deacon, 1995).

The Thai government has been using the traditional approach of command and control over many decades to protect and preserve natural forest, but there is ample evidence of its ineffectiveness. Government's failure to lift the majority of highland maize farmers out of poverty has indirectly increased the cost of forest law enforcement. Thus, the government has been unable to deliver sustainable protection of forest and failed to stop forest encroachment (Cropper et al., 2001).

Government's agricultural policies have been heavily based on the concept of crop subsidy. These policies raise returns from maize farming, which further incentivize farmers to expand maize farm areas. Theoretically, government can use this form of subsidy to negotiate with farmers so as to reclaim more land for reforestation by subsidizing only maize grown by farmers who agree to restrict forest conversion. However, it is very difficult and costly in practice to implement such a

---

[11]A synthesis work by Angelson and Kaimowitz (2001) discusses impacts of lowland irrigation on deforestation. On one hand, irrigation may reduce the degree of deforestation as it helps increase crop supply, resulting in falling prices, thus making the idea of farm expansion less attractive. Farmers can also cultivate more than one crop per year. This increases demand for labor and pushes up local wages. Hence, it makes returns from lowland cultivation more attractive than upland farming. On the other hand, irrigation in the lowlands could worsen the deforestation problem as technology improvement may reduce demand for labor input per cropping season. This may induce workers to migrate to frontier areas and continue their forest clearing. It is also possible that lowland farmers' returns are higher as a result of irrigation, and farmers have enough capital to cultivate more and expand farm area. However, data from the on-site survey and interviews in Nan show no evidence of the aforementioned drawbacks of irrigation provision in Nan.

scheme because it is an almost impossible task to differentiate maize from the qualified farmers from that grown by farmers who are not eligible for the scheme. These factors have partly contributed to the fast rate of forest clearing for maize farming in Nan.

Given that the majority of highland maize farmers remain in poverty, a natural adjustment to equilibrium, where people become more concerned about the environment and forests will be well protected, would seem like an endless wait. The current analysis highlights two possible government interventions that may bring about win—win outcomes for both farmers and forests in the shorter term: green subsidy and irrigation development. A full-fledged green subsidy can internalize external benefits of forest into farmers' consideration, and society can attain the optimal level of forest coverage. Even a partial green subsidy can help farmers make the important change from maize farming to reforestation and make their lives easier during the transition period. Irrigation development is a plausible solution, since water scarcity is one of the most important factors contributing to the large expansion of maize cultivation in the highlands. Theoretically, irrigation development works in the same way as crop subsidy by increasing the return from maize production. However, irrigation provides an easier channel for government to control and negotiate forest conversion. With irrigation, government can easily incentivize farmers to limit their farm area by scoping the irrigation area.

Having an irrigation system in highland areas offers more crop options—thus higher income—for farmers and can also be easily used by government as a bargaining chip to control farmland expansion. Thus, providing irrigation can be a means to achieving sustainable forest preservation. Along with findings from many previous works (e.g., Gunatileke and Chakravorty, 2003), this current study has demonstrated that a win—win solution to maize farming and reforestation in Nan, Thailand, can be attained for economic development and environmental viability, given a correct understanding of the situation and a well-designed policy.

REFERENCES

Angelson, A., Kaimowitz, D. (Eds.), 2001. Agricultural Technologies and Tropical Deforestation. CABI Publishing, Oxon, UK.

Berck, P., 1979. The economics of timber: a renewable resource in the long run. Bell J. Econ. 10 (2), 447—462.

Bhattarai, M., Hammig, M., 2001. Institution and the environmental Kuznets curve for deforestation: a cross-country analysis for Latin America, Africa and Asia. World Dev. 29 (6), 995—1010.

Bookaew, S., 2009. Value of non-timber forest products utilization in Khok Yai community forest in Can subdistrict, Wapi Pathum district, Maha Sarakham province (Master's thesis). Mahasarakham University, Thailand (in Thai).

Bull, G.Q., Bazett, T.M., Schwab, O., Nilsson, S., White, A., Maginnis, S., 2006. Industrial forest plantation subsidies: impacts and implications. For. Policy Econ. 9, 13—31.

Chaitup, D., 2003. Local Knowledge for Resources Manager, North Communal Forest Network, Chiangmai, Thailand (in Thai).

Cropper, M., Griffiths, C., 1994. The interaction of population growth and environmental quality. Am. Econ. Rev. 84 (2), 250—254.

Cropper, M., Puri, J., Griffiths, C., 2001. Predicting the location of deforestation: the role of roads and protected areas in North Thailand. Land Econ. 77 (2), 172—186.

Deacon, R.T., 1995. Assessing the relationship between government policy and deforestation. J. Environ. Econ. Manage. 28 (1), 1—8.

Deininger, K.W., Minten, B., 1999. Poverty, policies, and deforestation: the case of Mexico. Econ. Dev. Cult. Change 47 (2), 313—344.

Department of Forestry, 2014. Forestry Statistics Data 2014. Available from: <http://forestinfo.forest.go.th/55/Content.aspx?id=80>. (Retrieved 6 August 2014).

Ehui, S.K., Hertel, T.W., Preckel, P.V., 1990. Forest resource depletion, soil dynamics, and agricultural productivity in the tropics. J. Environ. Econ. Manage. 18 (2), 136—154.

Gunatileke, H., Chakravorty, U., 2003. Protecting forests through farming. Environ. Resour. Econ. 24, 1—26.

Hyde, W.E., Amacher, G.S., Magrath, W., 1996. Deforestation and forest land use: theory, evidence, and policy implications. World Bank Res. Obs. 11 (2), 223—248.

Meechoui, S., Surapornpiboon, P., 2008. Farmer participation in development of suitable alternate agricultural system replacing maize production on sloping land in Nan province. Area Based Cooperative Research Program Upper Northern Region. Thailand Research Fund (in Thai).

Nan Provincial Agricultural Extension Office, 2011. Agricultural Statistics in Nan Province (in Thai).

Nan Provincial Office of Natural Resources and Environment, 2012. A report on the haze situation in Nan province. Available from: <http://123.242.178.83/webjo/images/stories/mainmenu_images/air.pdf>. (Retrieved 2 December 2012).

Office of the National Economic and Social Development Board, 2011. Poverty statistics 1988—2010. Available from: <http://www.nesdb.go.th/Default.aspx?tabid=322>. (Retrieved 2 December 2012).

Roongtawanreongsri, S., 2006. Economic valuation of community forest in the south of Thailand: a case of Khaohuachang community forest, Tamod Sub-district, Tamod district, Pattalung. Available from: <http://kb.psu.ac.th:8080/psukb/handle/2553/4667>. (Retrieved 2 December 2012).

Shafik, N., Bandhyopadhyay, S., 1992. Economic growth and environmental quality: time series and cross section evidence. Working paper for the World Development Report 1992. The World Bank, Washington, DC.

Talerngsri, K., Pongkijvorasin, S., 2012. Mechanisms of inequality: a case study of maize farming in Nan Province. Paper presented in a conference at Chulalongkorn University, Thailand, August 15, 2012 (in Thai).

UNDP (United Nations Development Programme), 2012. Report of Sub-global Assessment in Nan. UNDP, Bangkok, Thailand (in Thai).

Vincent, J.R., 1992. The tropical timber trade and sustainable development. Science 256 (5064), 1651—1655.

Vincent, J.R., Gillis, M., 1998. Deforestation and forest land use: a comment. World Bank Res. Obs. 13 (1), 133—140.

INSTITUTIONS, GOVERNANCE, AND POLITICAL ECONOMY

THE ROLE OF INSTITUTIONS IN NATURAL RESOURCE USE

Nori Tarui

Department of Economics, University of Hawai'i at Mānoa, Honolulu, HI

JEL: Q20; O17; O13; D02

8.1 INTRODUCTION

Natural resource management takes various institutional forms, such as open access, common property, private property, and state property. Not only are there different institutional arrangements across resources (both over different types of natural resources and within a given type) but also changes in institutional setting for a given resource over time. What explains such institutional differences? Which institution prevails for each resource as its scarcity changes? Given the costs of institutional change and resource governance, how do resource institutions change over time? This chapter reviews studies that deal with these questions.

8.2 INSTITUTION, RESOURCE USE, AND RESOURCE SCARCITY: DEBATES IN THE LITERATURE

Existing studies have demonstrated how changes in the price of (or demand for) harvested products influence the extent and efficiency of resource use. I first review studies that take the relevant resource-use institutions as given.

8.2.1 GAME THEORY STUDIES ON COMMON PROPERTY RESOURCE MANAGEMENT

A large number of studies apply game theory to illustrate the outcome of strategic interactions among resource users in common property resources. Though most of them take common property institutions as given, they do illustrate whether or not cooperative (or efficient) resource use is supportable as an equilibrium outcome of the game. They address this issue by considering gains from both cooperative resource use and "cheating" from cooperation (by overharvesting) for each resource user. Many studies note that both gains increase with harvest price: whether or not cooperation remains (or emerges) to be an equilibrium outcome is not *a priori* certain.

Using a static game with a fixed number of resource users and costly sanctions on cheaters, McCarthy et al. (2001) explain that an increase in harvest price does not change the equilibrium degree of overharvesting because the gains from cooperation and deviation cancel each other. Considering dynamics in natural resource and resource users' interaction can change this prediction, however. Studies employing dynamic games characterize the conditions on relevant parameter values under which cooperative resource use is a subgame perfect equilibrium (Dutta, 1995a,b). Such conditions are more likely to be met when the number of resource users is smaller, the harvest price is lower, and the users are more patient (Polasky et al., 2006). Effects of resource users' inequality and their access to outside markets may be ambiguous. They depend on the degree of inequality in resource users' wealth or harvesting capacity, and the returns from access to outside markets that enhance harvesting capacity relative to the returns from access to outside labor markets, which tend to alleviate the pressure of harvesting efforts on the resource (Dayton-Johnson and Bardhan, 2002; Tarui, 2007).

Another strand of literature applies evolutionary game theory to study whether or not cooperative resource use prevails as an evolutionary stable equilibrium strategy. The studies describe the evolution of "social norms" in resource use by explaining how the shares of cooperators and defectors change as their relative profitability and resource stock change. They incorporate exogenous costs of punishment and penalty (Sethi and Somanathan, 1996); warm glow from cooperation (Osés-Eraso and Viladrich-Grau, 2007); and the possibility of resource collapse (Richter et al., 2013). Sethi and Somanathan (1996) describe what happens with an increase in harvest price when the equilibrium involves a high degree of social norms (with the majority of resource users cooperating). When the increase in harvest price is large enough, social norms break down and resource use transitions to a state with a lower steady state resource stock.

Though institutions are taken as given, the studies introduced above describe the effect of changes in parameter values on the supportability of cooperative resource use in equilibrium. Dynamic game approaches are suitable for illustrating resource users' strategic interactions as well as their interaction with changing resource scarcity. They can also incorporate inequality and earning opportunities outside the commons. Their limitation is that they usually admit multiple equilibria and, hence, do not generate a unique prediction in comparative dynamics (or comparative statics on the steady state). Evolutionary game theory approaches, in contrast, could generate a unique prediction, explaining the possible emergence of cooperative resource use. They assume homogeneous resource users in most cases. The replicator dynamics, typically embedded in evolutionary games, also assume that resource users are myopic (or nonoptimizing).

Though the studies cited above assume a fixed number of resource users, some studies investigated resource use in the presence of potential poaching (or illegal harvesting) by outsiders.

8.2.2 EFFECTS OF TRADE ON RESOURCE USE IN A RESOURCE-ABUNDANT ECONOMY

Studies with different assumptions of institutions help in understanding the mechanism through which trade liberalization (and an increase in the price of resource-intensive goods) improves or hinders welfare in a resource-abundant economy through changes in the harvest level (see Copeland, 2005 for a detailed review).

One strand of literature argues that trade enhances resource overuse. A simple two-country model explains the logic (Chichilnisky, 1994). Suppose there are two identical countries (the North and the South) with an identical renewable resource and endowment of a numeraire (i.e., labor for extracting harvest from the resource). Suppose the resource in the North is under private property while in the South it is open access. When the two countries trade, the South's resource stock decreases and that in the North increases; with a typical renewable resource model with logistic resource growth and Schaefer production function, the open-access harvest level is larger than the harvest under private property given any harvest price. Therefore, open access in the South generates "apparent comparative advantage" (Chichilnisky, 1994) that would have been absent if its resources were under private property, just as in the North.

"Resource curse," the focus of another large body of literature, refers to the detrimental effects on economic development of the windfall wealth coming from booming natural resources. The literature focuses on nonrenewable resources such as minerals, oil, and natural gas. It investigates the mechanisms through which such detrimental effects are realized upon trade liberalization or with an increase in the price of the resource-intensive goods. The main mechanisms include real exchange rate appreciation and political economy factors involved in rent-seeking from resource use. Consideration of the former implies that the general equilibrium effects (i.e., income effects from revenue increases in the resource-intensive sector and reallocation of inputs to the resource-intensive sector from other sectors), together with positive (growth) externalities associated with the sectors outside the resource-intensive sector, explain resource curse. The latter illustrates that rent-seeking activities attracted to high-potential rents from resource use trigger resource curse. Though the literature focuses on the curse associated with nonrenewable resources, the mechanisms of the curse described above are relevant to renewable resource management as well, to the extent that general equilibrium effects and rent-seeking activities are significant in renewable resource use.

The above studies assumed that institutional arrangements governing resource use are given and fixed (e.g., private property in the North, open access in the South). In fact, institutions are not given, as detailed in many case studies, including enclosure of open/common fields in England

(McCloskey, 1976; Allen, 1983), groundwater use in Southern California (from open access to restricted access) (Ostrom, 1965), use of forestland (*Iriaichi*) in rural villages in Japan (from commons to private) (McKean, 1986; Kijima et al., 2000), and lobster fisheries in Maine (from open access in colonial periods to group management) (Acheson, 1988). As another example, pest eradication programs such as the cotton boll weevil program in the United States have succeeded in organizing farmers to coordinate control efforts after individual farmers' control efforts failed in the beginning (Smith, 1998).[1]

Taking into account institutional change explains the possibility that trade may enhance efficient resource use (Margolis and Shogren, 2009; Copeland and Taylor, 2009). This view recognizes that institutions change depending on harvest price, which changes under trade liberalization. In general, the size of the resource rent available when restrictions on resource use are enforced—or the size of the gains from enforcement relative to the case of open access where rent is dissipated—is increasing in the harvest price. Thus costly enforcement may not be justified when the harvest price is low. However, the gains from enforcement may exceed its cost when the harvest price is high enough. Therefore, a resource price increase due to trade may induce more enforcement, leading to more efficient resource use. Margolis and Shogren's (2009) analysis treats institution as endogenous by assuming variable costs of enforcement. They focus on resource users' decentralized enforcement decisions as opposed to socially optimal enforcement. Their analysis implies that, though such efficiency enhancement is the case for a large enough price increase, opening up to trade deteriorates the South's welfare when the price change (starting from the autarky level) is sufficiently small. This is understood in light of the second-best theory (as explained in Margolis and Shogren, 2009; Copeland, 2005): given open access, an increase in harvest price reduces the South's welfare, and costly institutional change is not sufficient to compensate for welfare reduction when harvest price is sufficiently low.

The above discussion implies that whether or not trade enhances efficient resource use depends on the cost of enforcement (or of institutional change in general) and other parameter values that characterize the economic and ecological attributes of the resource. The next subsection further summarizes work that investigates this issue.

8.3 OPTIMAL INSTITUTIONS GIVEN THE COST OF INSTITUTIONAL CHANGE

8.3.1 STEADY STATE ANALYSIS

How does the cost of institution influence steady state resource use when the institution is endogenous? In studying this question, Hotte et al. (2000) assume that it is costly for the resource manager to deter poaching (harvesting by resource users outside the jurisdiction).[2] The second-best steady state resource stock, given costly enforcement, is lower than the optimal level when enforcement is costless. This is because poaching occurs at the optimal stock level; hence, the stock level must be

---

[1] The author thanks Brian Wright for introducing this example.

[2] Thus their model focuses on the cost of external enforcement (i.e., the cost of enforcing against outsiders) and assumes away the costs of internal enforcement (i.e., the cost of enforcing harvest by resource users in the context of common property).

kept lower to deter poaching. Hotte et al. (2000) also describe that, in equilibrium where resource users conduct decentralized enforcement, the social welfare may decrease when the resource-rich economy opens up to trade.

What is the form of optimal institutions given the costs of enforcement? Copeland and Taylor (2009) provide a comprehensive framework to address this question. Their model considers variable costs of enforcement by assuming that the resource manager incurs monitoring costs in enforcing harvest restrictions on resource users. They found that trade liberalization may improve welfare and resource use efficiency, depending on the extent of price changes and the type of economy: a "Hardin" economy where open access prevails regardless of harvest price; an "Ostrom" economy where limited governance is second best when harvest price is high enough (though open access prevails when harvest price is low); or a "Clark" economy where "complete" governance, supporting the first-best harvest, is second best when harvest price is high enough. This economy classification depends on parameters such as population size, discount factor, harvesting productivity, monitoring technology, and intrinsic growth rate of the resource. In particular, a larger labor force population (which determines the economy's harvesting capacity), a larger rate of time preference, a lower resource growth rate, higher harvest per effort (i.e., better harvesting technology), and less productive monitoring technology all lead a country to be a Hardin economy. Within Ostrom and Clark economies, what determines the threshold price levels above which the economy escapes open access and maintains positive rents in equilibrium? The thresholds are lower for an economy with a smaller labor force population, lower discount rates, more productive monitoring technology, lower harvest per effort, higher resource growth rate, and higher resource carrying capacity.

Thus Copeland and Taylor's study helps in understanding why open access persists upon trade liberalization in one case and why restricted access in the form of common or private property prevails in another.

8.3.2 INSTITUTIONAL CHANGE ON THE TRANSITION PATH

The above studies focus on the optimal institution in the steady state. How does institutional change occur in the transition to the steady state as resource scarcity changes? When is the optimal timing of institutional change? A classical conjecture to answer these questions postulates that an institution changes when the benefits of such change exceed the costs (Demsetz, 1967). A formal treatment of this conjecture describes the optimal timing of property rights enforcement (Anderson and Hill, 1990; Lueck, 2002). These studies consider a one-time, fixed cost of adopting an institution but without endogenous resource depletion.

Roumasset and Tarui (2010) endogenize resource depletion to analyze institutional change over the life cycle of a resource. They looked at the transition from one institution to another with (i) fixed and variable costs of governance (restricting harvest below open-access level) and (ii) endogenous timing of switching from open access to governance. In their framework, governance costs imply two effects. First, governance cost decreases the rent available from restricting harvest. Second, the marginal opportunity cost of harvesting is lower in the presence of variable governance costs; this is because restricting (or reducing) harvest is costly for the resource manager. The second effect implies that the optimal harvest level upon governance is larger than what would prevail in the absence of governance costs. When the resource stock is large enough, the optimal harvest level upon governance coincides with the level that would prevail under open access. Thus open access can be optimal when the resource stock is large. As the resource gets depleted and its scarcity increases, governance becomes justified if the gains from governance exceed its costs. Thus it can be optimal to allow open access when the stock is large and then adopt governance and restrict harvest as the stock gets smaller.

The model explains institutional change along the transition path (as well as possibly different institutions at the steady state). Roumasset and Tarui (2010) found that the optimal timing of institutional change is delayed if the harvest price is larger (steady state stock is smaller) and if the marginal and fixed governance costs are larger. When governance costs are large enough, open access remains optimal throughout the transition path (and at the steady state).

Analyzing the transition path also reveals that the optimal trajectory of the resource stock is not necessarily monotonic. Given a fixed cost of institutional change and a constant harvest price, overshooting becomes optimal: starting with open access, it is optimal to allow the resource stock to get depleted, start governance, and then build up the resource stock. This is because there are gains from delaying institutional change when it is costly. Similar nonmonotonicity applies when harvest price increases exogenously: though a small increase in the harvest price may not justify governance (and hence reduces the stock level), further increases in the harvest price could justify governance, which then increases the stock level. This nonmonotonicity (or overshooting) result is consistent with Libecap's (2008) observation that, in many cases, property rights are adopted to natural resource management only after the resource is depleted significantly.

To summarize the discussion so far, under exogenous institution, given open access, an increase in harvest price leads to further resource overuse and lower welfare. With an endogenous institution (given the costs of governance), an increase in harvest price may improve welfare (via institutional change). Institutional change occurs as resources become scarcer; overshooting of the resource stock is optimal given the cost of institutional change.

8.4 INSTITUTIONAL CHOICE IN EQUILIBRIUM

The majority of the studies described above focus on the second-best theory: in the presence of governance (monitoring and enforcement) costs, what is the socially optimal institutional change? They describe the second-best institutional choice for resource management given the costs of introducing and maintaining institutions. In practice, resource management involves decentralized decisions by individual resource users (with or without enforcement by a resource manager or a regulator). In other words, in many cases institutional change occurs as a result of resource users' deliberate decisions, as Ostrom (1990) describes.

Several studies investigate the outcome of decentralized decisions to enforce private property by resource users (de Meza and Gould, 1992; Hotte et al., 2000; Margolis and Shogren, 2009). One important insight from these studies is that decentralized enforcement decisions by resource users may be different from what is socially optimal. Their focus, however, is the steady state or static equilibrium. In addition, their frameworks do not apply to the case of common property, where individual resource users' interest may not be consistent with the group's interest. Under common property, enforcement among resource users (and cooperation among them) is a key challenge in resource management. Because of failure in cooperation among users, many common pool resources are overharvested to the extent close to that in open access (Ostrom, 1990). Given the challenges of internal and external enforcement, how would self-governance emerge as an equilibrium outcome of resource users' decisions?

Tarui (2004) presents a dynamic model of renewable natural resource use with entry deterrence by a group of resource users, where the harvest price is taken as given and fixed over time. The model—a simplified version of the Hotte et al. (2000) model of costly external enforcement—considers both external and internal enforcement. In the model, a coalition of agents self-governs the commons and monitors overharvesting by members and entry by outsiders. If the coalition chooses its group size in each period to maximize income per member, subject to costly internal and external enforcement, then a unique equilibrium is associated with smaller resource rents than the second-best level, given costly monitoring. The equilibrium steady state resource stock is larger than the second-best level—a result consistent with earlier studies (de Meza and Gould, 1992; Hotte et al., 2000; Margolis and Shogren, 2009) that describes the difference between decentralized enforcement decisions and the socially optimal enforcement level.

Though Tarui's analysis is limited to comparative statics of steady states, it illustrates that the steady state resource stock has a nonmonotonic relationship with harvest price—a result consistent with that of Roumasset and Tarui (2010). Given a low harvest price, self-governance does not emerge in the steady state because gains from enforcement are too low relative to the cost of enforcement. As harvest price increases, self-governance emerges and generates positive rents. Since the open-access stock level also decreases as harvest price increases, the model exhibits a U-share relationship between the steady state resource stock level and harvest price.

Tarui's (2004) model also describes the effects of monitoring productivity on steady state stock and rents. Monitoring productivity determines the effectiveness of monitoring efforts in detecting overharvesting by insiders and outsiders, thereby representing the cost of enforcement. Not surprisingly, with sufficiently high monitoring productivity (i.e., enforcement costs converging to zero), the second-best outcome converges to the efficient outcome. However, the equilibrium inefficiency remains: because of excessive enforcement by insiders to deter entry by potential resource users from the outside, the equilibrium resource stock level exceeds the Pareto efficient level when the monitoring productivity is large enough.

The studies cited above help in understanding how (steady state) equilibrium institutional arrangements based on decentralized decisions by resource users depend on the underlying parameter values (such as harvest price and costs of enforcement or resource governance). What has not been explored is how institutions emerge along the transition path in equilibrium as the number of resource users and the resource scarcity change.

8.5 RESEARCH OPPORTUNITIES ON RESOURCE GOVERNANCE

Many resources across countries have long been identified as overused and degraded, yet still many of them remain under imperfect enforcement. On the other hand, in some cases resource degradation has resulted in institutional change and improved enforcement of resource use restrictions (Ostrom, 1990, 1998, 2007; Libecap, 2008). The theoretical studies reviewed above indicate that whether or not institutional change (away from open access with no enforcement) occurs depends on various parameters that capture the ecological characteristics of the resource and the economic conditions surrounding the resource. Empirical studies to test the theoretical predictions of these studies will yield useful policy implications on how to improve resource use efficiency.

The review also indicates several directions in which research on the role of institution in natural resource use can be extended. What follows is a summary of such directions.

8.5.1 TRANSITIONS ACROSS DIFFERENT FORMS OF INSTITUTIONS

As mentioned before, many case studies document institutional change from open access to common property or private property. Further investigations that distinguish alternative institutions (e.g., common, private, government properties) will facilitate understanding of when an open-access resource transitions to common property and then to private property, or when an open-access resource transitions directly to private property.

8.5.2 GENERAL EQUILIBRIUM EFFECTS

The studies reviewed illustrate the importance of general equilibrium effects of changes in demand for harvest from the resource, such as allocation of efforts between the resource-intensive sector and other sectors and income effect. However, much of the analysis assumes identical resource users. Among potential resource users, who will participate in resource use? How do the incumbent resource users (insiders) and potential entrants (outsiders) interact, and how do such interactions determine endogenously the number of resource users? Several theoretical studies explicitly incorporated heterogeneity across resource users in terms of their harvesting productivity (Heaps, 2003; Tarui, 2007; Taylor, 2011). The frameworks used in these studies could be applied to investigate the above questions.

Incorporating heterogeneity across resource users is important for another crucial reason: institutional change (such as implementing individual transferable quotas, ITQs, in a fishery) may involve winners and losers as the gains associated with institutional change would be different for resource users with different productivity. (For example, highly productive fishers may be better off under open access than under ITQ if the quota given is sufficiently low.) Considering the distributional aspects of institutional change will help in understanding why institutional change is challenging politically in many cases and how an institutional reform could be arranged so that it is accepted by existing resource users.

The literature on resource curse and that on the role of institutions in resource use have not been synthesized in the existing studies. The general equilibrium effect investigated in the resource curse literature can also be incorporated further in the discussion of how an economy's resource abundance influences its development given endogenous institutional change.

8.5.3 THE ROLE OF GOVERNMENT AND ITS INTERACTION WITH RESOURCE USERS

The literature focuses on either the benevolent resource manger's problem (solving for the second-best institutional choice) or the outcome of decentralized enforcement decisions by resource users.

As described above, the latter could be explored further. Case studies and field research also point to the importance of "comanagement" by government (or resource managers) and resource users, as in Maine's lobster fishery (Acheson, 1988) and fisheries cooperatives in Japan (Platteau and Seki, 2001). This form of institutional arrangement has not been explored extensively in the literature. The interaction between government and resource users on institutional choice and resource management will yield useful policy implications on the efficacy of centralized versus decentralized resource management.

A recent study by Engel et al. (2013), which incorporates the endogenous nature of comanagement, is informative in this respect. Comanagement may be an outcome of negotiations between the state (or an authority regulating the use of a resource) and the local community that depends on the resource, which may have different interests in resource use. While the regulator may prefer preserving the resource, the community may benefit from harvesting it. Engel et al. (2013) recognize that the form of a state's (or a regulating authority's) intervention on local communities' resource use in such contexts may range not only from "fences and fines" by the authority to exclude the local community's access to resource use, to "paper tigers" where the regulator's protection of resources is not enforced (*de facto* open access), but also to comanagement, where the regulator negotiates with local groups to share the management of and benefits from the resources. Many case studies (Baland and Platteau, 1996) document how comanagement has been adapted in practice. As Engel et al. (2013) note, comanagement requires, as a precondition, that the regulator has sufficient capacity to enforce its resource-use regulation; otherwise "paper tigers" result. The case studies provide theoretical predictions and empirical tests on various factors that determine which regime prevails. Among other factors, a key parameter is the regulator's cost of monitoring and enforcement. When this cost is sufficiently high, *de facto* open access prevails. Moreover, an increase in the benefits from harvest (e.g., through price changes) to the local community tends to lead to *de facto* open access and resource overuse. When the regulator's cost of monitoring is sufficiently low, "fences and fines" or comanagement emerges. While fences and fines (state property with exclusion of local communities) result in equilibrium if the benefits from harvest to the community are low, comanagement is likely to occur otherwise. In this case, an increase in the benefits from harvest tends to result in comanagement.

Engel et al. (2013) informatively describe the strategic interaction between a regulating authority and local users of a natural resource. Their approach is promising as a baseline to exploring further the roles of government and local community in enhancing (or preventing) institutional change. It may be useful to study how their predictions would change with the incorporation of changing resource scarcity over time or the possibility of preexisting social norms in the community in the face of government intervention. Exploring what forms of interventions work for (or against) enhancing local resource use will provide useful insights for policymakers engaged in natural resource management.

8.5.4 INSTITUTIONS AND ECONOMIC DEVELOPMENT

Another (somewhat distinct) strand of literature has shed light on the role of government or political institutions on institutional change regarding natural resource management. For example, resource curse literature provides important insights on how rent-seeking may deteriorate the welfare of a resource-abundant economy when harvest price increases. These studies often assume rent-seeking (i.e., the underlying political institution) as a given. In the context of economic development, exploring the interaction between political and economic institutions has enabled researchers to understand the economic gap across nations (Acemoglu and Robinson, 2012). Such investigation has not been done in the context of natural resource use. How an economy's political institutions interact with resource-use institutions can be investigated further. Research along this line will also help in understanding the coevolution of governance/property and economic development.

There has been an extensive debate on the fundamental cause—institutions versus geography—of different economic performance in different countries (Acemoglu et al., 2001, 2005; Bloom and Sachs, 1998; Sachs, 2001; Diamond, 2005). Much of the discussion focuses on geographic conditions that are largely fixed over time. Unlike such geographical factors, natural resource abundance and scarcity change, as well as institutions for resource governance. Therefore, investigating the role of institutions in resource use provides a useful venue for demonstrating the interaction between institutions and changes in resource scarcity.

This chapter mainly discusses the impact of changes in harvest price on institutions and resource use. This focus is relevant when harvests are priced and sold to outside markets (e.g., forestry and fisheries), but not in cases where harvests are not priced or are used mainly within local communities surrounding the resources (e.g., water for irrigation). Recent studies have evaluated market-based reforms introduced to enhance efficient resource use in the latter context. For example, by assessing China's water management reforms, Wang et al. (2006) found that the creation of new management institutions that offer water managers monetary incentives enhanced water savings.[3] Moreover, farmers' participation in water management played no role in saving water. Detailed analysis of how institutional reforms work in such contexts would prove useful in drawing scalable policy implications in Asia.

Except in the discussion of evolutionary game theory applications, this chapter paid scant attention to the role of social capital in the efficiency of institutions and in institutional change. Recent empirical work combining experimental economics and empirical research illustrates that social capital perhaps plays a crucial role in these aspects (see Tsusaka et al., 2013 for findings of an experiment involving farmers cultivating rainfed and irrigated areas in the Philippines). Game theory offers various ways to incorporate social capital or other-regarding preferences in analyzing collective action (Rabin, 1993; Charness and Rabin, 2002). Further application of such theory in the context of institutions for natural resource use is another promising research venue.

ACKNOWLEDGMENTS

The author thanks the participants at the International Conference on Risk, Resources, Governance, and Development at the East-West Center, as well as a workshop on microeconomic analysis on the emergence of natural resource management institutions at Hitotsubashi University, for useful comments. The editors provided constructive suggestions for improving an earlier draft. The author is responsible for any remaining errors.

---

[3]The author thanks Kei Kajisa for introducing this work.

REFERENCES

Acemoglu, D., Robinson, J.A., 2012. Why Nations Fail: The Origins of Power, Prosperity, and Poverty. Crown Publishers, New York, NY.

Acemoglu, D., Johnson, S., Robinson, J.A., 2001. The colonial origins of comparative development: an empirical investigation. Am. Econ. Rev. 91 (5), 1369–1401.

Acemoglu, D., Johnson, S., Robinson, J.A., 2005. Institutions as a fundamental cause of long-run growth. In: Aghion, P., Durlauf, S.N. (Eds.), Handbook of Economic Growth, vol. 1A. Elsevier, Amsterdam, pp. 385–472.

Acheson, J., 1988. The Lobster Gangs of Maine. University Press of New England, Hanover, NH.

Allen, R.C., 1983. The efficiency and distributional consequences of eighteenth century enclosures. Econ. J. 92, 937–953.

Anderson, T.L., Hill, P.J., 1990. The race for property rights. J. Law Econ. 33 (1), 177–197.

Baland, J.M., Platteau, J.P., 1996. Halting Degradation of Natural Resources: Is there a Role for Rural Communities? Food and Agriculture Organization, Rome, Italy.

Bloom, D.E., Sachs, J.D., 1998. Geography, demography, and economic growth in Africa. Brookings Pap. Econ. Act. 2, 207–295.

Charness, G., Rabin, M., 2002. Understanding social preferences with simple tests. Q. J. Econ. 117 (3), 817–869.

Chichilnisky, G., 1994. North-south trade and the global environment. Am. Econ. Rev. 84 (4), 851–874.

Copeland, B.R., 2005. Policy endogeneity and the effects of trade on the environment. Agric. Resour. Econ. Rev. 34 (1), 1–15.

Copeland, B.R., Taylor, M.S., 2009. Trade, tragedy, and the commons. Am. Econ. Rev. 99 (3), 725–749.

Dayton-Johnson, J., Bardhan, P., 2002. Inequality and conservation on the local commons: a theoretical exercise. Econ. J. 112 (481), 577–602.

de Meza, D., Gould, J.R., 1992. The social efficiency of private decisions to enforce property rights. J. Polit. Econ. 100 (3), 561–580.

Demsetz, H., 1967. Toward a theory of property rights. Am. Econ. Rev. (Papers and Proceedings) 57, 347–359.

Diamond, J., 2005. Guns, Germs, and Steel: The Fates of Human Societies, second ed. W.W. Norton, New York, NY.

Dutta, P.K., 1995a. A folk theorem for stochastic games. J. Econ. Theory 66 (1), 1–32.

Dutta, P.K., 1995b. Collusion, discounting and dynamic games. J. Econ. Theory 66 (1), 289–306.

Engel, S., Palmer, C., Pfaff, A., 2013. On the endogeneity of resource comanagement: theory and evidence from Indonesia. Land Econ. 89 (2), 308–329.

Heaps, T., 2003. The effects on welfare of the imposition of individual transferable quotas on a heterogeneous fishing fleet. J. Environ. Econ. Manage. 46, 557–576.

Hotte, L., Long, N.V., Tian, H., 2000. International trade with endogenous enforcement of property rights. J. Dev. Econ. 62 (1), 25–54.

Kijima, Y., Sakurai, T., Otsuka, K., 2000. Iriaichi: collective versus individualized management of community forests in postwar Japan. Econ. Dev. Cult. Change 48 (4), 866–886.

Libecap, G.D., 2008. Open-access losses and delay in the assignment of property rights. Ariz. Law Rev. 50, 379–408.

Lueck, D., 2002. The extermination and conservation of the American Bison. J. Legal Stud. 31 (2), S609–S652.

Margolis, M., Shogren, J.F., 2009. Endogenous enclosure in north–south trade. Can. J. Econ. 42 (3), 866–881.

McCarthy, N., Sadoulet, E., de Janvry, A., 2001. Common pool resource appropriation under costly cooperation. J. Environ. Econ. Manage. 42 (3), 297–309.

McCloskey, D.N., 1976. English open fields as behavior towards risk. In: Uselding, P. (Ed.), Research in Economic History. Jai Press, London, pp. 124–170.

McKean, M.A., 1986. Management of traditional common lands (Iriaichi) in Japan. Proceedings of the Conference on Common Property Resource Management, National Research Council. National Academy Press, Washington, DC, pp. 533–589.

Osés-Eraso, N., Viladrich-Grau, M., 2007. On the sustainability of common property resources. J. Environ. Econ. Manage. 53 (3), 393–410.

Ostrom, E., 1965. Public entrepreneurship: a case study in ground water management (Ph.D. dissertation). University of California at Los Angeles.

Ostrom, E., 1990. Governing the Commons. Cambridge University Press, Cambridge, UK.

Ostrom, E., 1998. Reflections on the commons. In: Baden, J.A., Noonan, D.S. (Eds.), Managing the Commons. Indiana University Press, Bloomington, IN, pp. 95–116.

Ostrom, E., 2007. Challenges and growth: the development of the interdisciplinary field of institutional analysis. J. Inst. Econ. 3 (3), 239–264.

Platteau, J.-P., Seki, E., 2001. Community arrangements to overcome market failures: pooling groups in Japanese fisheries. In: Aoki, M., Hayami, Y. (Eds.), Communities and Markets in Economic Development. Clarendon Press, Oxford, pp. 344–402, Chapter 13.

Polasky, S., Tarui, N., Mason, C.F., Ellis, G., 2006. Cooperation in the commons. Econ. Theory 29 (1), 71–88.

Rabin, M., 1993. Incorporating fairness into game theory and economics. Am. Econ. Rev. 83 (5), 1281–1302.

Richter, A., Van Soest, D., Grasman, J., 2013. Contagious cooperation, temptation, and ecosystem collapse. J. Environ. Econ. Manage. 66 (1), 141–158.

Roumasset, J.A., Tarui, N., 2010. Governing the resources: scarcity-induced institutional change. Department of Economics Working Paper 10–15, University of Hawaii.

Sachs, J.D., 2001. Tropical underdevelopment. NBER Working Paper No. w8119, February, National Bureau of Economic Research, Cambridge.

Sethi, R., Somanathan, E., 1996. The evolution of social norms in common property resource use. Am. Econ. Rev. 86 (4), 766–788.

Smith, J.W., 1998. Boll Weevil eradication: area-wide pest management. Ann. Entomol. Soc. Am. 91 (3), 239–247.

Tarui, N., 2004. Essays on common-property resource management and environmental regulation. (Order No. 3142651, University of Minnesota). ProQuest Dissertations and Theses, 135-135 p. Retrieved from < http://search.proquest.com/docview/305156341?accountid=27140 >. (305156341).

Tarui, N., 2007. Inequality and outside options in common-property resource use. J. Dev. Econ. 83 (1), 214–239.

Taylor, M.S., 2011. Buffalo hunt: international trade and the virtual extinction of the North American Bison. Am. Econ. Rev. 101 (7), 3162–3195.

Tsusaka, T.W., Kajisa, K., Pede, V.O., Aoyagi, K., 2013. Neighbourhood effects and social behaviour: the case of irrigated and rainfed farmers in Bohol, the Philippines. MPRA Paper No. 50162, University Library of Munich, Munich.

Wang, J., Xu, Z., Huang, J., Rozelle, S., 2006. Incentives to managers or participation of farmers in China's irrigation systems: which matters most for water savings, farmer income, and poverty? Agric. Econ. 34 (3), 315–330.

PUBLIC CHOICE AND THE GENERALIZED RESOURCE CURSE

Majah-Leah V. Ravago* and James A. Roumasset*†

*University of the Philippines, Diliman, Quezon City, Philippines
†Department of Economics, University of Hawai'i at Mānoa, Honolulu, HI

CHAPTER OUTLINE

JEL: Q33; D72; D73; D61

9.1 INTRODUCTION

Sustainable development encompasses sustainable growth and dynamically efficient development patterns. Why do countries deviate from efficient development paths? We explore the role of a generalized resource curse as one possible reason for economic inefficiencies. The "resource curse" refers to the detrimental effects on economic development of the windfall wealth coming from booming natural resources such as minerals, oil, and natural gas. The *generalized resource curse*, on the other hand, refers to the negative effect promulgated not only by traditional extractive natural resources but also by foreign aid, remittances, tourism, and other activities that induce exchange rate appreciation.

Among several mechanisms that turn abundance into a curse, two have received ample but often separate attention in the literature. The first is commonly known as the Dutch disease or factor reallocation, which pushes up the real exchange rate and results in the contraction of nonbooming tradables, typically manufacturing, thus lowering the positive growth externalities of such activities. The second regards the political economy effects of either abundant resource endowments or the effect of an exogenous increase in resource abundance. The mechanisms by which abundance can induce a political economy curse in the sectors that lose from the boom, however, are unclear from the perspective of the three-sector "Australian model." We fill the lacuna by integrating the Dutch disease framework with a public choice explanation of how abundance, in its generalized form, can push the economy inside its welfare frontier.

A specific objective of this chapter is to seek a fundamental explanation of the stylized fact that resource-rich countries tend to have higher tariffs than nonresource rich countries (Collier and Venables, 2011b). We investigate postboom protectionist policies by extending the special interest model of Becker (1983) to include the rate of return to lobbying for political protection and the consequences thereof. To this end, the general equilibrium model of Corden and Neary (1982) is augmented to have four sectors: three traded and one nontraded good. A four-sector model illuminates the public choice aspect of abundance by identifying the losers and winners from the boom. In particular, we identify incentives to lobby by both advantaged and disadvantaged sectors, with a particular focus on the generation of protectionism. The deadweight loss arising from unproductive rent-seeking combined with the contraction of the growth-generating sector pushes the economy inside the welfare frontier.

We then explore another avenue of political economy wherein the government is active. The resource boom provides initial impetus to rent-seeking in the booming sector. Subsequently, there is learning-by-lobbying such that lobbying experience reduces the cost of political influence. This expertise can be readily transmitted to other rent-seeking coalitions. We provide a Philippine illustration of why generalized resource abundance calls for additional tools of government policy.

9.2 OTHER BOOM SOURCES

The term Dutch disease was coined by *The Economist* (November 1977) to describe the decline of the manufacturing industry of the Netherlands following its discovery of natural gas in the 1960s. The concept of Dutch disease was later expanded to include other extractive natural resources, such as oil, gas, agriculture, minerals, and fuel (Sachs and Warner, 1995, 2001; Bulte et al., 2005; Brunnschweiler and Bulte, 2008a,b) to explain the possible impediment to economic growth due to a *resource curse*. South Africa, Nigeria, Egypt, and Venezuela, among others, are sometimes

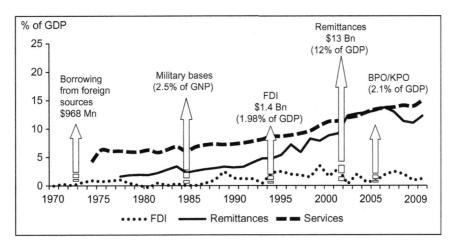

FIGURE 9.1

Sources contributing to exchange rate appreciation, Philippines.

Sources of data: Foreign borrowing is average of 1975–1976 from Bautista (1988). Military bases figure is an estimate of contribution to the Philippine economy sourced from Henry et al. (1989). Data on FDI and remittances are decadal averages from World Development Indicators (World Bank, 2011). Data on BPO are from the Philippine Statistics Authority-National Statistical Coordinating Board.

offered as examples of economies constrained by this curse of abundance. Symmetrically, the economies of Hong Kong and Singapore are said to have been blessed by a lack of natural resources.

Corden and Neary (1982) generalized the notion of "sectoral booms" to include nonextractive as well as extractive sectors. Tourism in Greece, Cyprus, and Malta (Copeland, 1991; Palma, 2008); the "export" of financial services in Switzerland, Luxembourg, and Hong Kong (Palma, 2008); and foreign aid in developing countries (Paldam, 1997; Rajan and Subramanian, 2009) exemplify nonextractive booms. These sectoral booms displaced economic activities of lower commercial value and increased foreign exchange into the economies (Torvik, 2009). Contrary to expectation, however, these windfall incomes have often decelerated growth. We refer to this negative effect as a *generalized resource curse*.

The Philippines has had several missed opportunities with episodes of abundance becoming an anti-development force rather than a propeller of growth (see also Chapter 21). Over the years, the country has experienced several episodes of boom that brought foreign exchange earnings into the country. These include foreign borrowings in the 1970s (Bautista, 1988); the United States military bases in the 1970s and 1980s; foreign direct investments (FDIs) in the late 1990s; and more recently the surge in remittances (Medalla et al., 2007; Tuaño-Amador et al., 2007; Bayangos and Jansen, 2010) and business and knowledge process outsourcing (BPO/KPO). Figure 9.1 shows a snapshot of these sources.

In the 1970s, the Philippines suffered from exchange rate appreciation due to the large increase in foreign borrowings used to finance its trade deficit.[1] The upsurge of FDIs in the 1990s had a similar

---

[1] After the 1974 oil price shock, the country ran huge current account deficits, which ballooned to 32% of the average of export and import of the Philippines in 1982 (Bautista, 1988).

effect. Long-term capital inflows have been on the rise[2] after the passage of the Foreign Investment Act of 1991.[3] The policy of deregulation and privatization of the service sectors (e.g., water, communications, and transport), especially during 1992−1998, attracted FDIs into the services sector, away from traditional manufacturing. More recently, efforts are made to attract FDI into BPOs.[4] The steady increase of remittances is likely to have the same effect of increasing exchange rate. Remittances accounted for an average of 12% of GDP during 2006−2009 (World Bank, 2008).

Like a booming mineral sector, these activities also bring in foreign exchange, resulting in appreciation of the real exchange rate, a contraction of the Philippines' manufactured exports.[5] Since manufactured exports serve as an engine of growth, economic development is likely to be adversely affected (World Bank, 1993) unless there are offsetting effects. Even the FDI in finishing-stage semiconductors and electronic equipment (accounting for more than 60% of merchandise exports) have had disappointing effects both on innovation and on the exploitation of backward and forward linkages (Canlas et al., 2011).

If, as suspected, the lagging sector is more labor intensive than the booming sectors, downward pressure on wages and increasing unemployment may also result. This effect has been exacerbated by the country's high population growth. While outward migration serves as a safety valve, higher skilled workers and entrepreneurs tend to leave the country, thus the moniker, "brain-drain." The unskilled workers who remain are more likely to live in poverty.

In what follows, some mechanisms by which abundance can become a curse are presented, focusing on the real exchange rate effect and a political economy explanation. A theory on the distribution of gains and losses from a boom is presented to motivate the public choice aspect of abundance. The implication of the model is then applied to the Philippine case, where the virulence of the political economy curse varies according to the source of the boom.

9.3 MECHANISMS BY WHICH ABUNDANCE CAN BECOME A CURSE

This section reviews the primary mechanisms of how abundance can become a curse. The first concerns boom-induced factor reallocation away from manufactured goods. The second mechanism relates to the *political economy* effects from either an initially abundant endowment or a boom. This mechanism carries with it another two curses: the emergence of distortionary tariff and its transmission effects. The latter pertains to the formation of the tariff coalition that contributes to a general rent-seeking coalition, which can distort other sectors (e.g., public utilities, contractors for infrastructure). Although it should be noted that this is not always harmful; for example, a subsidy to manufacturing exports may increase economic welfare. These mechanisms are introduced in this section and modeled in the subsequent section.

---

[2]Interrupted by the onslaught of the 1997 Asian Crisis.

[3]The Act allows for up to 100% foreign equity participation in all investment areas.

[4]The year 2003 marked a significant jumpstart of services export, most notably BPO, in the Philippines. This made the country a major contender for offshore providers in the Asia-Pacific region next to India, China, and Malaysia (NeoIT, 2004). The BPO sector accounted for 2.4% of GDP in 2005 and employed about 163,000 workers (Magtibay-Ramos et al., 2007).

[5]The country has low manufacturing exports by regional standards. Growth in the period of 2000−2005 was only 1.4% compared with 6.4% in Indonesia, 7.6% in Malaysia, and 11.8% in Thailand (Canlas et al., 2011).

9.3.1 CROWDING OUT MANUFACTURING

As discussed by Corden and Neary (1982) and Corden (1984), resource booms induce exchange rate appreciation via the *resource movement and spending effects*. The resource movement effect involves resources being pulled out from the manufacturing sector into the booming mining sector. The spending effect occurs when income generated from the booming traded sector is spent on consumption of nontraded goods, thereby exerting upward pressure on the price of nontraded outputs, increasing the real exchange rate, and further shrinking manufacturing.

The appreciation of real exchange rate and resource reallocation does not constitute a "disease" in itself. In an efficient economy, a change in endowments and/or opportunities results in reallocation of resources and changing relative prices. Due to the resource boom, the economy's comparative advantage shifts in favor of "production" of the booming traded sector. The boom becomes a disease when there is "something special" about the shrinking manufacturing sectors, e.g., manufactured exports and import substitutes. If these lagging subsectors have more potential growth externalities than the booming mining sector, the boom can negatively impact growth. Recall that the engine of growth in the East Asian miracle countries was manufactured exports with their abundant and never-ending possibilities for continuing vertical and horizontal specialization (Roumasset, 2014), production externalities (Sachs and Warner, 1995), and learning-by-doing induced technological progress (van Wijnbergen, 1984; Matsuyama, 1992; Gylfason, 2001). The resource boom discriminates against manufacturing, thus putting the brakes on the growth engine. Even if the overall effect of the boom is a small increase in growth, one could still identify a "disease" to the extent that the economy is not meeting its potential.

9.3.2 POLITICAL ECONOMY CURSES: DISTORTIONARY TARIFFS AND THE TRANSMISSION EFFECT

The second mechanism is the more pernicious form of the Dutch disease—the political effect of abundance on economic policy. A booming resource is a natural magnet of rent-seekers. The higher the inherent resource wealth, the greater are the returns on lobbying, the greater is lobbying, and the greater are the resulting policy distortions (Corden, 1984; Lane and Tornell, 1996; Tornell and Lane, 1999; Torvik, 2002; Humphreys et al., 2007; Kolstad and Søreide, 2009).

The political economy mechanism carries with it two curses. The first is the emergence of distortionary tariffs. Empirical evidence shows that governments in resource-rich countries are more likely to set higher tariffs than their counterparts (Collier and Venables, 2011b). The second political economy curse regards transmission effects. The formation of tariff-seeking coalitions lowers the cost of other rent-seeking coalitions.

Political economy models can be classified into two subclasses, depending on whether the government is viewed as a passive or active player. Where the government is passive, it is seen as an aggregator of special interests of two or more opposing groups (Tullock, 1967; Krueger, 1974; Becker, 1983). Corden (1984) gives the example of industrialists adversely impacted by the appreciation of the real exchange rate increasing their lobbying effort for tariff and nontariff protection. This type of political economy model is often referred to as "rent-seeking." Protectionism is perhaps the most commonly recognized type of rent-seeking inasmuch as industrialists are already well organized to lobby (Olson, 1982). Acemoglu et al. (2002) refer to this as the reaction of "entrenched interests" to the loss of rents.

Competition among latent groups in the society in the context of the resource curse was formally modeled by Tornell and Lane (1996, 1999), where the share obtained from available rents is endogenously determined by the existence of powerful groups, rates of return, and institutional barriers that constrain the power of each group. The rise in the numbers of rent-seekers in resource-rich economies is analytically shown by Torvik (2002). The higher the rents from resources the more entrepreneurs switch over to rent-seeking activities, where each rent-seeker gets an equal share of the pie. A similar setup is found in Mehlum et al. (2006), who showed that the higher the rents obtained, the greater the number of rent-seekers, albeit conditional on quality of institutions. More recent literature on rent-seeking in the context of the resource curse includes, among others, Baland and Francois (2000), Hodler (2006), and Bulte and Damania (2008).

In the second type of political economy modeling, the government plays an active role, for example using increased tax revenues to relax fiscal limits on wasteful government spending (Bulte and Damania, 2008). Wasteful spending may be motivated by *patronage* (Kolstad and Søreide, 2009), an implicit bargain whereby political support is exchanged for favors, such as pork-barrel projects, e.g., the *iron triangle* (Lowi, 1969).

9.4 MODELING THE CURSE OF ABUNDANCE

The political economy effects of the resource curse are not entirely clear, partly because previous models have been elucidated using only three sectors. While a three-sector model is sufficient to illustrate how resources may be allocated away from manufacturing, it does not allow for the transmission of protectionism. In the following sections, we first revisit the three-sector model of Corden and Neary (1982) and then expand it into four sectors to demonstrate how gains and losses are distributed, thus demonstrating the contagion effects of the Dutch disease.

9.4.1 THE THREE-SECTOR AUSTRALIAN MODEL

Figure 9.2 illustrates the Australian model of the three-sector disease. Each of the sectors in the small open economy is assumed to have a specific capital uniquely designed for its own use, while labor can freely move across the sectors (Jones, 1971; Snape, 1977; Corden and Neary, 1982). With this assumption, it has been shown that a boom in the traded sector induces real exchange rate appreciation and a contraction of the nonbooming traded sector.[6]

We illustrate the Australian model with the mineral sector as the primary source of the boom. There are two sectors: traded goods (T) and nontraded services (S). T has j subsectors: imported and domestic cars (c) and mineral exports (m), i.e., $j \epsilon T$ with $j = c$ and m. For simplicity, we assume that all minerals are exported. By means of the Hicks aggregation theorem (Clarete and Roumasset, 1987), the general equilibrium model is collapsed into a two-good model—the composite traded good (T) and nontraded services (S), as in Figure 9.2A. The figure illustrates a production

---

[6]These results may not extend to the case of complete factor mobility. For example, in the Samuelson-Komiya (Komiya, 1967) model of linearly homogeneous production functions for a nontraded good and two traded goods, the real exchange rate remains constant. See Corden and Neary (1982) for a rigorous derivation of the core model and the extensions on the different degrees of factor mobility.

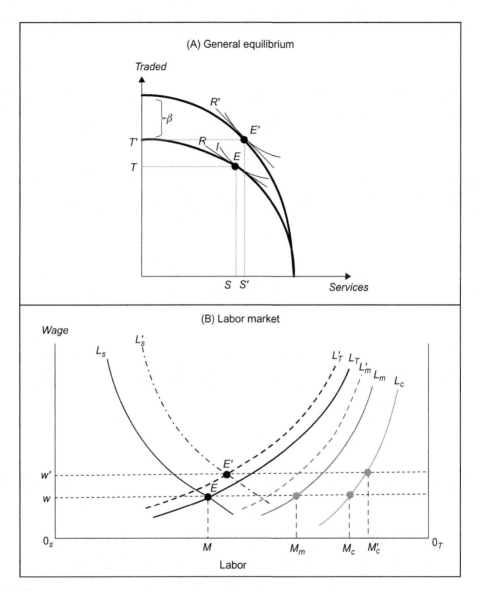

FIGURE 9.2

The three-sector model of Dutch disease: (A) general equilibrium and (B) labor market.

Note: Three sectors: Nontraded services (S) and two traded sectors j of T, i.e., $j \epsilon T$ with $j = c$ and m, where c = import and domestic cars and m = mineral exports. Subscripts refer to sectors.

possibility frontier (PPF) showing all the possible combinations of $j \in T$ and S that the economy can produce. Before the boom, consumption preference between T and S by a single representative consumer is plotted by the indifference curve I.

The small country assumption implies that all traded goods prices are fixed. Units can then be adjusted such that all traded goods prices and the price of the composite good, p_T, equal 1. The real exchange rate, R, is then given by the ratio of prices:

$$R = \frac{P_S}{P_T} = P_S \tag{9.1}$$

In Figure 9.2A, the tangency between the PPF, the consumer indifference curve (I), and line R gives the preboom equilibrium point E, with equilibrium aggregate output level at T and S. This gives an aggregate picture of the equilibrium in the economy.

The labor market is shown in Figure 9.2B. The vertical axis is the wage rate in terms of importable cars $j = c$. The horizontal axis $(0_S 0_T) = L$ is the total labor force in the economy. The quantity of labor that goes to the production of S is measured from left to right while the labor for $j \in T$ is measured from right to left. Labor demand is downward sloping due to diminishing returns. L_S is the labor demand of services. L_m and L_c are labor demands for traded minerals and cars, respectively. The aggregate labor demand for the traded sector (L_T) is obtained by horizontally adding the initial labor demand schedule of minerals and cars, i.e., $L_T = L_c + L_m$. While traded goods prices are fixed, the price of services (p_S) and the location of L_S are endogenously determined as part of the general equilibrium model in panel A.

Preboom equilibrium (E) in the labor market is given by the intersection of the aggregate labor demands L_T and L_S, where wage (w) is equalized across sectors. This equilibrium corresponds to the equilibrium shown in panel A where the price of services is determined. The allocation of labor corresponds to the output levels at E in panel A.

For example, M_c is the employment level for domestic car production. An increase or decrease in employment implies the same directional change in output level.

9.4.1.1 Crowding out of manufactured importables

The boom in exportable minerals, T, shifts the PPF outward, appreciates the real exchange rate from R to R', and crowds out manufacturing, the sector assumed to have the most growth externalities. β measures the potential increase in traded good production if all labor were employed therein. The maximum output of services remains unchanged. This results in the upward asymmetrical shift of the *PPF* in Figure 9.2A.

The reallocation of labor depends on both substitution and income effects. Consider first the substitution effect. The boom in mining shifts its labor demand from L_m to L'_m thus shifting aggregate labor demand of the traded sector from L_T to L'_T. This pushes up the wage rate, releasing labor from elsewhere in the economy. For example, labor in domestic car production moves up the demand schedule from M_c to M'_c. In the face of increased wages, returns to the fixed factors decrease in other industries, especially manufacturing.[7] Labor from the services sector also moves into mining, tending to decrease services output.

---

[7]See Corden and Neary, 1982 for a formal derivation.

Consumer real income increases as a result of the boom, thus shifting out the labor demand for services. If this income effect dominates, as shown, then output of services will increase after the boom. Labor demand for services shifts to L'_S in panel B further raising wages to w'.

The postboom equilibrium is at point E^1. The real exchange rate appreciates as shown by line R' in panel A. Output of services increases to S'. The output of the traded sector—mining and cars—increases to T'. However, domestic autoproduction contracts, corresponding to the fall in labor employed from M_c to M'_c in panel B. The appreciation of the real exchange rate lowers the cost of imported cars, and the additional imports crowd out domestic production. Domestic car producers are thus hit by two negative effects. First, labor gets pulled from production via higher wages. Second, the higher real exchange rate lowers the cost of imports. These negative effects on domestic car production outweigh the positive income effect on car production.[8]

9.4.1.2 Distortionary tariff after the boom

Losers from the Dutch disease wanting to restore their profits are likely to seek solace in the cloak of protection. Protectionism helps restore rents to protected industries but reduces total consumer and producer welfare. In the example above, the manufacturing industry lobbies for a protective tariff in response to their loss of sales to imports. Now the slope of the consumer price line or the marginal rate of substitution (MRS) is less than the marginal rate of transformation (MRT). Since the indifference curve is tangent to the consumer price line, MRS < MRT, excess burden is created.

While the overall static effects of the boom are likely to be positive, i.e., the welfare gain from the boom outweighs the negative effect of the import tariff, the long run effects may be negative. The loss of growth externalities means that the frontier is now shifting outward more slowly than before, such that the country may eventually be worse off than in the counterfactual no-boom case.

9.4.2 THE AUGMENTED DUTCH DISEASE: THE FOUR-SECTOR MODEL

While the three-sector model shows the contraction of the traded sector, it omits another source of loss—the growth externalities associated with *manufactured exports*. This is because the model has been elucidated using only three sectors. The aim of this section is to augment the model of Corden and Neary (1982) by adding a manufactured exportable electronics sector (e) to the analysis to facilitate understanding of how the curse of abundance affects another engine of growth. Hence, we now have a four-sector model with c, e, m, and S. The solid curves in Figure 9.3 show the preboom equilibrium in the labor market. We collapse the four sectors into two by combining the traded sectors, c, e, m into a composite traded good (T). Equilibrium is then determined, as in Figure 9.2, by the demands for the composite traded good (T) and nontraded services (S). The model is easily generalized to booms in other foreign exchange-generating sectors such as renting military bases, tourism, and remittances.

9.4.2.1 Crowding out of manufactured exportables

The same adjustment and assumptions as in the three-sector model are observed. The boom in the mineral sector shifts its labor demand, which also shifts the labor demand of the traded sector from L_T to L'_T. The combined substitution and income effect of a mineral (or other foreign exchange) boom creates an excess demand for services resulting in an increase in the price of services leading to a real

---

[8]These effects are also noted by Corden, 1984.

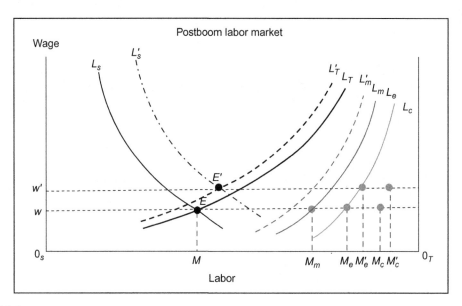

FIGURE 9.3

Augmented Dutch disease: The four-sector model.

Note: Four sectors: Nontraded services (S) and three traded sectors j of T, i.e., $j \epsilon T$ with $j = c,e$, and m where $c =$ import and domestic cars, $e =$ manufactured exportable electronics, and $m =$ mineral exports. Subscripts refer to sectors.

appreciation. Maintaining the assumption that income effect is greater, output of services increases. This is manifested in the shift of labor demand from L_S to L'_S. The postboom equilibrium E' is shown in Figure 9.3.

The service sector again gains from the boom and is considered as the *advantaged* sector. On the other hand, the outputs of the traditionally traded sectors—cars and exportable electronics ($j = c, e$)—decline from M_c to M'_c and from M_e to M'_e, respectively. While the nonmineral subsector loses as a whole, the losses are not spread evenly. In the context of the model sketched above, the labor employed in the manufactured export electronics (e) subsector declines more than the contraction in the manufactured cars (c) subsector. Export manufactures are hit by a double whammy. In addition to the resource movement effect, the appreciation of the exchange rate (the inverse of the price of foreign exchange) lowers the value of the foreign exchange that exports earn. Unlike manufactured imports, there is no income effect to cushion the negative effects. Since manufactured exports are a key engine of growth (World Bank, 1993), their inherent potential for specialization and learning-by-doing is partially lost, thereby stifling economic development.

9.4.2.2 Distortionary tariff and the distribution of gains and losses

Figure 9.4 illustrates the distribution of gains and losses in each of the four sectors. Panel A shows the boom in the mining sector. The newly discovered resource acts similarly to a reduction in the cost of production. The supply shifts to the right (from S to S') and producers' surplus increases (gray area). Panel B shows the market for domestic cars. Due to decreasing relative prices,

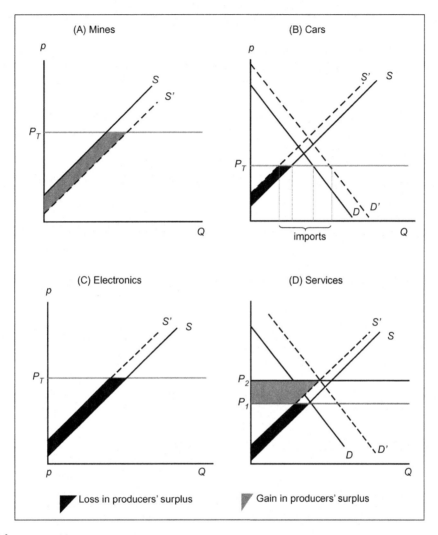

FIGURE 9.4

Distribution of gains and losses after the boom: (A) mines, (B) cars, (C) electronics, and (D) services.

imported cars become cheaper and demand for cars increases (from D to D'). On the other hand, the cost of production has increased due to higher wages. Since capital is immobile, its returns have gone down. This combined effect is shown as a leftward shift of the supply curve (from S to S'). The net effect is a decrease in producers' surplus by the amount of the black area.

Electronic producers lose the surplus area shown in panel C. Panel D shows the gains and losses of the producers' surplus in the services sector. The black area is lost due to higher labor costs, while the (larger) gray area is the gain due to the income effect.

The preceding analysis describes the economy's statically efficient response to the boom. Static efficiency, however, is not everything. The core model above omits dynamics, including considerations of investment in sectors with future comparative advantage, growth externalities, and political economy effects. For example, since future prices do not exist for most goods and services, markets cannot internalize spillover effects among interdependent investments such that market equilibrium fails to achieve dynamic efficiency.

9.5 RENT-SEEKING EFFECTS ON PUBLIC POLICIES

In the preceding section, three traded sectors and one nontraded sector augment the core model of Dutch disease. With four sectors (importable cars, exportable electronics, exportable minerals, and nontraded services), the distribution of gains and losses from the boom can be identified. This allows us to examine the effects of rent-seeking on public policy.

The existing political economy models of the resource curse focus on explaining how political economy effects lower total income (Auty, 2001; Humphreys et al., 2007; Bulte and Damania, 2008; Kolstad and Søreide, 2009; Kolstad and Wiig, 2009). This set of models focuses on the political economy concerning the source of the boom, in this case—the mining sector. The abundant resource begets a state that is "predatory" in nature, dragging the economy into an antidevelopmental state. Where the rents obtained from mining are easily appropriable, government implicitly colludes with the mineral sector in extracting and dividing rents.

However, while these models demonstrate the increase in unproductive activities after the boom, they ignore rent-seeking in sectors that lose from the boom—the exportables and the importables. Gaining a complete picture of the political economy emanating from a boom is critical in explaining why abundance can lead to curse and ultimately in the search for fundamental sources of institutional effectiveness.

The lacuna regarding public policies augmented by the resource boom calls for an inquiry into its fundamental causes, particularly the provision of optimal amount of political influence within the groups of proponents and opponents of protection. To recover part or all of the losses in producer surplus, producers of domestic cars do not remain passive; instead they increase their investment in political influence to lobby for tariff protection. On the other hand, while exportable electronics may also lobby for an export subsidy, the threat of additional losses when importables succeed in lobbying may serve as their incentive to lobby against tariff protection. The producers of exportable electronics find consumers and mine producers as allies in opposing the tariff because trade taxes reduce their surpluses. In the context of the present model, the proponents of protection are the import-competing domestic car producers; the opponents are the exportable electronics, mine producers, and consumers.

9.5.1 MODELING RENT-SEEKING AND THE POLITICAL ECONOMY EFFECTS OF THE BOOM

The presence of rent-seeking behavior has been modeled in the tradition of Becker (1983) with one group being favored at the expense of lowering the welfare of another group. Our strategy is to augment the Becker framework by investigating the mechanisms that provide the optimal amount of political influence within a group. In doing so, we apply the framework of Balisacan and Roumasset

(1987) who model the optimal investment in political influence as the Lindahl equilibrium expenditure of an endogenous coalition, where coalition size is determined where the marginal benefits of adding a coalition member are equal to the additional organizational costs. Since optimal investment for or against protection is contingent on investment by the opposing group, the theory yields the optimal reaction functions for the advocates and opponents of protection. By then solving for the Nash equilibrium investment by proponents and opponents of a particular policy, they provide a positive theory of public choice of economic policy (e.g., the degree of agricultural protection).[9]

This theory can be applied to elucidate the rent-seeking aspect of the *resource curse*. The exogenous resource boom impacts the benefit−cost calculus of investment in political influence by special interest groups. Political influence is treated as a public good to members of interest group coalitions. Investment in political influence is modeled as *within group* cooperative provision of public good while the equilibrium level of protection is modeled as a noncooperative equilibrium *between groups*. Time and money spent on political campaigns, advertising, grooming of bureaucrats and politicians, retaining the services of party organizers, bribery, and political demonstrations are among the different forms of investment in political influence.

The benefits of collective action to each group can then be defined as the proponents' gain in producers' surplus and prevention of surplus losses by opponents.

The benefits of collective action accrue to "club" members of the proponents and to opponents. In the following model, proponents of protection (*d*) represent the import-competing domestic car industry and the opposing group (*l*) represents consumers and producers of electronics and minerals.

The benefit to the coalition is given by:

$$B_i = B_i(F, N_i; x_i) \quad \text{where } i = d, l \tag{9.2}$$

where *F* is the level of protection (e.g., effective rate of protection, including tariff and nontariff barriers), *N* is the size of the coalition or the number of group members, and *x* represents a vector of other variables. The subscript *i* refers to the two opposing groups: the domestic car industry (*d*) and the coalition of the electronics sector, the mining sector, and the consumers (*l*). Consumers oppose tariff protection because it increases the price of cars. The mining and electronics sectors are opposed because the tariff decreases the price of foreign exchange, meaning their earnings decrease. A rise in *F* increases the surplus of the import-competing domestic car producers (B_d) but lowers the surpluses (B_l) of electronics producers, the miners, and the consumers, all other things constant. The other vectors of variables for the producers of cars (x_d) include the elasticity of supply and the share of market surplus in the domestic market. For the consumers and the producers of electronics and minerals, (x_l) includes the elasticity of demand and the importance of protected commodity in the consumer's budget.

In our application, *F* represents the level of protection of domestic autos. Investments in political influence by the competing groups are assumed to directly influence the function *F*. Commodity protection (I_d) benefits the members of the *d* group, but members of the *l* group are hurt by protection (I_l). The protection formation function[10] is given by:

$$F = F(I_d, I_l; z); \frac{\partial F}{\partial I_d} > 0, \frac{\partial F}{\partial I_l} < 0 \tag{9.3}$$

---

[9]See also Gardner (1988).

[10]The protection formation function is analogous to the "political influence function" in Becker (1983).

where z represents all other variables including the prevailing social norms of granting protection and a country's commitment, if any, to international trade agreements. The marginal product of investment by interest groups depends on the ideology of the government. From Eq. (9.3), commodity protection can be increased by the domestic car producers through increased investment in political influence. On the other hand, the group of consumers and producers of exportable electronics can alleviate their loss by increasing investment in opposing commodity protection. Given the relative investment of interest groups, the government provides protection F. As in Becker (1983), regardless of its form, government responds to special interests.

There are costs incurred in collective action: opportunity costs of investment in political influence (V) and organizational costs (G). The opportunity costs increase as the level of resources siphoned off from economic to political activities increases, i.e., V is an increasing function of I. Furthermore, since I is just the summation of individual investments by members of the coalition, the opportunity costs are also a rising function of N, all other things constant. The second component is organizational costs, including information and enforcement costs. As in the canonical public choice literature, G is an increasing function of N (Buchanan and Tullock, 1965; Olson, 1965). As members of the coalition increase, so does the cost of organization and enforcement, including decision-time costs. The cost of collective action can be written as:

$$C_i = V_i + G_i = V_i(I_i, N_i) + G_i(N_i, v_i), \quad \text{with } i = d, l \tag{9.4}$$

where v is a vector of other variables, including group heterogeneity, geographical dispersion, communication costs, and transportation costs.

Each group tries to maximize its net benefits from investing in political influence by finding the optimal level of investment and coalition size given the level of investment made by the other group. The necessary conditions for the optimal values of I_i and N_i are thus obtained by maximizing $B_i - C_i$ with respect to I_i and N_i:

$$\text{Max}_{I_i, N_i} NB_i = B_i - C_i = B_i(F, N_i, x_i) - V_i(I_i, N_i) - G_i(N_i, v_i) \tag{9.5}$$

which gives the following first-order conditions:

$$\frac{\partial NB_i}{\partial I_i} = \frac{\partial B_i}{\partial F}\frac{\partial F}{\partial I_i} - \frac{\partial V_i}{\partial I_i} = 0 \quad \text{with } i = d, l \tag{9.6}$$

$$\frac{\partial NB_i}{\partial N_i} = \left(\frac{\partial B_i}{\partial N_i} - \frac{\partial V_i}{\partial N_i}\right) - \frac{\partial G_i}{\partial N_i} = 0 \quad \text{with } i = d, l \tag{9.7}$$

Equation (9.6) states that the optimal I_i is obtained when the marginal benefit of investment in political influence is equal to the marginal cost of investment, i.e., the opportunity cost of investment. The condition for optimal N_i is given by Eq. (9.7). The term in parentheses, marginal net benefit less change in investment cost with respect to coalition size, is offset by the marginal cost of organization. The optimality conditions are ensured due to the following: protection formation function is concave at the origin; the cost functions are convex and increasing with I_i and N_i, i.e., $C'(I), C'(N) > 0$ and $C''(I), C''(N) > 0$; and the marginal benefit with respect to N_i is constant or diminishing.

Figure 9.5 illustrates the results. Panel A gives the Lindahl equilibrium amount of collective good attained without organizational costs at I_i^0. At this point, the sum of the marginal benefits of group members ($D_{N_i}^{max}$) is equal to the marginal investment cost MIC_i. In Eq. (9.6), the marginal

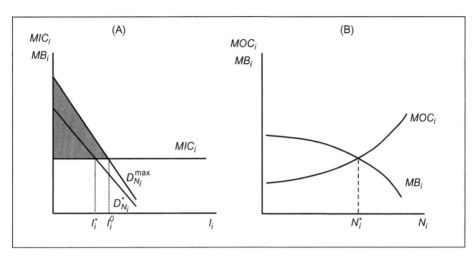

FIGURE 9.5

Collective provision of investment in political influence.

Source: *Adapted from Balisacan and Roumasset (1987).*

benefits are $(\partial B_i/\partial F)(\partial F/\partial I_i)$ and the marginal investment costs $(\partial V_i/\partial I_i)$ are equalized. The shaded gray area is the maximum net benefit obtained by the members of the group. The figure also shows the net benefit accruing to smaller groups with a corresponding smaller aggregate demand schedule, e.g., at $D^*_{N_i}$, when organizational costs are included.

Panel B shows a falling marginal benefit schedule (MB_i) reflecting the smaller benefit of "fringe" members. On the other hand, the marginal cost schedule MOC_i is increasing, indicating that higher organizational costs are required as the size of the coalition gets bigger. It is assumed that the marginal benefit schedule intersects the marginal cost schedule from above. The intersection gives the optimal coalition size N^*_i, which corresponds to optimal investment I^*_i in panel A. Moreover, given the costs, optimal coalition size is less than the total number of members in the group, i.e., $N^*_i < N^{max}_i$.

The simultaneous solutions of Eqs. (9.6) and (9.7) show that investment in political influence depends on each coalition's conjecture regarding the amount of investment by the opposing group. Figure 9.6 shows a potential outcome of a noncooperative Cournot-Nash process of investment. I_d is the investment by the protection-seeking group (import-competing producers) and I_l is the investment by the opposing coalition. Reaction functions of the two groups, proponents of protection (d) and opponents (l), are given by the lines dd and ll, respectively. The solid line F, passing through the Nash equilibrium at e_1, is the locus of I_d and I_l combinations that give rise to the same level of economic protection.

Factors that determine the level of investment in political influence are also the shifters of the marginal benefit and marginal cost schedules. Factors on the cost side include Olson's (1982) "individualized selective incentives" that increase group-oriented behavior in an individual. Examples of such incentives are fines and rewards to members behaving against or in accordance to the accepted

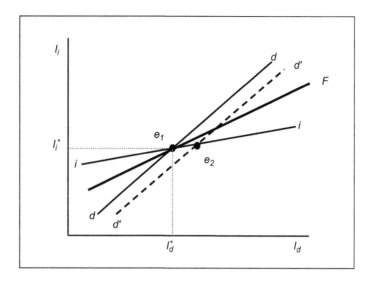

FIGURE 9.6

Cournot-Nash investment equilibrium.

group norm. Moral sanctions, such as ostracism, also mitigate violations of the group norm and help in enforcing group contracts (Hirshleifer and Rasmusen, 1989). Such factors can lower the marginal cost curve in panel B of Figure 9.5, thereby contributing to a larger optimal coalition size and a larger investment in political influence. On the organizational cost side, a critical determinant is the cost of communication, which increases with the geographical and social distance between members and decreases with improved information technology. Organization costs also include the difficulty of negotiating and enforcing the group agreement and increase with factors such as member heterogeneity (Ostrom, 2010).

In the case at hand, the impact of the boom causes a reconfiguration of the investment in political influence by special interest groups. Without an explicit rent-seeking model of protection, Corden (1984) noted that the government would be urged to protect the lagging import-substituting sector by raising tariffs or tightening import quotas. This can be explained in the context of our rent-seeking model.

The proponents of tariff protection are the import-competing domestic car producers (d). The opponents taken collectively as (l) are: electronics (e), the mineral producers (m), and consumers (r). Suppose that auto imports before the boom were small. Then the corresponding benefits of import protection would also be small. It is plausible therefore that greater competition from imports would increase the benefits from protection. The domestic autoproducers would also be keenly aware of their own losses, and this may lower the costs of organization. In terms of Figure 9.6, these factors shift dd to $d'd'$ and to the new Nash equilibrium at e_2. Since the new political equilibrium is below isoprotection line F, a higher level of protection is implied.

Consider the contending groups and how their investments influence F. The benefit of collective action by the import-competing car producers (d) is primarily determined by the returns to specific factor (r). In the face of cheaper imports due to the appreciation of the exchange rate, auto-producer real incomes will be lower when the substitutability between foreign and domestic cars is greater.[11] On the other hand, for the producers of exportable electronics (e) and producers of minerals (m), the benefits of political investment in opposing tariff protection depend on the sensitivity of their profit to changing terms of trade. For consumers, the main determinant of opposition to tariff protection is the share of manufactured goods to total expenditure (n). To the extent that manufactured goods have higher income elasticities than nontraded goods, their share in total expenditure increases as development proceeds.

The primary determinants of the costs of collective action are communication and enforcement costs. These costs are influenced by the technological advances in transportation, communication, group heterogeneity, production location, and proximity or dispersal of group members. In particular, the concentration of production (t) and the proximity (p) of producers of cars are low relative to those of consumers and producers of electronics. For consumers, urbanization (u) also reduces the cost of organizing and lends a voice to exert political influence. Advances in the level of communication and transportation technology (k) bring down the organizational cost for the three groups. Information and technological advances, such as social networking, may also reduce the difficulty of organizing large groups, even with members in remote areas.

The determinants of optimal investment in political influence of the four groups are identified: import-competing domestic car producers (I_d^*), producers of exportables (I_e^*), producers of minerals (I_m^*), and consumers (I_r^*). This is given by:

$$I_i^* = I_i^*(r, n, t, p, u, k) \quad i = d, e, m, r \tag{9.8}$$

where,

r—returns to specific factor;
n—share of manufactured goods to total expenditure;
t—concentration of production;
p—proximity;
u—urbanization; and
k—communication and transportation technology.

To recover losses from real exchange rate appreciation due to the boom, the import-competing domestic car producers may increase their investments in political influence. The greater are t, p, and k, the higher the investments in lobbying for tariff protection. Their investment in political influence is also affected by r, which depends in turn on the substitutability between foreign and domestic cars. Anticipating such actions from the domestic car producers, the producer of exportable electronics, along with the producer of minerals and consumers, would also increase their investments in political influence. The higher n, u, and k, the greater will be the investments by these groups.

Protection of the import-competing domestic car producers is at the expense of the (already squeezed) exportable electronics sector, the booming mining sector, and consumers. Thus, the resource boom may initially increase tariff protection. The following section explores a potential

---

[11]In the Australian model of Figures 9.2 and 9.3, foreign and domestic cars are assumed to be perfect substitutes in order to facilitate the graphics.

reason why this initial protection may lead to subsequent rounds of protectionism, helping to explain the stylized fact of persistent protectionism in resource-rich economies (Arezki and van der Ploeg, 2011; Collier and Venables, 2011b).

9.5.2 LEARNING BY LOBBYING

Another mechanism for explaining persistently high tariffs is the active-government version of political economy, sometimes associated with the iron triangle influence on projects and policies. The iron triangle (McConnell, 1966; Lowi, 1969) represents the alliances between politicians, bureaucrats, and special interests. Politicians form alliances with bureaucrats and special interest groups, who enjoy more sharply focused objectives and ease of organization (Olson and Zeckhauser, 1966; Olson, 1982). Special interest groups benefit from lax regulations and special favors, which are forms of political payoffs. Politicians influence the legal, policy, and fiscal environment, conferring rents in return for campaign contributions, favors, and other support. Bureaucrats implement rent-capturing agreements and broaden their power base. With these dynamics, the minority coalition readily tyrannizes the needs of the unorganized majority of taxpayers and consumers.

The conceptual foundations behind the iron triangle have not been well developed, although a close cousin uses voting theory. For example, in capture theory (Stigler, 1971; Peltzman, 1976), politicians maximize "votes," which depend on the benefits and deadweight losses from particular policies. Since politicians are maximizing votes, the solution is found by appeasing losers from the boom in order to keep their votes. Tariffs and other indirect measures of transfer are often politically favored over direct cash transfers inasmuch as the transparency of the latter would induce too much opposition (Becker, 1976; Magee et al., 1989).

By focusing only on politicians, however, capture theory leaves out other potential coalitions. Dixit (1996) gives an example of new policies and regulations generating the formation of strong special interest groups who perceive themselves as potentially disadvantaged and fight hard to maintain the status quo. In other words, the high returns of maintaining current policies trigger the formation of the strong constituencies. The returns are net of any costs of organizing the coalition, but the returns accruing to each member may be high enough to motivate collusion. In our example, domestic auto producers may be sufficiently motivated by their collective loss to organize additional tariff-seeking action.

This explanation holds promise for explaining Adam Smith's observation that "people of the same trade seldom meet together, even for merriment and diversion, but the conversation ends in a conspiracy against the public, or in some contrivance to raise prices" (Smith, 1776, Book I, Chapter X, para 82). In this perspective, special interest groups choose policy-altering influence levels to maximize their net returns. The size and strength of the coalition are endogenous. How do such coalitions arise, and how is cooperation within each coalition assured? Problems of free-riding and externalities necessitate setting up a constitutional infrastructure to govern the coalition. The constitutional infrastructure provides explicit and/or implicit rules, regulations, sanctions, and other incentives for motivating compliance. Membership of the coalition and maintenance of the same entails costs, which can be viewed as investments in the coalition. As the coalition seeks to maximize net returns to its investment, it is always vigilant regarding rents that can be seized. In other words, there is learning-by-lobbying. Once social capital in predation is created, it can be deployed to other rent-seeking opportunities. Once the resource boom forms a predatory group to seize part of the new rents, the same predatory infrastructure and knowledge can be used to distort other policies. Since protection by one group disadvantages others, thus begetting more investment in influence, there is an escalating arms race of lobbying (Magee et al., 1989).

9.6 ALL THAT CURSES IS NOT GOLD: IMPLICATION FOR THE PHILIPPINES

In the past, the Philippines has been called a historical underachiever (Briones, 2009), laggard among flying geese (Intal and Basilio, 1998), and stray cat amongst economic tigers (Vos and Yap, 1996). Policy distortions have been well documented (see, e.g., Balisacan and Hill, 2003; Lucas, 1993; Bautista and Power, 1979; Clarete and Roumasset, 1987; Power et al., 1971; Chapter 21 of this book). The resource curse model provides a possible reason for some of these distortions. The generalized resource boom creates economic inefficiencies, which, if not mitigated, undermine economic growth. Inasmuch as the Philippines has experienced several episodes of foreign exchange booms, the adverse effects above may be useful in explaining the lack of sustained growth, relative to other Southeast Asian economies. While minerals were used above to explain these "curses," other sources of foreign exchange may be easily substituted, e.g., tourism, remittances, and renting military bases.

The curse from a mineral boom has two notable features: the reallocation of resources and changes in relative prices and the rent obtained from minerals, which is easily appropriable. The rents are easy to appropriate because the industry is small and concentrated; the boom is akin to *manna* from heaven that is ready for the taking.

Any exogenous development that brings foreign exchange earnings into the country will have adverse affects on the "lagging sector" whose exports will earn a lower price of foreign exchange. If the lagging sector is an engine of growth, such as manufactured exports, there may be a negative effect on economic development. If the exogenous boom increases the rate of return to rent-seeking, e.g., through discovery of improved methods of organizing and political influence, there will be an additional negative effect on development, depending on the size and nature of the rents obtained.

Other foreign exchange booms (e.g., remittances) may generate different degrees of rent-seeking. For booming natural resources, the exploitation of the resource is easily conferred on a "natural" monopolist (concessionaire or license) due to the high fixed costs. Remittances, in contrast, tend to come from many relatively small sources. However, the huge inflow of remittances alters the domestic terms of trade due to appreciation of real exchange rate (Amuedo-Dorantes and Pozo, 2004; Acosta et al., 2009; Lartey et al., 2012; Bayangos and Jansen, 2011), which, in turn, changes the behavior of government and their cronies in the private sector. Since remittances are often channeled into real-estate investments, these become an easy target for taxation or extortion. For example, development of residential and commercial property garners huge rents that can easily be shared between developers, politicians, and bureaucrats.

The sources of boom also have different effects in terms of growth externalities. FDIs and business processing outsourcing ventures (BPOs) may generate growth externalities, though perhaps not in the same degree as manufactured exports. For example, FDI in banking and telecommunication can generate some innovations, spillover effects, and learning-by-doing. For BPOs, it is possible for institutions, human capital, and knowledge provided by BPOs to create a comparative advantage in knowledge process outsourcing (KPO). The provision of business and technical analysis, animation services, pharmaceuticals and biotechnology, and architectural design may be similarly advantaged. These opportunities present policy challenges to facilitate warranted investments to overturn the curse.

In addition to the negative effects of the boom on manufacturing, booms may also increase the returns to rent-seeking. It is plausible that import-competing domestic car producers would discover successful lobbying techniques, e.g., using well-known infant industry arguments and the "Filipino first" arguments embedded in the Philippine constitution. On the other hand, manufactured export producers are likely to be more diffused and possibly not even in business yet, making it more costly for them to organize. The losses to the manufactured exports may be accordingly more permanent. Even if producers of domestic manufactures recover some rents, the losses to the "black hole" of lobbying (Magee et al., 1989) cannot be recovered.

Rent-seeking coalitions may also find ways to diversify. In the Philippines, big Taipans are divesting from their core business and moving into public utilities. In this subsector of services, government typically sets the price as well as the rules and regulations governing the provision of services. The services subsector includes roads and highways, electricity provision, and increasingly nontraditional services such as hospitals, schools, and tourism development.

Sometimes, competition for rents occurs prior to entry into the industry, e.g., obtaining concessions. Given that big conglomerates are likely to have an established political infrastructure, these resources can be deployed in lobbying for special concessions and price-setting policies. There may or may not be a limit to predatory behavior. In Becker (1983), deadweight losses from rent-seeking limit the resources that can be used for rent-seeking, thus putting a ceiling on levels of protection. In Magee et al. (1989), however, there are no such limits, and the economy descends into a black hole of lobbying.

9.7 CONCLUSION

Sustainable development should include the dynamic patterns of sectoral development as well as the more traditional aspects of sustainable growth (Ravago et al., 2010). In examining the stylized patterns of development, we focus on the mechanics of how an economy can deviate from its optimal trajectory. In particular, we investigate a generalized version of the resource curse as a possible cause of economic inefficiencies. That is, the mechanics of the resource curse can apply to other foreign exchange booms such as foreign aid and borrowing, tourism, remittances, FDIs, and BPOs.

Any exogenous development that brings foreign exchange earnings into the country has the potential to move the economy inside its welfare frontier. We have focused on two primary mechanisms: the discrimination against imported and exported manufactures and the political economy consequences. Political economy effects may differ according to the size and nature of rents. For example, rents from remittances may not be as easily appropriable as minerals. Since resource extraction typically involves large fixed costs, it may seem "natural" to grant monopolies to concessionaires. In contrast, remittances come from many relatively small sources. Another difference between resource extraction and other sources of foreign exchange regards the degree of growth externalities generated by the booming sector. In particular, FDIs and BPOs may be able to generate growth externalities that overcome the negative pull of the curse.

The implication is not to eschew boom opportunities any more than one would ban mining a new gold discovery. Even if natural resources have the least redeeming features compared to other sources of foreign exchange, they do not have to impose a curse. Getting the right institutions is a prerequisite but not a sufficient condition for overturning the curse. If earnings from the minerals are properly invested to sectors that induce specialization, increase productivity of human capital, and accelerate technological progress, then the curses can be turned into blessings. Investments can also be put into infrastructure, such as roads, power, and communication, that increase the productivity of workers and the nontradable sector (Sachs, 2007; Collier and Venables, 2011a).

In summary, we provide an integrated model of the sources of the resource curse.

The core model of Corden and Neary (1982) has been augmented to include the losing export sector. This in turn facilitates an inventory of gains and losses from the boom and how those may influence rent-seeking activity. Rather than focus on the empirical question of whether the benefits of booms are outweighed by the costs, we have focused attention on the induced inefficiencies, i.e., sources of diversion between actual and potential performance.

Given its many foreign booms over the years, we have afforded particular attention to the Philippines. If not countered by prudent investments and actions to reduce rent-seeking, booms can deepen fragmentation and economic stagnation. One particular policy challenge relating to outmigration and foreign remittances is how to productively employ more of the country's human capital at home. For example, investments to stimulate and capture growth externalities of BPOs may help to overturn the curse of abundance. Improved mechanisms of transparency and accountability can also be deployed to counter the temptations of rent-seeking.

There are several promising avenues for further research. One would be the calibration of general equilibrium models such that the gains and losses to various groups can be quantified. The greater challenge is to formulate political economy models that go beyond the two-player Nash equilibrium rent-seeking models and political economy models of citizen voting. To capture the spirit of the iron triangle, models need to allow parts of government to be active players. Furthermore, coalitions should be genuinely endogenous. This may require applications of matching theory (Shapley and Scarf, 1974; Roth, 2002) where participants weigh the costs and benefits of joining alternative coalitions and the strategies that coalitions could employ. Formally, modeling the conceptual foundations of "learning-by-lobbying" would also help to illuminate the transmission effects of the resource curse.

ACKNOWLEDGMENTS

This research was supported by the PhD Incentive grant of the University of the Philippines, University of Hawaii Economics Department, and the Southeast Asian Graduate Study for Research in Agriculture (SEARCA). The authors are grateful to the participants of the UH Honolulu Conference and the FAEA Annual Meeting in Singapore for useful comments. Any errors of commission or omission are our responsibility and should not be attributed to any of the above.

REFERENCES

Acemoglu, D., Johnson, S., Robinson, J., 2002. Reversal of fortune: geography and institutions in the making of the modern world income distribution. Q. J. Econ. 117 (4), 1231−1294.

Acosta, P.A., Lartey, E.K., Mandelman, F., 2009. Remittances and the Dutch Disease. SSRN eLibrary.

Amuedo-Dorantes, C., Pozo, S., 2004. Workers' remittances and the real exchange rate: a paradox of gifts. World Dev. 32 (8), 1407−1417.

Arezki, R., van der Ploeg, F., 2011. Do natural resources depress income per capita? Rev. Dev. Econ. 15 (3), 504−521.

Auty, R.M., 2001. The political economy of resource-driven growth. Eur. Econ. Rev. 45 (4−6), 839−846.

Baland, J.-M., Francois, P., 2000. Rent-seeking and resource booms. J. Dev. Econ. 61 (2), 527−542.

Balisacan, A.M., Hill, H. (Eds.), 2003. The Philippine Economy: Development, Policies, and Challenges. Oxford University Press, New York, NY, pp. 283−310.

Balisacan, A.M., Roumasset, J., 1987. Public choice of economic policy: the growth of agricultural protection. Rev. World Econ. 123 (2), 232−248.

Bautista, R.M., 1988. Foreign borrowing as Dutch disease: a quantitative analysis for the Philippines. Int. Econ. J. 2 (3), 35−49.

Bautista, R., Power, J. and Associates, 1979. Industrial Promotion Policies in the Philippines. Philippine Institute for Development Studies, Manila.

Bayangos, V., Jansen, K., 2011. Remittances and competitiveness: the case of the Philippines. World Dev. 39 (10), 1834−1836.

Becker, G.S., 1976. Toward a more general theory of regulation. J. Law Econ. 19 (2), 245−248.

Becker, G.S., 1983. A theory of competition among pressure groups for political influence. Q. J. Econ. 98 (3), 371−400.

Briones, R.M., 2009. Asia's underachiever: deep constraints in Philippine economic growth. In: Carnaje, G.P., Cabanilla, L.S. (Eds.), Development, Natural Resources & The Environment, pp. 226–239. College of Economics and Management, UPLB College, Los Baños, Laguna.

Brunnschweiler, C.N., Bulte, E.H., 2008a. Economics: linking natural resources to slow growth and more conflict. Science 320 (5876), 616−617.

Brunnschweiler, C.N., Bulte, E.H., 2008b. The resource curse revisited and revised: a tale of paradoxes and red herrings. J. Environ. Econ. Manage. 55 (3), 248−264.

Buchanan, J.M., Tullock, G., 1965. The Calculus of Consent; Logical Foundations of Constitutional Democracy. University of Michigan Press, Ann Arbor, MI.

Bulte, E.H., Damania, R., 2008. Resources for sale: corruption, democracy and the natural resource curse. B.E. J. Econ. Anal. Policy 8 (1): 1935−1682. Available from: http://dx.doi.org/10.2202/1935-1682.1890.

Bulte, E.H., Damania, R., Deacon, R.T., 2005. Resource intensity, institutions, and development. World Dev. 33 (7), 1029−1044.

Canlas, D., Khan, M., Zhuang, J. (Eds.), 2011. Diagnosing the Philippine Economy: Toward Inclusive Growth. Asian Development Bank, Mandaluyong City, Philippines.

Clarete, R.L., Roumasset, J.A., 1987. A Shoven-Whalley model of a small open economy: an illustration with Philippine tariffs. J. Public Econ. 32 (2), 247−261.

Collier, P., Venables, A., 2011a. Plundered Nations?: Successes and Failures in Natural Resource Extraction. Palgrave Macmillan, New York, NY.

Collier, P., Venables, A.J., 2011b. Illusory revenues: import tariffs in resource-rich and aid-rich economies. J. Dev. Econ. 94 (2), 202−206.

Copeland, B.R., 1991. Tourism, welfare and de-industrialization in a small open economy. Economica 58 (232), 515−529.

Corden, W.M., 1984. Booming sector and dutch disease economics: survey and consolidation. Oxf. Econ. Pap. 36 (3), 359–380.

Corden, W.M., Neary, J.P., 1982. Booming sector and de-industrialisation in a small open economy. Econ. J. 92 (368), 825–848.

Dixit, A.K., 1996. The Making of Economic Policy: A Transaction-Cost Politics Perspective. MIT Press, Cambridge, MA.

Economist, 1977. The Dutch Disease. November 26, pp. 82–83.

Gardner, B., 1988. The Economics of Agricultural Policies. Macmillan, New York, NY.

Gylfason, T., 2001. Natural resources, education, and economic development. Eur. Econ. Rev. 45 (4–6), 847–859.

Henry, D.P., Crane, K., Webbe, K.W., 1989. The Philippine Bases: Background for Negotiations. RAND Publication Series R-3674/1-USDP/DOS. RAND National Defense Research Institute, Santa Monica, CA.

Hirshleifer, D., Rasmusen, E., 1989. Cooperation in a repeated prisoners dilemma with ostracism. J. Econ. Behav. Organ. 12, 87–106.

Hodler, R., 2006. The curse of natural resources in fractionalized countries. Eur. Econ. Rev. 50 (6), 1367–1386.

Humphreys, M., Sachs, J., Stiglitz, J.E., 2007. Escaping the Resource Curse. Columbia University Press, New York, NY.

Intal, P.J.S., Basilio, L.Q., 1998. The International Economic Environment and the Philippine Economy. Philippine Institute for Development Studies.

Jones, R., 1971. A three-factor model in theory, trade, and history. In: Bhagwati, J.N., Jones, R.W., Mundell, R.A., Vanek, J. (Eds.), Trade, Balance of Payments and Growth. North-Holland Publishing Company, Amsterdam and London.

Kolstad, I., Søreide, T., 2009. Corruption in natural resource management: implications for policy makers. Resour. Policy 34 (4), 214–226.

Kolstad, I., Wiig, A., 2009. It's the rents, stupid! The political economy of the resource curse. Energy Policy 37 (12), 5317–5325.

Komiya, R., 1967. Non-traded goods and the pure theory of international trade. Int. Econ. Rev. 8 (2), 132–152.

Krueger, A., 1974. The political economy of the rent-seeking society. Am. Econ. Rev. 64 (3), 291–303.

Lane, P.R., Tornell, A., 1996. Power, growth, and the voracity effect. J. Econ. Growth 1 (2), 213–241.

Lartey, E.K.K., Mandelman, F.S., Acosta, P.A., 2012. Remittances, exchange rate regimes and the dutch disease: A panel data analysis. Rev. Int. Econ. 20 (2), 377–395.

Lowi, T.J., 1969. The End of Liberalism: The Second Republic of the United States. Norton, New York, NY.

Lucas Jr., R.E., 1993. Making a miracle. Econometrica 61 (2), 251–272.

Magee, S., Brock, W., Young, L., 1989. Black Hole Tariffs and Endogenous Policy Theory: Political Economy in General Equilibrium. Cambridge University Press, Cambridge.

Magtibay-Ramos, N., Estrada, G.E.B., Felipe, J., 2007. An Analysis of the Philippine Business Process Outsourcing Industry. Asian Development Bank, Manila.

Matsuyama, K., 1992. Agricultural productivity, comparative advantage, and economic growth. J. Econ. Theory 58 (2), 317–334.

McConnell, G., 1966. Private Power & American Democracy. Knopf, New York, NY.

Medalla, P., Fabella, R., de Dios, E., 2007. The Remittance Driven Economy. Paper written for the Center for National Policy and Strategy (CNaPS).

Mehlum, H., Moene, K., Torvik, R., 2006. Institutions and the resource curse. Econ. J. 116 (508), 1–20.

NeoIT, 2004. Mapping Offshore Markets Update Offshore Insights White Paper.

Olson, M., 1965. The Logic of Collective Action; Public Goods and the Theory of Groups. Harvard University Press, Cambridge, MA.

Olson, M., 1982. The Rise and Decline of Nations: Economic Growth, Stagflation, and Social Rigidities. Yale University Press, New Haven, CT.

Olson Jr., M., Zeckhauser, R., 1966. An economic theory of alliances. Rev. Econ. Stat. 48 (3), 266–279.

Ostrom, E., 2010. Beyond markets and state: polycentric governance of complex economic systems. Am. Econ. Rev. 100, 1–33.

Paldam, M., 1997. Dutch disease and rent seeking: the Greenland model. Eur. J. Polit. Econ. 13 (3), 591–614.

Palma, J.G., 2008. De-industrialization, 'premature' de-industrialization and the Dutch disease. In: Durlauf, S. N., Blume, L.E. (Eds.), The New Palgrave Dictionary of Economics. Palgrave Macmillan, Basingstoke. Available from: http://dx.doi.org/10.1057/9780230226203.0369.

Peltzman, S., 1976. Toward a more general theory of regulation. J. Law Econ. 19 (2), 211–240.

Philippine Statistics Authority-National Statistical Coordinating Board. Available from: <http://www.nscb. gov.ph/factsheet/pdf07/FS-200711-ES2-01_BPO.asp>.

Power, J., Sicat, G., 1971. The Philippines: Industrialization and trade policies. Oxford University Press, London.

Rajan, R.G., Subramanian, A., 2009. Aid, Dutch Disease and Manufacturing Growth. SSRN eLibrary.

Ravago, M., Roumasset, J., Balisacan, A., 2010. Economic Policy for Sustainable Development vs. Greedy Growth and Preservationism. In: Roumasset, J.A., Burnett, K.M., Balisacan, A.M. (Eds.), Sustainability Science for Watershed Landscapes, pp. 3–45. Institute of Southeast Asian Studies, Los Baños, Philippines: Southeast Asian Regional Center for Graduate Study and Research in Agriculture, Singapore.

Roth, A., 2002. The economist as engineer: game theory, experimentation, and computation as tools for design economics. Econometrica 70 (4), 1341–1378.

Roumasset, J., 2008. Population and agricultural growth. In: Durlauf, Steven N., Blume, Lawrence E. (Eds.), The New Palgrave Dictionary of Economics, Second ed. Palgrave Macmillan, The New Palgrave Dictionary of Economics Online. Palgrave Macmillan. August 09, 2014. Available from: http://dx.doi.org/ 10.1057/9780230226203.1308.

Sachs, J., 2007. How to handle the macroeconomics of oil wealth. In: Humphreys, M., Sachs, J., Stiglitz, J.E. (Eds.), Escaping the Resource Curse. Columbia University Press, New York, NY, pp. 193–213.

Sachs, J.D., Warner, A.M., 1995. Natural Resource Abundance and Economic Growth. Available from: <http://ideas.repec.org/p/nbr/nberwo/5398.html> (December).

Sachs, J.D., Warner, A.M., 2001. The curse of natural resources. Eur. Econ. Rev. 45 (4–6), 827–838.

Shapley, L., Scarf, H., 1974. On cores and indivisibility. J. Math. Econ. 1, 23–28.

Smith, A., 1776. An Inquiry into the Nature and Causes of the Wealth of Nations. Methuen & Co., Ltd., London.

Snape, R.H., 1977. Effects of mineral development on the economy. Aust. J. Agric. Econ. 21 (3), 147–156.

Stigler, G.J., 1971. The theory of economic regulation. Bell J. Econ. Manag. Sci. 2 (1), 3–21.

Tornell, A., Lane, P.R., 1996. Power, growth, and the voracity effect. J. Econ. Growth 1 (2), 213–241.

Tornell, A., Lane, P.R., 1999. The voracity effect. Am. Econ. Rev. 89 (1), 22–46.

Torvik, R., 2002. Natural resources, rent seeking and welfare. J. Dev. Econ. 67 (2), 455–470.

Torvik, R., 2009. Why do some resource-abundant countries succeed while others do not? Oxf. Rev. Econ. Policy 25 (2), 241–256.

Tuaño-Amador, M.C.N., Claveria, R.A., Co, F.S., Delloro, V.K., 2007. Philippine Overseas Workers and Migrants' Remittances: The Dutch Disease Question and the Cyclicality Issue, *Bangko Sentral Review* (Manila: Bangko Sentral ng Pilipinas) pp. 1–23.

Tullock, G., 1967. The welfare costs of tariffs, monopolies, and theft. West. Econ. J. 5 (3), 224–232.

van Wijnbergen, S., 1984. The 'Dutch Disease': a disease after all? Econ. J. 94 (373), 41–55.

Vos, R., Yap, J.T., 1996. The Philippine Economy: East Asia's Stray Cat?: Structure, Finance and Adjustment. Macmillan Press Ltd. and St. Martin's Press, London and New York, NY.

World Bank, 1993. The East Asian Miracle: Economic Growth and Public Policy. Oxford University Press, New York, NY.

World Bank, 2008. World Development Report 2008: Agriculture for Development. Washington, DC.

World Bank, 2011. World Development Indicators 2011. Washington, DC.

GOVERNING COMMERCIAL AGRICULTURE IN AFRICA: THE CHALLENGES OF COORDINATING INVESTMENTS AND SELECTING INVESTORS[1]

10

Marilou Uy

World Bank Group, Washington, DC

CHAPTER OUTLINE

---

[1]While the author is a senior staff member of the World Bank, the views expressed in this chapter are solely the author's and are independent of World Bank policies or practices.

10.1 INTRODUCTION

There is a great need and also strong potential for increasing agricultural productivity in Sub-Saharan Africa.[2] Agriculture contributes one-third of the continent's GDP. More than 70% of its people, especially the poor, are engaged in the sector. Agricultural transformation could have far-reaching developmental impact, by reducing poverty,[3] providing better food security, and creating a better foundation for urbanization and overall economic transformation.

Recent reports highlight the solid potential for agriculture and agribusiness to improve productivity and contribute to economic growth in Africa (McKinsey, 2010; World Bank, 2013a,b). Strong global demand and growing urbanization throughout Africa present compelling market opportunities for developing a more competitive and commercial agricultural sector[4] that has so far been lethargic. International investor interest in Africa's agricultural possibilities has picked up dramatically since the international food price increases of 2008.

However, it is difficult for many African countries to make the transition to commercial agriculture, a form that has typically evolved in countries where land is relatively scarce and production more input intensive than has been the case in much of Africa. While the discussion about Africa's future agricultural development has focused mainly on the role of small farms, there has been a surge of large land deals across the continent. This surge of investor interest suggests there could be real potential in larger scale commercial investments in Africa, but the evidence shows that many of these land deals have not resulted in the expected levels of investment. Some appear to be no more than speculative ventures—pay a little bit now for the option of developing something later—and the downside of speculative leases is that they tend to discourage serious investors. Some investors have also either discontinued or slowed their investments in Africa because of social conflicts or poor infrastructure. There is increased doubt that African countries will be able to translate the stronger investor interest into investment opportunities capable of yielding sustainable benefits and poverty reduction.

A key question, therefore, is how to govern commercial agricultural interests in Africa and to determine what policies and institutions are necessary to enable countries to get the most out of these investments in terms of economic development and food security. Experience shows a need to address the key constraints to increased productivity and agricultural development. In broad terms: What are the bottlenecks that inhibit successful commercial agriculture? How can one contain speculation? What can be done to reduce the risks of investing when the requisite infrastructure that needs to complement farm investments is lacking? How can one manage the entry of private investors where there is ambiguous definition of land-use rights?

This chapter focuses on how private investors and governments in Africa can work with each other to overcome bottlenecks in commercial agriculture. It focuses on the need to coordinate interdependent private and public investments, especially in infrastructure, in order to reduce investment risks. Governing this coordination requires effective mechanisms for investor selection, such through auctions, and enforcement of requirements for keeping land leases. The chapter also describes emerging mechanisms for information sharing between the public and private sectors, in order to better

---

[2]This will be referred to simply as Africa in this chapter.

[3]Increasing agricultural productivity is one of the fastest means to reducing poverty (World Bank, 2008).

[4]Commercial agriculture is defined as that part of the agricultural sector producing primarily for the market (Kaldor, 1967).

understand the bottlenecks for investments and improved productivity and to provide solutions to these, and for coordination within spatial or sectoral development strategies. It highlights the importance of engaging communities and local stakeholders to manage the social risks associated with land transfers and ensure that they benefit from the commercial investments. There is clearly room to learn from global experiences in governing commercial agriculture, both successes and failures, and to adapt lessons to the country context, as shown by some emerging practices in governing commercial agriculture in parts of Africa.

The chapter also explores ideas for governance mechanisms beyond governments. These include the appropriate use of global standards and frameworks for better governance of investments in natural resources. While these frameworks have been used mainly in mining and natural resource management, they also provide relevant lessons and guidance in governing agricultural investments by both multinational and local investors.

A sub-theme of the chapter is the importance of infusing the field of agricultural development with analytical approaches to industrial organization, public policy, and natural resource management.[5] Drawing on the field of industrial organization leads us to replace the question of *whether* small or large farms are more efficient with *what causes* small and large farms to emerge and coexist and how to work strategically with private investors to improve the development gains from investments in commercial agriculture. Successful industrial policies provide lessons in the value of greater and effective coordination between governments and private investors. Experience in governing mining rights and contracts, in turn, provides useful guidance in designing the process of investor selection and the models for coordinating public and private investments.

Section 10.2 discusses the potential for increased productivity growth in Africa and reviews the small versus large farm debate for the future of African agriculture, taking lessons from experiences in commercial agriculture on the continent. Section 10.3 focuses on the challenge of selection and design of incentives for investors and on mechanisms to coordinate public and private investments. Section 10.4 discusses the importance of engaging local communities and other stakeholders who would be affected by the commercial investments and examines the scope for leveraging their comparative advantage in local knowledge and monitoring. Section 10.5 examines the potential use of collective initiatives at the international level—specifically of international producer associations, civil society organizations, and international development institutions—in influencing the governance of commercial agricultural investments. Section 10.6 sets out concluding remarks and recommendations for further research.

10.2 CAPTURING THE PRODUCTIVITY GROWTH POTENTIAL THROUGH COMMERCIAL AGRICULTURE

Agricultural productivity in Sub-Saharan Africa has lagged behind many other regions despite a strong comparative advantage in natural resources. With this loss in competitiveness over the past four decades, Africa's share of world agricultural exports declined steadily (Figure 10.1).

---

[5]Roumasset (2010) discusses the potential of "new tools of analysis to explain empirical patterns in behavior and organization in developing agriculture and to build the foundations of a public microeconomics of development" (ibid, p. 17).

(A)

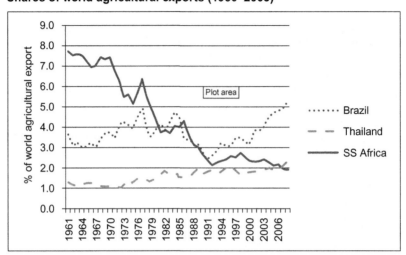

Shares of world agricultural exports (1960–2008)

(B)

Current Yield Relative to Estimated Potential Yield

Country/Region	Maize	Oil Palm	Soybean	Sugarcane
Sub-Saharan Africa	0.20	0.32	0.32	0.54
Asia	0.62	0.74	0.47	0.68
South America	0.65	0.87	0.67	0.93

FIGURE 10.1

(A) Decline in competitiveness of Africa's agricultural exports and (B) with crop yields far below potential.

(A) FAO STAT and (B) Deininger and Byerlee, 2011.

Actual yields in key agricultural commodities are way below estimated potential yield, ranging from 20% of potential yields for maize to 54% for sugarcane (Deininger and Byerlee, 2011). Studies nevertheless point to substantial, unexploited potential for higher yields and the strong potential comparative advantage in agriculture of most African countries. Within the Africa region itself, stronger economic growth, a growing middle class, and increased urbanization are creating new market opportunities (McKinsey, 2010; World Bank, 2013a,b). There is also scope for greater utilization of farmland: various studies estimate that Africa has close to half of the world's agriculturally suitable yet unused land (McKinsey, op.cit.; Fischer and Shah, 2010).[6]

In recent years, especially after the 2008 food price increases, global investors have shown strong interest in investing in agriculture and agribusiness. Half of the land deals[7] reported

---

[6]As cited in Deininger and Byerlee (op. cit.).
[7]These are documented large-scale land acquisitions.

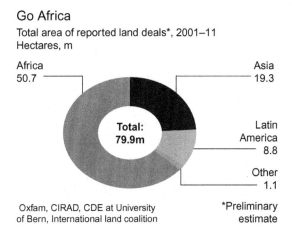

Go Africa
Total area of reported land deals*, 2001–11
Hectares, m

Africa 50.7

Asia 19.3

Total: 79.9m

Latin America 8.8

Other 1.1

Oxfam, CIRAD, CDE at University of Bern, International land coalition

*Preliminary estimate

FIGURE 10.2

Africa's share of reported land deals, 2001–2011.

The Economist, 2011.

between 2001 and 2011 have been in Africa (Figure 10.2). The increased investor interest in agricultural land, however, has raised concerns as to whether the transition to more commercial agriculture in Africa would indeed bring much-needed economic benefits that also benefit farmers and communities. The reality to date is not encouraging. Deininger and Byerlee, drawing on a database of media reports on land transactions, estimated that land deals in 2011 involved more than 400 million hectares, about 2% of Africa's land area. Most were very long leases with minimal rents, and the median size was estimated at 40,000 hectares. Many of these projects, however, have not led to actual commercial operations, prompting concerns that many of these deals were speculative and not commercially motivated. This has led observers to question whether large-scale commercial agriculture in Africa is capable of yielding sustainable and shared benefits in the countries concerned.

10.2.1 SMALL VERSUS LARGE FARMS?

The poor outcomes of many large land investment projects have reinforced the traditional view that the future development of Africa's agriculture lies in focusing primarily on smallholders (World Bank, 2008; Binswanger, 2009). The discussion about how best to develop Africa's agricultural sector has been dominated by the "small-versus-large farms" debate. Support for targeting small farms as the key to Africa's future agricultural development is based largely on empirical evidence of the inverse relationship between farm size and production per unit of land that has been observed in many countries, including in Africa (Larson et al., 2012; Otsuka and Larson, 2013; Ayalew and Deininger, 2014). African agriculture is dominated by small farms: about 80% of the continent's farms are smaller than two hectares (Nagayets, 2005). When land is abundant and the population is growing, farms will take advantage of the low shadow price of labor by using family labor more intensively and avoiding the relatively high transaction cost of recruiting and supervision hired labor. Since farm size adjusts to the availability of family labor, small farms are thought to have advantages over large ones. Critics assert that this empirical work does not consider the effect of land quality but attempts to include it in country level regressions have been hampered by poor proxies of land quality.

It is also argued that small farms are much better than large farms for improving equity and reducing poverty (Hazell et al., 2007; World Bank, 2008). Smaller farms tend to use more labor per unit area, and their expenditure patterns tend to favor spending on rural nontradables compared to large farms, thus creating more income earning opportunities (Hazell and Roel, 1983). The World Bank's 2008 World Development Report recommended "improving the productivity, profitability, and sustainability of small holder farming as the main pathway out of poverty in using agriculture for development" (The World Bank, 2008, p. 10). This could be achieved by improving price incentives, more and better public investments, better access to markets and credits, effective producer organizations, and promoting innovation through science and technology. Binswanger (2009), in his book *Awakening Africa*, argued that the "smallholder-led revolution" of Northeast Thailand, which led to major increases in productivity with desirable distributive effects, offers lessons on the comparative advantages of small-scale agriculture that dominates in the Africa setting. Otsuka and Larson further noted that an encouraging lesson for Africa from East Asia's green revolution is that rapid growth in agriculture is feasible without a change in farm structure. The limited success of large-scale farming in Africa, according to Binswanger-Mkhize et al. (2010) in any case, further reinforces the view that the family farm model is an appropriate model for most of Africa's agricultural development.

Counter arguments point to the need for farms in Africa to adopt more capital intensive technologies and marketing practices if they are to improve productivity and compete in world markets. Modernization of supermarkets makes it more difficult for small farms to survive. Small farms are likely to be marginalized as new technologies require more capital and specialization more suited for larger farms (Hazell et al., 2007). The reality of Africa's agriculture clearly suggests that opportunities for greater commercialization that can exploit the benefits of scale are still to be captured. The growth of Africa's aggregate agricultural output has been slower compared to Latin America and developing Asia. Africa's output growth has also been accompanied by slow growth in labor and land productivity and has drawn primarily from area expansion (Haggblade and Hazell, 2010). In fact, farms in Africa have become smaller over recent decades (Eastwood et al., 2010), and large-scale commercial farming has scaled back over the past 40 years (Collier, 2010). This suggests that African agriculture could be foregoing the potential advantages of scale economies in access to finance, marketing, technology, and specialization.

An alternative to the "small-versus-large farm" paradigm could start with the premise that the structure of the agricultural organization and the choice of farming techniques are endogenous. For a particular location that is associated with certain economic and agroclimatic conditions, one determines how best to farm if organized as a family farm and how best to do so if organized as a commercial venture. Comparative advantage is determined by the arrangement that will yield the highest rent per hectare. In this way, farming technique and organization are determined jointly. The inverse relationship between farm size and yield per hectare may prevail among family farms, while a positive relationship simultaneously prevails among commercial farms (Uy, 1979). Both can be efficient outcomes. Since the organizational form and farming techniques are determined simultaneously, it is also misleading to attribute differences in productivity to different forms of agricultural organization (Roumasset and Uy, 1986).

Where management is less important, family labor combined with family management may enjoy a comparative advantage in exploiting the lower shadow price of family labor and lower supervision costs. For these farms, to the extent that farm size is adjusted to the family labor force, there will be an inverse relationship between farm size and yield per hectare. For the many family farms that hire labor at the margin, the inverse relationship will be augmented by the higher shadow price of labor inclusive of recruiting and supervision costs.

For land that is suited to commercial crops, including locational advantages, more management per hectare is warranted, including the organization of hired labor. Since hired labor affords economies of specialization, the optimal size will tend to increase. The optimal size of a commercial farm can vary widely. Some could function as nucleus farms that provide support and services to their numerous suppliers that could be much smaller farms. Some could be associated with extensive vertical integration or horizontal specialization where teams specialize in different tasks. The extent and limits of these economies are not clear. If farm size is limited instead by imperfections in the land and credit markets, one expects to see a positive relationship between farm size and productivity.

Family farms can also evolve into more commercial undertakings. As the economy develops, wages rise relative to effective interest rates and capital is substituted for labor, which increases optimal farm size, especially for family farms. This could prompt some degree of consolidation (unless otherwise blocked by land policies).

Under this framework of endogenous institutions, one may observe an increasing diversity of farm sizes and organizational forms with varying degrees of specialization in Africa. Family farms are likely to dominate in areas where land is relatively abundant, or otherwise has a low shadow price such that land-using technologies that economize on other expenses are favored. In some other settings, such as those that will produce for global markets, large-scale farming would be favored. Scale economies would be needed to use the right technology, access financing, and access markets. The "supermarket revolution" (Reardon and Timmer, 2007), which is now taking root in many African countries, is leading to more direct links between farms and wholesalers that in turn work with large retailers (World Bank, 2013a,b). These linkages could be carried out through either large-scale, vertically integrated farms or through partnerships between large retailers and small-scale farmers. In either case, the links with supermarkets have meant greater attention to compliance with quality standards at the farm level and most likely greater tendency to increased specialization in how labor is organized. These trends could be associated with a diversity of organizational arrangements in commercial farming in Africa. Commercial farming has not been limited to plantations. It has also been associated with a better integration of large farms and small farms, such as through "contract farming" practices and "nucleus farm arrangements" (World Bank, 2013a,b). These arrangements tap the scale economies of large-scale farming and the comparative advantages of smallholder farms. Africa's commercial agriculture of the future is likely to consist of small and large farms that are viable and that coexist.

When farm organizations are viewed as endogenous, policy makers will be concerned less with choosing between small and large farms and more with providing the necessary enabling environment to address bottlenecks and market failures that constrain the efficient functioning of farms, large and small. While much has been written about how to improve the productivity of small farms, more work is needed to understand the enabling policy environment that will enable large-scale commercial agricultural investments to achieve the best development outcomes.

This alternative view therefore leads to us to ask what can be done to reduce speculation in large land deals, coordinate public and private investments in light of highly inadequate infrastructure and missing markets, and better manage the social risks associated with the ambiguous state of land governance. These are key areas for policy makers to tackle in governing the transition toward more commercial agriculture with the goal of ensuring that investors invest responsibly so as to discover the true value of the land and thereby enable countries to achieve and capture the benefits of agricultural transformation.

10.2.2 THE CHALLENGES OF GOVERNING LARGE-SCALE COMMERCIAL FARMING

The experience so far in commercial agriculture in Africa has yielded useful lessons for putting in place sound policies and institutions to improve the governance of large-scale investments in commercial agriculture. Deininger and Byerlee's case studies of 14 countries in Africa showed that different country policies tended to determine the scale of recent land transfers. In Tanzania, for example, <50,000 hectares were transferred to investors between January 2004 and June 2009, while in Mozambique 2.7 million hectares were transferred. Many countries that accelerated land transfers, however, showed a lack of readiness and serious weakness in their institutional capacity to manage large-scale private investments. Mozambique's land audit in 2009, for example, found that about half of the transferred land was unused or not fully used. Many countries had limited due diligence in screening project proposals, and there was an "air of secrecy" around the process of land transfers that was conducive to weak governance. Incomplete official land records and widespread neglect of social and environmental norms were pervasive. The process led to failure to recognize and protect land rights, to compensate communities when there were voluntary transfers, and to conduct participatory consultations to clarify agreements. Neglect of existing land rights prompted social conflicts that further slowed progress of investments. These were evident in the reports of displacement of small farmers without appropriate compensation that led to accusations of "land grabs" by large investors from local holders of farming rights. Taking commercial agriculture to scale will need a better system of governing land deals, alongside a longer term agenda to clarify land rights and ownership in Africa.

While data on the provisions of large-scale land transfers are not readily available, there is qualitative evidence that in many African countries, the entry of large-scale investors in agriculture has been seen as an opportunity to introduce new technology and financing, facilitate infrastructure development, and create employment. In most large-scale land transfers, land has been sold cheaply but the contracts include commitments on investment, infrastructure development, and employment (Cotula et al., 2009). Actual investments, however, have been well below expectations, either because of speculation or because investors eventually found the risks to their investments to be too high. The risks of social conflicts from issues related to land claims are well known. Other investors cite the poor state of public infrastructure. Expected public sector investments to upgrade infrastructure services have not been forthcoming, raising risks to the profitability of the private investments. The availability of major road networks, power, irrigation, and infrastructure for wholesale and retail markets is essential to improve the viability of and to reduce the risks inherent to agricultural projects. Truck freight rates in Africa can be twice as high as those of Latin America and Asia (World Bank, 2013a,b). About 75% of farmers in Africa are more than four hours away from the nearest market by motorized transport, compared to 45% in Asia, and poor rural roads lead to much higher transport costs and postharvest losses (World Bank, 2013a,b).[8] Less than 5% of the land under cultivation in Africa is irrigated. Developing commercial agriculture will clearly need a sharp increase in infrastructure investments and workable models to coordinate public and private investments.

The experiences of international development financial institutions that have invested in agribusiness in Africa and elsewhere provide further lessons from their successes and failures. Tyler and Dixie's (2013) review of the development impact of large investments made by the Commonwealth Development Corporation (CDC) in commercial agriculture in Africa, Southeast Asia, and the Pacific between 1948 and 2000 showed that more successes were observed in established business models than in startup investments. However, successful pioneers—after a period of learning—attracted further investments that brought transformative results. The study covered 179 projects in 32 countries and a variety of organizational forms, such as plantations, nucleus estates, and outgrower schemes. Controversial land acquisition issues were not prevalent in these projects so that the study was able to focus on operational and market issues affecting investment outcomes. Successes included some highly profitable investments in palm oil production in Southeast Asia and tea in Kenya and investments involving nucleus states with outgrower arrangements. Startups had a high failure rate, and many of these, especially in Africa, were judged to have been highly uncertain at the time of their appraisal.[9] Most of these failures were due mainly to flawed project concepts that identified the wrong crop or the wrong location. The CDC's success rates improved over time through more rigorous preparation that took account of lessons learned from previous projects, such as the importance of gathering sufficient initial knowledge of agronomic conditions in order to weed out unviable projects.

Clearly, there are difficult challenges to governing the growth of commercial agriculture in Africa, but there is scope to learn from experience so as to achieve the large development potential from these investments. Some countries in Africa are putting in place more structured processes to select investors and govern commercial contracts and are encouraging more diverse models to coordinate public and private investments. Countries increasingly are exploring a wider range of business models that better integrate large farming operations with local communities and surrounding small farms, such as nucleus farms and outgrower schemes. Some of these are more successful, not least because they tend conform more to existing land-use rights. Governments are also starting to build capacity to manage private sector investments in large-scale commercial agriculture and to calibrate their investment promotion initiatives with their capacity to implement them.

The World Bank's recent 2013 publication on unlocking the potential of agribusiness in Africa recommends focusing initially on a few locations, preferably where more is known about the commercial potential. The spatial focus will enable governments to build the capacity to design, implement, and coordinate difficult reforms and demonstrate success that could be scaled up later. It can also facilitate these efforts to clarify and implement their public investment program. Governments

---

[8]Poor infrastructure is not the only cause of the high transport costs: high truck freight rates in Africa are due to high fuel costs, uneconomic one-way payloads, logistical delays, and rents extracted at various checkpoints (World Bank, 2013b).
[9]Some private companies manage the risks of early investments in Africa by working with local firms that have established management capacity. A Harvard Business School case study of Olam, a private sector company that has been successful in sourcing agricultural products from a number of countries in Africa for global distribution, documents its success in investing with local companies and managers that have strong local knowledge and management expertise (Bell and Shelman, 2008).

and investors should work together to provide information and evidence on market opportunities and constraints to investments that should inform the policies and steps needed to manage the high risks in the early stages of investments. The study also proposes engaging only with firms that have established reputations and capacity to undertake the desired investments and approving investments that are viable and in accord with responsible business practices. It proposes a much stronger focus on alleviating key constraints to the proper functioning of a few value chains[10] and locations as a way of harnessing scarce resources so as to capture the opportunities from increased investor interest in commercial agriculture.

10.3 COORDINATING INVESTMENTS AND SELECTING INVESTORS FOR BETTER GOVERNANCE OF COMMERCIAL AGRICULTURE

The experience so far of large land deals in Africa illustrates the challenge that governments need to manage to ensure that increased demand from investors translates into actual investments that increase productivity. It points to the importance of a much more strategic approach to developing commercial agriculture by engaging serious investors and discouraging speculators and mitigating the high risks of pioneer investments. This approach also envisages diverse models of coordinating interdependent public and private investments. One model consists of very large farms where the private investor internalizes many of the externalities from large-scale investments, including infrastructure. Another model involves a much greater role for governments in coordinating interdependent investments and investing in infrastructure. The latter may provide more opportunities for greater integration of large-scale commercial investments with the production capacity of small farms. An effective enabling environment for commercial agriculture will need to address the governance challenges of each model. We discuss below approaches to investor selection that could guard against speculation and improve incentives to invest and discover the value of the land. We also discuss additional emerging mechanisms for public—private coordination that will improve information sharing and cooperation aimed at identifying the constraints to investment and productivity growth and defining effective solutions that could be an essential part of an agricultural development strategy.

10.3.1 COORDINATING INVESTMENTS AND SELECTING INVESTORS FOR "VALUE" DISCOVERY

In most African countries, the potential value of the land that investors might be interested in is not known *ex ante*. In many settings, there is uncertainty about which products and production processes will be commercially viable, and experimentation will be needed to discover them. Speculation is a concern. Early investors could purchase leases cheaply but will not have the

---

[10]Typically, "value chain" describes activities that add value by bringing a product or a service through the different phases of production to bring the product to the consumer (Kaplinksy and Morris, 2002). Innovations in products and processes improve the function of value chains. By contrast, the term "supply chain" is used to describe the logistical and procedural activities involved in taking the product from supplier to customer. Reduced delays will improve the functioning of the supply chain.

incentive to invest quickly because if they succeed they will share the benefits with new entrants, while they will bear the full cost of failure. Their incentive would be to wait until others succeed, or until there is more information available, before they invest or sell the land when its (higher) value is known. The downside of speculation, however, is that serious investors are precluded.

Collier and Venables (2012) present a compelling framework that builds incentives in agricultural land deals to encourage investors to invest rather than speculate. Holding a lease on land would be akin to holding an option on an asset whose future value is uncertain and could have an upside value to both governments and investors. While land prices will be very low initially, both will stand to gain if investors succeed in discovering the value of the land. Governments could gain from the increased value of the land by phasing land transfers, starting with a few and expanding as the land gains more value. Governments could also encourage early investors to invest and experiment by selling them more land than investors would initially cultivate. If the investments succeed, investors stand to gain from the increased value of the "excess" land. Governments gain from the higher value of the land they will sell in subsequent phases. If the investments fail, governments still have the flexibility to work with the investors on alternative business plans or with other investors on the later phases of the land transfers. This framework implies a much more measured approach to commercialization by phasing in some private investments and scaling up at later stages when there is more information about their commercial value.

Another concern is the high risks of investing that could deter serious investors or lead to major delays in investments. Poor infrastructure is a key factor. If interdependent investments in farming, processing, production of inputs and infrastructure are coordinated effectively, the probability of success of projects will be greater. In the early development literature, public provision of infrastructure as well a strong public intervention in planning and investments have been the preferred solutions to the coordination problem, but these have met with little success. The experience of East Asia's industrial transformation provides some lessons from successful experiences: World Bank (1993) and Roumasset and Barr (1992) attribute an important part of the East Asian economic miracle to the governments' roles in solving "coordination problems" that arise with interdependent large-scale investments, including in infrastructure, and externalities needed for development. Governments played a key role in putting in place mechanisms for coordinating investment decisions and sharing of information with the private sector.

In many of the very large land deals for commercial agriculture in Africa, governments have sought to negotiate private investments on infrastructure and other public services as part of the contractual agreements. With sufficiently large project sizes, investors could realize the scale economies from large-scale investments and internalize the spillovers from learning and value discovery. These arrangements suggest a strategy where governments take on less, and the private investors more, of the investment coordination role.

This approach poses risks from governance challenges that could lead to poor selection of investors and projects. One of the more important lessons from the experience so far in many of the land deals in Africa is the need for greater oversight and discipline to ensure the choice of responsible investors and viable commercial transactions. "The deals are too opaque, too large, and too long" (Collier, 2010, p. 218) is a complaint often heard. Land deals need to be made through a structured auction process, with greater sharing of information. An auction will ensure competition among several bidders.[11] Making information on investor interest, their project proposals, and the selection process available to all stakeholders will also be important measures within the governance process. It will be necessary also to build in periodic reviews to determine impact, identify weaknesses, and address them in a timely way. The participation of external stakeholders, such as development finance institutions and civil society organizations, could strengthen the oversight of the process.

The auction process would need to build in mechanisms to select investors that have the capacity to manage and finance commercial agricultural projects and track records in investing responsibly. Peru's successful auction of 235,000 hectares of public land for agricultural investments prequalified potential bidders that deposited with the government an amount equal to a certain percentage of the bid price plus the intended investment amount, which provided the comfort that the bidders consisted only of investors who had the financial capacity to undertake the project. Before the contract was signed, successful bidders had to pay for the land and guaranteed their commitment to invest by depositing with the government a letter of credit covering the amount of the proposed investment over a period of three years (Hernandez, 2010). Investors could also be prequalified based on a broader set of criteria that included financial capacity and technical expertise as well as a record of investing responsibly. For example, the "Principles for Responsible Investment in Agriculture," which were formulated through a process led by FAO, IFAD, UNCTAD, and the World Bank,[12] could inform the criteria for investor selection. These seven principles draw attention to the responsibilities of countries and investors to ensure that rights to land and associated natural resources rights are respected, food security is strengthened, good governance is practiced, and investments contribute to sustainable development and desirable social and distributional effects. Commercial agricultural projects that the World Bank Group and other development institutions are supporting in Africa are increasingly using the Principles to guide investor choice. The participation of external stakeholders, such as development finance institutions and civil society organizations, could strengthen the oversight of the investor selection process.

Adequate preparation would involve making available upfront relevant information that would help inform investors' investment proposals. This would include clear plans on the governments' investment programs on infrastructure that are relevant to the private investments. An effective process to clarify land tenure needs to be in place. Governments, possibly in collaboration with international development institutions, have a role to play in conducting social, agronomic, and environmental assessments that are useful inputs to designing viable projects. Investors, on the other hand, should provide an assessment of market opportunities and the technical and financial viability of their project plans. For example, Peru's auction built in a process of vetting the viability of the investors' project proposals, which was assigned to a group of credible technical experts. Information on project proposals was made publicly available so that other potential investors also benefited from the assessments of early movers.

Most of the large commercial agriculture contracts involve promises of future investments, infrastructure development, and employment creation over long terms. Without periodic reviews and the enforcement of performance requirements for keeping a lease, there is a serious risk that failed investments will not be detected and resolved in a timely way, thereby precluding the entry

---

[11]Even if there is only one expression of interest and bid, the process should still be transparent by prequalifying the investor, making information regarding the project proposal available publicly, and subjecting the project proposal to a rigorous technical review.
[12]See FAO, IFAD, UNCTAD, and the World Bank Group (2010).

of new investors. Evidence suggests that many of the early large-scale contracts that have experienced major delays in investments have not captured the complexity associated with coordinating investments and enforcing responsible business practices (Cotula et al., 2009). While many of the large land deals in Africa have been conditioned on a timeline of delivery of investments according to their projects' plans, enforcement has been lacking (Cotula et al., 2009). Governments, possibly with the support of international development institutions, will need to build the capacity to monitor and put in place institutions to enforce these long-term contracts. For example, Senegal's recent agribusiness development project supported by the World Bank provides assistance to rural communities in putting in place a legal mechanism for expeditious cancellation of dormant leases.

The experience of many countries in designing and enforcing mining leases provides useful lessons to public–private coordination in commercial agriculture. Investors in mining development projects often invest in large-scale infrastructure. Some of these investments are dedicated to the project, but in some cases the investments in roads, power, and possibly education and health facilities could also benefit the broader community and other investors. Sharing this infrastructure would have the additional benefits of catalyzing new investments and economic activity. One of the best practice recommendations in the mining literature, which could apply to large-scale commercial agriculture projects, is to reduce the risk of enclaves by fostering workable public–private agreements to broadly share the use of this infrastructure (Stanley and Mikhaylova, 2011).

Most of the African countries' programs to increase commercial agriculture will not rely on mega-farm arrangements, however. There is an increasing trend for governments to encourage diverse institutional arrangements that combine the benefits from larger scale farming with smallholder farming. This tendency is observed in a number of more recent spatial and product development strategies in Tanzania, Ghana, and Senegal where governments are taking the lead in coordinating, with strong private sector participation, the promotion of commercial agriculture. These examples are discussed in the following section. In Ghana, for example, nucleus farms are increasingly viewed as attractive options for the future, largely because they do not alter existing patterns of land rights (Throup et al., 2014).

Ensuring a rigorous process of selecting investors and screening investment plans is important and consistent with a variety of organizational forms. Governments that intend to encourage "nucleus" farms, contract farming, or other outgrower arrangements may include scrutiny of relevant track records in such operations as an added criterion for prequalifying investors. Governments' infrastructure investments will also need to reflect what will be needed to make outgrower arrangements work. Greater interaction between the investors and local communities or representatives of holders of land rights may also be an important part of the contracting process to define the scope of land deals and to verify the viability of the business proposals. In this context, more will need to be done to build the capacity of communities and smallholder farms to interact and negotiate with investors.

Finally, there are a variety of models for public–private coordination of investments in the development of large-scale commercial agriculture. In some models, investors are provided a greater role in coordinating interdependent investments, including in infrastructure. In others, governments play a greater role in catalyzing private investments and in ensuring greater links between larger commercial investors and the local community. Each one presents specific governance challenges, but broadly effective auctions, more explicit agreements on requirements to keep leases, and effective enforcement mechanisms will introduce greater transparency in the selection of investors and incentives to meet the promised development objectives.

10.3.2 INSTITUTIONAL ARRANGEMENTS FOR PUBLIC AND PRIVATE COORDINATION

One of the ways to address the challenge of coordination would be to put in place mechanisms for consultations and cooperation with the private sector to gain a better understanding of obstacles to firm performance and to devise effective solutions to address them. These mechanisms have been associated with the practice of industry policy: Rodrik (2007) portrays industrial policy as a process that is "a more flexible form of strategic collaboration between public and private sectors, designed to elicit information about objectives, distribute responsibilities for solutions, and evaluate outcomes" (ibid, p. 112). Appropriate institutions will need to be built to make this process work. Governing the development of commercial agriculture in Africa could draw on the industrial policy literature, particularly in finding effective mechanisms for more systematic dialogue among governments, private investors, and other stakeholders, recognizing that there is no single model of public−private dialogue that will fit all settings.

There have been initiatives within some African countries to set up mechanisms for improved public and private consultations and coordination. Three of these mechanisms are Presidential Investor Advisory Councils, commodity/industry-level consultations, and spatial development strategies that develop and coordinate investments in commercial agriculture in selected areas. The discussion below does not attempt to capture the full range of emerging practices but discusses a few examples and the elements that lead to more effective communication and collaboration between the public and private sectors.

10.3.3 PRESIDENTIAL INVESTOR ADVISORY COUNCILS

Over the past decade, a number of African countries have introduced Presidential Advisory Councils to promote a structured dialogue between governments at the highest levels and the private sector on investment opportunities and the constraints entrepreneurs face to capture them. They were also intended to be a sounding board for the President to get feedback on design and implementation of policies of concern to the private sector. The councils were also seen as a vehicle for removing obstacles to investments and a source of pressure on government entities to deliver on implementation. They were not focused solely on commercial agriculture but most promoted agriculture as a national development priority. Between 2002 and 2004, Ghana, Tanzania, Senegal, Mali, and Uganda created such councils; in 2010 Ethiopia created a public−private consultative forum to carry out similar functions.

These investor councils drew from the design of "deliberation councils" that a number of fast-growing East Asian countries used to facilitate communication and cooperation between public and private sectors on industrial development (Rodrik, 2007; World Bank, 1993). There is broad agreement on the key elements of success of these coordination processes: a strong commitment from their governments at the highest level, focus on alleviating specific obstacles to achieving competitiveness in key areas, a keen sense of experimentation, and emphasis on feedback (Page, 2013). Japan's deliberation councils were structured forums between a powerful technocratic bureaucracy and private sector and civil society to discuss policy issues and exchange information as well as to obtain feedback on specific policies under consideration. Korea's most prominent communication channels with the private sector until the early 1980s were the monthly export promotion meetings that were chaired by the President of Korea himself. More recently, the development of light manufacturing and the evolution of industrial parks in China show a history of dynamic interaction between private entrepreneurs and public officials at the national and local levels (Dinh, 2013).

The councils today differ in the scope of their reform agenda and the degree to which they take up and act upon reforms (Page, 2013). Sustained commitment of senior policy makers has been important to sustaining the councils' momentum for reforms. Uganda's Presidential Investors Roundtable has met twice a year with the President and Cabinet Ministers. Its private sector membership has expanded over the years. The roundtable has taken up a large reform agenda that includes endorsing broad-based development initiatives and sector-specific reforms in areas such as agribusiness, information technology, and business process outsourcing. They have been credited with eliminating export taxes on agribusiness, allowing duty free imports of packing materials for agribusiness products, and supporting the infrastructure for internet connectivity and improved energy services. The Tanzania National Business Council (TNBC) has also been a sounding board for national development strategies such as *Kilimo Kwanzaa*, its strategy to focus on agriculture. In agriculture, it has endorsed specific recommendations to reform land tenure and promote the development of an agriculture value chain, although many of these reforms have been delayed. Senegal's *Conseil Presidential d'l'Investissement* (CPI) meets once a year with the President, who also chairs an interministerial council to monitor and guide the implementation of CPI's decisions. The CPI is well regarded by the business community. It has been associated mainly with reforms on broad-based regulations, such as the reform of Senegal's Investment Code, and administrative procedures governing business.

When viewed through the lens of the East Asian experience of public–private coordination, the councils' performances differ in a number of ways (Page, 2013). They brought to bear private sector development policy experience on policy reforms, and both government entities and the private sector gained valuable experience by working together on policy issues.[13] In most cases, however, follow-up actions on the recommendations of the councils were found wanting. The agenda of the councils has also largely tackled broader investment climate reforms in contrast to East Asia's strong focus on coordination at the industry-, sector-, and firm-specific constraints. The councils' agendas "fall more in the rubric of public–private dialog than coordination" (Page, 2013, p. 21). In part because of the broad agenda and also because of the absence of systematic monitoring and evaluation, the councils received limited feedback on the impact of the reforms—what worked and what did not—on investor performance, growth in investments, and employment.

10.3.4 INDUSTRY-LEVEL PUBLIC–PRIVATE DIALOGUE AND COORDINATION

Beyond the Presidential Advisory Councils, other forums for public–private coordination have emerged at the commodity- or industry-specific level. These often have involved producer or professional associations, which have been created to enhance the growth and prosperity of a particular agribusiness activity (Lamb, 2004). These associations have contributed to public and private dialogue and coordination in the agribusiness area by expressing the views of those in the industry, undertaking collective action, and enabling the flow of communication between members and government.

The development of Ethiopia's cut-flower industry provides one such illustration. It emerged in the late 1990s, grew very rapidly after 2003, and is now the second-largest source of flower exports

---

[13]See also World Bank (2005).

in Africa. This growth has been attributed largely to the joint efforts of a private association of cut-flower producers and the government of Ethiopia that pushed a 5-year plan to promote the industry in late 2002 (Gebreeyesus and Iizuka, 2010). The association has taken collective steps to develop the industry, such as through training and certification. It also raised awareness in government about the obstacles exporters face, notably the need for reliable and affordable air cargo, access to land, and availability of financing. The government has since made available public land near the airport at Addis Ababa and has played an important role in resolving the air transport problem by initiating discussion between the flower producers and Ethiopian airlines. The government's strong support for this public–private coordination was evident from the frequent interactions between producers and the Prime Minister and from the follow-up steps to remove obstacles to exporting that had been identified during the process.

Over the past few decades, some countries in Africa have worked closely with private sector associations to support sector-specific strategies to develop their agricultural exports with strong participation of the private sector. To reinvigorate the growth of its rubber exports, for example, Cote D'Ivoire continues to draw on the strengths of its well-established interprofession[14] organization in the rubber industry (World Bank, 2012a,b). The rubber interprofession organization coordinates the determination of domestic rubber pricing and extension services provided by the larger plantations, while the government regulates and sets standards for the industry. Because of its positive experience with interprofession groups, Cote D'Ivoire is setting up a similar group in the cocoa industry as part of its national strategy to revitalize its cocoa exports. In Botswana, the association of beef producers catalyzed discussions with the government on policy reforms needed to reverse the sharp decline of beef exports between 1998 and 2004, resulting in the change of the export pricing policy (Weber and Labaste, 2010).

While stakeholder associations have brought successful outcomes for some agroindustries in several countries, there also have been instances where associations have not been effective in delivering on long-run development objectives. While they act for the benefit of members and lobby governments for actions that will help members or against actions that hurt them, they also face risks of capture by a few dominant groups who use the association as vehicles to lobby for subsidy or rents. So long as associations are able to manage this governance risk, they can be effective partners with responsive governments in developing the commercial potential of agribusiness industries.

10.3.5 COORDINATING PUBLIC–PRIVATE INVESTMENTS WITHIN SPATIAL DEVELOPMENT PLANS

More recently, some countries have begun to develop spatial corridors with strong private sector participation as part of their strategies to develop commercial agriculture. Two ongoing initiatives by Tanzania and Senegal highlight common elements. First, these are primarily government-led efforts to coordinate public and private investments in public goods and firm-specific activities in order to lower the risk and improve profitability of investments within a specific corridor. Corridors with high commercialization potential, either because they have the basic infrastructure

---

[14]See Teyssier (2008) for a brief review of the agricultural interprofessional organizations in West Africa. These have been established in part to facilitate interactions and coordination among the government and various stakeholders—producers, processors, and distributors—to ensure the development of a specific agricultural product.

in place or because initial experiments to commercialize targeted products have had positive results, have been selected. Second, within this coordination framework, governments invest heavily in infrastructure but they also leverage private investments in public goods from large investors. Third, these location-specific strategies aim to identify obstacles to investments—often at the level of value chains of specific products—and devise solutions to remove these investment bottlenecks. Fourth, within these corridors, there is an enabling environment for both large and small farms to be commercially viable. Diverse models of commercial agriculture, such as plantations, nucleus farms, and contract farming, are expected to emerge. In this context, land leases are not limited to land concessions made directly with governments but also include voluntary contracting between investors and local communities. Fifth, the location-specific strategies facilitate efforts to build the capacity of public agencies and communities to negotiate and contract with private investors and monitor progress and outcomes of commercial investments.

The investment blueprint of the Government of Tanzania's Southern Growth Corridor (SAGCOT) initiative has been developed jointly with the private sector.[15] The blueprint describes the plan to develop commercially an initial six clusters of potential groupings of farming and agribusiness processing along the southern corridor of Tanzania that links the port of Dar es Salaam to landlocked countries in southeastern Africa. The blueprint describes this corridor as passing through some of the richest farmland in Africa, but agricultural productivity has been low so far. Commercial farming in this corridor has been limited, and there has been little interaction between existing large farms and the smallholder farmers. The SAGCOT corridor is considered to have strong commercial potential, however, in part because of the agroclimatic conditions and because the basic infrastructure encompassing port, rail, road transport, and water are in place, which, with some improvements, would provide strong incentives for the private sector to invest. Developing this corridor represents a pragmatic starting point in Tanzania to boost commercial agriculture.

Within the SAGCOT corridor, it is envisaged that the government and private firms will coordinate their infrastructural investments to bring road access, better logistics, power, and water to farms, as well as processing and storage facilities and research institutions that are located within the close proximity of investors within a cluster. Other policies could include facilities to improve access to finance. Firms benefit from the externalities associated with geographic concentration of suppliers and downstream firms, and shared access to research, finance, and other services. Furthermore, within this corridor, small farms obtain access to affordable infrastructure services that they would not have been able to afford by themselves. Within this corridor, large and small farms can be better integrated by encouraging nucleus farm hubs and contract farming arrangements that link the farms to better effect. Coordinated investments in the supporting infrastructure for agriculture could catalyze a rapid increase in private investments from both small and large farms. The estimated total investments within SAGCOT of more than US$3 billion include public and private investments in infrastructure as well as onfarm private investments that would bring more than 350,000 hectares under commercial production.

---

[15]See SAGCOT Investment Blueprint, January 2011, at www.sagcot.com. The SAGCOT initiative expects to expand the land under cultivation to 350,000 hectares. Now, out of Tanzania's 7.5 million hectares of arable land, <2% is farmed with irrigation. The total area under cultivation is below 2000 hectares, of which 95% is farmed by smallholders (ibid., p. 27).

As part of its national strategy to diversify agricultural exports and increase export revenues, Senegal has sought to develop the agribusiness industry within the St. Louis/Senegal River Region (The World Bank, 2013a). Its good agroclimatic conditions, relatively good logistics to serve regional and international markets, locational advantage to export to Europe, strong investor interest, and positive experience so far in increasing horticulture exports underpinned the choice to develop six "growth pole" areas within the region. Within these growth poles, the government, local authorities, and the private sector could focus their efforts to address the constraints faced along the value chain of the main products to be developed to improve their competitiveness. For example, the two most important binding constraints for investors in the Ngalam Valley are the lack of access to water and land degradation. The left bank of Lac de Guiers, on the other hand, lacks primary road infrastructure and secured access to land for investors seeking larger landholdings. The regional development plan, which the World Bank is supporting, involves public and private investments in the irrigation infrastructure in the Ngalam Valley and around the Lac de Guiers. Public financing will focus on construction of primary and some secondary irrigation canals, the tertiary irrigation of smallholders within the Region's perimeter, and the rehabilitation of access roads. Large-scale investors will finance the secondary and tertiary irrigation canals that directly affect their farms. The government has also put in place a framework for securing land tenure for large investors.

In both initiatives discussed above, governments have also invested in building their capacity, including at the local level, to govern the entry of investors and monitor their performance. SAGCOT's implementing agencies involve government entities, private sector representatives, and development partners. The Senegal project has set up the local implementing agencies in partnership with development partners, such as the World Bank. The good initial knowledge of the potential value of the investment within the selected locations has helped governments to plan their investments so as to reduce obstacles to investments. The investment frameworks within the corridors involve the participation of both large and small farms and a diverse set of agricultural organizational forms. Both projects have introduced systems for screening investors and identifying viable business proposals. The capacity of agencies as well as local communities to negotiate and administer land deals and other business arrangements with investors is also being developed.

The East Asian Miracle study emphasized the importance of the coordination of investments in effecting industrial transformation and for such coordination to include effective instruments of communication between the public and private sectors. The discussion above draws from this experience and highlights some of the initiatives on public and private collaboration in the early stages of commercializing agriculture in some countries in Africa. It is evident that there are different ways to conduct this coordination in different settings; one size does not fit all. We have focused the discussion above on the mechanisms for information sharing and cooperation. There are still, however, unanswered questions on the nature of the subsidies and incentives, if any, and in what form that could induce early investors to invest. Much better understanding of the elements of effective public and private collaboration in the investment in and provision of public goods will be a useful area of future investigation.[16] Successful initiatives could provide large demonstration effects for further similar efforts in other countries in Africa.

---

[16]For example, Donahue and Zeckhauser (2011) discuss cases in the United States where the public sector has harnessed private expertise to achieve public goals.

10.4 COORDINATING WITH COMMUNITIES AND LOCAL STAKEHOLDERS IN GOVERNING COMMERCIAL AGRICULTURE

A key lesson of the global experience on land transfers is that they run the risk of social conflicts when land governance is weak and fails to recognize, protect, and compensate local communities with existing land rights. "Clarifying land rights, titling land, and creating land markets is the biggest agricultural challenge for African governments in the next 40 years and will be critical for realizing the benefits from Africa's currently uncultivated arable land." (noted Ahlers et al., 2014, p. 63). While this is a long-term agenda,[17] more could be done now to strengthen the policy, legal, and institutional frameworks for land governance as countries embark on scaling up commercial agriculture, and Deininger and Byerlee recommend a number of steps to improve land governance over the shorter term. Rights to land and natural resources, including customary rights, need to be recognized and transparent systems for their registration put in place. Transfers of rights should be based on users' voluntary and informed agreement and fair level of proceeds. Local communities will therefore need to be aware of their use rights and the potential value of the land and have the ability to negotiate with investors. They also need access to accurate and up-to-date information on large-scale investments and land transfers, among others, so that they have the means to monitor the performance of commercial agricultural projects. In addition, mechanisms to assess the technical and economic viability of projects, as well as their environmental and social sustainability, will need to be in place.

Given the ambiguity and complexity of land rights, clarifying them for new investors would require participatory and inclusive approaches involving investors and local stakeholders. These would have the dual advantage of effecting buy-in of communities and exploiting the comparative advantages of investors and communities alike. Ghana's experience in encouraging private investors in agriculture show that while "the village chief may be the entry point for negotiating access to land, it is imperative also to negotiate and agree on appropriate compensation directly with clan and family heads", highlighting the ambiguities of control that make land transactions complex (Throup et al., 2014, pp. 162–163). Thus the ambiguity of land tenure cannot be resolved by "fiat." It requires consultations and direct negotiations between rural communities and private sector investors with regard to land and water allocation. Senegal has made local consultations a key part of its agribusiness development strategy. In partnership with the World Bank and other development partners, Senegal is providing technical assistance to rural communities in identifying the land to be made available to private investors, selecting investors, and designing the lease contracts for investors. It is also supporting rural communities in monitoring the investors' implementation of their planned investments and the application of sustainable land and water management practices.

Creating an enabling environment for commercial agriculture may also call for institutional arrangements that mitigate land transfer risks. Throup et al.'s (2014) analysis of Ghana's political economy of commercial agriculture at the local level concludes that for the long term, there is a need to pursue reforms to clarify and formalize property rights in land. In the short term, however, more feasible institutional arrangements should be the strategic focus. Models that complement

---

[17]Byamugisha (2013) provides a comprehensive long-term program to scale up reforms for improving land administration in Africa.

commercial investments by outgrowing arrangements are attractive options especially as they build on existing land-use patterns. Nucleus outgrowing arrangements could offer a solution as they address financial and market access constraints and harness the production capacity of participating small-farm holders. Governments need to play a supportive role by providing communities with information and assistance to ensure a transparent process of negotiations and to build their capacity to work effectively with commercial investors.

It is evident that expanding commercial agriculture in many countries in Africa entails drawing on the strengths of many stakeholders, including governments, investors, and communities. Learning from the experience in Africa and elsewhere, drawing on the knowledge of local stakeholders and obtaining their agreement as holders of land rights and potential beneficiaries is an important element to creating an enabling environment for commercial agriculture. In this regard, greater efforts to understand the political and social environment would help to better inform the efforts to increase private investments in agriculture.

10.5 GOVERNANCE BEYOND GOVERNMENTS

An added dimension in the governance of the use of natural resources is the emergence of global nongovernmental entities concerned with setting standards and regulations around corruption, environmental concerns, conflict, corporate responsibility, and other areas of global concern. These bottom-up mechanisms, beyond governments, that represent the diverse interests of global stakeholders, such as investors, consumers, and civil society, have led to collective pursuits to influence investor behavior.[18] These have complemented efforts of countries hosting global investors to regulate investor behavior (Moore, 2013). While these mechanisms have been largely applied to mining and extractive industries, they provide useful guidance to African governments embarking on building local capacity to manage large investments in commercial agriculture. International development agencies are also global stakeholders and partners that can support countries in designing and implementing their strategies to further commercial agriculture.

10.5.1 VOLUNTARY INDUSTRY STANDARDS

When the behavior of an investor within an industry exerts negative spillovers, producer organizations might impose performance standards and sanctions on its members. This is the case with "industry roundtables" that have emerged, motivated largely by the desire to sell in major markets that exert pressure for sustainable practices in areas such as forest products, palm oil production, and soybean production. These roundtables demand that member investors are certified according to a set of standards and principles. The Roundtable for Sustainable Palm Oil (RSPO), for example, was formed in 2004 by commercial participants and the World Wildlife Fund (WWF) to improve the environmental sustainability of palm oil operations globally through the development and application of global standards and a code of social and environmental practices within the production process.

---

[18]Deininger and Byerlee call for various stakeholders—government, investors, civil society, and international organizations—to have a shared responsibility in managing the risks associated with the rising global investor interest in farmland.

The roundtable brings together stakeholders from seven sectors of the palm oil industry: producers, processors or traders, consumer goods manufacturers, retailers, banks and investors, and nongovernmental organizations concerned with nature conservation and social development. Through engagement with these stakeholders, the RSPO developed its global standard and code of practice (called Principles and Criteria) that are implemented through an (auditable) certification system.

Will private voluntary standards influence and lead to sustainable and responsible business practices? The RSPO has raised awareness among governments to environmental and social issues around palm oil production and the nature and scope of regulatory interventions. The impact of these standards will depend partly on demand and the willingness of consumers to pay for the certification premium. While the Forest Steward Council's (FSC) certification is demanded by almost all retail markets for wood and associated products, market demand for sustainable palm oil is still limited, however, including in the major markets of India and China. By 2011, about 7.5% (about 3.4 million tons) of global palm oil supply was RSPO certified (World Bank, 2011). Nevertheless, almost all major international trading companies are expected to require all of their palm oil suppliers to be RSPO certified by 2016, which will broaden the reach of these standards.

The effectiveness of these industry standards will also depend on what governments do to implement them and the local context. The implementation of standards by the RSPO on palm oil producers investing in Liberia, Sierra Leone, and Gabon could avoid the mistakes of previous expansions (Byerlee, 2013). De Man (2010) notes that global standards still need to be strengthened on matters related to land issues in agricultural investments, such as in recognition of informal land rights in the conversion of traditional farmland to large-scale commercial undertakings, which are relevant in the African context. Similarly, extending certification to smallholders may require additional capacity building as they will not have the same capacity as large investors to meet the requirements for certification.

10.5.2 CIVIL SOCIETY ORGANIZATIONS AND STANDARDS FOR TRANSPARENCY AND GOOD GOVERNANCE

There have been calls to draw on the experience of the Extractive Industry Transparency Initiative (EITI) to improve the governance of large-scale land deals related to agriculture in Africa. The EITI framework involves principles related to transparency and government accountability, and it works with countries, private sector enterprises, and civil society organizations that would like to adopt and implement these principles. Participating countries[19] require companies engaged in extractives to report the payments they make to the government and to reveal the amounts they receive from these companies, thus making this information transparent and available for monitoring by citizens. A key strength of EITI is its emphasis on country driven processes for implementation, so that countries committed to this process put in place the needed legislation, institutions, and a process for multistakeholder consultations. A more recent global multistakeholder initiative, the Natural Resources Charter (NRC), provides a broader framework for governments to manage their oil, gas, and natural resources. Its intention is for governments to use the charter voluntarily as a tool for self-assessment of their capacity to conduct the chain of decisions needed to manage

---

[19]Forty-one governments have committed to implementing the EITI framework, and more than 80 companies involved in oil, gas, and mining support the EITI (see www.eiti.org).

natural resources, from the discovery of the resource to investing for development (Cust, 2013). EITI and NRC have been used so far only for extractive industries, but the principles contained in them could be usefully applied to large-scale investments in agriculture.

Yet another initiative arising from civil society's collective efforts, particularly the Publish What You Pay (PWYP) coalition, is the provision[20] in the US Dodd-Frank Act of 2010 to introduce mandatory disclosure requirements for extractive companies listed in the United States' stock exchanges. These listed extractive companies are required to provide details of their payments, including taxes, royalties, fees, bonuses, etc., to US and foreign governments. This requirement, especially if adopted by other financial centers, provides an additional source of monitoring by the financial sector. In order to tap into this additional requirement for transparency from the mainstream financial sector, host governments, including those in Africa, would need to put in place complementary domestic mechanisms for public accountability, including disclosure rules and their enforcement, stronger regulatory capacity, and transparent reporting to the public that covers extractives and agricultural firms.

10.5.3 INTERNATIONAL DEVELOPMENT INSTITUTIONS

What might be the role of international lending and development agencies? A key role would be to continue their support of policies and institutions that would enable countries to achieve and sustain productivity enhancing investments in agriculture. This would include maintaining strong support for the governments' long-term efforts to put in place effective land governance systems and to build capacity of their institutions to implement them. The World Bank, for example, has had a longstanding engagement with many countries in Africa to address issues of land tenure and governance. In partnership with other donors, it has also been supporting countries, such as Cote D' Ivoire, Senegal, Ghana, and Tanzania, which are cited in this chapter, in creating an enabling environment for commercial investments in agriculture that also benefits local communities and small farms.

The World Bank and other multilateral lending institutions provide a neutral, development-oriented perspective to public—private initiatives and have the expertise for working on public sector management. In supporting countries in scaling up commercial agriculture, the World Bank has supported efforts to clarify responsibilities of oversight institutions within governments, scale up infrastructure investments, provide technical support to inform public—private dialogue, oversee the preparation of environmental risk assessments and analysis of obstacles to competitiveness, and manage studies on impact evaluation. More recently, demand has increased for their assistance to facilitate public—private dialogue and build governments' and communities' abilities to frame and negotiate contracts on large-scale commercial agriculture. Many of the emerging practices in these areas cited in this chapter benefited from the support of the World Bank and other donors.

Development partners could also facilitate global knowledge sharing and/or be the source of knowledge about best practices. The World Bank, IFAD, FAO, and IFPRI, among others, have been rich sources of research in, and policy work on, land governance and the development of the agricultural sector more broadly. Financial and advisory support for the provision of other public goods, such as research, would also be important areas of emphasis.

At the global level, development partners can be effective catalysts for building greater awareness of the objectives and principles of global standards that could support countries in ensuring

---

[20]North-South Institute (2013) provides a brief policy description of this provision.

good governance and sustainable use of resources. Development partners are among EITI's partner organizations in supporting capacity building at the country level to implement the EITI framework. The International Finance Corporation (IFC) led the effort for banks to use the Equator Principles for Environmental Sustainability in assessing their transactions with borrowing companies. The IFC is also an active participant in RSPO (World Bank, 2011). The World Bank, in cooperation with others, supports country level implementation of the FSC principles.

Following the surge of investor interest in large-scale agriculture in 2008, FAO, IFAD, UNCTAD, and the World Bank Group collaborated to develop The Principles for Responsible Agricultural Investments (PRAI), which we discussed earlier. The PRAI complement frameworks for responsible investing involving land that were also developed by FAO, IFPRI, and Germany's BMZ. According to Deininger and Byerlee (2011, p. xxvi) "the principles have already served a useful purpose in reminding countries and investors of their responsibilities and in drawing attention to situations where they were not adhered to." The World Bank and other donors have helped embed these principles in the selection of investors and in designing approaches to commercialization of agriculture in developing countries. There is emerging interest among private financiers: Campanele (2013) notes that African investment funds for agriculture have also expressed adherence to these principles for responsible investment.

10.6 CONCLUSION

Increased investor interest in commercial agriculture presents opportunities as well as challenges for many African countries. Much of the literature on Africa's agricultural development has revolved around the small versus large farm debate and on policies and institutions that would improve the productivity of small farms. We argue in this chapter that, if one views farm size as endogenous, an efficient and productive agriculture sector should consist of small and large farms, and both should contribute their respective comparative advantages to the sector's development. From this perspective, the field of agricultural development will need to look at agriculture differently, as consisting not only of small family farms but as a set of small and large businesses that are linked to domestic and global markets and are potential sources of new growth.[21] Increasing their productivity requires an appropriate enabling environment for investment. Policy makers can draw on lessons from industrial policy and development to address the question of how governments and private investors might work to overcome the bottlenecks in investments in commercial agriculture in order to get the best out of them in terms of economic development and food security.

Coordinating interdependent investments, especially because of the large infrastructure needs, is important in the early phases of commercializing agriculture in Africa. Reviews of the early experience of large land deals in Africa show that governments sought from the private sector a large role in performing this coordinating role, including investing in infrastructure. This chapter has emphasized, therefore, the importance of addressing associated governance issues by designing auction mechanisms to improve the process of selecting investors and to enforce the contractual terms as requirements for keeping leases over the long term. Building institutions for public—private

---

[21]See also Weber and Labaste (2010).

coordination and sharing of information to focus efforts on identifying bottlenecks to investments and devising solutions to address them is an important policy and institutional challenge of building the enabling environment for private sector investments. There are emerging practices of greater public—private coordination in Africa. The Presidential Advisory Councils in some countries have provided the opportunity to improve public—private dialogue especially on broader policy issues. Some countries are also implementing spatial development strategies and sector level approaches that are more focused efforts to improve the functioning of value chains in selected agricultural products as means to spur investments in commercial agriculture and meet desirable development goals. This shift in agricultural development strategies reflects greater coordination roles for governments and offers more opportunities to integrate large-scale investments with developing opportunities for small farms.

Improving land governance is crucial to agricultural development in Africa. A number of studies have provided very useful guidance for pursuing this long-term agenda. Meanwhile governments need to manage the social risks associated with the complexity of defining land-use rights and irresponsible business practices with the increase of private sector investments in large-scale agriculture. Engaging local communities in more participatory and bottom-up approaches in defining land rights, setting contractual terms, and monitoring the implementation of investment plans has the double dividend of improving their buy into private investments and of capturing their comparative advantage in their knowledge of the local context and ability to monitor.

Global governance mechanisms that have been developed by nongovernmental stakeholders could also support countries' efforts at governing investments in commercial agriculture. Voluntary industry standards and certification could offer additional sources of pressure for responsible business practices. Governments can draw on partnerships with global initiatives to improve disclosure and transparency of companies' payments and activities, thus allowing more effective public monitoring. International development institutions, too, have an important role to play by continuing to support policies and institutions to improve land governance and to develop the agricultural sector more broadly. They are important partners in governments' efforts to build their capacity to coordinate public and private investments to ensure that commercial agriculture leads to economic growth, reduced poverty, and improved food security.

Finally, there is scope for further research to increase the understanding of a number of policy issues. More can be done to understand how governments might harness effectively the expertise and financing of private sector entities in meeting the objectives of public goods. What elements determine effective public—private collaboration? What subsidies could catalyze investments and value discovery among early movers, and why? Additional research could also provide more guidance on policies for commercial agriculture that could lead to greater equity. There is also ample scope to study the political economy of reforms in agricultural organizations to determine the constraints many organizations and countries face in adapting to market developments.

REFERENCES

Ahlers, T., Kato, H., Kohli, H., Madavo, C., Sood, A. (Eds.), 2014. Africa 2050: Realizing the Continent's Full Potential. Oxford University Press, Oxford and New York, NY.

Ayalew, D.A., Deininger, K., 2014. Is there a Farm-Size Productivity Relationship in African Agriculture? Evidence from Rwanda. World Bank Policy Research Working Paper 6770, January.

Bell, D.E., Shelman, M.L., 2008. Olam International. Harvard Business School Case 509-002, December 2008. (Revised October 2009.) http://www.hbs.edu/faculty/Pages/item.aspx?num=36686.

Binswanger, H., 2009. Awakening Africa's Sleeping Giant: Prospects for Commercial Agriculture in the Guinea Savannah Zone and Beyond. World Bank, Washington, DC.

Binswanger-Mkhize, H., McCalla, A., Patel, P., 2010. Structural transformation and African agriculture. Global J. Emerg. Market Econ. 2 (2), 113–152.

Byamugisha, F., 2013. Securing Africa's Land for Shared Prosperity: A Program to Scale Up Reforms and Investments. Africa Development Forum Series, World Bank, Washington, DC.

Byerlee, D., 2013. Are we learning from history? In: Kugelman, M., Levenstein, S. (Eds.), The Global Farms Race. Island Press, Washington, DC, pp. 21–44.

Campanele, M., 2013. Private investment in agriculture. In: Alan, T., Keulertz, M., Sojamo, S., Warner, J. (Eds.), Handbook of Land and Water Grabs in Africa. Routledge, Abingdon, United Kingdom.

Collier, P., 2010. The Plundered Planet: Why We Must — and How We Can — Manage Nature for Global Prosperity. Oxford University Press, Oxford and New York, NY.

Collier, P., Venables, A.J., 2012. Land deals in Africa: pioneers and speculators. J. Global. Dev. 3 (1), 1–22, June.

Cotula, L., Vermeulen, S., Leonard, R., Keeley, J., 2009. Land Grab or Development Opportunity? Agricultural Investment and International Land Deals in Africa. IIED/FAO/IFAD, London/Rome.

Cust, J., 2013. The Natural Resources Charter in Africa: A Tool for National Strategy and Evaluation. Presentation at the North-South Institute, Governing Natural Resources for Africa's Development, Ottawa, Canada.

De Man, R., 2010. Land Issues in Voluntary Standards for Investments in Agriculture. Paper submitted to The World Bank Annual Conference on Land Policy and Administration, Washington, DC.

Deininger, K., Byerlee, D., 2011. Rising Global Interest in Farmland: Can it Yield Sustainable and Equitable Benefits? World Bank, Washington, DC.

Dinh, H., 2013. Tales from the Development Frontier: How China and Other Countries Harness Light Manufacturing to Create Jobs and Prosperity. World Bank, Washington, DC.

Donahue, J.D., Zeckhauser, R.J., 2011. Collaborative Governance: Private Roles for Public Goals in Turbulent Times. Princeton Collaborative Press, Oxford and Princeton, NJ.

Eastwood, R., Lipton, M., Newell, A., 2010. Farm size. In: Pingali, P.L., Evenson, R.E. (Eds.), Handbook of Agricultural Economics. Academic Press, Burlington, MA, pp. 3323–3397.

The Economist, 2011. The Surge in Land Deals. March.

FAO, IFAD, UNCTAD, and the World Bank Group, 2010. Principles for Responsible Agricultural Investment that Respects Rights, Livelihood, and Resources, a discussion note to contribute to an ongoing global dialogue, Rome.

Fischer, G., Shah, M., 2010. Farmland Investments and Food Security: Statistical Annex. Report prepared for the World Bank and the International Institute for Applied System Analysis, Luxembourg.

Gebreeyesus, M., Iizuka, M., 2010. Discovery of the Flower Industry in Ethiopia: Experimentation and Coordination. Working Paper #2010-025, Maastricht, UNU-MERIT.

Haggblade, S., Hazell, P.B.R., 2010. Successes in African Agriculture, Lessons for the Future, published for International Food Policy and Research Institute (IFPRI). Johns Hopkins University Press, Baltimore, MD.

Hazell, P., Roel, A., 1983. Rural Growth Linkages: Household Expenditure Patterns in Malaysia and Nigeria. IFPRI Research Report 41, International Food Policy Research Institute, Washington, DC.

Hazell, P., Poulton, C., Wiggins, S., Dorward, A., 2007. The Future of Small Farms for Poverty Reduction and Growth. Discussion Paper 42, International Food Policy Research Institute, Washington, DC.

Hernandez, M., 2010. Establishing a Framework for Transferring Public Land: Peru's Experience. Paper presented at the Annual Bank Conference on Land Policy and Administration, World Bank, Washington, DC.

Kaldor, D., 1967. Policy for Commercial Agriculture, Increasing Understanding of Public Problems and Policies. Farm Foundation. <http://ageconsearch.umn.edu/bitstream/17580/1/ar670059.pdf>.

Kaplinksy, R., Morris, M., 2002. A Handbook for Value Chain Research. IDRC. <http://www.value-chains.org/dyn/bds/bds2search.details2?p_phase_id=395&p_lang=en&p_phase_type_id=1>.

Lamb, J., 2004. Establishing and Strengthening Farmer, Commodity, and Inter-Professional Associations. Agriculture and Rural Development, World Bank, Washington, DC.

Larson, D.F., Otsuka, K., Matsumoto, T., Kilic, T., 2012. Should African Rural Development Strategies Depend on Smallholder Farms? An Exploration of the Inverse Productivity Hypothesis. Policy Research Working Paper Series 6190, World Bank, Washington, DC.

McKinsey Global Institute, 2010. Lions on the Move: The Progress and Potential of African Economies. McKinsey Global Institute.

Moore, B., 2013. The Continental Land Rush: Canada's Role in Advancing the Land Rights of Africa. Presentation to the North-South Institute, Governing Natural Resources for Africa's Development, Ottawa, Canada.

Nagayets, O., 2005. Small Farms: Current Status and Key Trends. Information brief prepared for the Future of Small Farms Research Workshop, Wye College, Ashford, United Kingdom.

North-South Institute, 2013. Policy Brief on Dodd-Frank 1504 and Extractive Sector Governance in Africa.

Otsuka, K., Larson, D. (Eds.), 2013. An African Revolution: Finding Ways to Boost Productivity on Small Farms. Springer, London and New York, NY.

Page, J., 2013. Industrial Policy in Practice: Africa's Presidential Investors' Advisory Councils. Paper Presented at the UNU-Wider Conference "Learning to Compete: Industrial Development and Policy in Africa", Helsinki, Finland.

Reardon, T., Timmer, C.T., 2007. The supermarket revolution with Asian characteristics. In: Balisacan, A., Fuwa, N. (Eds.), Reasserting the Rural Development Agenda: Lessons Learned and Emerging Challenges in Asia. SEARCA/ISEAS, Los Banos, Philippines/Singapore, pp. 370–391.

Rodrik, D., 2007. One Economics Many Recipes. Princeton University Press, Oxford and Princeton, NJ.

Roumasset, J., 2010. Wither the Economics of Agricultural Development? Working Paper, University of Hawaii Economic Research Center, University of Hawaii, Manoa.

Roumasset, J., Uy, M., 1986. Agency Costs and the Agricultural Firm. Yale Economic Growth Center Discussion Paper 501, May.

Roumasset, J., Barr, S. (Eds.), 1992. The Economies of Cooperation: East Asian Development and the Case for Pro-Market Intervention. Westview Press, Inc., Oxford and Boulder, CO.

Stanley, M., Mikhaylova, E., 2011. Mineral Resource Tenders and Mining Infrastructure Projects Guiding Principles, Extractive Industries for Development. Series #22, Oil, Gas and Mining Policy Unit Working Paper, World Bank, Washington, DC, September.

Teyssier, J., 2008. Agricultural Interprofessional Organizations in West Africa. Farming Dynamics, Issue No. 17, June.

Throup, D.W., Jackson, C., Bain, K., Ort, R., 2014. Developing commercial agriculture in Ghana. In: Fritz, V., Levey, B., Ort, R. (Eds.), Problem-Driven Political Economy Analysis: Directions in Development (Public Sector Governance). World Bank, Washington, DC, pp. 145–174.

Tyler, G., Dixie, G., 2013. Investments in Agribusiness: A Retrospective View of a Development Bank's Investments in Agribusiness in Africa and East Asia. World Bank, Washington, DC.

Uy, M.J., 1979. Contractual Choice and Internal Organization in Philippine Sugarcane Production (M.A. thesis). University of Philippines, Diliman, Quezon City.

Weber, C.M., Labaste, P., 2010. Building Competitiveness in Africa's Agriculture: A Guide to Value Chain Concepts and Applications. World Bank, Washington, DC.

World Bank, 1993. The East Asian Miracle: Economic Growth and Public Policy. Oxford University Press, New York, NY.

World Bank, 2005. Presidential Investors' Advisory Councils in Africa: Impact Assessment Study, Washington, DC, May.

World Bank, 2008. Agriculture for Development, World Development Report, Washington, DC.

World Bank, 2011. The World Bank Group Framework and IFC Strategy for Engagement in the Palm Oil Sector, Washington, DC, March.

World Bank, 2012a. Cote D'Ivoire: The Growth Agenda: Building on Natural Resources and Exports. Report No. 62572-CI, Washington, DC, March.

World Bank, 2012b. Africa Can Help Feed Africa: Removing Barriers to Regional Trade in Food Staples, Washington, DC.

World Bank, 2013a. Project Appraisal Document of the Sustainable and Inclusive Agribusiness Development Project for the Republic of Senegal, Washington, DC, November.

World Bank, 2013b. Growing Africa: Unlocking the Potential of Agribusiness, Washington, DC, January.

LAND CONFISCATIONS AND LAND REFORM IN NATURAL-ORDER STATES

11

Sumner La Croix

Department of Economics, University of Hawai'i at Manoa, Honolulu, HI

CHAPTER OUTLINE

JEL: Q15; N41; N43; N51; N53

11.1 INTRODUCTION

Throughout his illustrious career at the University of Hawai'i, Jim Roumasset has emphasized to multiple generations of graduate students the possibility of identifying and implementing Pareto-improving policy changes for a wide range of important national policies. Of course, not all policies implemented by national governments are Pareto-improving. Land reform stands out as a leading example, as landowners have rarely been fairly compensated by land reform programs.

Because of this, social scientists' discussions of land reform have generally focused on whether the reform benefits the relatively poor recipients of the confiscated land or contributed to the country's economic growth instead of on whether a land reform is Pareto-improving. For example, Lipton (2009) in his encyclopedic review of modern land reform programs concludes that "at least 1.5 billion people today have some farmland as a result of land reform, and are less poor, or not poor, as a result" (p. 8). Virtually all economic historians and development economists studying the rise of the economies of South Korea, Taiwan, and Japan after World War II have listed land reform as a critical ingredient in each economy's success. Land reform may also have been necessary for the political success of post-war governments, as it helped them fend off potential challenges from communist or socialist parties with land reform at the center of their agendas.

One less emphasized feature of the East Asian land reforms is that landlords throughout the region resisted reform, in part because they anticipated that they would ultimately receive little compensation (Otsuka, 2012). In fact, all land reform programs carried out in China, Taiwan, South Korea, North Vietnam, and Japan in the decade following the end of World War II involved massive land confiscations, with the vast majority of owners receiving little or no compensation. Land reform programs in both Taiwan and Korea initially confiscated agricultural lands controlled by Japanese colonizers and corporations and then transferred property rights to tenant farmers. In Taiwan, the land reform program redistributed agricultural lands from Taiwanese and Chinese landlords to tenant farmers. The government paid compensation to landlords by tapping proceeds from the government's sale of confiscated Japanese corporate assets. The 1949 communist victory in China's civil war led to the largest land reform in modern history in the early 1950s. The new national government confiscated all agricultural lands from landowners in China, paid no compensation to them or their families, and executed between 1 and 3 million landowners. Land reform in North Vietnam proceeded in a similar fashion. In Japan, the Supreme Command of the Allied Powers initiated a land reform program in 1947 in which the national government seized agricultural lands from both resident and absentee landlords and resold them to tenant farmers. Compensation was determined by capitalizing the annual rents paid in 1938, payable to the landlord with 30-year fixed-rate government bonds. Not only was the ex ante compensation specified by the law inadequate (given the substantial inflation in Japan that had occurred since 1938), but the ex post compensation was even lower given the unexpected high inflation that prevailed after the land reform measure was enacted.[1]

From the perspective of early modern and modern global history, the post-war land confiscations in East Asia were far from extraordinary. Confiscation of land and other assets from the losers of international and civil wars has long been the historic norm; winners typically redistributed land to build a new governing coalition that reflected the new distribution of power in the society at the end of the war. Analysis of these types of redistributions is facilitated by the distinction made by North et al. (2009) between a limited-access order ("the natural state") and an open-access order. Limited-access orders are the norm in history and exist to solve the problem of violence in society. They do so "by forming a dominant coalition that limits access to valuable resources ... to elite groups. The creation of rents through limiting access provides the glue that holds the coalition

---

[1]See Dorner and Thiesenhusen (1990) for an overview of land reform in East and Southeast Asia; North et al. (2012) for an analysis of post-World War II land reform programs in South Korea and the Philippines; and Ramseyer (2012) for new perspectives on post-World War II land reform in Japan.

together, enabling elite groups to make credible commitments to one another to support the regime, perform their functions, and refrain from violence" (p. 30). The distribution of rents is heavily influenced by groups' and individuals' violence potential and by established networks of unique personal, family, and group relationships. A change in a country's ruler often necessitates that additional rents be created for or transferred to the ruler's supporters. In preindustrial revolution societies, this was usually accomplished by transferring property rights to valuable lands.

The transition to an open-access order occurs when the personal privileges of elites are transformed into impersonal rights (p. 27). In natural states, formation of economic, social, or political organizations is restricted to elites. In an open-access order, any group may form an organization. "The ability to form organizations at will without the consent of the state ensures nonviolent competition in the polity, economy, and indeed in every area of society with open access" (p. 22). The hallmark of an open-access society is its capability of "sustaining impersonal relationships on a large scale through their ability to support impersonal perpetually lived organizations, both inside the state and in the wider society" (p. 23).

The contrast between a limited-access order and an open-access order is most clearly seen during political transitions due to the death of a ruler or a change in the core of support for a ruler. In an open-access order, the rulers are perceived to be temporary and the governance institutions are perceived as perpetual. For instance, the transition to a new US president takes place within the perpetual institutions arising from the US Constitution. In a limited-access order, a transition to a new president or prime minister immediately raises questions on which aspects of the system of rights and privileges supporting the current order will survive and which aspects will change to reflect changes in the sources of support for the new ruler and changes in the underlying distribution of violence potential.

A critical question emerges regarding the effects of confiscations of land in a natural state: How can a system of de jure property rights in land be sustained after widespread confiscations? As a general rule, state confiscation of private property should lead remaining property owners to reason that if the state has seized property from one group of owners, then their properties could be next. But in the natural state, a government's confiscation of some property, if properly executed, will increase the likelihood that other property owners' de jure rights will be maintained and enforced in the future. This case occurs when the government of a natural-order state redistributes confiscated lands in a manner calculated to bolster the strength of its governing coalition. When done following a process perceived to be legitimate, such redistributions signal to all other property owners that the government has established sufficient order to enforce their de jure rights for a longer period of time. This chapter examines this proposition, focusing on government confiscations of land in early modern Europe, Britain's rebellious North American colonies, and Hawai'i.

The early modern period also saw the emergence of limitations on executive power—defined by Acemoglu et al. (2001) as binding constraints on the ability of the executive to confiscate assets from the public. Such limitations are at the core of the political and economic competition that characterizes an open-access state; they include limitations on land confiscation. The chapter concludes with a brief discussion of two important cases—Britain immediately after the Glorious Revolution and France after the restoration of the monarchy—in which a cycle of land confiscations was broken, thereby signaling an emergence of new limitations on executive power.

11.2 CONFISCATIONS IN EARLY MODERN EUROPE AND ITS OFFSHOOTS

Established monarchs, fragile monarchs, and revolutionary parliaments confiscated lands during the early modern period (1500−1800) in Europe and North America (Reynolds, 2010). Revolutionary parliaments are likely to be particularly interested in redistributing rents and property, as they usually face immediate problems with reestablishing order. Most are interested not just in bolstering the strength of their new coalition but also in reducing the strength of supporters of the just overthrown government who, given the chance, could be waiting on the sidelines to violently overthrow the new regime. The interaction between the 1789−1793 revolutionary governments in France and royalist émigrés organizing in neighboring states to restore a royal government provides a clear example (White, 1995). Monarchs consolidating power tended to confiscate church lands whereas revolutionary parliaments in Northern Europe and North America confiscated lands from a wider variety of owners—the Catholic Church, the crown, and royalist supporters.

11.2.1 CONFISCATIONS OF CHURCH LANDS BY ESTABLISHED GOVERNMENTS

Church properties were confiscated by rulers throughout Northern Europe during the Reformation as cities and states acted to reduce the power of the Catholic Church, by then both an ecclesiastical and political opponent.[2] Confiscated properties were transferred to the ascendant Protestant sect in their jurisdiction. Rulers confiscated Jesuit properties in several Catholic countries after members became entangled in the losing side of political intrigues. Governments in countries with Catholicism as the state religion confiscated church and monastic properties when it became in their interest to weaken the Vatican's influence within their countries. Consider now a few specific examples of each type of confiscation.

11.2.1.1 Henry VIII's monastic confiscations

Henry VIII's confiscation of monastic properties (1536−1540) was triggered by his 1534 confrontation with the Pope over his divorce. However, the confiscations should be viewed more properly as a fundamental component of his drive to consolidate his government's power. The likelihood of opposition by members of contemplative orders to his policies of church reorganization was clearly a factor behind the confiscations. On the other hand, the attack on the special corporate privileges of the church was far more expansive than could possibly be explained by potential opposition from the orders' members. Henry's commission of Richard Cromwell to compile the *Valor Ecclesiasticus*, a report containing an exhaustive inventory of church assets, not only served to establish their value and identify which assets might be sold, but also the full scope of the corporate body that Henry was reconstituting.[3] The abolition of the papal tax, the imposition of a new direct tax to support the clergy, and the sale of the confiscated monastic properties served not just to sever the Church's corporate bodies from the Vatican, but also to fundamentally reintegrate them within the corpus of the several British crowns.

---

[2]Some early confiscations in England during the fifteenth century may be traced back to the Black Death and the subsequent sharp decline in the number of people residing in some monastic houses.

[3]Their sale also enabled Henry VIII to bypass Parliament's consent for additional tax revenues to fund his wars. See also Habakkuk (1958).

11.2.1.2 Joseph II's monastic confiscations

While some state-led reorganizations of Catholic corporate bodies were not founded on severing their linkages with the Vatican, they were ultimately just as far-reaching as the British reorganization. Upon assuming the Hapsburg throne in 1780, Emperor Joseph II began a comprehensive program to reorganize the operations of the Roman Catholic Church and its properties, both in central Austria and Hungary.[4] The emperor confiscated and resold the majority of Austria's monasteries during the 1780s and reorganized church corporate bodies to weaken their linkage with the Vatican and tie them more closely to the Austrian state. While revenues from the sale of the monasteries contributed substantially to the Austrian fisc, the restructuring of the church also had roots in the consolidation of military power around the monarch during both Marie-Theresa's and Joseph's reigns as well as in the constant strategic interaction between cities and regions in Italy controlled by Austria and the Vatican. Like Henry VIII's confiscations, Joseph II's takings were preceded by an exhaustive inventory of church personnel and assets that not only facilitated confiscations but also identified the full scope of the corporate body being reconstituted.

11.2.2 CONFISCATION OF LANDS BY REVOLUTIONARY PARLIAMENTS

van Zanden et al. (2012) charted the rise of parliamentary governments in Northern Europe and their fall in Southern Europe during the seventeenth and eighteenth centuries. Some economists view the establishment of parliamentary representation in Northwest Europe as arising from the demands of newly enriched Atlantic Coast merchants for limitations on executive power (Acemoglu et al., 2005). In numerous instances, however, revolutionary parliaments were noted for their seizure of crown, royalist, and church lands. The most prominent examples are England in 1649–1651, France in 1789–1793, and Britain's rebellious North American colonies in 1775–1786.

11.2.2.1 The interregnum confiscations in Great Britain

The Long Parliament's confiscation of crown, church, and royalist lands came near the end of a decade of civil war. The victors in England's Civil War tried and executed King Charles I and confiscated properties of the crown and of the crown's prominent supporters. Three successive legislative measures in 1649, 1650, and 1651 put into force the confiscations. Lands were sold at auction to a restricted group of buyers, and proceeds were primarily used to pay wage arrears of the large standing army present at the end of the war and to retire some of the growing national debt. The authorities found the lands difficult to sell, however, in part due to a perception of a nontrivial probability that a Stuart might return as monarch and attempt to return the confiscated lands to their owners.[5]

---

[4]Confiscation of church properties in the central lands of Austria in the late 1780s was preceded in the 1770s by the dissolution of a number of religious orders, including the Jesuits. The Portuguese government confiscated Jesuit monastic lands in 1761 after Jesuits were implicated in an assassination attempt on King Joseph (Maxwell, 1995).
[5]See Habakkuk (1962) for a fascinating discussion of "doubling" debt and the use of doubled debt to secure land purchases during the Interregnum.

11.2.2.2 The loyalist confiscations in North America

After declaring their independence from Great Britain in 1776, the revolutionary state governments quickly moved to confiscate lands of Loyalists and all who would not pledge allegiance to the revolution. The extent of confiscations varied across states, with all states confiscating some property (Allen, 2010; Moore, 1994; van Tyne, 1959). Most states sold confiscated properties at auction, with proceeds dedicated to replenishing their war-depleted treasuries. The 1783 peace treaty between Britain and its former colonies provided that the Congress of the Confederation (of the newly independent states) "shall earnestly recommend" that each state legislature "provide for the Restitution of all Estates, Rights, and Properties, which have been confiscated belonging to real British Subjects." All state governments ignored the treaty's recommendation. They neither returned confiscated lands nor allowed Loyalists to continue in residence unless they took a loyalty oath.

11.2.2.3 Confiscations during the early French revolution

During the early phases of the French Revolution (1789–1791), the National Assembly, composed of representatives from each of the three estates, passed a series of confiscatory measures. On August 3, 1789, the Assembly eliminated all remaining feudal privileges and obligations, including the tithe dedicated to supporting the church clergy and the pope; the clergy, nobles, and the pope were not compensated for the lost stream of revenues. From the middle of August through October 1789, the Assembly debated whether or not it should confiscate church lands in order to remedy France's deteriorating public finances.[6] Doyle (2002, pp. 131–132) noted a fierce debate on whether or not the sale of such vast lands would depress prices. Proponents argued that the sale of church lands would adhere buyers to the new regime, thereby lending it some stability. Opponents countered that the sale of such lands violated Article Two of the Assembly's earlier *Déclaration des Droits de L'homme et du Citoyen*, which proclaimed property as one of the "natural and imprescriptible rights of man," and that such sales would destabilize the new government.[7]

On November 2, 1789, the Assembly passed and the King authorized a law mandating the confiscation of all properties belonging to the Catholic Church in France. Clergy were to be partly compensated for their loss of revenues by becoming salaried employees of the state. Church lands and monasteries were to be sold at auction, with proceeds to be used to pay off a portion of the national government's large public debt. A law of May 14, 1790, established a secure title for these lands and rules for their auction by governments of the newly constituted *Départments*. Sales commenced in August 1790 and were made at robust prices to a "large cross section of society," thereby providing "a political base to reduce the future danger of expropriation" (White, 1995, pp. 235–241).[8]

---

[6]See White (1989) for a sweeping discussion of the *ancien régime*'s finances.

[7]"The aim of all political association is the preservation of the natural and imprescriptible rights of man. These rights are liberty, property, security, and resistance to oppression."

[8]White (1995) provides a lucid discussion of the economics underpinning the confiscation of land to finance France's national debt as well as an analysis of the *assignat* currency issued against the confiscated lands.

11.2.2.4 Confiscations by ruling chiefs in Hawai'i

Now consider an example from an entirely different cultural context: the redistribution of land by new rulers of polities in Hawai'i. La Croix and Roumasset (1984, 1990), Roumasset and La Croix (1988), and Kameʻeleihiwa (1992) provide accounts of property rights redistributions in pre- and post-contact Hawai'i, in which rights to rents from some productive agricultural lands were directly redistributed to a group of ranking chiefs who could constitute a stable core of support for a ruling chief (*mōʻi*) who had recently assumed power. Fornander (1969, p. 300) related that "[i]t has been the custom since the days of Keawenui-a-Umi on the death of a Moi (King) and the accession of a new one, to redivide and distribute the land of the island between the chiefs and favorites of the new monarch." Following this ancient practice, King Kamehameha (who united all but one of Hawai'i's competing island polities in 1795) extensively redistributed rights in land to the chiefs in his conquering army after taking control of islands governed by rival chiefs. The lands were redistributed after elaborate discussions and ceremonies conducted by *kahuna*, the priests of the state religion.

After the death of Kamehameha in the spring of 1819, the structure of property rights evolved to reflect fundamental shifts in the underlying power structure. Power transferred to Kamehameha's young son, King Liholiho, and Kamehameha's wife, Queen Ka'ahumanu. The two faced competition from another ranking chief, Kekauaokalani, the son of Kamehameha's brother, who had been named as the second heir. The scenario that two factions would form and a civil war would ensue was not improbable. To consolidate their power base among the ranking chiefs, Liholiho and Ka'ahumanu redistributed rights to earn income from the sandalwood trade to ranking chiefs. The transfer of property rights to sandalwood income further secured Liholiho's coalition, as the chiefs supporting him now had bigger stakes in maintaining him in power. When the revolt by Kekauaokalani came, it was quickly crushed.

Even prior to the revolt, the king and his family had moved to dismantle the cultural and physical manifestations of the state religion (La Croix and Roumasset, 1984). The introduction of Western military technology to the Islands rendered the support offered by the state religion (*'aikapu*) less important to the existing political order, and the resources used to maintain the network of temples and the voluminous sacrifices to the Hawaiian gods loomed as more of a burden. Violations by the monarch, his family, and ranking chiefs of religious rules mandating that males and females eat separately provided signal to chiefs and common people that the king no longer considered the state religion to be a key element in the political, economic, and cultural mechanisms that supported his dominant coalition. Within a few weeks, Hawai'i's common people (*makaʻāinana*) reacted to the news by burning the wooden temple statues (*akua*) and plundering the archipelago's vast network of large and small stone temples (*heiau*). The religious vacuum would prove fortuitous for the Protestant missionaries from New England who arrived in Hawaii in March 1820 with a new religion to support the monarch's rule.

11.3 ORIGINS OF EARLY MODERN CONFISCATIONS

The cases discussed above are somewhat distinct in their origins yet similar in two respects. First, natural states engineered land confiscations to reduce the wealth and presence of an interest that was a vital part of the old but not the new coalition. The Dutch were rebelling against the occupation of Holland by Spain, and their confiscation of Catholic Church properties was due not only to the emergence of Protestantism in Holland but also to the close links between the Catholic Church and the state in Spain. The revolution in France was as much against the Catholic Church as the monarch, given the status of the Church as France's privileged state religion. Second, the revolutionary parliaments did not redistribute the confiscated church properties directly to supporters but rather sold them at auction. This has its roots in the growing importance of land markets during this period, the diminishing role of land in holding together the natural state's coalition, and in the universal need of new revolutionary parliaments for revenue (North et al., 2009, p. 99).

Confiscations to reduce the wealth of ecclesiastic and political opponents are closely related to the state's recognition that these opponents do not recognize the legitimacy of the government, that they will use their wealth to sponsor its violent overthrow, and that they have enough wealth for the state to consider them to have some reasonable chance of success. Violent opposition to the state's existence is defined as a crime—treason—in both natural-order and open-access states, the penalty for which typically involves some combination of imprisonment, deportation, and execution as well as confiscation of assets. In all five cases considered herein, the Church was viewed either as a direct opponent of the government (England 1536–1540 and Austria 1780s) or as an ally of the deposed government (Holland 1570s, France 1789–1793, and England 1649–1651) whose rule had been sanctioned by divine authority, as represented temporally by the Church.[9]

11.4 REDISTRIBUTION AND SALE OF CONFISCATED LANDS

Another factor common to the three-century-long confiscation of Catholic Church lands in Europe was that governments often opted to resell confiscated lands at auction rather than to redistribute them directly to supporters of the government. In other times and places, rights to residual revenues from confiscated lands have often been directly redistributed to supporters while leaving tenants in place on the land. In some instances, governments chose direct redistribution due to the lack of a well-functioning or thick enough market in land. In other cases, land redistribution took place to reflect changes in the composition of the ruling coalition. William the Conqueror's redistribution of lands in England to his Norman supporters after the 1066 conquest, while leaving Anglo-Saxon landlords in place with some residual claims, is one widely analyzed example (see North et al., 2009, pp. 79, 96, 105).

On the other hand, the state's use of land confiscations could undermine its support in the longer run. For instance, in the early 1680s, opponents of the Duke of York (James II), on his assumption of the English throne, worried that he would restore the central role of the Catholic Church in Britain and that purchasers of confiscated monastery lands—taken more than 130 years earlier(!)—would be forced to return them. Harris (2006, p. 149) noted that "[t]he reason why William Lord Russell, Whig MP for Bedfordshire, was such a staunch supporter of Exclusion

---

[9]Confiscation of the lands of nobles who fled France after the National Assembly voted in August 1789 to eliminate all feudal obligations in France followed the same logic.

was because most of his estate was from the lands of dissolved monastic institutions, such as the abbeys of Tavistock (Devon) and Woburn (Bedfordshire) and the monastery of Thorney (Cambridgeshire)."[10]

One of the signs of a transition to an open-access society is that a new ruler refrains from explicitly adjusting the wealth of his or her governing coalition when assuming power. Bogart's (2011) study of investment in transportation infrastructure in seventeenth- and eighteenth-century England identified the Glorious Revolution—an event widely associated with the emergence of an open-access order in England—as a critical juncture for confiscation of property rights in England. Bogart shows that prior to the Glorious Revolution, rights granted to developers to undertake new transportation projects were susceptible to expropriation, "especially if they were associated with the losing side" of Britain's Civil War or subsequent Restoration (p. 1081). After the Glorious Revolution, a striking change took place: in most cases, Parliament did not violate developer rights to undertake transportation projects that had been granted by the Stuarts. Bogart found that when rights were violated, "it was linked to [developers'] failure to complete the navigation improvements" (p. 1084).

A second example comes from France's transition at the end of the Napoleonic Wars to a monarchy, albeit one with more restricted powers than those possessed by the prerevolutionary seventeenth- and eighteenth-century monarchs. Nobles whose lands were confiscated at the start of the French Revolution petitioned King Louis XVIII to restore the lands to them. His successor, King Charles X, resolved the issue by paying 988 million francs in compensation, with the quid pro quo being relinquishment of claims to the confiscated lands. (No compensation for confiscated lands, however, was ever paid to the reorganized Catholic Church in France.)

11.5 CONCLUSION

Land redistribution was regularly observed in developing countries during the twentieth century to benefit particular groups within a ruling coalition. The major reason underlying such reforms is that natural state governments needed to transfer wealth to various members of their dominant coalition as their influence or the violence potential increased. The bottom line is that such adjustments are *never meant to be Pareto improving* by their very nature. Solidifying a governing coalition often means weakening opponents as well as benefiting allies. Even when the veneer of the law allows for compensation, it is rarely the intent of the natural-order ruler to make such compensation.

Natural-order governments constantly have to negotiate with corporate bodies that have special privileges and will only retain them when doing so is advantageous to the dominant coalition. The above examples have emphasized situations in which a reforming monarch or parliament has taken

---

[10]In 1686, James established the "Dominion of New England," which administratively merged colonies from Delaware to Massachusetts. The new entity threatened to restructure property rights in New England and led to a vigorous colonial opposition to James.

action to reduce the power of state religions, not only by confiscating their lands but also by reorganizing the relationships between the church's corporate bodies and the state. Such reorganization of religious corporate bodies facilitates the transition to an open-access order by limiting the potential losses that religious organizations might face if a new ruling coalition was to assume power. The Philippines provides a contemporary example: the Catholic Church resists political reforms because the transition from today's natural state to a limited-access order would entail the loss of its special privileges and influence.

If history is to teach us anything about property rights, it is that property rights in land are never fully secure in the natural state and that land reform programs implemented in a natural state are motivated by consolidating the durability of the ruling coalition rather than implementing Pareto-improving policies that would benefit all parties.[11] Lamoreaux (2011) shows, however, that the same property rights are also not always secure in open-access societies. Their governments regularly encounter situations in which they need to restructure and reallocate property rights in order to accommodate technological and other changes (Bogart, 2011; Bogart and Richardson, 2011). Lamoreaux catalogued numerous examples in the United States in which property has been involuntarily reallocated and owners have received inadequate compensation. For property rights to be perceived as both secure and occasionally confiscated, Lamoreaux argues that the widespread ownership of land in the United States provides a self-enforcing mechanism that "prompts voters to mobilize whenever they think redistribution in favor of the top is getting out of hand" (p. 301).

Top-down land reform in East Asian developing countries is another matter. It would be naïve to believe that when victors in international or civil wars proposed land reform, they thought they were establishing a self-enforcing mechanism to secure property rights in land and fundamentally alter the nature of their social orders. Their goal was to bolster the newly established natural states to ensure that they could effectively maintain order and stay in power in the chaotic political environment seen in most Asian states in the late 1940s and early 1950s.

ACKNOWLEDGMENTS

Comments from John Wallis, Naomi Lamoreaux, Eugene White, Jim Roumasset, Brian Hallet, and participants in the 2012 Conference on Risk, Resources, Governance, and Development were exceptionally helpful. Thanks also to John Lynham for presenting the paper for me at the 2012 Conference and to the American Business School of Paris, the London School of Economics, the University of Arizona, and the University of Hawai'i for providing quiet times to finish the article. I am responsible for all errors of both omission and commission.

---

[11]One only has to look at a recent study of World Bank-funded land reform efforts in Cambodia to understand this (Thin, 2012). The land reform programs were designed to dismantle a traditional system of property rights in land and to replace it with a de jure system in which properties are surveyed and registered, thereby facilitating their rent, sale, and taxation. The new system suffered from one particularly prominent defect: common people holding lands rights were not confident that they would receive the same treatment from Cambodian courts as the elites holding land rights. In response, a shadow system of property rights emerged that bypassed Cambodian courts and the newly established land registry and reestablished the traditional village system of property rights specification and enforcement.

REFERENCES

Acemoglu, D., Johnson, S., Robinson, J.A., 2001. The colonial origins of comparative development: an empirical investigation. Am. Econ. Rev. 91 (5), 1369−1401.

Acemoglu, D., Johnson, S., Robinson, J.A., 2005. The rise of Europe: atlantic trade, institutional change, and economic growth. Am. Econ. Rev. 95 (3), 546−579.

Allen, T.B., 2010. Tories: Fighting for the King in America's First Civil War. HarperCollins, New York, NY.

Bogart, D., 2011. Did the glorious revolution contribute to the transport revolution? Evidence from investment in roads and rivers. Econ. Hist. Rev. 64 (4), 1073−1112.

Bogart, D., Richardson, G., 2011. Property rights and parliament in industrializing Britain. J. Law Econ. 54 (May), 241−274.

Dorner, P., Thiesenhusen, W.C., 1990. Selected land reforms in East and Southeast Asia: their origins and impacts. Asia Pac. Econ. Lit. 4 (1), 65−95.

Doyle, W., 2002. The Oxford History of the French Revolution. second ed. Oxford University Press, New York, NY.

Fornander, A., 1969. An Account of the Polynesian Race: Its Origins and Migrations, and the Ancient History of the Hawaiian People to the Times of Kamehameha I. C.E. Tuttle, Co., Rutland, VT.

Habakkuk, H.J., 1958. The market for monastic property, 1539−1603. Econ. Hist. Rev. 10 (3), 362−380.

Habakkuk, H.J., 1962. Public finance and the sale of confiscated property during the Interregnum. Econ. Hist. Rev. 15 (1), 70−78.

Harris, T., 2006. Restoration: Charles II and His Kingdoms, 1660−1685. Allen Lane, New York, NY.

Kameʻeleihiwa, L., 1992. Native Land and Foreign Desires. University of Hawaii Press, Honolulu, HI.

La Croix, S.J., Roumasset, J., 1984. An economic theory of political change in Pre-Missionary Hawaii. Explor. Econ. Hist. 21 (1), 151−168.

La Croix, S.J., Roumasset, J., 1990. The evolution of private property in nineteenth-century Hawaii. J. Econ. Hist. 50 (4), 829−852.

Lamoreaux, N.R., 2011. The mystery of property rights: a U.S. perspective. J. Econ. Hist. 71 (2), 275−306.

Lipton, M., 2009. Land Reform in Developing Countries: Property Rights and Property Wrongs. Routledge, London.

Maxwell, K., 1995. Pombal, Paradox of the Enlightenment. Cambridge University Press, New York, NY and London.

Moore, C., 1994. The Loyalists: Revolution, Exile and Settlement. Macmillan and Stewart, Toronto.

North, D.C., Wallis, J.J., Weingast, B.R., 2009. Violence and Social Orders: A Conceptual Framework for Interpreting Recorded Human History. Cambridge University Press, New York, NY.

North, D.C., Wallis, J.J., Webb, S.B., Weingast, B.R. (Eds.), 2012. In the Shadow of Violence. Cambridge University Press, New York, NY.

Otsuka, K., 2012. Book Review of Michael Lipton, Land Reform in Developing Countries. EH.Net Book Review, November. http://eh.net/book_reviews/land-reform-in-developing-countries-property-rights-and-property-wrongs/.

Ramseyer, J.M., 2012. The fable of land reform: expropriation and redistribution in occupied Japan. Harvard Law School, John M. Olin Center Faculty Discussion Paper No. 733, October.

Reynolds, S., 2010. Before Eminent Domain: Toward a History of the Expropriation of Land for the Common Good. University of North Carolina Press, Chapel Hill, NC.

Roumasset, J., La Croix, S.J., 1988. The coevolution of property rights and the state: an illustration from nineteenth-century Hawaii. In: Ostrom, V., Picht, H., Feeny, D. (Eds.), Rethinking Institutional Analysis and Development. Institute for Contemporary Studies Press, San Francisco, CA, 315−336.

Thin, K., 2012. Essays on land property rights in Cambodia: empirical analysis (Unpublished Ph.D. dissertation). University of Hawai'i at Manoa.

van Tyne, C.H., 1959. The Loyalists in the American Revolution. P. Smith, Gloucester, MA.

van Zanden, J.L., Buringh, E., Bosker, M., 2012. The rise and decline of European parliaments, 1189–1789. Econ. Hist. Rev. 65 (3), 835–861.

White, E.N., 1989. Was there a solution to the ancien régime's financial dilemma? J. Econ. Hist. 49 (3), 545–568.

White, E.N., 1995. The French Revolution and the politics of government finance, 1770–1815. J. Econ. Hist. 55 (2), 227–255.

REGIONAL INTEGRATION AND ILLICIT ECONOMY IN FRAGILE NATIONS: PERSPECTIVES FROM AFGHANISTAN AND MYANMAR

12

Manabu Fujimura

Aoyama Gakuin University, Shibuya-ku, Tokyo, Japan

CHAPTER OUTLINE

JEL: O17

12.1 ECONOMIC FRAMEWORK FOR ILLICIT ACTIVITIES AND ITS CROSS-BORDER CONTEXT

While illegality is defined in strictly legal terms, illicitness can be ambiguous as its definition may include social values that may not coincide with legality. In economic terms, a reasonable criterion for illicit as opposed to licit activities would be those that impose significant costs on the society in question that outweigh the total private benefits, provided that both social costs and private benefits are valuable across all stakeholders in a consistent way.

Applying such criterion, illicit goods include addictive drugs, alcohol, gambling, etc., which induce irrational behaviors in individuals that cause harm to them and the society as a whole. Production and use of addictive drugs may be the most significant illicit activity because it is often complementary to other illicit activities, such as human trafficking, bonded labor, and forced prostitution.

An economic analysis of criminal activities generally involves interaction between offenders (supply), consumers (private demand), and victims (derived demand for protection) in a market

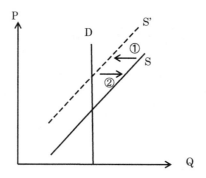

FIGURE 12.1

Effect of supply-side intervention when demand is inelastic.

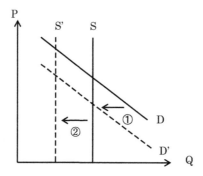

FIGURE 12.2

Effect of demand-side intervention when supply is inelastic.

setting (Ehrich, 2010). In this setting, government intervention works either as a supply or demand control through different degrees of deterrence measures. In the market for drugs, supply-side interventions are intended to raise the cost of producing drugs (leftward shift in the supply curve). Demand-side interventions are intended to raise the expected punishment for drug use, therefore lowering the users' willingness to pay (leftward shift in the demand curve). The relative effectiveness of supply-side versus demand-side controls depends on price elasticities. Supply-side controls are ineffective to the extent that demand is inelastic: interventions lead to a higher market price without reducing the equilibrium quantity. Demand-side controls are ineffective to the extent that supply is inelastic: interventions lead to a lower market price without reducing the equilibrium quantity. However, in the long run, asymmetric outcomes would be expected. A higher market price as a result of supply-side controls would invite new entries, potentially pushing the supply curve back to the original equilibrium in the long run (Figure 12.1). In contrast, a lower market price as a result of demand controls would discourage new supply entry as well as induce existing producers to put their resources to other productive use, shifting the supply curve to the left in the long run (Figure 12.2). Therefore, aside from relative intervention costs, in principle, demand-side control seems superior in reducing the equilibrium quantity. In the context of the United States, Caulkins (2006) provides a good discussion of the benefit–cost comparison and limitations of such

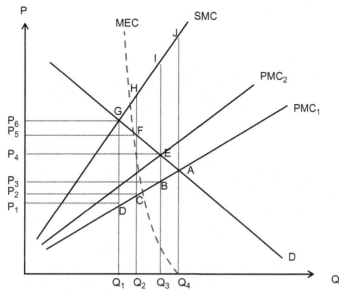

PMC$_1$: private marginal cost under unregulated legalization
PMC$_2$: private marginal cost under regulated legalization
SMC: social marginal cost that incorporates all possible negative spillovers
MEC: marginal enforcement cost

FIGURE 12.3

Effects under different policy settings in an illicit drug market: a possible illustration.

comparison for various types of drug control interventions, such as supply control, prevention, treatment, harm reduction, and legalization.

Despite failed experiments on alcohol and tobacco prohibitions, by the last half of the twentieth century, virtually every country in the world had adopted some form of drug prohibition. Critics of prohibition maintain that prohibition/eradication strategies could prove to be ineffective, only increasing the profits accruing to the illicit drug industry (Shepard and Blackley, 2010). However, available studies estimating the elasticity of demand for illicit drugs have found the numbers too wide-ranging to be useful for policymaking (Winter, 2008, pp. 89−91). Estimating supply elasticities would also be a daunting task as suppliers of illicit drugs are in the underground economy, thus hard to observe. Practical policymaking should probably compare marginal enforcement cost between supply-side and demand-side interventions and then apply the best guess on the elasticities for the specific drug in question and its associated market.

Figure 12.3 depicts economic implications under four different policy settings in an illicit drug market. The first setting, represented by the equilibrium quantity Q_1, is associated with an optimal prohibition with no net negative spillover to the society but rent accruing to drug traffickers corresponding to area P_1DGP_6, and probably infinite marginal enforcement costs, thus infeasible in practice. The second setting, represented by the equilibrium quantity Q_2, is associated with an imperfect prohibition and enforcement (presumably the status quo) with relatively large marginal enforcement costs, relatively small negative spillover corresponding to area FGH, and rent accruing to drug traffickers corresponding to area P_2CFP_5. The third setting, represented by the equilibrium quantity Q_3, is associated with a regulated legalization with relatively small marginal enforcement costs, potentially (subject to empirical testing) larger negative spillover corresponding to area EGI, and rent accruing to regulators corresponding to area P_3BEP_4. The fourth setting, represented by the equilibrium quantity Q_4, is associated with unregulated legalization with no enforcement costs and the largest negative spillover corresponding to area AGJ. If the last option of unregulated legalization would inflict on the society an overwhelming cost, especially in the case of "dangerous" drugs, such as amphetamine type stimulants (ATS) including methamphetamine, then a practical policy choice would be narrowed down to the second and third options for most drugs. The main task of policy analysts would be to weigh relative magnitudes among consumer surplus (private benefit), producer surplus (private benefit), traffickers' profits (private benefit), negative spillovers (social costs of treating addicts and crimes induced by drug use), and enforcement/regulation costs between imperfect prohibition/enforcement and regulated legalization.

In trying to apply the above framework to an international context surrounding the two largest sources of illicit drugs in Asia, Afghanistan and Myanmar, the factors to incorporate into practical policy analysis become very complicated. The link between regional economic integration and cross-border illicit drug economy would most likely work through both lower transportation and transaction costs, and, therefore, higher marginal enforcement costs.

The broad questions to be asked toward improving the status quo include: can there be an internationally tolerable size of an illicit drug economy (like the case of greenhouse gas emission); and who are the winners and losers from the status quo and from the changes to it? As a step toward tackling these daunting questions, this chapter proposes an analytical requirement in comparing the status quo and a potential policy alternative surrounding narco-nations.

12.2 AFGHANISTAN

Since 1980, Afghanistan has emerged as the world's largest producer of opium. Such development was catalyzed by stricter drug control laws imposed on neighboring Iran, Pakistan, and India at the end of the 1970s. Other catalysts include the *mujahideen* groups' search for military funds and increased cross-border movement by Pashtun tribes involved in drug production. When the Taliban regime shifted its drug policy from tolerance to an enforced total ban by the end of 2000, one of the reasons was that they had their opium warehouses full and needed to reduce their stocks to keep opium prices high. The Taliban collected up to 20% of the value of all drug shipments made by traffickers, supposedly as a form of *zakat* (a practice of charitable giving by wealthy Muslims). After the demise of the Taliban regime, regional warlords and militia commanders continued to engage in the drug business but refused to send in fees levied on imported goods at border posts to the newly established central government. For ordinary Afghans, armed robbery, kidnapping, and intimidation have displaced the Taliban as the principal security threat. Guns have been a way of life in Afghanistan for much longer than the last few decades of conflict and war. Disarming men who have made their living by the gun and have no other trade or skills is thus a daunting task (Macdonald, 2007).

The strong link between insecurity and opium cultivation is obvious: 78% of the opium cultivated in 2011 was concentrated in the southern and western regions where security conditions are classified

Table 12.1 Continued Importance of Drug Economy in Afghanistan		
	2010	**2011**
Net opium poppy cultivation after eradication (ha)	123,000	131,000
Weighted average dry opium yield (kg/ha)	29.2	44.5
Potential production of opium (mt)	3600	5800
Average farm-gate price of fresh opium (USD/kg)	128	180
Average farm-gate price of dry opium (USD/kg)	169	241
Current GDP (USD)	12.7 billion	16.3 billion
Potential gross export value of opiates (USD)	1.4 billion	2.6 billion
Share in GDP (%)	11	16
Potential net export value of opiates (USD)	1.2 billion	2.4 billion
Share in GDP (%)	9	15
Farmers' net income from opium per ha (USD)	2900	6400
Ratio of farmers' net income from wheat to opium	1:4	1:8

Source: *UNODC, 2011b*

as high or extreme risk by the United Nations (UN). Opium continues to be an important part of the Afghan economy, "guesstimated" to account for 16% of the gross domestic product (GDP) and the growing farmers' income source as eight times as valuable as wheat production (Table 12.1).

Afghanistan meets over 80% of the world's heroin demand today. Globally, around 16.5 million people use illicit opiates. In 2009, Afghan farmers earned an estimated USD 440 million, the Afghan Taliban and other insurgency groups around USD 155 million, Afghan drug traffickers USD 2.2 billion, and non-Afghan traffickers at least USD 7 billion from the market. Profits from heroin trafficking increase rapidly as heroin travels closer to consumers. The process of heroin production involves extracting morphine from opium and then mixing it with acetic anhydride (AA) and other chemicals. The heroin industry relies on AA availability. About 475 tons of AA is required annually to manufacture the current volume of Afghan heroin. In licit trade, a liter of AA costs no more than a dollar, but in Afghanistan it costs an average of USD 350 in the retail market. Given such markups, the incentives for criminal groups are obvious. Upwards of 2 million tons of AA are produced annually, out of which only 0.02% would need to be diverted and trafficked into Afghanistan for heroin processing. AA trafficking routes appear to roughly overlap with major Afghan opiate routes in a reverse direction (UNODC, 2011a, pp. 91−92).

Heroin flows out of Afghanistan in three main directions: Pakistan, Iran, and Central Asia. The 2430-km border between Afghanistan and Pakistan, known as the Durand Line drawn when the British left, artificially divides an area inhabited by an ethnically coherent Pashtun tribe. The strong tribal ties make it easy to run cross-border smuggling and illicit activities between Afghanistan, Iran, and Baluchistan (Pakistan). Along the 930-km Baluchistan−Afghanistan border are numerous routes through which drug traffickers can easily pass undetected. In the Northwest Frontier Province (NWFP) and the Federally Administered Tribal Area (FATA) to the north of Baluchistan, the lack of employment opportunities and licit livelihood have created a conducive environment for militant groups to engage in illicit activities. Many Pashtun people in the FATA live and work in both Afghanistan and Pakistan, regularly crossing the border without any identification documents. Moreover, with the Afghan Transit Trade Agreement (ATTA), signed in 1965 between Pakistan and Afghanistan, Afghan imports arriving in Karachi port pass free of duties. A similar arrangement is in place between Iran and Afghanistan. According to estimates, 60–80% of Afghan imports are smuggled back into Pakistan. The ATTA similarly works for drugs that are smuggled into Pakistan from Afghanistan, hidden in containers arriving in Karachi (UNODC, 2011a, pp. 75–90).

Afghan opiates that enter Central Asia mainly travel through the porous Tajikistan–Afghanistan border. Observers believe that drug money is fueling abnormally high property prices in Dushanbe and in the provinces. The combination of poverty, weak governance, and extremist violence makes the situation of Tajikistan virtually identical to that of neighboring Kyrgyzstan. Most of the drug flow crosses into Batken province in Kyrgyzstan. Alternatively, the drugs are transported to Bishkek along the country's main road. Once in Bishkek, the drugs blend into the major flows traveling into Kazakhstan and further to Russia. Law enforcement officials suspect that traffickers are increasingly using the International Road Transport Convention system (TIR) to move narcotics through Kazakhstan in sealed trucks.

Profits made from trafficking Afghan opiates into Central Asia, estimated at USD 344 million in 2010, are small compared with the net profit pocketed by criminals trafficking onward to Russia, which was around USD 1.4 billion, equivalent to one-third of Tajikistan's GDP. The markup on heroin further goes up as it nears the final consumer market in Russia: 1 kg of 70% purity fetches USD 3000 in Afghanistan, USD 4000 in Tajikistan, USD 10,000 in Kazakhstan, and USD 22,000 in Russia. Organized crime groups controlling the drug trade in Central Asia have developed intimate ties with the region's power structure. As an indication of their entrenched power, no major drug kingpin has been arrested in Central Asia since the region's independence in 1991. Initiatives of regional cooperation in counter-narcotics measures discussed in venues such as the Shanghai Cooperation Organization (SCO), the Collective Security Treaty Organization (CSTO), and the Central Asia Regional Information and Coordination Centre (CARICC) tend to be confined to rhetoric when implementation is needed. The unfortunate truth is that criminals have better cross-border coordination than law enforcement agencies (UNODC, 2012a, pp. 44–50).

Domestic drug use in Afghanistan is also becoming increasingly problematic. Norms on substance abuse have been eroded due to intense social disruptions from war and conflict and the influence of returning refugees. Many excombatants throughout the country started their drug use during the fight against the Soviets. According to a UNODC report, nearly 50% of heroin users interviewed in Afghanistan in 2003 first started using heroin in either Iran or Pakistan. In 2005, the Pakistani government estimated there were still over 3 million Afghan refugees in the country, with 1.8 million of them in NWFP. It ordered the closure of all refugee camps in the tribal areas of NWFP due to security concerns. The return of a massive number of drug-using refugees to Afghanistan has had a significant impact on the extent of drug use within contemporary Afghanistan (Macdonald, 2007).

12.3 MYANMAR

The so-called Golden Triangle encompassing northeastern Myanmar, northern Thailand, and northwestern Laos has been reputed for its long history of opium production and trade. The main actors in the drug trade changed hands from the British East India Company in the eighteenth to nine-teenth century to China's Kuomintang (KMT) forces fighting the communists in the 1950s–1980s, and then to various ethnic minority groups in Myanmar seeking to establish their autonomy. By the late 1970s, almost 80% of the world's supply of opium and heroin was originating from the Golden Triangle. By 1990, Khun Sa, who transformed himself from a KMT soldier to a local warlord and opium tycoon, was controlling half of the global heroin trade. While in recent years Afghanistan has taken over the status of being the top opium producer by a large margin, still the Golden Triangle continues to produce drugs, shifting from opium to heroin, and then to methamphetamine.

Mae Sai, Thailand's northernmost border town, is lively with various shops, restaurants, and vendors. In fact, there seems to be an excessive number of shops in Mai Sai relative to its small population of <30,000 and apparently no major industries. It is not difficult to imagine that the town has been thriving on border trade, including illegal trafficking of drugs and gemstones from Myanmar. Likewise, Tachileik on the Myanmar side is lively with border markets selling all kinds of items, most likely smuggled from southern Chinese provinces. After a few decades of an unstable border situation, including occasional closure of the border gate due to skirmishes between the central government and ethnic minority armies in Myanmar, the trade through this border has been robust in recent years.

Traveling from Tachileik northward to the Mongla (Myanmar)/Daluo (Yunnan Province of China) border, one enters the Special Region IV of Shan State. As part of the ceasefire arrangement since the late 1980s, Myanmar's central government granted autonomy to the region composed of local ethnic minority people. The region is led by a local tycoon named Lin Mingxian, who runs a business conglomerate and heads the 2000-strong National Democracy Alliance Army (NDAA) that rules the region. Lin Mingxian is a former Chinese Red Guard volunteer and originally came from Yunnan in the late 1960s. With his old contacts in China, he was able to solicit investments in building a gambling city out of Mongla. In January 2005, however, responding to growing gang-related violence and gambling debt borne by some eminent individuals, Chinese security squads crossed the border and raided the casinos in Mongla. They also pressured Lin to move his casinos deeper (16 km) inside the border (Swe and Chambers, 2011). When this author visited Mongla in September 2012, the new casino complex in the middle of a farm village had at least 10 gaming houses and looked busy with Chinese customers betting Chinese money and served by Chinese staff. There were no obvious signs of opium poppy cultivation or heroin trade, but Chin's (2009) surveys suggest that heroin refineries and methamphetamine laboratories are located in Mongla.

Despite Khun Sa's surrender to the Myanmar government in 1996, which subsequently announced in 1999 its 15-year plan to eliminate illicit crop production, many ethnic minority communities in Myanmar's hilly parts continue to cultivate opium poppy, which remains as an important source of their livelihood and food security. Particularly, farmers living on poor soil conditions have no reasonable alternatives for cash income. As such, Myanmar still accounted for 23% of the world's poppy under cultivation in 2011. Opium yields in Myanmar are much lower than in Afghanistan due to poor soil conditions and the absence of irrigation facilities in hilly areas. Nonetheless, the cultivation persists and appears to have regained momentum in the past several years since 2006 (Table 12.2). The market value of the average per-hectare output of dry opium is some 19 times that of upland rice. Many households that do not cultivate poppy on their own land still benefit from working as wage laborers on poppy fields owned by others.

Shan State, a core of the Golden Triangle, accounts for 90% of opium production in Myanmar (Table 12.3). Its Wa region (North Wa borders on China and South Wa borders on Thailand) is

Table 12.2 Drug Economy's Importance in Myanmar

	2010	2011	2012
Opium poppy cultivation (ha)	38,100	43,600	51,000
Weighted average dry opium yield (kg/ha)	15.2	14.0	13.5
Potential production of opium (mt)	580	610	690
Opium poppy eradication (ha)	8268	7058	23,718
Average price of opium (USD/kg)	305	450	520
Total potential value of opium production (USD million)	177	275	359

Source: *UNODC, 2012b*

Table 12.3 Opium Poppy Cultivation Myanmar, by Region (ha)

	2010	2011	2012	Share
East Shan	12,100	12,200	14,200	28%
North Shan	3700	4300	6300	12%
South Shan	19,200	23,300	25,400	50%
Shan State total	35,000	39,800	46,000	90%
Kachin	3000	3800	5100	10%
National total	38,100	43,600	51,000	100%

Source: *UNODC, 2012b*

currently the largest producer of opiates in the country. Opiates from this region are transported to the world market mainly via two routes: the Thai route and the Chinese route. After Khun Sa's downfall, many heroin factories were set up in North Wa. The first step in heroin travel entails transportation from refineries in North Wa to the Thai-Myanmar border. Residents of a large number of Chinese villages located near the border are heavily involved in carrying the drug across to Thailand. While the Thai route is believed to be on the decline, it is by no means obsolete. Once in Thailand, heroin can be transported overland to Southeast Asian countries and onward to international markets. As cross-border trade between China and Myanmar took off in the 1980s and expanded rapidly in the 1990s, drug trafficking along the Chinese border also flourished. Increasingly large amounts of opiates have been smuggled from Shan State to Yunnan Province and onward to the southeastern provinces of China and Hong Kong. The booming drug trafficking business in China can be attributed to a now sizable domestic drug-consumption market. The number of officially registered drug addicts in China increased from 148,000 in 1991 to more than 900,000 in 2001 and 1.44 million in 2004. For drug traffickers in the Golden Triangle, who are mostly ethnic Chinese, the Chinese route is far more convenient than the Thai route because of geographical proximity to North Wa, better road conditions, and familiarity with the language and customs (Chin, 2009).

In the face of international pressure against opium cultivation, unlike Lin Mingxian in Special Region IV who shifted to the gambling business, Wa leaders turned to the methamphetamine business. Their plan was to allow Chinese and foreigners to mass produce methamphetamine pills in their territory, collect fees from them, and then let them ship the products out of their territory to be consumed elsewhere—a variant of an export processing zone. To prevent consumption by their own people, Wa authorities penalized those who retailed or consumed the drug. However, the plan failed. First, it was not possible to prevent their people from gaining access to the drugs produced in their backyards. Soon large numbers of Wa students and soldiers were addicted to the drug. Second, Wa leaders could not hide their role in the methamphetamine trade, as hundreds of millions of the tablets were smuggled into Thailand, causing a strong reaction in Thailand as indicated by former Prime Minister Thaksin's war on drugs in 2003. Still, many former opium and heroin traders shifted to the methamphetamine business, which is more lucrative than the heroin trade; the pills (known as *yaba* in Thailand) are also much easier to transport. Easier connectivity across international borders involving Myanmar facilitates importation of ingredients for methamphetamine production. Myanmar's army became involved in the drug trade because in 1996 the army was forced to institute a self-support system for troops in the fields and their families. The army force had expanded to 400,000 soldiers by the 1990s and the central coffers were insufficient to maintain such force. Therefore, the soldiers resorted to collecting their own military taxes on opium production and also allegedly cultivated their own poppy fields in remote areas. A creeping criminalization eroded Myanmar's economy: narco-capitalists and their close associates became involved in running ports, toll roads, airlines, banks, and industries, often in joint venture with the government (Lintner and Black, 2009).

It appears that the drug issue in Myanmar cannot be tackled in a practical way unless the political issues, especially the unstable relationship between the majority Burmans and ethnic minority groups in border areas, are sorted out first. Leaders of local armed groups may be more concerned about maintaining their power and autonomy than about social spillovers affecting their own people and neighboring countries. For local leaders, being drug entrepreneurs is one way of securing funds for their autonomy. Gaining revenue from drugs is a natural extension of their historical use of opium in their livelihood. For them, there may be no moral distinction between state builders and drug lords.

12.4 **A WAY FORWARD FOR POLICY ANALYSIS**

While there is a large amount of literature on economic benefits derived from integration due to easier movements of goods and people, analyses and policy discussion are scant on negative spillovers associated with integration, particularly in the context of fragile nations. While how best to go about regional integration for fragile nations is beyond the scope of this chapter, it wishes to provoke analyses and policy discussions on how to deal with such negative aspects. As described above, the state of drug prohibition and enforcement surrounding Afghanistan and Myanmar is far from perfect. It seems to benefit disproportionately traffickers and criminal organizations and to perpetuate a vicious circle of illicit money, rule of guns, and fragile nations, without containing the negative spillovers. Afghanistan represents a global problem of the post-Great Game world, mired in a global network of criminal organizations, requiring policy solution at a global level. Myanmar represents a relatively regional problem of a long history of ethnic conflicts intertwined with a regional illicit network, requiring a solution at a relatively regional level. In both cases, however,

Table 12.4 Possible Benefit–Cost Distribution of a Shift from the Status Quo to a Regulated Legalization: Afghanistan's Case

	Producers	Consumers	Traffickers/Organized Crime Groups	Government	General Public
Afghanistan	Farmers' income (+)	Consumer surplus (+)	Anti-government elements (AGEs): decreased profit (−)	Saved eradication cost (+) New regulatory cost (−)	Saved public health cost due to safer drug use (+)
Pakistan			Organized crime groups: decreased profit (−)	(Trafficking countries) Saved enforcement cost (+)	Negative spillovers due to increased drug use (−)
Iran			Organized crime groups along the Balkan route: decreased profit (−)	New regulatory cost (−)	
Turkey—Central and Western Europe			Organized crime groups along the Northern route: decreased profit (−)		
Central Asian countries					
Russia			Organized crime groups: decreased profit (−)	(Consumer countries) Saved enforcement cost (+) New regulatory cost (−) New rent to regulators (+)	Less armed conflicts and violence due to less financing capacity of illicit economy (+)
China					
Other heroin user countries					

Note: " + " indicates presumed net benefit and " − " indicates presumed net cost.

Main beneficiaries from drug trafficking evidenced in recent drug seizures (UNODC, 2011a) include the following:

Afghanistan: Afghan Taliban, Hezb-i-Islam, Afghan public servants

Pakistan: Pakistan Taliban, Al-Qaida, Haqqani network, Lashkar-e-Islam, Baluchistan's organized crime groups

Iran: Organized crime groups working with Turkish and Kurdish crime groups through the Iran–Turkey border

Turkey: Kurdish Workers' Party (PKK), organized crime groups, Nigerian networks buying heroin from Turkish ringleaders. Citizens arrested for drug trafficking in Turkey include those from Albania, Bulgaria, Macedonia, and Iran.

Central Asian countries: Criminal groups in Tajikistan, Kyrgyzstan, and Uzbekistan; West Africans (especially ethnic Nigerians) operating in Kazakhstan

Russia: Criminal groups working with Tajik criminal groups in trafficking Afghan heroin

China: Some 1600 drug traffickers from 50 different countries were arrested in China in 2009, including those from Myanmar, Pakistan, Nigeria, Hong Kong, etc. An increasing share of these seizures is considered to be Afghan opiates rather than Myanmar's.

coordination failure across stakeholder governments in prohibition and enforcement is inevitable, as entrepreneurs coordinate better and faster than bureaucrats.

An alternative policy setting to be considered would be some form of regulated legalization, probably combined with a stronger demand control. For example, the Netherlands allows adults to take marijuana in small quantities at licensed coffee shops; some states in America allow use of medical marijuana prescribed or recommended by doctors; and Swiss hospitals distribute pharmaceutical grade heroin to addicted patients (Shepard and Blackley, 2010). Heroin and methamphetamine, with their higher potency, may require stronger regulations than cannabis-based substances (see the Appendix for the characteristics of the different types of harmful drugs). Table 12.4 illustrates the possible benefit–cost distribution associated with a shift from the status quo to a regulated legalization. Institutional design should target a Pareto improvement: for example, opium farmers are better off with lower distributor margins and some diversion of opium to alternative crops, heroin processors and traffickers are diverted to licit businesses, anti-government elements and warlords are guided into peace-making deals due to squeezed financial capacity, and drug users and patients have better access to safer drugs.

REFERENCES

Caulkins, J.P., 2006. Cost–benefit analyses of investments to control illicit substance abuse and addiction. Research Showcase Working Paper, Carnegie Mellon University.

Chin, K.-L., 2009. The Golden Triangle: Inside Southeast Asia's Drug Trade. Cornell University Press, Ithaca, NY.

Ehrich, I., 2010. The market model of crime: a short review and new directions. In: Benson, B.L., Zimmerman, P.R. (Eds.), Handbook on the Economics of Crime. Edward Elgar, Celtenham, UK, pp. 3–23.

Lintner, B., Black, M., 2009. Merchants of Madness: the Methamphetamine Explosion in the Golden Triangle. Silkworm Books, Chiang Mai, Thailand.

Macdonald, D., 2007. Drugs in Afghanistan: Opium, Outlaws and Scorpion Tales. Pluto Press, New York, NY.

Shepard, E.M., Blackley, P.R., 2010. Economics of crime and drugs: public policies for illicit drug control. In: Benson, B.L., Zimmerman, P.R. (Eds.), Handbook on the Economics of Crime. Edward Elgar, Celtenham, UK, pp. 249–275.

Swe, T., Chambers, P., 2011. Cashing in Across the Golden Triangle. Mekong Press, Chiang Mai, Thailand.

UNODC (United Nations Office on Drugs and Crime), 2011a. Global Afghan Opium Trade: A Threat Assessment, July 2011, New York, NY.

UNODC, 2011b. Afghanistan Opium Survey, December 2011, New York, NY.

UNODC, 2012a. Opiates Flow through Northern Afghanistan and Central Asia: A Threat Assessment, May 2012, New York, NY.

UNODC, 2012b. South-East Asia Opium Survey, October 2012, New York, NY.

Winter, H., 2008. The Economics of Crime. Routledge, New York, NY.

APPENDIX: NOTES ON HARMFUL DRUGS

Harmful drugs can be broadly defined as those that show one or more of the following characteristics:

— Psychological dependence: Development of a strong desire to continue its use;
— Physical dependence: Physical inability to do without it, usually revealed by withdrawal symptoms; and
— Development of a resistance to an increasing amount of dosage.

Different types of drugs have different characteristics, as presented in Table A.1. Some, such as heroin and cocaine, are processed from existing plants, while others are synthesized chemically, such as ATS, including methamphetamine, and LSD (Lysergsäure Diäthylamid in German or lysergic acid diethylamide).

Table A.1 Characteristics of Major Harmful Drugs

	Effect on the Nerve System	Psychological Dependence	Withdrawal Symptom	Resistance Development	Major Method of Intake
Opiate	Depressant	Very strong	Very strong	Very strong	Smoke, inject, eat
Cannabis	Depressant	Strong	None	None	Smoke, eat
Cocaine	Stimulant	Very strong	None	None	Sniff, inject
ATS	Stimulant	Very strong	None	Strong	Inject, eat
LSD	Stimulant	Exists	None	Strong	Eat, inject
Substance with similar characteristics					
Alcohol	Depressant	Very strong	Strong	Strong	Drink
Tobacco	Stimulant	Very strong	Strong	Strong	Smoke, chew
Caffeine	Stimulant	Strong	Moderate	Moderate	Drink

Source: *Author, compiled from various sources.*

The history of opium use by humans goes as far back as sixteenth century BC. The oldest medical document in ancient Egypt indicates some 700 kinds of drugs, including opium, which was recorded as having a sedative effect on crying babies. In many parts of the world, opium has been a household drug for pain sedation and recovery from fatigue. Arabian merchants and crusade soldiers brought it to Europe. For Europeans in the nineteenth century, opium use was socially acceptable; the mix of opium and flavored liquids was a popular "energy drink." In America, opium use spread as a sedative for soldiers during the Civil War. In 1805, a German pharmacist named Friedrich Sertuner succeeded in extracting morphine ($C_{17}H_{19}NO_3$) from opium. Morphine is 10−20 times more potent than opium. Later, in 1878, a British chemist named Alder Wright discovered enhanced potency of morphine by adding acetic anhydride [$(CH_3CO)_2O$], making diacetylmorphine ($C_{21}H_{23}NO_5$). In 1898, Bayer, a German pharmaceutical company, introduced the new substance as heroin. Heroin is three to four times more potent than morphine.

Cocaine ($C_{17}H_{21}NO_4$) is extracted from coca plants, which originally grew in the Andean mountains. The local Indio people chewed coca leaves to recover from fatigue and to suppress hunger. In 1859, a German chemist named Albert Friedrich Niemann first succeeded in extracting cocaine from coca leaves. A kilogram of coca paste of 40–90% cocaine purity can be extracted from about 200 kg of coca leaves. Coca paste is further processed using hydrochloric acid, calcium carbonate, ammonia, etc., into cocaine hydrochloride for general consumption. Cocaine was a popular party commodity in the late nineteenth to early twentieth century in European upper class societies. Sigmund Freud, an Austrian psychologist, tried cocaine on himself and his friends and promoted its use, even publishing an academic paper on cocaine, until one of his patients died from overdose. In America, cocaine-containing Coca Cola was popular until harmful side effects became clear; cocaine was removed from its ingredients in 1903. Cocaine continued to be marketed in the underground economy. It is suspected that during the Vietnam War, American soldiers used it heavily, leading to today's market demand in America.

The origin of ATS goes back to ephedrine, an asthma drug developed in 1885 by a Japanese doctor named Osayoshi Nagai. Ephedrine is an alkaloid extracted from euphedria plants originally found in northern China and Mongolia. In 1887, a Romanian chemist named Lazăr Edeleanu at the University of Berlin succeeded in synthesizing amphetamine ($C_9H_{13}N$) from ephedrine. Amphetamine was marketed in America and Britain in 1935 as "Benzedrine," becoming popular first among day laborers and long-distance truck drivers and later among soldiers, pilots, and factory workers in many countries during World War II. In the 1950s, amphetamine use spread among athletes, particularly bicycle racers, until a Danish cyclist died from its abuse in the 1960 Rome Olympic Games. Regarding methamphetamine ($C_{10}H_{15}N$), Dr. Nagai first succeeded in synthesizing it from ephedrine in 1893. Later in 1919, a Japanese pharmacist named Akira Ogata succeeded in its crystallization. Methamphetamine is known to be twice as potent as amphetamine. It was first marketed in Japan as "Philopon," a fatigue-recovery medicine, and later put to use during the war, including as a ration to *kamikaze* (suicide attack) pilots. Its continued distribution after the war had caused enormous social ills, thus Philopon was classified as a dangerous drug and banned in 1949. ATS does not generally cause withdrawal symptoms but has a very strong psychedelic effect, causing a strong psychological dependence.

CORRUPTION, TRANSACTIONS COSTS, AND NETWORK RELATIONSHIPS: GOVERNANCE CHALLENGES FOR THAILAND

13

Sirilaksana Khoman

*Chair, Economic Sector Corruption Prevention, National Anti-Corruption Commission,
Nonthaburi, Thailand*

CHAPTER OUTLINE

JEL: K4

13.1 INTRODUCTION

> When plunder becomes a way of life for a group of men living together in society, they create for themselves, in the course of time, a legal system that authorizes it and a moral code that glorifies it.
>
> **Frederic Bastiat (1801–1850), French Economist**

Knowledgeable observers nowadays wonder what has become of Thailand. From a promising high growth country that routinely addressed conflicts through conciliation and tolerance and that welcomed religious, cultural, and social diversity, it is now deep in the kind of polarity that threatens to derail the economy and worse.

To understand the situation, one needs to delve into the intricate economic, political, and social structures that make up the Thai society, a Herculean task that will not be attempted here. This chapter focuses only on corruption and network relationships that have polarized Thai society in the context of structural shifts that occurred after the Asian economic crisis in 1997.

Corruption in many parts of the world has steadily evolved from simple administrative transactions involving straightforward embezzlement or bribes in return for favors to sophisticated and complex activities, often labeled "regime corruption" or "political corruption," more serious and pernicious, and made to appear legal. It may become entrenched when the network of beneficiaries widens and the cost of leaving a network becomes prohibitive. Under these circumstances, the governance challenges are formidable, and limited features of democracy and decentralization can exacerbate the situation.

This chapter draws on Thailand's experiences in order to pinpoint and analyze the causes of the malaise. It examines the underlying structural shifts as a result of the 1997 economic crisis to the present time, the irony of the promising governance trends in which political corruption had been brought more into the open, the awareness-raising campaigns organized by local and international civic organizations, the country's increasingly dynamic media, and the reformist constitution that encompasses most of the standard provisions said to underpin good government and good governance, as well as specific provisions for increasing transparency and probity. It tackles network relationships that underlie governance failures in Thailand and possible ways and means for combating the new forms of corruption.

Many aspects of the problems facing Thailand today can be traced back to the well-intended reform movements in the mid-1990s and the Asian economic crisis in 1997. Political and administrative reforms were in full swing in the mid-1990s with concrete plans for decentralization of political and budgetary control and the drafting of the reform-minded Constitution that was ratified by a national referendum—the first of its kind in Thailand. This Constitution was promulgated in 1997. Unfortunately this coincided with the Asian economic crisis, precipitated by the financial crisis in Thailand. The crisis transformed Thailand's economic landscape and power structures even as the political and administrative reforms were being ushered in.

Section 13.2 briefly describes the setting: the effects of the economic crisis and the salient features of the 1997 Constitution, which shaped Thailand's political structure thereafter. The role of network relationships and corruption is discussed in Section 13.3 and cases analyzed. Section 13.4 talks about measures undertaken to address corruption. Section 13.5 concludes with lessons that may be applicable to other developing countries.

13.2 THE SETTING: POLITICAL AND ADMINISTRATIVE REFORMS AND THE ASIAN ECONOMIC CRISIS

"The End of Brand Thailand" announces a *Newsweek* article (June 3, 2010), detailing "how mismanagement and mistakes turned a high growth democratic paradise into a violent mess."

This chapter argues, however, that errors and mismanagement are the least of Thailand's problems, and that intentional plunder of the economy occurred because of the entrenched network of strategically placed corrupt individuals who had taken advantage of the key administrative reforms.

The 1997 Constitution embodies the philosophical frameworks for political and administrative reform, as well as protections of human rights; it contains all of the features of good governance and probity. It embraces empowerment and protection of the disadvantaged. Gradually, the Constitution itself was seen as the cause rather than the cure.

The coincidence of the reformist Constitution and the economic crisis in 1997 had the following consequences.[1] First, the economic crisis of 1997 transformed Thailand's economic landscape. High debt companies were the hardest hit, causing a permanent change in Thailand's capitalist structure. The banking sector was the most badly affected; many banking families had their power diminished and some have not recovered. The other side of this structural shift in economic power was that government banks gained greater prominence and quickly became important instruments in implementing government policy. At the same time, the face of cronyism in Thai society changed, as economic power shifted to the telecommunications sector.

While the country was still reeling from the economic downturn and the accompanying hardship on the population, the 1997 Constitution ushered in a new era of strong government, and a new charismatic and innovative leader was elected in a landslide victory in 2001. The outcome of the election combined with the provisions of the Constitution that favored strong government soon led to a concentration of political power.

As the economy began to rebound, the government adopted a dual-track program of development, ostensibly to build resilience and reduce vulnerabilities from exposure to external shocks from the global economy that had precipitated the crisis. Track One concentrated on domestic demand-led growth—that is, stimulating grass roots activity and small- and medium-scale enterprises (SMEs). Huge funds were injected to the rural areas under various schemes. Track Two was for global engagement, with the government handpicking industrial sectors to be promoted, identifying new clusters of industries, and following an export-led growth path. However, the more interventionistic policies, greater micromanagement of the economy, and the practice of picking winners paved the way for a different kind of vulnerability by allowing cronyism to occur on a greater scale than had hitherto been seen.

The higher propensity to consume and the lower propensity to import of the rural community with its local traditional wisdom were seen as a new source of growth. Measures used for Track One included credit extension and asset capitalization to increase purchasing power, debt moratorium, the village fund, conversion of assets into capital, a people's bank, SMEs, and other populist measures like the 30-baht healthcare scheme. These programs had mixed success in terms of feasibility and sustainability, but were hugely popular.

The consumption stimulus with greater reliance on domestic demand and high household spending led to increased indebtedness among rural households. Table 13.1 shows that household debt increased from 370% of the income in 1994 to 640% in 2004, which worried analysts (Government of Thailand, 1992; Government of Thailand, 1994, 1996, 1998, 2000, 2002, 2004; Government of Thailand, 1999). The households living on debt became strong supporters of the populist regime.

---

[1]A great deal of literature on the Asian economic crisis exists and details will not be discussed here.

Table 13.1 Household Debt (1994–2004) During the Period Before Press Allegations of Corruption by the Prime Minister Surfaced

Year	2004	2002	2000	1998	1996	1994
Debt (THB/household)	110,133	84,603	70,586	72,345	55,300	31,079
% Debt for consumption	65.3%	64.1%	61.0%	61.2%	50.8%	59.7%
% of monthly income	640	610	570	570	500	370

Source: *Government of Thailand, National Statistical Office, Socio-Economic Surveys, various years.*

At the same time, technocrats in public agencies were discredited because the crisis occurred under their watch. The reputations of revered institutions, like the Bank of Thailand and the Ministry of Finance, were tarnished. This paved the way for a host of vulnerable economic policies, including the paddy pledging program that continues to haunt Thailand and threatens both its budgetary and overall macroeconomic stability.

As administrative reforms were embraced, new types of government organizations were set up, which had greater flexibility in personnel management, rules and regulations, and budgetary processes. The intention was to reduce bureaucratic red tape and provide better services to the public. However, vulnerabilities also emerged. As controls were eased, opportunities for nepotism and corruption increased.

Cases of alleged corruption and conflict of interest mushroomed in both Track One and Track Two programs.[2] The business-politician nexus, supported by grass roots mobilization, dominated the scene—a recent phenomenon that claims legitimacy because of popular voting. Even on taking office, the charismatic leader was found to have concealed his assets, violating the law that allows high-level politicians to hold no more than 5% of the shares in any private company. This law aims to prevent conflict of interest, create a culture of transparency, and promote the practice of placing shares in blind trusts. This was not the case with Thaksin Shinawatra; instead, he placed the shares in the names of his household staff. With the evidence clear, he was found culpable, but the Constitutional Court voted by a narrow margin to drop the charges in view of his election win.

Populist policies have many well-known problems. Foremost is that political objectives take precedence over economic or social returns. There is a tendency to circumvent proper evaluation, feasibility studies, and the normal processes of scrutiny. Distortions in the economy are likely to be created, since necessary projects and programs may be crowded out. Riskiness is compounded by the likely allocation of funds to high-risk groups and projects, leading to widespread moral hazard. The inclination to emphasize political gains often leads to a disregard for the fiscal burden and the consequences for macroeconomic stability.

---

[2]Among the notable (alleged and concluded) cases against Prime Minister Thaksin Shinawatra are: (i) concealment of shares and tax evasion, (ii) Independent Television (ITV), (iii) Petroleum Authority of Thailand (PTT) shares, (iv) conflict of interests in airline purchase, (v) land purchases from government agencies, e.g., the THB 772 million deal, at prices five times lower than the estimated market price, (vi) Channel 7, (vii) EXIM Bank loan to Myanmar of THB 4 billion and conflict of interest, (viii) failed schemes at taxpayers' expense, e.g., Elite cards (THB 1 billion), and (ix) kickbacks from procurement projects. As to Track One programs, the National Anti-Corruption Commission had received complaints regarding the Housing program and the Village Fund, among others. (THB 31 = USD 1 approximately.) See Government of Thailand, Assets Examination Committee (2007) and Bangkok Post (2008).

"Quasi-fiscal" measures (requiring only Cabinet approval and thereby circumventing normal budgetary processes) were used by this (strong) government under the new Constitution, more than any other previous government. This has become its legacy. Populist combined with quasi-fiscal measures tend to result in overspending by government, budget deficits, increased borrowing, current account deficits, and balance of payments problems, as witnessed in other countries in the past.

The new Constitution provides for the decentralization of political, administrative, and budgetary power. It contains provisions for local administration to be elected and not appointed from the central government, as had been the case in the past. Moreover, the government budget was devolved with 40% of the national budget allocated to local administration. Ironically, while the decentralization process was under way, the government devised the concept of CEO provincial governors (i.e., acting like Chief Executive Officers in the corporate sector), with full powers and answerable directly to the Prime Minister. This increased the concentration of power with the Prime Minister.

The Constitution was amended to provide for strong government because of the country's weariness with the large number of political parties in the past vying for power, resulting in coalition governments that constantly shifted blame and accountability. One may ask, what about checks and balances? Indeed, the Constitution provides for several independent agencies and vests powers in the people to provide the checks and balances. In addition, there are the Cabinet and Parliament.

These avenues, however, are negated by the details of the Constitution. According to the 1997 Constitution, members of Parliament (MPs) have to belong to a political party; no independents are allowed to contest the national elections. Moreover, a candidate has to be a party member for at least 90 days before standing for election. On the other hand, if Parliament is dissolved, new elections have to be held within 60 days. This means that if MPs play a conscientious role and vote against their own party—and succeed in forcing dissolution—they will not be eligible to stand in the next election. This effectively deters them from crossing party lines. The Constitution also stipulates that Cabinet ministers cannot be MPs. This fosters solidarity because any dissention could result in dismissal and loss of employment.

The technocrats were largely discredited after the Asian economic crisis which was precipitated by the financial crisis in Thailand and could not aggressively intervene with proper analysis or cautionary advice.

The newly created independent bodies (Election Commission, Constitutional Court, National Anti-Corruption Commission, Human Rights Commission, Office of the Auditor-General, Ombudsman) experienced interference from corrupt politicians at varying degrees. For instance, after the January 2001 election, the Election Commission accused 100 political candidates of fraud. It subsequently disqualified only a handful of MPs, however. The main reason for this leniency seems to be due to the election law requirement that the commission be unanimous in its decision to disqualify candidates. In addition, these independent agencies need Parliamentary approval of their budgets, which a hostile Parliament could routinely slash.

13.3 NETWORK RELATIONSHIPS, TRANSACTIONS COSTS, AND CORRUPTION

In just over a decade after Thailand's path-breaking, people-centered Constitution was promulgated, the country's legal infrastructure has been transformed. The 1997 Constitution has been widely regarded as containing all of the underpinnings for good government: promoting transparency in government, fostering improvement in the quality of public services, strengthening integrity in public life, preventing corruption and misconduct for personal gain, as well as creating a sense of mutual responsibility, participation, and accountability in society.

However, Thailand continues to grapple with legal interpretations, behavioral consequences, and loopholes in the implementation of the new legal infrastructure. As such, judicial rulings have assumed an increasingly important role in the recent volatile years. Since many laws are path-breaking, hence without legal precedence, legalities are subject to differing interpretations, and questions of partisanship or at least partiality have been raised. Corruption and malfeasance issues, as well as court decisions, are hotly debated. With the different views on legal interpretations, coinciding with the unprecedented concentration of political power, the role of "network" membership and "connections" has come to the fore. "Connected dealings" and "procurement conspiracies" are difficult to tackle. Even when the National Anti-Corruption Commission arrives at a finding regarding corruption and malfeasance, legal wrangling often ensues, undermining anti-corruption efforts. Consequently, rent-seeking and "plunder" have approached crisis proportions.

The term "connected dealings" is used here to refer to situations of conflicts of interest, where at least one party to a contractual arrangement, who is in a position of authority to (legally) make decisions on the use of public funds regarding that contractual arrangement, and at least one other party, who is in a position to benefit from the use of such public funds, are connected through business relations, family ties, school or institutional affiliations, or other previous dealings. Contracts resulting from connected dealings are deemed to be part of a procurement conspiracy, because they are won due to undisclosed network relationships rather than objective criteria. When relationships are formally disclosed and criteria are objective, clearly spelled out, and satisfied, the resulting contracts are not considered a connected dealing or procurement conspiracy even if the parties involved are connected. This is different from other forms of corruption, such as simple extortion, that can occur between individuals unknown to each other.

Network relationships are a cornerstone of human interaction. Worldwide social networking, loose or tight, has exploded within the last decade. By creating trust, network relationships facilitate efficient interaction by reducing the three components of transactions costs: information and search costs, negotiating and contracting costs, and policing and enforcement costs. Whether in business or politics or any other area of interaction, network relationships play an important role and help create efficiency. Indeed investments in creating trust, brand loyalty, recognition, and reputation, whether in personal or business relationships, can be seen as networking. Trust implies confidence that some persons or institutions will behave in an expected way (Rose-Ackerman, 2001). However, there are built-in dangers to maintaining the status quo and creating situations of power concentration if the network is large enough, as well as facilitating transactions with corrupt intent.

A network may be regarded as a set of contracts, which can be loose or tight, formal or informal, that establishes the internal rules of exchange and cooperation. In some cases, the set of contracts gives the group a collective identity vis-a-vis others, replacing individual identity. An organization, or clan, is thus a set of contracts and rules defining roles and establishing their relationships within the network, in which individuals play or exit roles that have been defined. Investment in identity takes place in the selection for roles and in the process by which individuals select the organization/clan that they join. This network may start out being an innocuous social network where members assist each other and some kind of reciprocity is the norm. It later can transform into one with a patron−client relationship, paving the way for the formation of a more pernicious network, whereby the patron provides resources and protection to the clients who, in return, provide services, rent collection, and other forms of support to the patron, including facilitating corrupt acts.

Since there are many competing networks, each patron needs to accumulate a large amount of resources to feed the needs of the clan. Corruption then becomes a method for accomplishing this task, allowing the network to accumulate sufficient funds and attract a large number of members to compete successfully against other clans. Members recruited into the corruption ring may actually not be aware of the ring in the beginning, but the cost of leaving the network becomes prohibitive and the option of moving to another network may not be available due to mutual distrust and possible hostility between clans/networks. Figure 13.1 illustrates the situation.

People with independent sources of power obtain it from knowledge, expertise, wealth, ethical values, religious values, patriotism, prudence in business, and so on. They have social networks as well. Because of their independence, they do not necessarily fall prey to corrupt networks, unless they want to achieve more power.

People without independent sources of power tend to be the poor and disenfranchised, although this is not a necessary nor sufficient condition. Many who cannot rely on the State to provide basic services seek assistance by affiliating themselves with groups. Groups, clans, or

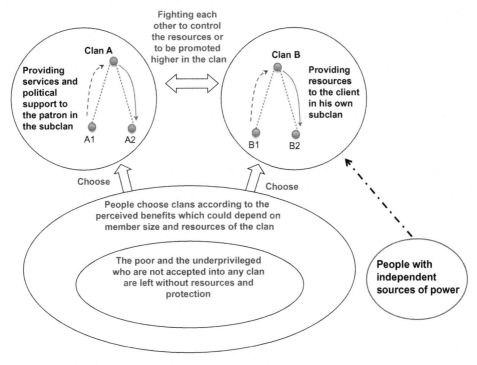

FIGURE 13.1

Rivalry between clans ("puak") or subclans, choosing clan affiliation. Independent sources of power include knowledge, expertise, wealth, morality, religious values, patriotism, business ethics, and so on. Cost-benefit calculations preclude leaving a clan for those without independent sources of power.

"*puak*" are collective entities whose members may include local and national politicians and officials, supporters, and followers. They share an identity. "*Puak*" consists of personal relationships and is integrated by a leader who has been instrumental in building up the group (and who can "deliver" resources, including a certain number of votes at election time or number of people for *ad hoc* protest rallies). Knowing that one belongs to the same *puak* encourages communication and cooperation. Belonging also provides safety, being under the protection of the "mosquito net" that the group provides. Conversely, there is suspicion of people from other *puak*, as well as people whose affiliation is not clear. Groups can be unstable as some clients may make their own contracts with the patron; this can even breed hostility and rivalry between subgroups. In all, the cost and benefit calculations favor remaining in a group when one does not have an independent source of power.

In order for the group leader to deliver on his/her promises, key players are placed in the bureaucracy to carry out key tasks. Thus, provincial political groups and influential families not only try to capture local political positions but also try to build bases of support in the bureaucracy, in the police force, and even probably also in the judiciary.[3] This is one of the reasons why being a government MP is so important: MPs are able to lobby ministers and other ministry officials to put their people in important administrative positions. Bureaucrats also lobby their local MPs to use their influence to obtain transfers and promotions. Such a service is reciprocated at a later point, perhaps during election campaigns.

Corollary to this, group members evaluate actions or policies to determine whether these are beneficial or detrimental to the group or subgroup's interests. They generally have no patience for complex issues, such as fiscal discipline, sophisticated siphoning, or macroeconomic stability. To a large extent, political loyalties are not directed toward ideas but toward leading personalities. Therefore, for the most part, it is not ideological persuasion but the leadership qualities of individual politicians that determine political structures and affiliations.

13.3.1 CONNECTED DEALINGS: CASES FROM THAILAND

Corruption risks were examined using case studies to highlight the connection between corruption and network relationships. These are selected cases of alleged corruption as well as completed cases in Thailand. Only corruption involving networks of connected persons was examined; these included "political corruption," where projects were designed and initiated to benefit specific suppliers within connected networks, possibly identified prior to project initiation, and collusion among networks of private sector suppliers. Even "simple kickbacks" from suppliers to government officials can involve a network of players and brokers to reduce the possibility of detection.

Thailand is a country of contradictions. Religious piety predisposes government officials to honesty. However, a culture of kinship and caring for friends and relatives reinforces the tendency toward "prebendalism," whereby state offices are regarded as "prebends that can be appropriated by officeholders, who use them to generate material benefits for themselves and their constituents and kin groups" (Joseph, 1987). The cases examined here explore these influences.

---

[3]Newspapers reported paper bags containing millions in cash left at the courts. http://www.transparency-thailand.org/thai/index.php/2012-05-07-10-50-47/83-2 (in Thai).

Case 1: Intervention in the Longan Market: Strategic (Corrupt) Partners

Thailand has a range of market intervention schemes to assist agricultural producers. One of the most widely used is the "pledging scheme," which was intended to reduce the effects of seasonal price variation on agricultural producers. This case presents the intervention in the longan fruit market.

In the pledging scheme (which was also used in the rice market), a government agency operates like a pawnshop in that just after harvesting, when prices of produce are low, producers can pledge or pawn their produce at a government agency and redeem it after prices improve. For longan, the scheme was started in the year 2000. Initially, the buying price was set at 70% of the market price. In 2002−03, however, this scheme was hijacked by politicians and turned into a populist scheme. The buying price was jacked up higher than the market price, encouraging excessive expansion of production. As a result, the level of intervention (defined as the quantity purchased divided by total production) increased from 4.4% in 2000 to 62.2% in 2002 (Jaruk and Issariya, 2010). The successive intervention measures are summarized in Table 13.2.

The corruption ring during the 2002−03 intervention scheme operated at every stage of the process (Figure 13.2). The pledge prices of THB 72/kg for grade AA, THB 54/kg for grade A, and THB 36/kg for grade B were well above market prices, attracting wrongdoers to extract undue benefits from the scheme (USD 1 = THB 30). Starting from the producers and cooperatives, estimates of production were inflated to take advantage of the high prices. At this stage, "substitution of rights" also occurred, whereby legitimate longan producers were asked/coerced/recruited to sign documents falsifying production figures. Nonproducers subsequently used these documents to obtain payment for nonexistent stock.

Table 13.2 Interventions in the Longan Market

Year	Intervention Measure
2000	Bank of Agriculture and Agricultural Co-operatives (BAAC) — accepts pledging documents to dry longan — pays 70% of target price — receives target of 30,000 tons of longan
2002−03	— BAAC pays a price higher than market price — huge increases in size of scheme
2004	— longan market management through such schemes as output distribution outside production location, further processing, such as canning, promotion of fresh longan export, domestic consumption, and purchase of fresh longan for processing into dried longan
2005−present	— limited intervention and greater reliance on market mechanism and consumption boosting measures

Source: Adapted from Jaruk and Issariya (2010).

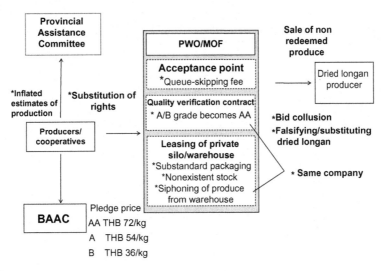

FIGURE 13.2

Corruption ring in the longan pledging scheme in 2002−03.

At the level of government agencies (i.e., Public Warehouse Organisation and Marketing Organisation for Farmers), illegal "queue-skipping" fees were charged at the acceptance point, where longan producers were supposed to submit their pledge. At the quality verification stage and leasing of private sector silos and warehouses, the same company was awarded the contract for both quality verification and warehousing, so there were no checks and balances. Investigation by the National Anti-Corruption Commission uncovered fraud in the grading process, substandard packaging resulting in product deterioration, siphoning of produce from the warehouse, and falsification of records (i.e., nonexistent stocks).

The high purchase price also led to low redemption rates, as there was no incentive to redeem the pledged produce. The stock in the warehouse therefore deteriorated and had to be sold at a loss to dried longan producers, a process that again involved bid collusion and fraud.

Consequently, in 2004 the intervention scheme was modified to focus more on market management, output distribution outside the production location, promotion of domestic consumption and fresh longan export, further processing (canning), and purchase of fresh longan for processing into dried longan. This led to another systemic pattern of corruption in procurement. Figure 13.3 shows that inflated production estimates and substitution of rights also occurred at the producer level. Krungthai Bank was brought in to replace the Bank of Agriculture and Agricultural Co-operatives (BAAC), and Chiangmai University (CMU) was recruited for quality inspection and grading. However, grading fraud continued, and Por Heng Inter Trade Company was awarded the contract for drying fresh longan even though the company did not possess any drying facilities and had no experience in agricultural processing. That Por Heng did not have the qualifications yet obtained the contract shows the collaboration of the network members: procurement officers, politicians, and private business. Its scandalous failure to deliver nearly 50,000 tons of dried longan exposed the corruption ring. The poor quality of dried longan produced as a result of corrupt practices further damaged the export market.

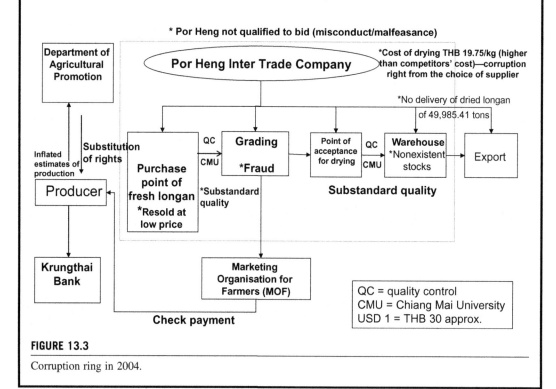

FIGURE 13.3

Corruption ring in 2004.

Case 2: Intervention in the Rice Market: Insider Information About Policy

The intervention in the rice market, which was implemented prior to longan's, is similar to the description above and so will no longer be discussed here, except for one phenomenon. The pledging scheme for rice, introduced in 1995, was intended to operate like a pawnshop; it provided low-interest loans to rice farmers to tide them over during the harvest season until market prices increased. Thus the pledging price was, in principle, always lower than the market price, so that farmers would redeem their produce when market prices increased. This had been the norm until the 2001–02 season, when the Thaksin government unprecedentedly announced a pledge price (or price at which rice was "pawned") **above** market prices at THB 7000 per ton (Figure 13.4). This move resulted in falsification of documents and substitution of rights to take advantage of the high government price, and nonredemption of pledged output, leading to enormous stockpiles, storage of which cost hundreds of millions of Baht per month. Worse, only the wealthy farmers were able to take advantage of the government's high price.

More significantly, in early 2004, President Agri Trading (PAT) Company, a newcomer in rice exporting, inexplicably made a bid to buy 1.68 million tons of rice from the government at prices *above* the market price, thereby knocking out all competition and subsequently possessing the largest amount of rice among all the exporters—2.2 million tons. A few months later, the government announced the pledging price for the new season paddy at THB 10,000 per ton *(higher than market price and higher than PAT's purchase price)*. Consequently, the market price shot up, other exporters could not compete with PAT, and many had to resort to buying rice from PAT. The Ministry of Commerce later backlisted PAT, citing other reasons; however, it was reinstated during the time of a "friendly" government.

This case strongly suggests a connection between this company and government policymakers.

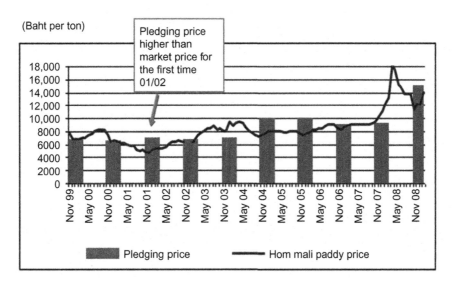

FIGURE 13.4

Comparison of pledging price and market price of jasmine (hom mali) rice.

Case 3: Highway Department: Collusion and Bid Rotation

The close ties between political parties and road projects have long been known in Thailand. In fact, many of the key politicians made their fortunes by owning construction companies whose bulk of business was undertaken with government agencies.

Table 13.3 provides a glimpse of the close connections. All of the top 10 road projects during fiscal years 2007–11 went to companies known to have political affiliations or to have made large overt campaign contributions to political candidates and parties.

The Office of the Auditor-General had found that collusion and favoritism can occur in various ways during the awarding of contracts (Pan et al., 2012). Risks are present at each procurement stage: project initiation or site of construction, technical specifications and reference prices, canvassing of suppliers, tendering process, contract design and management, and verification and acceptance of work.

Table 13.3 Top 10 Road Projects During Fiscal Years 2007–11, Contract Winners, and Political Affiliation

Highway Department Contracts from 1 October 2006 to 30 September 2011 (Fiscal Years 2007–11)

	Contract No.	Project	Value of Contract (THB)	Contractor	Political Affiliation
1.	SRN 27/2551 LW 23 Sept 2008	Rehabilitation of asphalt surface, number 2081 CS 0100 at intersection of Highway 214 (Taha)—and Highway 219 (Stuk)	19,657,158,361	Yingcharoen Construction Buriram Limited Partnership	Bhumjaithai Party
2.	SKM 2/2552 LW 6 May 2009	Construction of bridge across Mekong River at Nakorn Panom	1,723,234,000	Italthai PLC	Pueathai Party + contributes to Democrat Party
3.	SPB 6/2551 LW 23 Sept 2008	Construction of overpass at Sukapiban intersection 2–3	913,988,982	Prayoonwit Company Limited	Chart Thai Pattana Party
4.	SPB 6/2551 LW 30 Sept 2008	Construction of overpass at Sukapiban intersection 2 and 3 (West Bridge and East Bridge), casting and installing of Precast Segment Box Girder, 3235 m	913,988,982	Prayoonwit Company Limited	Chart Thai Pattana Party
5.	STI 1/5/2551 LW 23 Sept 2008	Road widening from 4 to 8 lanes, and rehabilitation of surface, Highway 9 Bang-pa-in to Bangplee (section 9)	800,038,725	Chainan Construction Supplies Company Limited (2524)	Pueathai Party
6.	STI 1/1/2552 LW 11 Dec 2008	Rehabilitation of Highway 3256 intersection of Highway 3 (Bangpu) Ban Klong Krabue section 1 between km 2 + 200.00 and km 5 + 209.329 (west) 3009 km	776,442,243	See-sang Yotha Company Limited	Chart Thai Party

Table 13.3 Top 10 Road Projects During Fiscal Years 2007–11, Contract Winners, and Political Affiliation *Continued*

Highway Department Contracts from 1 October 2006 to 30 September 2011 (Fiscal Years 2007–11)

	Contract No.	Project	Value of Contract (THB)	Contractor	Political Affiliation
7.	STI 1/2/2551 LW 22 Sept 2008	Road widening from 4 to 8 lanes, rehabilitation of surface, Highway 9 Bang-pa-in to Bangplee (section 2)	766,881,047	Roj-sin Construction Company Limited	
8.	SPB 5/2551 LW 23 Sept 2008	Construction of Highway 9 Bang-pa-in to Bangplee (section 8)	712,899,338	Krungthon Engineer Company Limited	Chart Thai
9.	SPB 4/2551 LW 23 Oct 2008	Construction of Highway 9 Bang-pa-in to Bangplee (section 6)	683,946,431	Prayoonwit Company Limited	Chart Thai Pattana Party
10.	STI 1/4/2551 LW 23 Sept 2008	Road widening from 4 to 8 lanes, rehabilitation of surface, Highway 9 Bang-pa-in to Bangplee (section 4)	680,209,428	Kampaengpet Wiwat Construction Company Limited	Bhumjaithai Party
11.	STI 1/1/2551 LW 22 Sept 2008	Road widening from 4 to 8 lanes, rehabilitation of surface, Highway 9 Bang-pa-in to Bangplee	658,338,675	M.C. Construction Company Limited (1979)	Democrat

Sources*: Contracts selected from http://procurement-oag.in.th, public procurement of supplies and services, accessed on June 1, 2012. Political affiliations from various news; party contributions from www.ect.go.th. USD 1 = THB 30 approximately.*

Interviews with contractors who asked to remain anonymous indicate that bid collusion is the norm rather than the exception. Collusion can take several forms. There could be the formation of a bidding ring, whether *ad hoc* or longstanding, among the private contractors. Bid rigging occurs when members of the bidding ring offer prices within a narrow (inflated) range. Bid rotation often happens, whereby suppliers take turns at offering the lowest bid and therefore getting the contract. This is also called complementary bidding. In some cases, bid suppression occurs, where competitive suppliers are coerced into withdrawing from the bidding process or suppliers in the bidding ring abstain. Any of these practices could be combined with subcontracting, whereby members of the bidding ring receive subcontracts from the winning bidder. In many cases, the procurement officer is complicit and can exclude nonfavored potential bidders (who are not part of the network) through various ways, such as sending the project announcement to select groups only. Network relationships are built up through time and often succeed in eliminating competition. Many corruption cases are currently being investigated by the NACC.

Case 4: Thai Oil Public Company (TOP): Outside Interference and Intimidation

The case of TOP involved the purchase of a Distributed Control System (DCS), which controls the production processes of the refinery, and an Instrumented Protection System (IPS), which is the emergency shutdown system for the refinery.

The procurement irregularities are summarized in Figure 13.5.

A search of the business registration records and board membership shows cross-membership among them (Table 13.4). It remains to be determined, however, whether or not bid rigging and collusion occurred. If so, the companies can be indicted for collusion.

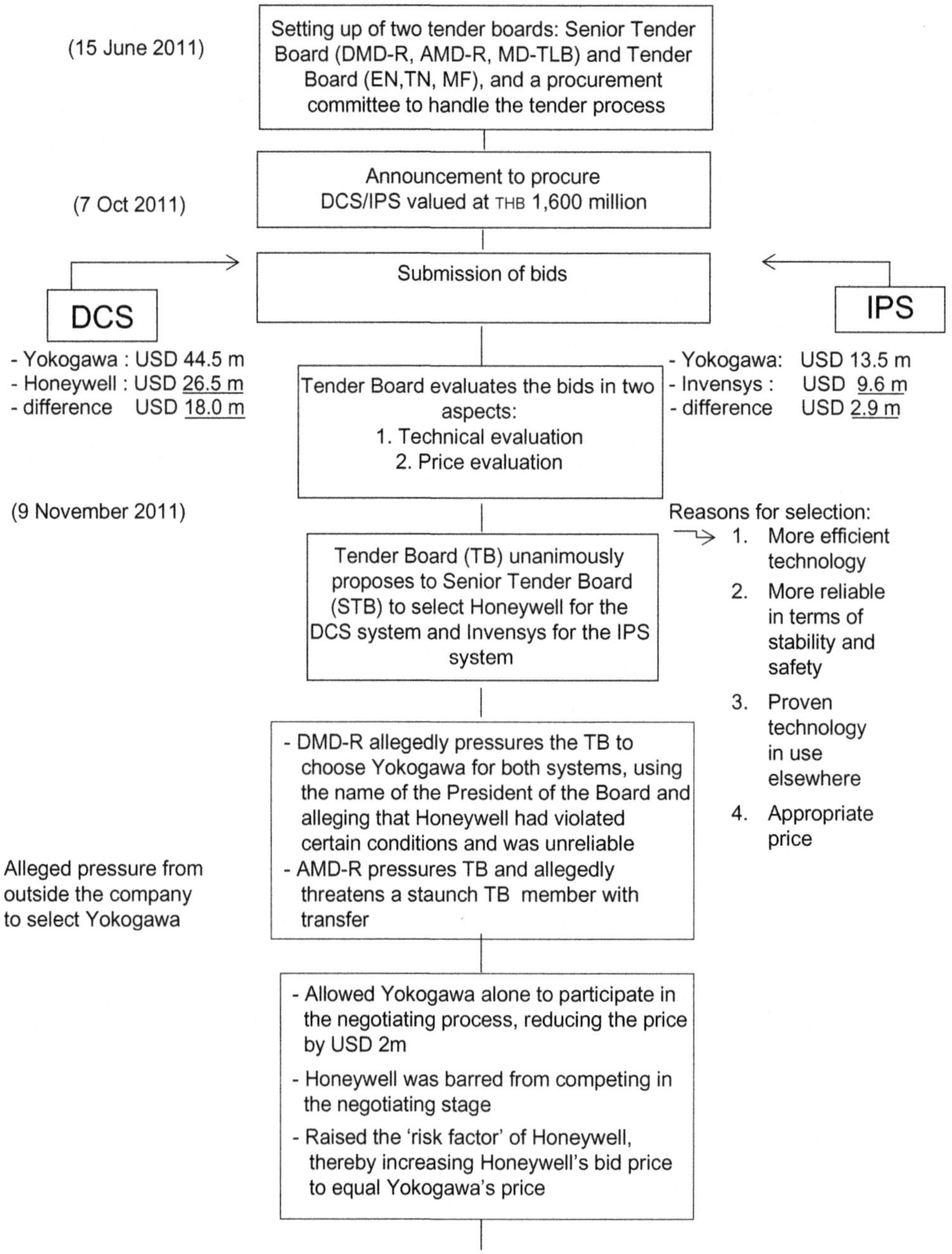

(15 June 2011)

Setting up of two tender boards: Senior Tender Board (DMD-R, AMD-R, MD-TLB) and Tender Board (EN,TN, MF), and a procurement committee to handle the tender process

(7 Oct 2011)

Announcement to procure DCS/IPS valued at THB 1,600 million

Submission of bids

DCS

- Yokogawa : USD 44.5 m
- Honeywell : USD 26.5 m
- difference   USD 18.0 m

IPS

- Yokogawa: USD 13.5 m
- Invensys :   USD 9.6 m
- difference   USD 2.9 m

Tender Board evaluates the bids in two aspects:
1. Technical evaluation
2. Price evaluation

(9 November 2011)

Tender Board (TB) unanimously proposes to Senior Tender Board (STB) to select Honeywell for the DCS system and Invensys for the IPS system

Reasons for selection:
1. More efficient technology
2. More reliable in terms of stability and safety
3. Proven technology in use elsewhere
4. Appropriate price

- DMD-R allegedly pressures the TB to choose Yokogawa for both systems, using the name of the President of the Board and alleging that Honeywell had violated certain conditions and was unreliable
- AMD-R pressures TB and allegedly threatens a staunch TB member with transfer

Alleged pressure from outside the company to select Yokogawa

- Allowed Yokogawa alone to participate in the negotiating process, reducing the price by USD 2m
- Honeywell was barred from competing in the negotiating stage
- Raised the 'risk factor' of Honeywell, thereby increasing Honeywell's bid price to equal Yokogawa's price

FIGURE 13.5

Thai Oil Public Company: procurement process of a computerized DCS and IPS.

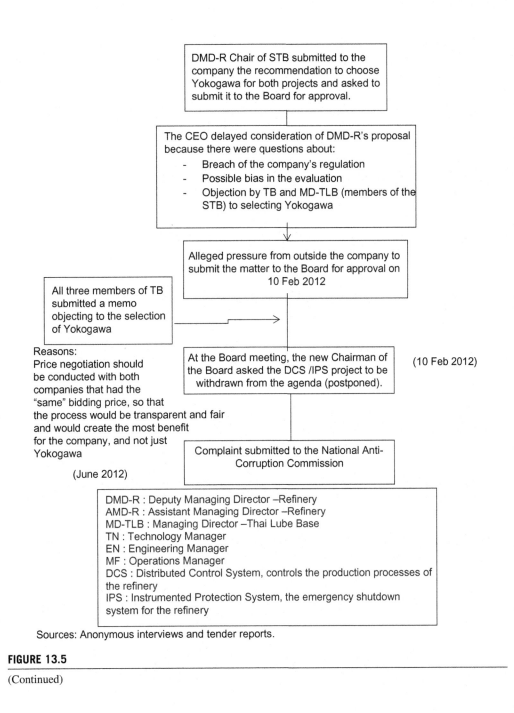

DMD-R Chair of STB submitted to the company the recommendation to choose Yokogawa for both projects and asked to submit it to the Board for approval.

The CEO delayed consideration of DMD-R's proposal because there were questions about:
- Breach of the company's regulation
- Possible bias in the evaluation
- Objection by TB and MD-TLB (members of the STB) to selecting Yokogawa

Alleged pressure from outside the company to submit the matter to the Board for approval on 10 Feb 2012

All three members of TB submitted a memo objecting to the selection of Yokogawa

Reasons:
Price negotiation should be conducted with both companies that had the "same" bidding price, so that the process would be transparent and fair and would create the most benefit for the company, and not just Yokogawa

(June 2012)

At the Board meeting, the new Chairman of the Board asked the DCS /IPS project to be withdrawn from the agenda (postponed).

(10 Feb 2012)

Complaint submitted to the National Anti-Corruption Commission

DMD-R : Deputy Managing Director –Refinery
AMD-R : Assistant Managing Director –Refinery
MD-TLB : Managing Director –Thai Lube Base
TN : Technology Manager
EN : Engineering Manager
MF : Operations Manager
DCS : Distributed Control System, controls the production processes of the refinery
IPS : Instrumented Protection System, the emergency shutdown system for the refinery

Sources: Anonymous interviews and tender reports.

FIGURE 13.5

(Continued)

Table 13.4 Related Companies and Board Members

Company Name	Mr. Somkhit Theeraboonchaikul	Mr. Takeshi Aoyama	Mr. Wichai Suwannavasit	Mr. Somchart	Mr. Thee	Mr. Masatochi Nakahara
1. Yokogawa (Thailand) Ltd.	✓	✓	✓	✓	✓	✓
2. E&I Solution	✓	✓	✓			
3. Advance Computer Machinery Co., Ltd.			✓			
4. Bilonker Company Limited		✓				
5. Boonkij Development Ltd., Partnership			✓			
6. Computer and Communication Dynamic Co., Ltd.			✓			
7. Direction of Technology Co., Ltd.			✓			
8. Dynamic Computer Co., Ltd.			✓			
9. Dynamic Engineering System Co., Ltd.			✓			
10. Dynamicsupply Engineering Registered Ordinary Partnership		✓				
11. Edynamic Co., Ltd.			✓			
12. River Process Engineering Co., Ltd.			✓			
13. Thai International Foam Co., Ltd.			✓			
14. Wood Talk Co., Ltd.			✓			

Source: *Government of Thailand, Database of Registered Companies*

13.4 **REDUCING CONNECTED DEALINGS AND IMPROVING PROCUREMENT IN THAILAND**

To counter corruption networks, it is important to note that certain types of social action, even if effective, may change the types of connections created but not necessarily reduce their number or importance. The threat of punitive social action on certain transactions induces connections of mutual dependence within the network at different stages. High penalties on crimes in general increase the mutual dependence of the criminals, but not necessarily their number, if the network is strong.

The difficulty with network relationships and social capital is that they can benefit or harm society. Close-knit, trusting criminal groups may create networks based on a mixture of empathy, threats, and shared goals that negate law enforcement efforts (Gambetta, 1988). Trust among law enforcers facilitates processing of cases; trust motivated by moral values, such as respect, even when extended altruistically, can mitigate against effective law enforcement. Both kinds of organizations can exhibit interpersonal solidarity, and can create benefit or harm; a critical mass of desirable networks needs to be created because people are affected by their perception of what others are doing (Fehr and Gachter, 2000).

What is perhaps alarming in Thailand is that harmful networks are being created and perpetuated. Massive proceeds of corruption can be used to mobilize supporters to protest court verdicts, indictments, and anti-corruption efforts. Network members show solidarity with the group or network leader, whatever the issue, by their readiness to protest, block roads, put up barricades, and even openly threaten the lives of judges and their families. The creation of networks is grafted on the "democratic" machinery, and right and wrong become a popularity contest.

Further complicating the situation is the possibility that a person helps another or fails to act against certain persons without any corrupt intent or even expectation of reciprocity. A person's good reputation or affiliation (e.g., being a child of a respected person) can cause others to act in his or her favor without expecting any personal gain. Deference to an "institution" can also be the motivating factor. For example, Thais often advise children to respect "the monk's cloth" and ignore the monk's behavior.

Accusations of "double standard" float around every day. The question on many people's minds is whether "good" people should allow themselves to be punished, knowing that the "bad" people would get off scot free because of their connections. There are a few examples of "good officials" resigning their posts when faced with corruption charges; the really corrupt, however, tenaciously cling on to their positions. Both good and bad people have connections. "Everybody knows everybody" and is related to everybody—this is the key to Thailand's political turmoil. When suspicions about corrupt acts arise concerning "someone known for a long time," even uncorrupt academics would refuse to cooperate even when they know the facts. The Thai word *khon gun eng* and Chinese word *kaki nung*—meaning, "familiar people"—are often evoked. Indeed interpersonal relationships play an important role in every aspect of life, including securing procurement contracts.

The returns to corrupt acts depend on how diffused corruption is in society—that is, how much corruption is inherited from the past. The larger the share of corrupt agents, the higher the returns to corruption, because of several reasons. First, auditing corrupt officials is not easy when corruption is widespread in society. Second, the expected profitability of corruption from an individual viewpoint is a positive function of the degree to which a society as a whole is corrupt (Andvig and Moene, 1990).

Third, corrupt officials have an incentive to establish communication or networking among themselves and fuel the corruption ring (Sah, 1988). In addition, in societies where rent-seeking and bribery abound, the return to rent-seeking relative to entrepreneurship is high (Murphy et al., 1991, 1993; Acemoglu, 1995). Fourth, individuals may be discouraged from fighting corruption when it is widespread, even if everybody would be better off if corruption was eliminated.

To tackle corruption in procurement, reforms are needed in the following areas: (i) legal infrastructure, (ii) corruption-friendly economic policies, (iii) upgrading of the database, and (iv) social mobilization for enhanced transparency. This chapter argues that membership in international conventions, such as the World Trade Organization's Government Procurement Agreement (WTO-GPA), could help alleviate current problems.

13.4.1 IMPROVING THE LEGAL INFRASTRUCTURE

The legal infrastructure needs to be reformed in many ways. This section discusses two key points.

First, even though Thai procurement regulations emphasize openness and transparency as the main principles, there is room for many improvements. In terms of openness and transparency, announcements and dissemination of information are required to go through the Public Relations Department, Mass Communication Organisation of Thailand, the G-Procurement website, etc. Procurement committees have to be formed, often with citizen participation. Contracts worth more than THB 1 million have to be sent to the Office of the Auditor-General and Revenue Department within 30 days of signing. Regulations for e-procurement include additional criteria: value for money, transparency, efficiency and effectiveness, and accountability and responsibility for completion. At least three tenders (in the case of standards license or meeting quality control systems) are required. Indeed, Cabinet decisions are announced on the website but in very brief form only. In the past, many cases that had led to corruption cases were approved via Cabinet decisions. More detailed disclosures should be required.

Second, "locking of technical specifications," which favors certain suppliers, is practiced. The dilemma is how to specify enough details so as to verify suitability and at the same time avoid such specificity that may exclude some suppliers, particularly for complex, sophisticated procurements that need customized designs. Often the suppliers themselves are consulted for expert knowledge, even though this is prohibited.

On the other hand, *functional requirements* are not specified in Thai regulations, and technical specifications are invariably related to physical characteristics. This area could be the focus of reforms; it would require nationwide training of procurement officers.

The NACC recently amended the Anti-Corruption Law, allowing it to closely monitor large procurement projects and requiring procuring agencies to publish and explain how the reference prices are calculated. Protection for whistle blowers has also been introduced. Moreover, a new Integrity Index has been constructed to evaluate all government agencies at the departmental level. It gives marks for procurements done using proper procedure and with transparency.

13.4.2 TARGETING CORRUPTION-FRIENDLY ECONOMIC POLICIES

The NACC now plays an increasingly proactive role in scrutinizing economic policies and measures that open up opportunities for corruption. Several pre-emptive interventions had been made, such as in the scheme to lease 4000 natural gas operated city buses and the auction of 3G telecommunications licenses, which subjected the projects to greater scrutiny. Subsequently, the Cabinet asked the proposers of the bus scheme to withdraw the project so that further studies could be done on aspects questioned by the NACC technical committee.

Reform is also needed in the intervention schemes for the agricultural sector. At the behest of the NACC, reforms in this area have been started; however, the new government in 2011 seems to be bringing back some of the old risky policies.

13.4.3 UPGRADING OF THE DATABASE

Proactive approaches to countering corrupt networks require knowing the identity of the network members. Efforts are currently underway to improve the anti-corruption database, with linkages to information from financial institutions, the Land Department, vehicle registry, business ownership, and tax returns.

Nongovernment groups are also collecting information on and analyzing the behavior of elected officials. For the 2011 election, a civil society group collected information on Parliament members' absences from parliamentary meetings and distributed copies of each member's "report card." Regretfully, in spite of the dismal scores given to the MPs who failed to attend any parliamentary sessions or were absent for more than 50% of the time, they were all reelected.

13.4.4 INCREASED SOCIAL MOBILIZATION FOR ENHANCED TRANSPARENCY

Transparency in procurement is vital to prevent wrongdoing. The Thai procurement regulations actually require publication of procurement information. In cases of complaints and appeals, the regulations state that aggrieved suppliers or contractors may lodge complaints directly with the procuring agency, the Committee in Charge of Procurement, or the Petition Council. If with the Petition Council, the petitioner must lodge the complaint within 90 days of knowledge of wrongdoing. The Council is to consider the petition "without delay" and recommend remedial measures, if any, within seven days to the Prime Minister. Such measures may include overturning acts that are inconsistent with the law or cannot be supported by "justifiable reason." The Council itself may issue an interim remedy when appropriate.

However, transparency remains a problem, and efforts are needed to mobilize stakeholders in the society. This may sound like a broken record already, but the means of mobilization itself have to be overhauled so that citizens could obtain some benefits from their involvement that would make it worth their while. Given the network connections and the possibility of retaliation, Thailand continues to grapple with the means to mobilize and incentivize citizens in the fight against procurement corruption.

Khoman et al. (2009) and Tangkitvanich et al. (2009) have proposed that Thailand's membership in the WTO-GPA could be a tool for increasing transparency in government procurement. Aside from increasing transparency, WTO-GPA membership could also lead to more efficient use of government budgets, as it would stimulate fair competition, help honest and efficient suppliers, and may foster industrial growth and development. Greater foreign involvement and competition can thus help uncover or prevent wrongdoing. There is some apprehension about becoming a WTO-GPA member, however. First, opportunities for Thai suppliers to access WTO-GPA procurement markets remain limited, and domestic suppliers will face stiffer competition from foreign competition. This is the familiar "infant industry" argument. However, if long-run efficiency is the goal, then gradual expansion of competition may help attain it. Second, there are some concerns that opening up may lead to international collusion instead of greater competition and efficiency. Third, if foreign governments subsidize their service sectors, particularly construction, the WTO-GPA does not have any provision for countervailing action or remedy, unlike the case of subsidies under WTO's General Agreement on Tariffs and Trade (Khoman et al., 2009).

13.5 CONCLUSION

In a society dominated by interpersonal relationships for social, business, and other activities, understanding these relationships is key to understanding corruption. Networks cut across the usual socioeconomic characteristics because members with different skills and characteristics are required for a corrupt network to be effective. On the other hand, designing a procurement system is a big challenge, particularly since government procurement usually involves multiple objectives, with efficiency being just one of them. In addition, procurement is often used as a means to effect a geographical redistribution of income or to favor underprivileged groups (e.g., people with disabilities). The system must also align public benefit with personal incentives, as the same observed behavior could be motivated by opposite motives. Strict conformity to rules sometimes results in less efficiency; the "special method" could reflect either a sinister motive or a desire to be efficient; the lowest price may involve sacrifice of quality; and detailed specifications could limit competition.

At the societal level, social enforcement of private contracts, ready access to adjudication, morality, and religious pressure for generalized honesty (in contrast to "contextual morality") cannot be overlooked. These elements tend to reduce the importance of identity (thus facilitating transactions between strangers) and the need for specific mutual investment by connected parties. Procurement also needs to be accompanied by effective monitoring systems (e.g., corruption report, witness protection, etc.) and sufficiently stringent penalties for the wrongdoers and conspirators.

Thailand's experience shows that developing countries dominated by personal networks can fall prey to so-called democratic reforms that usher in elections and prebendal politics without proper infrastructure to provide checks and balances. Blatant cases of connected dealings reveal that even when the legal framework exists, rules can be violated with impunity if the connected persons feel "protected" by their corrupt network. The larger the network of corruption rings, the larger the returns to corrupt acts. Thus, in addition to the legal infrastructure, the creation of a certain critical mass of networks of clean officials is absolutely vital to counter the corrupt networks.

REFERENCES

Acemoglu, D., 1995. Reward structures and the allocation of talent. Eur. Econ. Rev. 39, 17−33.
Andvig, J.C., Moene, K.O., 1990. How corruption may corrupt. J. Econ. Behav. Organ. 13 (1), 63−76.
Bangkok Post, June 16, 2008. New Thaksin charges. (Accessed May 20, 2012.)
Fehr, E., Gachter, S., 2000. Cooperation and punishment in public goods experiments. Am. Econ. Rev. 90 (4), 980−994.

Gambetta, D., 1988. Can we trust trust? In: Gambetta, D. (Ed.), Trust: Making and Breaking Cooperative Relations. Basil Blackwell, Oxford, UK, pp. 213–237.

Government of Thailand, 1999. Act on Offences Relating to the Submission of Bids to State Agencies, Bangkok, Thailand.

Government of Thailand. Assets Examination Committee, 2007. Corruption Cases.

Government of Thailand. Database of Registered Companies. Department of Business Development, Ministry of Commerce, Bangkok, Thailand.

Government of Thailand. National Statistical Office, Socio-Economic Surveys, 1994, 1996, 1998, 2000, 2002, 2004, Bangkok, Thailand

Government of Thailand. Regulation of the Office of the Prime Minister on Procurement, 1992, amendment No. 6, 2002, Bangkok, Thailand.

Jaruk, S., Issariya, B., 2010. The Political Economy of Longan Market Intervention. Research Report Submitted to the National Anti-Corruption Commission, Bangkok, Thailand.

Joseph, R.A., 1987. Democracy and Prebendal Politics in Nigeria. Cambridge University Press, Cambridge, UK.

Khoman, S., Mingmaninakin, W., Thosanguan, V., Tantivasadakarn, C., Chetsumon, C., Suntharanurak, S., et al., 2009. The World Trade Organisation's government procurement agreement: a study of Thailand's preparation for accession. Research Report Submitted to the Comptroller-General's Department, Ministry of Finance, Bangkok, Thailand (in Thai).

Murphy, K.M., Shleifer, A., Vishny, R.W., 1991. The allocation of talent: implications for growth. Q. J. Econ. 106, 503–530.

Murphy, K.M., Shleifer, A., Vishny, R.W., 1993. Why is rent seeking so costly to growth? Am. Econ. Rev. 83, 409–414.

Pan, A., Withee, P., Sutthi, S., 2012. Thai Budgetary Reform to Counter Corruption. Research Report Submitted to the National Anti-Corruption Commission, Bangkok, Thailand (in Thai).

Rose-Ackerman, S., 2001. Trust, honesty, and corruption: reflection on the state-building process. John M. Olin Center for Studies in Law, Economics, and Public Policy Working Papers. Paper 255. <http://digital commons.law.yale.edu/lepp_papers/255>. (Accessed January 20, 2013.)

Sah, R.K., 1988. Persistence and pervasiveness of corruption: new perspectives. Yale Economic Papers. Yale University, New Haven, CT.

Tangkitvanich, S., Intrawitak, C., Rattanakum, S., 2009. Evaluating Costs and Benefits of Joining the Government Procurement Agreement (GPA). Research Report Submitted to the Comptroller-General's Department, Ministry of Finance, Bangkok, Thailand (in Thai).

FURTHER READING

Bangkok, Thailand. Available from: <http://www.transparency-thailand.org/thai/index.php/2012-05-07-10-50-47/138-2012-05-07-07-08-28> (in Thai).

<http://www.moj.go.th/Law/MojLaw/EngLaw/Act%20Con>. (Accessed June 1, 2012.)

<www.ect.go.th>. (Accessed June 1, 2012.)

. (Accessed June 1, 2012.)

<www.sec.gov/news/press/2010/2010-144.htm>. (Accessed June 1, 2012.)

<http://www.transparency-thailand.org/thai/index.php/2012-05-07-10-50-47/83-2> (in Thai). (Accessed June 1, 2012.)

<www.weforum.org/en/initiatives/gcp/Global%20Competitiveness%20Report/index.htm> (Accessed June 1, 2012.)

The Plurilateral Agreement on Government Procurement (GPA). <www.wto.org/english/tratop_e/gproc_e/gp_gpa_e.htm>. (Accessed March 14, 2014.)

THE NATURE, CAUSES, AND CONSEQUENCES OF AGRICULTURAL DEVELOPMENT POLICY

THE ROLE OF AGRICULTURAL ECONOMISTS IN SUSTAINING BAD PROGRAMS

14

Brian Wright

*Department of Agricultural and Resource Economics, UC Berkeley, Berkeley, CA and
Member, Giannini Foundation, San Francisco, CA*

JEL: Q1

Agricultural economics has a long and proud tradition of developing and disseminating practical tools that can quantify the gains and losses from changes in policies or programs and show the net change in efficiency. This type of analysis drives home the point that the redistribution achieved usually dominates the efficiency effect, at least when decisions are made on a case-by-case basis, and that the natural result is that political forces tend to reflect the concentrated interests in prospective gross benefits or losses more effectively than the diffused interests in net gains.

Land grant universities in the United States and public universities in Australia, New Zealand, and the United Kingdom have produced generations of applied economists who have worked on policy analysis mostly in public and nonprofit institutions. They have helped stave off numerous bad policy initiatives and initiated some very useful reforms.

Even those who attained their greatest achievements elsewhere were significantly influenced by United States land grant economists. In Australia, starting in the 1970s, many such economists contributed to the work of the Tariff Board as it morphed from granting protection to all comers into an organization that reformed not only tariff rates but also the prevailing attitudes of economists, journalists, and the general public to industrial and agricultural policy interventions that were strangling national economic progress (Productivity Commission, 2003; Corden, 1996). In New Zealand, a small group of agricultural economists spearheaded the world's fastest and most complete transformation of a national agricultural policy. They seized upon a unique political opportunity to turn the country's highly protected agricultural sector into an efficiency leader in global markets that now dominates international dairy trade and is a major player in many specialty products (Sandrey, 1990). In China, Chicago-trained economists influenced by T.W. Schultz, whose doctorate was from the University of Wisconsin, a land grant college, guided the path of reform that since 1978 has transformed the lives of many who were born near the bottom of the world's income distribution.

It is, I posit, a mistake to frame these achievements as proof of the benefits of introducing innovative new economic theories into policy discussions. The innovations were largely in insights, rhetoric, and persuasion, drawing on new capabilities in mathematical modeling and empirical testing; the theories are basically those outlined by Adam Smith two centuries ago, which had been largely neglected by *dirigistes* who had done so much damage. The effects were largely corrective, diminishing the influence of some arguably innovative and certainly creative economic theories regarding sources of economic growth and the role of industrial policies. These theories had neglected (or even assumed the absence of) private saving, private investment responses to incentives, and private capacity to bear risk.

The battle was fought in the public eye; combatants on both sides were generally academics, social philosophers, or employees in the public or nonprofit sectors. Few participating economists were full-time consultants or employees in the private sector, largely because the latter perceived little need for them. The dominance of nonprofit employment of economists probably reflected a general lack of appreciation of the ability of individuals and firms to take care of their economic challenges without public intervention; it frequently and understandably tended to persuade the profession to focus on policy and programs favoring public and nonprofit institutions. The battle was in large part about the means, not the ends; all parties were pursuing what they conceived to be the pubic interest.

Recently, the balance seems to have shifted. Private sector employment or income has become more important to the profession. Corporations have learned that consultants and in-house economists can add to their bottom line and that commercial or academic consultants can be worth their significant fees. In many cases the results have been impressive, refreshing, or at least interesting. Some have helped separate the wheat from the chaff (much of the latter generated by some financial economists who really seem to believe that prices are true "fundamentals") regarding the evils of speculation and the need for further market regulation. They have argued that the welfare benefits of large national retailers, which follow a strategy of low pricing, can be huge and have shown that patents and venture capitalists seem important for the success of innovative startups. They have estimated how much genetically modified crops have raised yields and/or reduced farm workers' exposure to harmful chemicals, when prominent nongovernment organizations have been loudly proclaiming that such innovations were useless, if not harmful. They have showed that export bans can have devastating consequences for global commodity markets and for farmers. They improve insurance pricing with better loss forecasting, and they develop means of identifying arbitrage opportunities that improve the functioning of markets.

Most of the applied economists I know work, as I do from time to time, as expert witnesses in litigation; others are engaged in policy analyses designed to maximize the profits of their clients or employers. Others consult for public sector regulators who attempt to control rampant moral hazard in, for example, subsidized crop insurance programs marketed by private firms. A few work on fashioning internal corporate strategies or advise on investment decisions. Many others work for the private sector indirectly through grants sourced via their academic or nonprofit employers. Other colleagues are in-house employees of private firms, but these still seem to be a small minority; I personally know only a handful of these.

No doubt the shift toward private sector employment (or, more importantly, private sector consulting) has introduced some useful knowledge of the needs and capabilities of private firms into the work of our discipline. However, anyone who fondly anticipated that the new private sector orientation would foster the advocacy of development of fierce competition and productive efficiency, promoting a new and fruitful appreciation of the true lessons of Adam Smith, will have already been disappointed. Such misguided souls would, like the rest of us, be well advised to pay closer attention to what Smith actually said about the kinds of policies promoted by private industry groups.

This short commentary's sole instructive, but not necessarily representative, example is the evolution of multiperil crop insurance in the United States. These comments would be more or less irrelevant had the World Trade Organization (WTO) not outlawed many forms of support for agriculture but allowed the proliferation of subsidized insurance programs.

About 10 or 15 years ago at a beautiful banquet in Paris, I found myself sitting next to the Secretary of a French Wheat Growers Association. This was not too long after the WTO agreement and I remarked, "So, is your association interested in programs for crop insurance?" He responded, "Yes, we are very interested and, in fact, we are now trying to design some programs. We have only one problem." "What is that?" I asked. "We do not have much yield variation," he replied.

The United States has finally caught up with, if not surpassed, France. Just last year, peanut farmers began complaining that the crop insurance program in the proposed new farm bill was unfair. Because their production is quite stable, they did not stand to participate in the benefits of the grotesquely costly and inefficient transfer program.

The US crop insurance started as a pilot program during the New Deal era. Prior to the early 1980s, US crop insurance, though wasteful, was not very expensive. It was sensibly a modest pilot project. It did not apply to many crops and payouts were not set at very high levels, so the subsidies were modest. In the 1980s and 1990s, much larger transfers evolved in the form of "*ad hoc* disaster assistance" (Wright and Hewitt, 1994). (As observed during the last US presidential election campaign, disaster payments are politically popular. People obviously in trouble, and others plausibly in trouble, get significant payments, especially in parts of the country likely to determine the outcome of an imminent election.) These disaster payments were increasing and becoming quite expensive. Billions of dollars were paid almost every year—quite a lot of money in those days.

What economists call a time inconsistency problem was evident to all, and Congress (or its attendant lobbyists) found an ingenious way to exploit it. They argued that they recognized a problem in granting too many disaster payments, but that they could not stop themselves from doing it. Consequently, farmers know they are going to get disaster payments, which lowers their demand for insurance and the incentive for protecting themselves against risk. Government's promises to discontinue disaster payments are just not credible. Even constituencies stridently opposed to interference from Washington and disgusted with excessive handouts to the unworthy turned out to be quite happy to take the disaster payouts. (Both senators from Oklahoma, who had adamantly opposed emergency support for the state of New Jersey after Hurricane Sandy, later enthusiastically supported disaster relief after a huge tornado hit their home state.)

Investment in mitigation is low and competitive insurance that could bring the private cost of rebuilding closer to the social cost is underdeveloped. Hence public intervention after a disaster is all the more necessary. This reminds me of a story (too good to investigate too critically) that some sugar substitutes with zero calories are appetite stimulants that help ensure their own continued demand.

Arguing that it could not resist the pressure for disaster payments each year, the US Congress voted to expand subsidized crop insurance, making the plausible prediction that the pressure for disaster payments would subside. The persuasive spin was that offering crop insurance of the multiple peril type, or increasing its coverage, would decrease the political pressure for disaster payments, which were obviously very wasteful, unpredictable, and growing out of control. The crop insurance program would be more predictable and generate less moral hazard and adverse selection; it would be less wasteful than the disaster payment program.

Uncritical economists, thrilled to be able to demonstrate the value of their training by applying risk models in a seemingly sophisticated second-best analysis, lent ample support to this cover story. (Remember, there was a time when some Congress members were still interested in economists' welfare analyses of costs and benefits of policy proposals.) Congress was delighted to be able to increase transfers to its rural supporters and the insurance industry, while looking responsible at the same time. Thus, crop insurance expanded in the 1980s, not because it was thought of as really good public policy on its own, but on the argument that it would mitigate a worse policy.

The solution turned out to be far more costly than the original problem. Disaster payments increased and crop insurance expenditures soared; now we even have officially designated "permanent" disaster payments!

If one is looking for good government, this is not an example. However, if one is a farmer or an insurance agent looking for generous handouts, the outcome can be wonderful: high payments generate rents to both along the way. Of course, some provisions specified that farmers who wanted to be eligible for crop insurance would have to waive claims for disaster assistance and that enrollment in a minimal program was a necessary condition for receipt of disaster relief. When the next disaster came along, the government duly waived these provisions. They actually gave disaster assistance to people who also had crop insurance—a case of double handouts.

Thus the US Federal crop insurance program has become one of the most expensive ways of transferring income to farmers. It provides insurance products that, absent subsidies, most farmers would never buy. Even Blanche Lincoln, then Chair of the Senate Agriculture Committee, pushed for *ad hoc* disaster assistance to farmers in her own state, which would also receive substantial payments from crop insurance.

Some of the very authors of legislation introducing insurance programs promoted as mitigating disaster payments (Bruce Babcock is an admirable, and vocally critical, exception) clearly were actually pushing for the controls to be vitiated and payments increased. One may support these programs because they are politically expedient, but to pay a proper regard for the virtue of hypocrisy, it is necessary to make them seem respectable. Academic economists stand ready to help with "second-best" political economy rationalizations. If only they were merely second best!

In the naïve economic theories used to buttress the case for the expansion of crop insurance, the (generally implicit) assumption has been that variations in crop yield coincide with fluctuations in farmers' consumption and that crop insurance stabilizes consumption. These assumptions did not come from direct measurement of the correlation of crop yields and farmer consumption but from experiments designed to test for risk aversion.[1] These experiments were almost always based on atemporal, one-shot games, wherein changes in outcomes were implicitly equated to changes in consumption.

Occasionally, agricultural economists would do something in the field to actually test these theories beyond, say, risk premia for gambles of two dollars up or two dollars down. When I was an undergraduate at New England in the 1960s, Dillon and Anderson were leaders in introduction and critical discussion of expected utility analysis, in work which culminated in Anderson, Dillon and

---

[1]Testing farmers' marginal propensity to consume out of current income received much less attention than crop insurance. A useful example is Langemeier and Patrick (1990).

Hardaker (1977). Officer and Halter (1968) were surveying and then interviewing farmers northern New South Wales, testing farmers' attitudes toward risk and then measuring how much hay they stored against the next drought to see if there is any relation between the two. This work, which raised questions about the von Neumann approach to risk preference elicitation, was quite novel for economists at that time. A decade later Hans Binswanger (1981), who, like my mentors Jock Anderson and John Dillon, was more sophisticated than typical risk aversion modelers of the era, played von Neumann-Morgenstern games with farmers in India at economically meaningful levels, up to a month's income. He found that the (reasonable) proposition that risk premia should respond to wealth changes did not explain the farmers' choices. I was at Yale by that time and Binswanger's results were noted by theorists in passing as a curiosity produced by a person with a quirky fascination with actual evidence in a real-world setting. This was, as far as I know, the first evidence of people playing von Neumann-Morgenstern games with amounts of money large enough to reflect the implications of serious farm-income fluctuations, and the theory did not work out too well. More scholars are now realizing that such real-world evidence has its place in economics; extrapolating results from small-stakes, one-shot classroom games to farmer behavior is too big of a stretch. That realization has been a long time in coming.

George Patrick (1988) subsequently pursued in Australia another line of research on farm decision making under uncertainty, directly addressing crop insurance. If one wants risk, Australia is the place to find it. In the Mallee region of Australia, people grow wheat where the average rainfall is 8 inches a year and the coefficient of yield variation is about 0.4. They grow wheat on the mere chance that it will rain enough, at the right time, in a given year. If these farmers are asked what their major risk is, they will say that it is rainfall risk, but most of them many years ago were still unwilling to buy index insurance based on rainfall at rates calculated to be actuarially fair (and therefore less than the competitive rates that would cover the minimum load of perhaps 0.2 or 0.3 and would account for administrative costs).

Given their reactions, one might conclude that these farmers were totally ignorant of insurance or just philosophically independent, but most of them also had other forms of insurance, such as car insurance in excess of the required minimum, and substantial life insurance. They graduated from school at an average age of about 15 and had wives who tended to be better educated, often with jobs such as school teaching. They were hardly averse to education. Nor were they unsuccessful. They typically started farming as teenagers, but their average age was around 50. This was all inconvenient evidence that, in fact, farmers did not see a value proposition in unsubsidized crop insurance or even index insurance. This inconvenient evidence did not have any effect on support by many empirical economists, including prominent development economists, for crop insurance. One reason, I discovered, is that they are not aware of (or at least not citing) this old empirical work.

How can we explain the lack of demand for this kind of insurance? I think farmers understand index insurance quite well, and after much of a lifetime in farming they understand weather risk very well. Farmers' reluctance to buy multiperil crop insurance does not imply that they avoid any kind of insurance; many have ample life and automobile insurance. Many farmers who will not buy multiperil crop insurance will, when necessary, insure their crops against fire at the end of the season. In the Australian context, when ripened wheat is exposed to fire, there is a limit to what farmers can do to reduce the risk. So private insurance for this well-defined and verifiable hazard can be feasible and profitable. The coverage is very specific, the hazard is easily observable, and there is minimal moral hazard. In many countries, farmers support a market in hail insurance for the same reasons. Nowhere in the world have farmers supported a sustained market for unsubsidized multiple-peril crop insurance or revenue insurance.

Several years ago, my students were frequently telling me about multiple-peril insurance programs in India and China with 10,000 farmers in the first year and 20,000 in the second. It sounded very plausible that the programs were working well, but I happen to know the person in charge of running many of these programs at the World Bank. It turns out that the 10,000 farmers in the first year were different from the 20,000 farmers in the second year. Times were trying, and if it was not sufficiently heavily subsidized, they would not keep on buying the insurance offered.

How do the Australian Mallee farmers stabilize their consumption? First, if their spouses work in the local town in jobs such as teaching, their family income is substantially diversified already. However, they also do much to mitigate rain and price risks. Initially they accumulate substantial savings in liquid form, which means they pay an opportunity cost for this. Next, they diversify. For example, they raise sheep on their wheat farms. The sheep can graze on postharvest stubble. They produce wool and can produce fat lambs, too. So farmers can diversify their income sources via the meat and fiber markets when the crop is not very good. The sheep can eat what is left on the ground, even if it cannot be harvested profitably as grain.

Farmers are also very careful to coordinate household expenditures and their major capital expenditures. They buy cars and other major household items in good years when they also buy combined harvesters and tractors and use the income tax advantages when they are most valuable. What these farmers are really good at is risk management. They are able to smooth consumption without insurance. They pay some opportunity cost, but it is far less than what the unsubsidized cost of crop insurance would be.

Now what is the fundamental reason why crop insurance does not work well? The simple answer is the size of the administrative load. Even index insurance tends to cost at least 20% of the expected indemnities. That is just too much for a reasonable risk premium for one crop in one year. The second reason is that the original theory was based on the assumption that fluctuations in consumption are equal to fluctuations in income from a single crop, but since farmers actually save and invest more when crop income is high and income tax is progressive, consumption fluctuates less than crop incomes. (Check back and see the stack of studies that specify utility as $U(Y)$ and define Y as net crop income.) For index insurance, there is still moral hazard in rain measurement, especially if measuring stations are spatially dispersed. (I once interviewed a weather risk expert who had a consulting contract to design a rain gauge not subjected to manipulation. He eventually, and reluctantly, concluded that the task was impossible.) Furthermore, rainfall often varies greatly on a local scale, so the basis risk is large. The worst outcome for risk-averse farmers who buy index insurance is that the crop fails, but their local index does not reflect the low rainfall received by their crops.

At this point one might well be asking, "Why is index insurance so popular as a research topic for agricultural and development economists?" One reason already mentioned is a lack of awareness of the historical evidence. Another is that there is plenty of money for research and pilot programs. Why do insurance companies support experiments in this area? I found one answer at the FarmD conference in Zurich a few years ago. The head of index insurance at a major Swiss reinsurer assured me his corporation was very interested in index insurance programs, "but only in the pilot stage." At the pilot stage, relevant distributions of weather and yield are not yet well established. This just might plausibly offer a smart Swiss reinsurer some opportunities for supernormal profits.

Let me now comment on promising future research opportunities. Remarkably little is known about fluctuations of household consumption and their relation to economic factors. There is little beyond anecdotal and survey evidence for it, although economists studying microfinance projects seem to have learned quite a bit about the hitherto undocumented consumption smoothing capacities of seriously poor people.

Second, it would be really interesting to know what sort of a minimum level of plausibly reputable theory lobbyists now need in order to maintain sufficient support for subsidized crop insurance. Do they actually have no need of theory at all, just a PhD and an intuitively plausible rationale? As long as insightful, tough and independent economists like Jim Roumasset are around, their task will be harder.

Third, it would be interesting to know the opportunity cost of the resources recently expended on crop insurance studies by economists and economic students. Economists who are among the major architects of new multiple peril crop insurance or revenue insurance contracts have been openly and admirably critical of government policy in this area, but donors continue funding projects. These projects create the impression that these policies are respectable interventions, needing just a few more clever design modifications. Economists go where the money is, and too little of the latter is directed to empirical analysis that can clearly document the social costs of bad programs, like United States crop insurance.

REFERENCES

Anderson, J.R., Dillon, J.L., Hardaker, J.B., 1977. Agricultural Decision Analysis. Iowa State University Press, Ames, pp. x + 344.

Binswanger, H.P., 1981. Attitudes toward risk: theoretical implications of an experiment in rural India. Econ. J. 91 (364), 867−890.

Corden, W.M., 1996. Protection and liberalization in Australia and abroad. Aust. Econ. Rev. 114, 141−154.

Langemeier, M.R., Patrick, G.F., 1990. Farmers' marginal propensity to consume: an application to Illinois grain farms. Am. J. Agric. Econ. 72 (2), 309−316.

Officer, R.R., Halter, A.N., 1968. Utility analysis in a practical setting. Am. J. Agric. Econ. 257−277.

Patrick, G., 1988. Mallee wheat farmers' demand for crop and rainfall insurance. Aust. J. Agric. Econ. 32 (1), 37−49.

Productivity Commission, 2003. From Industry Assistance to Productivity: 30 Years of 'The Commission'. Productivity Commission, Canberra.

Sandrey, R., 1990. Farming without Subsidies: New Zealand's Recent Experience. Ministry of Agriculture and Fisheries, New Zealand.

Wright, B.D., Hewitt, J.A., 1994. All-risk crop insurance: lessons from theory and experience. In: Hueth, D.L., Furtan, W.H. (Eds.), Boston Economics of Agricultural Crop Insurance: Theory and Evidence. Natural Resource Management and Policy Series. Kluwer Academic Publishers, Waltham, MA, pp. 73−112.

AGRICULTURAL R&D POLICY AND LONG-RUN FOOD SECURITY

15

Julian M. Alston[*], **William J. Martin**[†], **and Philip G. Pardey**[‡]

[*]*Department of Agricultural and Resource Economics, University of California-Davis, Davis, CA*
[†]*Agriculture and Rural Development, Development Research Group of the World Bank, Washington, DC*
[‡]*Department of Applied Economics, University of Minnesota, St. Paul, MN*

CHAPTER OUTLINE

JEL: O30; Q16; Q18

15.1 INTRODUCTION

Agricultural science made great strides in the twentieth century. Between 1960 and 2012, the world's population more than doubled, from 3.1 billion to over 7 billion. Over the same period, growth in the global food supply significantly outpaced growth in demand, arising mainly from growth in population and income, to the extent that since 1975, real prices of cereals have fallen by roughly 60% (Alston et al., 2012). These gains are largely attributable to improvements in agricultural productivity achieved through technological changes, enabled to an increasing extent over time by investments in agricultural research and development (R&D). The benefits have accrued through enhanced farm incomes, lower food production costs, reduced stress on the natural resource base, and the release of resources for nonagricultural uses. The poor, in particular, have benefited from lower food prices.

The prospects for the next 50 years are less sanguine, however, given the new demands for bioenergy, emerging challenges of climate change, and limited scope for expanding the agricultural use of land and water, combined with generally slower growth in agricultural productivity in many countries and the world as a whole. In the face of these prospects, some substantial shifts have recently been observed in funding patterns for agricultural R&D around the world— but sometimes in the wrong direction! Even though rates of return to agricultural research are demonstrably very high, spending growth has slowed in numerous countries and funds have been diverted away from farm productivity enhancement, especially in the richest countries. The middle income countries, especially China, India, and Brazil, have risen in importance both as producers and consumers of agricultural output and as (public sector) producers of agricultural scientific knowledge. The private sector's presence in food and agricultural R&D has also grown, but this research is still overwhelmingly done by companies based in rich countries. The world's poorest countries may increasingly come to depend on the middle income countries and private sector participants as sources of farming (and other agricultural and food related) technologies.

Research-induced changes in farming technology and productivity can benefit the poor by reducing their costs of producing staple food items and driving down the market prices of food, thereby increasing their real incomes and making them less vulnerable to the consequences of production shocks or price changes. Reduced rates of investment in agricultural R&D, and consequently slower rates of farm productivity growth, can be expected to result in larger numbers of poor people who are more vulnerable to external shocks, like the price spikes for staple grains or the effects of unfavorable weather experienced by the US corn industry in 2012. This chapter presents simple analytical models and results from a computable general equilibrium (CGE) model of the world economy to demonstrate the important links between agricultural productivity and long-term food security of the poor. Before presenting these models and results, it first summarizes key points from the literature on the returns to agricultural R&D and recent evidence on the shifting global patterns of agricultural research investments and agricultural productivity and prices.

15.2 RETURNS TO AGRICULTURAL R&D

Hurley et al. (2014) recently updated the InSTePP (International Science and Technology Practice and Policy Center) metadata files used by Alston et al. (2000) and reconsidered the evidence on the returns to agricultural research published over the past half century. As discussed by Alston et al. (2011), the conventional internal rate of return (IRR) presumes implicitly that, as they accrue, all benefits are reinvested and earn the same rate of return. This presumption is implausible, however, for public research investments generating rates of return that are much higher than can be earned by the farmers and consumers who reap the returns. Alston et al. (2011) argue that a modified internal rate of return (MIRR) measure, which allows private beneficiaries to reinvest at a different rate, is more appropriate for publicly performed (and largely publicly financed) research; they illustrate its application to US public agricultural research. In keeping with that view, Hurley et al. (2014) developed an algorithm to transform published evidence on IRRs into counterpart MIRRs.[1]

Based on 2242 evaluations from 372 separate studies published between 1958 and 2011—almost half of which were published since 1990—the average reported IRR is 67.6% per year.[2] The distribution of IRRs is heavily right skewed, with a median of 42.6% per year. Using the MIRR to recast previous estimates of the IRR, Hurley et al. (2014) found much more muted returns to public agricultural R&D: a median of 9.8 versus 39% per year. These recalibrated estimates are still substantial compared with the opportunity cost of the funds used to finance the research.[3] They suggest that society has continued to persistently underinvest in public agricultural R&D, notwithstanding the distorted view of the evidence created by reliance on the IRR to represent the returns to this investment, which has characterized the literature for the past 50 years.

15.3 A NEW WORLD ORDER FOR AGRICULTURAL R&D SPENDING

In spite of these past high returns and clear evidence of persistent underinvestment, public support for agricultural R&D has waned in recent decades for roughly half the countries of the world (and two-thirds of the rich countries). Overall, the trends look encouraging at first blush. Worldwide, public investment in agricultural R&D increased from just USD 5.4 billion (2005 PP prices) to USD 33.7 billion in 2009—an average growth rate (inflation-adjusted) of 3.31% per year (Pardey et al., 2014).[4] However, big and, of late, accelerating geographical shifts in the location of performance of agricultural R&D have been observed.

In 1960, high income countries—classified as such according to average per capita incomes in 2009—accounted for 58% of the world's total: almost 50 years later, in 2009, that share had dropped to 48%. The United States has lost significant global market share, falling from 21% of the total in 1960 to just 13% in 2009. Sub-Saharan Africa has also lost market share, falling from 10% of the world's total in 1960 to 6% in 2009. So too has the Latin America and Caribbean region, except for Brazil whose share increased from 2% to 5%. The notable expansion in market share was in the Asia and Pacific region: China's share grew from 13% in 1960 to 19% in 2009, and India's, from 3% to 7%. In 2009, 31% of the world's public agricultural R&D took place in the Asia and Pacific region, compared with just 21% in 1960. Strikingly, in 2009 the middle

---

[1]In recalibrating the reported rate of return evidence, Hurley et al. (2014) used revealed US savings rates to estimate that portion of the benefit stream likely to be reinvested and also accounted for the deadweight loss of taxation incurred in publicly funded R&D.

[2]Half of the IRR observations are for research done in high income countries (with almost one-third of the estimates being for the United States), about 8% report on research that had global or multinational impacts, and the rest deal with research of direct consequence for low and middle income countries. Just 57% of the estimates are for research done on specific crops, 10% are for livestock research, 23% report the returns to an aggregate measure of agricultural R&D investments, and the rest deal with an assortment of other agricultural commodities.

[3]For example, in the United States, the real rate of return on long-term US treasuries averaged 2.96% per year over the period 1969–2010 and 2.33% per year for the years 1919–2010.

[4]Pardey et al. (2014) report a preliminary estimate for public agricultural R&D conducted in the former Soviet Union (FSU) and Eastern European countries of USD 1.14 billion (2005 PPP prices) in 2008. This figure includes estimates for the Czech Republic, Hungary, Poland, Romania, the Russian Federation, and 23 other countries from this region.

income countries (which include Brazil, India, and China) collectively invested as much in public agricultural R&D as all the high income countries combined.

These changes in the global structure of public agricultural R&D reflect a shifting pattern of growth in R&D spending among countries and regions of the world. Figure 15.1 shows the average growth rate by decade since 1960 for the United States, the high income countries as a group, and the low and middle income countries as a group further split into five regions. In the United States, spending grew faster than in most other high income countries, but the growth rate of spending by the rich country group (and by 29 of the 33 countries in this group) has slowed markedly over the years and is now well below the corresponding low and middle income rates (1.4% vs. 3.6% per year for the period 1990−2009). Among the low and middle income countries, the slower overall pace of spending growth is especially evident in Latin America and Sub-Saharan Africa compared with the Asia and Pacific region. The Asia and Pacific region sustained a long-run (1960−2009) growth rate of 5.1% per year, compared with 3.8% for Latin America and 1.8% for Sub-Saharan Africa. The 1990s was an especially dismal decade for Sub-Saharan African research: the region spent less on agricultural R&D in 1999 than it did at the dawn of that decade. The recent recovery in Sub-Saharan Africa appears fragile and not widespread: over half the increase in spending from 2000 to 2009 came from just two countries—Nigeria and Angola.

Public spending on agricultural R&D is highly concentrated, with the top 10 (of 126 countries) in the data set accounting for two-thirds of the total in 2009. The lowest ranked 100 countries accounted for just 14% of the total spending. This includes mostly low income countries whose economies are still heavily reliant on agriculture and whose populations still look to agriculture as a major source of employment and income.

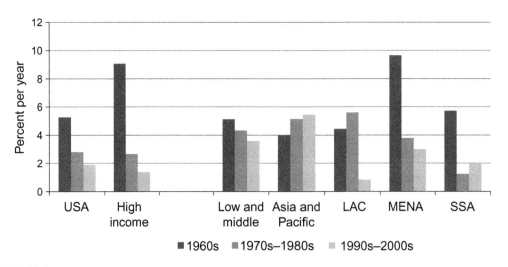

FIGURE 15.1

Rates of growth in public agricultural R&D spending by decade, 1960−2009.

Pardey et al., 2014.

15.4 PRICES AND PRODUCTIVITY: LONGER TERM PATTERNS AND PROSPECTS

The World Bank (2012, p. 1) noted that "(i)n 2011 international food prices spiked for the second time in 3 years, igniting concerns about a repeat of the 2008 food price crisis and its consequences for the poor." Table 15.1 presents measures of rates of change in real and nominal prices of maize,

Table 15.1 Average Annual Percentage Changes in US Commodity Prices, 1950–2011

Period	Commodity			Commodity		
	Maize	Wheat	Rice	Maize	Wheat	Rice
	(Average Annual Percentage Change)			(Trend Growth Rate, Percent per Year)		
Nominal Prices						
1950–2011	2.25	2.15	1.59	1.73	1.79	1.26
				(8.78)	(8.86)	(6.21)
1950–1970	−0.67	−2.04	0.08	−1.53	−2.65	−0.07
				(−3.71)	(−7.99)	(−0.36)
1975–2005	−0.87	−0.20	−0.29	−0.49	0.07	−0.90
				(−1.48)	(0.22)	(−1.82)
1975–1990	−0.72	−2.05	−1.47	−0.61	−0.19	−2.68
				(−0.61)	(−0.19)	(−2.06)
1990–2011	4.62	4.99	3.32	2.78	3.23	3.03
				(3.07)	(3.98)	(2.99)
2000–2011	10.70	9.70	7.86	9.75	8.71	11.27
				(6.37)	(6.20)	(6.35)
Deflated Prices						
1950–2011	−1.63	−1.73	−2.29	−2.46	−2.40	−2.94
				(−15.85)	(−15.00)	(−14.58)
1950–1970	−2.67	−4.04	−1.92	−3.10	−4.22	−1.64
				(−8.96)	(−11.30)	(−8.55)
1975–2005	−4.32	−4.07	−4.94	−3.61	−3.04	−4.01
				(−11.41)	(−9.09)	(−7.95)
1975–1990	−5.89	−7.22	−6.64	−5.44	−5.02	−7.51
				(−6.66)	(−6.11)	(−6.56)
1990–2011	1.19	1.56	−0.10	−0.48	−0.03	−0.23
				(−0.65)	(−0.04)	(−0.27)
2000–2011	5.92	4.92	3.08	4.76	3.71	6.27
				(3.41)	(2.86)	(3.76)

Notes: *Values in parentheses are* t-*statistics. Deflated prices were computed by deflating nominal commodity prices by the consumer price index.*

wheat, and rice over the period 1950–2011, which includes three distinct subperiods. First, over 1950–1970, deflated prices for rice and maize declined relatively slowly, while wheat prices declined fairly rapidly. Next, following the price spike of the early 1970s, over 1975–1990, prices for all three grains declined relatively rapidly and more or less in unison. Finally, over 1990–2011, prices increased for all three commodities, especially toward the end of that period. This reflects a generally slowing rate of price decline throughout the period prior to the price spike in 2008—in fact, essentially from 2000 onward, prices increased in real terms.

Growth in agricultural productivity, fueled by investments in agricultural R&D, has been a primary contributor to the long-run trend of declining food commodity prices, and the slowdown in the decline of real commodity prices since 1990 reflected a slowdown in the growth rate of crop production and yields, among other things. Global annual average rates of crop yield growth for maize, rice, wheat, and cereals are reported in Table 15.2, which includes separate estimates for various regions and for high, middle, and low income countries, as well as for the world as a whole, for two subperiods: 1961–1990 and 1990–2010. In both high and middle income countries—collectively accounting for 78.8–99.4% of global production of these crops in 2007—average annual rates of yield growth for cereals were lower in 1990–2010 than in 1961–1990. The growth of wheat yields slowed the most and, for the high income countries as a group, wheat yields

Table 15.2 Global and Regional Yield Growth Rates for Selected Crops, 1961–2010						
	Maize		**Wheat**		**Rice, Paddy**	
Group	**1961–90**	**1990–2010**	**1961–90**	**1990–2010**	**1961–90**	**1990–2010**
	(Percent per Year)					
World	2.33	1.82	2.73	1.03	2.14	1.09
Geographical Regions						
North America	2.19	1.75	1.38	0.98	1.22	1.33
Western Europe	3.73	1.32	3.21	0.83	0.62	0.70
Eastern Europe	2.54	1.93	3.19	0.18	0.51	3.49
Asia & Pacific (excl. China)	1.96	2.88	2.96	1.39	1.83	1.49
China	4.39	0.81	5.76	2.05	3.06	0.64
Latin America and Caribbean	2.01	3.22	1.67	1.52	1.39	3.10
Sub-Saharan Africa	1.30	1.70	2.88	1.84	0.83	1.03
Income Class						
High income	2.24	1.68	2.02	0.68	1.03	0.79
Upper middle (excl. China)	1.85	3.04	2.22	1.19	0.99	2.23
China	4.39	0.81	5.76	2.05	3.06	0.64
Lower middle income	1.79	3.06	3.27	1.42	2.36	1.36
Low income	1.19	0.36	2.08	2.02	1.50	2.18
Source: *Pardey et al., 2013.*						

grew by just 0.68% per year over 1990–2010. Global maize yields grew at an average rate of 1.82% per year during 1990–2010 compared with 2.33% for 1961–1990. Likewise, rice yields grew at 1.09% per year during 1990–2010, less than half the average growth rate in 1960–1990. These slower growth rates of crop yield are reflected broadly in other partial productivity measures and in multifactor productivity measures for countries where suitable data are available (see Alston et al., 2009, 2010), with China standing out as a significant exception.

"Green Revolution" varieties of wheat and rice (and other crops) combined with complementary fertilizer and irrigation technologies contributed to very significant growth of grain yields in the latter part of the twentieth century. Some writers have suggested that these technologies also contributed to greater variability of yields, production, and prices; however, the most compelling evidence indicates that the predominant effect has been to reduce variability of yields and production (Gollin, 2006). Alston et al. (2012) found that, for the world as a whole, variability of both production and yields trended down, but the converse was true in the low income countries, especially since 1990. The reasons for this dichotomy remain uncertain, but a significant factor might have been the slower growth of the means of yield and production in the low income countries. The pattern of production variability everywhere changed toward the end of the series, probably as a result of supply response to commodity prices that became more variable in the same period.

15.5 IMPLICATIONS OF ALTERNATIVE PRODUCTIVITY PATHS FOR THE WORLD'S POOR

Agricultural technology induces changes in the distribution of income among households through a multitude of direct and indirect effects and the optimizing responses of the households. Even if agricultural technology has no direct effect on household incomes, it affects food security or poverty through its effects on food prices. Benefits accrue to agricultural households both through reductions in costs of production (for those who adopt the technology) and through reductions in net costs of food purchases (the difference between their expenditure on consumption and the value of their production) resulting from the fall in price. Hence, for food-deficit agricultural households, the fall in price means a benefit; for food-surplus agricultural households, it means a loss. In addition to these two direct sources of benefits, households may gain (or lose) indirectly from induced adjustments in factor markets in which they participate as well as broader, general equilibrium impacts.

Figure 15.2 compares two stylized distributions of household income across households, conditional on the state of technology and a given draw of exogenous factors that gives rise to particular price outcomes. The income distribution across households, given technology τ_0, is denoted as Y_0^e. Associated with this distribution and defined by the corresponding prices is a "poverty line," reflecting the cost of a minimal quantity of food (or food calories) and other necessities, drawn at L_0^e. Under the alternative technology scenario, τ_1, but for the same draw of exogenous factors, food prices are lower and the poverty line shifts to L_1^e, reducing the portion of the population living in poverty for a given income distribution. This effect can be big if the change in food price is considerable, even with no direct changes in household incomes of the types that result from the effects of adopting new technologies on reducing production costs and the effects on revenues induced by price changes. When the income distribution shifts to the right from, say, Y_0^e to Y_1^e, as a result

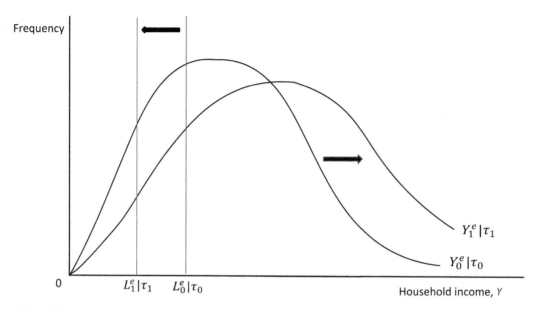

FIGURE 15.2

Agricultural technology and household income distributions.

of shifting from technology regime τ_0 to τ_1, then the portion of the population living in poverty is further reduced.[5]

Richer people are affected less by a given shock to prices of staple grains. When the income distribution has shifted substantially to the right, many fewer people will suffer severe consequences from a given price shock. This idea is illustrated in Figure 15.3, which shows the distribution of household income, Y_0^e and Y_1^e, under two alternative technologies, τ_0 and τ_1, with the corresponding poverty lines, L_0^e and L_1^e—all conditional on a particular draw of exogenous environmental factors that gives rise to particular price outcomes, P_0^e and P_1^e. The corresponding numbers of people below the poverty line are N_0 and N_1, with $N_0 > N_1$. Suppose a substantial negative environmental shock befalls the agricultural economy, such as a widespread drought, which under either technology scenario shifts the income distribution to the left, to $Y_0^{\sim}$ and $Y_1^{\sim}$, and shifts the poverty line to the right, to $L_0^{\sim}$ and $L_1^{\sim}$. Intuitively, the consequences are expected to be much smaller under technology τ_1 because (i) a smaller number of people were already poor, (ii) staple food commodities generally represent a smaller share of incomes such that the proportion of the population driven into poverty is smaller under technology τ_1, and (iii) farmers represent a smaller share of the population such that the direct effects on farm incomes from the shock are less important for the overall picture.

---

[5]Even though some farmers will be made worse off (if, for instance, they are surplus producers and do not adopt the new technology), the distribution generally shifts to the right, as drawn, reflecting the general improvement in incomes for households, although some have shifted to the left within the distribution.

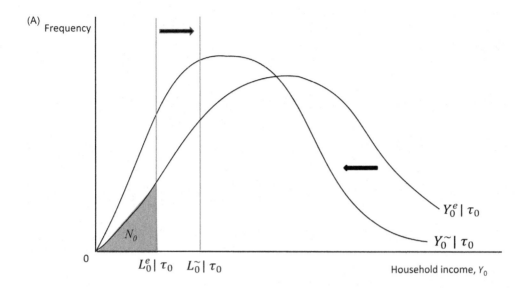

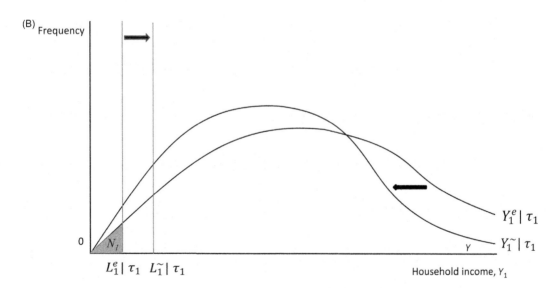

FIGURE 15.3

(A) Consequences of a shock under alternative technology scenarios (Technology Scenario τ_0) and
(B) Consequences of a shock under alternative technology scenarios (Technology Scenario τ_1).

15.6 IMPLICATIONS OF ALTERNATIVE PRODUCTIVITY PATHS: QUANTITATIVE ILLUSTRATION

These possibilities were explored by conducting simulations using a CGE framework. Using a model and approach developed and applied by Ivanic and Martin (2014) to evaluate the impacts of agricultural productivity growth on poverty, their analysis was extended to evaluate the effect on the vulnerability of the poor to food price shocks. To do this, the global economy from 2010 to 2050 was simulated under two alternative agricultural technology scenarios: (i) a pessimistic (slower growth) scenario, with equal productivity growth rates in agriculture and other sectors, and (ii) an optimistic (faster growth) scenario, with agricultural productivity growing by 1% per year faster than the rest of the economy. Then, the effects of a negative agricultural shock were simulated for each scenario, and the impacts on the number of people in poverty in a selection of less-developed countries were compared between scenarios.

The complete details of the simulation model are documented by Ivanic and Martin (2014) and summarized by Alston et al. (2012). Poverty assessment was based on the household survey datasets compiled by the World Bank for 29 developing countries that span the developing world, but notably exclude China. When a change in prices was simulated, the first step was to identify those households whose cost of living less any income changes moved them across the poverty line, defined as a particular level of utility. The poverty rate for each country was then recalculated following each simulation. Of specific interest is the difference in the effects of a commodity supply shock on poverty outcomes between the optimistic and pessimistic productivity growth scenarios.

Under the pessimistic scenario of uniform productivity growth across the agricultural and non-agricultural sectors, the prices of many foods rise substantially: food prices at the household level increase by an average of 48% by 2050 (63.3% in developing countries). Under the optimistic scenario, with productivity growing 1% per year faster in agriculture than in other sectors, food prices rise by a modest 1.4% over the same period (8% in developing countries).

Table 15.3 shows the total population (column 1) and the initial baseline percentage poverty rate (at USD 1.25 per person per day) in each of the 29 countries of interest (column 2). The next two columns show the effects of 1% per year higher productivity growth over the next 40 years (2010−2050) on reducing the poverty rate (column 3) and the number of people in poverty (column 4). The new poverty rate under the high-productivity growth scenario is shown in column 5. Thus, for example, in India the initial poverty rate of 43.83% applied to a population of 1.17 billion implies a total of some 513 million people in poverty. If global agricultural productivity grew by 1% per year faster for 40 years, this number would be reduced by 89 million and the poverty rate would be reduced by 7.6 percentage points. The reductions in poverty rates are even more pronounced in some countries. Under the faster productivity growth scenario, across all of the countries in this sample, poverty rates would be reduced by an average of 4.75 percentage points and a total of more than 135 million people would be lifted above the poverty line.

Table 15.4 shows the impacts of a substantial externally generated (say drought- or crop-pest-induced) price shock on poverty rates under the pessimistic agricultural productivity scenario (columns 1 and 2) and the optimistic scenario (columns 3 and 4). In most cases, the price shock increases the poverty rate (positive signs on entries in columns 2 and 4) but in other cases—where there are many poor net-selling households—the price shock decreases the poverty rate (negative signs on entries in

Table 15.3 Baseline Scenario: Changes in Poverty from 1% per Year Higher Agricultural Productivity Growth Over 2010–2050

Country	Population	Initial Poverty Rate, Percent	Change in Poverty Percentage Points	Change in Poverty Headcount	New Poverty Rate, Percent
	(1)	(2)	(3)	(4)	(5)
	Number	Percent	Percent	Number	Percent
Albania	3,204,284	0.85	−0.13	−4104	0.72
Armenia	3,092,072	10.63	−1.27	−39,176	9.36
Bangladesh	148,692,100	50.47	−4.29	−6,372,561	46.18
Belize	344,700	33.50	−1.73	−5962	31.77
Cambodia	14,138,260	40.19	−18.96	−2,680,020	21.23
Cote d'Ivoire	19,737,800	23.34	−3.94	−777,204	19.40
Ecuador	14,464,740	15.78	−3.27	−473,067	12.51
Guatemala	14,388,930	12.65	−5.02	−722,634	7.63
India	1,170,938,000	43.83	−7.59	−88,868,501	36.24
Indonesia	239,870,900	7.50	−1.54	−3,682,462	5.96
Malawi	14,900,840	73.86	−12.71	−1,894,637	61.15
Moldova	3,562,062	8.14	−4.04	−143,983	4.10
Mongolia	2,756,001	22.38	−6.30	−173,642	16.08
Nepal	29,959,360	55.12	−4.46	−1,337,469	50.66
Nicaragua	5,788,163	45.10	−5.62	−325,177	39.48
Niger	15,511,950	65.88	−2.10	−326,292	63.78
Nigeria	158,423,200	64.41	−3.47	−5,493,147	60.94
Pakistan	173,593,400	22.59	−6.97	−12,094,064	15.62
Panama	3,516,820	9.48	−1.94	−68,181	7.54
Peru	29,076,510	7.94	−1.77	−514,516	6.17
Rwanda	10,624,010	76.56	−2.26	−239,671	74.30
Sri Lanka	20,859,950	14.00	−3.20	−668,386	10.80
Tajikistan	6,878,637	21.49	−8.67	−596,488	12.82
Tanzania	44,841,220	67.87	−3.62	−1,621,932	64.25
Timor-Leste	1,124,355	52.94	−3.29	−37,033	49.65
Uganda	33,424,680	51.53	−6.78	−2,267,582	44.75
Viet Nam	86,936,460	13.70	−2.10	−1,824,816	11.60
Yemen	24,052,510	17.53	−5.25	−1,263,621	12.28
Zambia	12,926,410	61.87	−5.30	−684,590	56.58

Notes: In the "low-productivity" scenario, productivity grows at the same rate in agriculture as in the rest of the economy; in the "high-productivity" scenario, productivity grows 1% per year faster in agriculture than in the rest of the economy in all countries. The changes in poverty in this table reflect 49% higher productivity in agriculture as a result of 1% higher growth over 40 years.

Table 15.4 Changes in Poverty Rates Resulting from a Supply Shock in the Industrial Countries Causing Agricultural Commodity Prices to Double

	Low-Productivity State of the World		High-Productivity State of the World		Reduction in Poverty Impact: High- Versus Low-Productivity State	
	Initial Rate	Change	Initial Rate	Change	Rate	Headcount
	(1)	(2)	(3)	(4)	(5) = (2)−(4)	(6)
	Percentage Points					*Thousands*
Albania	0.85	0.11	0.72	−0.26	0.37	11.9
Armenia	10.63	0.92	9.36	0.14	0.78	24.1
Bangladesh	50.47	1.74	46.18	0.06	1.68	2498.0
Belize	33.50	2.43	31.77	0.44	1.99	6.9
Cambodia	40.19	−2.85	21.23	−3.09	0.24	33.9
Côte d'Ivoire	23.34	−0.26	19.40	−0.63	0.37	73.0
Ecuador	15.78	2.25	12.51	0.19	2.06	298.0
Guatemala	12.65	6.59	7.63	0.42	6.17	887.8
India	43.83	4.70	36.24	1.74	2.96	34,659.8
Indonesia	7.50	0.77	5.96	0.15	0.62	1487.2
Malawi	73.86	1.14	61.15	−0.59	1.73	257.8
Moldova	8.14	3.99	4.10	0.55	3.44	122.5
Mongolia	22.38	2.31	16.08	0.57	1.74	48.0
Nepal	55.12	−0.67	50.66	−1.27	0.6	179.8
Nicaragua	45.10	3.16	39.48	−0.35	3.51	203.2
Niger	65.88	−0.75	63.78	−1.29	0.54	83.8
Nigeria	64.41	0.32	60.94	−0.10	0.42	665.4
Pakistan	22.59	3.02	15.62	0.73	2.29	3975.3
Panama	9.48	1.20	7.54	−0.42	1.62	57.0
Peru	7.94	0.93	6.17	−0.50	1.43	415.8
Rwanda	76.56	0.49	74.30	0.21	0.28	29.7
Sri Lanka	14.00	2.45	10.80	0.72	1.73	360.9
Tajikistan	21.49	6.14	12.82	0.37	5.77	396.9
Tanzania	67.87	1.61	64.25	0.05	1.56	699.5
Timor-Leste	52.94	0.00	49.65	−0.43	0.43	4.8
Uganda	51.53	−0.07	44.75	−0.95	0.88	294.1
Viet Nam	13.70	−0.58	11.60	−0.84	0.26	226.0
Yemen	17.53	3.35	12.28	0.33	3.02	726.4
Zambia	61.87	0.77	56.58	−0.27	1.04	134.4
Average	**34.18**	**1.56**	**29.43**	**−0.15**	**1.71**	**39,460.4**

Notes*: In the "low-productivity" scenario, productivity grows at the same rate in agriculture as in the rest of the economy; in the "high-productivity" scenario, productivity grows 1% per year faster in agriculture than in the rest of the economy in all countries. The external price shock is represented by a 100% increase in prices of all agricultural commodities. The numbers in column 6 were derived by applying the rates in column 5 of Table 15.3 to the total population given in column 1 of Table 15.3.*

columns 2 and 4). However, in every case the entry in column 2 is more positive than the corresponding entry in column 4, such that the difference (in column 5, given by column 2 minus column 4) is positive—the poverty rate increases less (from a lower base) or decreases more in the high-productivity scenario than in the low-productivity scenario. This means that the effect of price shock on poverty is always more favorable given the high-productivity scenario compared with the low-productivity scenario. On average across countries in the high-productivity scenario, the external price shock reduces poverty by 0.15 percentage points; in the low-productivity scenario, the poverty rate increases by 1.56 percentage points. The difference reflects a benefit from higher productivity in providing some insulation against the impoverishing effects of price variability and—in most cases—reductions in the proportion of the population vulnerable to poverty.

In general, the high-productivity scenario makes households less vulnerable to price shocks. Higher productivity growth lowers real prices and—given the small price elasticities of demand for staple foods—enables households to spend less of their income on food. The numbers are large. As shown in column 6 of Table 15.4, in the high-productivity growth scenario 39.5 million fewer people across the 29 countries would be cast into poverty by a doubling of food commodity prices. This total benefit—i.e., reduced poverty impact of the price change in the high-productivity growth scenario—reflects the effects of (i) having a smaller shift of the income distribution induced by the price change in the high-productivity state and (ii) generally having a smaller share of the population close to the poverty line, as illustrated in the heuristic analysis using Figures 15.1 and 15.2.

15.7 **CONCLUSION**

This chapter examines the role of agricultural technology in contributing to or mitigating the consequences of variability in agricultural production. Technological changes in agriculture can affect food price variability both by changing the sensitivity of aggregate farm supply to external shocks and by changing the sensitivity of prices to a given extent of underlying variability of supply or demand. At the same time, by increasing the general abundance of food and reducing the share of income spent on food, agricultural innovation makes a given extent of price variability less important.

The direct effects of agricultural innovation on variability and its consequences are generally favorable. Since World War II, technological changes have contributed significantly to growth of yields and production and to reducing real prices, but probably not to increased price variability. Rather, it seems more likely that technological changes in agriculture have contributed to an underlying trend of production, yield, and prices that was generally less variable, with other factors giving rise to periodical increases in variability, such as in the early 1970s and in the late 2000s.

The indirect effects are also generally favorable. Agricultural technology reduces the importance of food price variability for food security of the poor by reducing the number of farmers, the number of poor, and the importance of food costs in household budgets. An illustrative analysis uses simulations of the global economy to 2050. The results also show that the vulnerability of households to poverty is lower following a sustained period of higher agricultural productivity growth.

Agricultural productivity growth rates have slowed, especially in the higher income countries, while growth rates of investment in productivity-enhancing agricultural R&D that slowed earlier have turned negative in numerous (especially high income) countries, suggesting a worsening of the agricultural productivity slowdown in the years to come, given the long R&D lags. Both the slowdown in agricultural productivity patterns generally, and the divergent patterns among countries in rates of research investments and productivity, will have implications for future paths of agricultural prices, price variability, and consequences of variability. A restoration of research investments and a refocusing of these investments on productivity may well preserve past productivity gains in the face of ever-evolving pests and diseases and climate change, which act to undercut these past gains, while also spurring the additional productivity growth required to meet the increased demand for agricultural output envisaged in the decades ahead.

ACKNOWLEDGMENTS

This work was partly supported by the University of California, Davis; the University of Minnesota; the HarvestChoice initiative, funded by the Bill and Melinda Gates Foundation; and the Giannini Foundation of Agricultural Economics. The authors gratefully acknowledge excellent research assistance provided by Jason Beddow, Maros Ivanic, and Connie Chan-Kang, and helpful comments from Jim Roumasset.

REFERENCES

Alston, J.M., Marra, M.C., Pardey, P.G., Wyatt, T.J., 2000. A meta analysis of rates of return to agricultural R&D: ex pede Herculem? IFPRI Research Report No. 113, IFPRI, Washington, DC.

Alston, J.M., Beddow, J.M., Pardey, P.G., 2009. Agricultural research, productivity and food prices in the long run. Science 325 (4), 1209—1210.

Alston, J.M., Beddow, J.M., Pardey, P.G., 2010. Global patterns of crop yields, other partial productivity measures, and prices. In: Alston, J.M., Babcock, B.A., Pardey, P.G. (Eds.), The Shifting Patterns of Agricultural Production and Productivity Worldwide. CARD-MATRIC Electronic Book. Center for Agricultural and Rural Development, Ames, IA. Available from: <http://www.matric.iastate.edu/shifting_patterns/>, May 2010.

Alston, J.M., Andersen, M.A., James, J.S., Pardey, P.G., 2011. The economic returns to U.S. public agricultural research. Am. J. Agric. Econ. 93 (5), 1257—1277.

Alston, J.M., Martin, W.J., Pardey, P.G., 2012. Influences of agricultural technology on the size and importance of food price variability. Paper presented at NBER Conference on Economics of Food Price Volatility, Seattle, Washington, August 15—16.

Gollin, D., 2006. Impacts of International Research on Intertemporal Yield Stability on Wheat and Maize: An Economic Assessment. CIMMYT, Mexico, DF.

Hurley, T.M., Rao, X., Pardey, P.G., 2014. Recalibrating the reported rates of return to food and agricultural R&D. Am. J. Agric. Econ. Available from: http://dx.doi.org/10.1093/ajae/aau047 (First published online: May 31, 2014).

Ivanic, M., Martin, W.J., 2014. Sectoral productivity growth and poverty reduction: national and global impacts. Mimeo. The World Bank, Washington, DC.

Pardey, P.G., Alston, J.M., Chan-Kang, C., 2013. Public Food and Agricultural Research in the United States: The Rise and Decline of Public Investments, and Policies for Renewal. AGree Report. AGree, Washington, DC.

Pardey, P.G., Chan-Kang, C., Dehmer, S., 2014. Global Food and Agricultural R&D Spending, 1960—2009. InSTePP Report. University of Minnesota, St. Paul, MN.

World Bank, 2012. Food Prices, Nutrition, and the Millennium Development Goals: Global Monitoring Report 2012. World Bank, Washington, DC.

ENERGY AND AGRICULTURE: EVOLVING DYNAMICS AND FUTURE IMPLICATIONS

16

Siwa Msangi and Mark W. Rosegrant

Environment and Production Technology Division (EPTD), International Food Policy Research Institute (IFPRI), Washington, DC

CHAPTER OUTLINE

JEL: Q49; Q27; Q37; Q17

16.1 INTRODUCTION

As global energy markets become ever more dynamic in the face of new resource discoveries, continual technology improvements, and ever-increasing demands, the connections to critical sectors of the economy are becoming more evident and important. Agriculture is a key sector of the global economy that has critical interactions with energy on both the supply and demand sides of the market.

On the supply side of agriculture, the linkages with energy are most strongly felt through the use of energy-intensive inputs in production, such as the use of nitrogen or phosphate-based fertilizers, and in the extraction of irrigation water from groundwater aquifers with electric or diesel pumps. The other important dimension of energy usage in agriculture has to do with the application of traction in the process of cultivation, which draws from either mechanical sources of power (in the case of more modern and intensive systems) or animal-based draft power (in the case of more traditional systems).

On the demand side, agriculture's interactions with the energy sector are increasingly felt through the demand for feedstocks used in the production of crop-based biofuels. As the demand for renewable fuels continues to grow in OECD[1] countries (as well as large non-OECD countries like Brazil), so does the strong need for first-generation biofuels from starch- or sugar-based feedstock sources. If we consider the energy content embodied in the consumption of energy-rich foods like meat, starches, sugar, and oils, we would see an increasing flow of energy demands as the growth of population and incomes lead to dietary changes and demands for these products.

The growth of demand for agricultural and energy-based products can be complementary, as in the case of biofuels or in cases where hydropower generation creates facilities that can help to better manage the water resources used for irrigation. There can also be trade-offs, as in cases where land areas exploited for energy-rich minerals overlap with areas that are suitable for cultivation.

To better understand the nature and strength of these interactions, a comprehensive analytical framework is needed to rationalize and explore the complex linkages that tie agriculture to energy production and demand systems, through the natural resource base. Policymakers need to be better informed about the potential trade-offs that might be embodied in policy goals, like enhancing energy security or improving environmental quality, and understand where synergies can be created.

This chapter draws the key linkages between energy and agriculture in terms of their critical market interactions and highlights some key uncertainties for policy. It first motivates the linkage between energy, agricultural markets, and the environment with a simple theoretical framework. It draws important illustrations of these interactions resulting from a stylized quantitative model-based analysis using groundwater, fertilizer use, and biofuels as examples. The analysis allows for a better understanding of the dynamics of these interactions and the drawing of lessons for policy design. The results generated within this framework illustrate the impact that technology- and policy-oriented interventions might have on future food security, which could help in prioritizing areas of scientific uncertainty that need to be addressed by future research.

16.2 KEY LINKAGES BETWEEN ENERGY AND AGRICULTURE

This section discusses the key linkages between energy and agriculture within the context of groundwater, fertilizers, and crop-based biofuels as renewable energy sources. Each of these cases

---

[1]Organisation for Economic Cooperation and Development.

represents an important set of relationships and interactions that tie the supply and demand of energy to the production or consumption patterns of certain agricultural products and that have implications for natural resource availability and environmental quality.

In cases where certain productive inputs are lacking in the natural environment for agricultural production, energy and energy-rich products can be used as a supplementary input to the cultivation process. In the case of soil nutrients, for instance, the lack of certain key minerals and nutrients needed for vigorous crop growth (such as phosphate or nitrogen) would need to be supplemented with energy-rich products such as potash- or urea-based fertilizers. In cases where surface water availability is limited and insufficient for achieving desired yield levels, energy-intensive extraction of groundwater may be necessary, where such groundwater resources exist. In the case of transportation energy demands, the rising prices of fossil-based fuel products have encouraged the development of alternative biofuel supplies, which rely heavily on agricultural feedstocks.

In each case, the exploitation of energy and energy-rich products for agricultural production could lead to resource constraints that might represent serious limiting factors to their continued future use. In the case of groundwater, exploitation often leads to its depletion over time, especially in cases where there is no natural recharge or when the extraction rate consistently exceeds its natural recharge rate. In the case of fertilizers, some important fertilizer feedstocks, such as phosphate rock, are in limited supply and will increasingly become the drivers of fertilizer prices in the future, as their scarcity deepens. In the case of biofuels, the key constraint to their production in the future might eventually arise in the form of trade-offs in land availability for production of agricultural crops that are mostly used for food and feed, as opposed to dedicated energy crops (e.g., switchgrass, miscanthus, or jatropha). In such cases, a country-level deficit in land availability could be compensated for by imports, as would be the case with fertilizers. In the case of water, imports of this resource cannot be directly done, unless implicitly through negotiated reallocations of transboundary resources or (more plausibly) through the import of those water-intensive goods themselves.

These cases are summarized in Table 16.1, which shows how the dimensions of scarcity of key inputs to agriculture intersect with the use of energy as a supplementary source, which, in turn, might be drivers of other kinds of scarcity or resource trade-offs.

Table 16.1 enables a better appreciation of some trade-offs and complementarities associated with agricultural production and consumption and the use of (and demand for) energy.

Such trade-offs and complementarities also convey the close connection between resource scarcity and technological change and transformation. Often the initial scarcity of a productive factor induces a change in the production or extraction technology, which serves to relieve the initial constraint. In the course of time, the substitution toward a new resource could lead to an increasing scarcity level, which has to be addressed through further measures or innovations in the productive or transformative process. In cases where additional energy inputs can substitute for the scarce resource, energy intensity and overall usage may increase. At the point where the limiting factors of production or

Table 16.1 Illustrative Linkages Between Agriculture, Energy, and Resource Scarcity

Underlying Scarcity	Resource Use	Limiting Factor
Water scarcity	Groundwater	Aquifer capacity
Fossil fuel price	Renewable energy	Trade-off in land
Soil nutrient deficiency	Fertilizer use	Phosphate rock availability

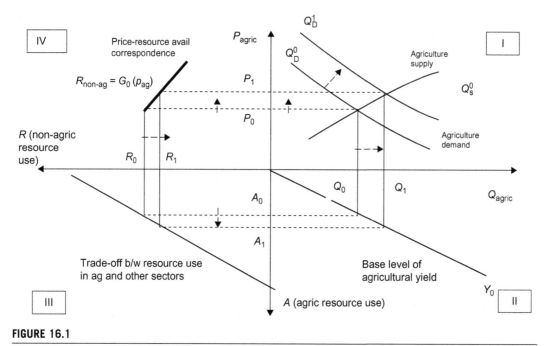

FIGURE 16.1

Linkages between demand-driven market shifts and agricultural land use in base case.

transformation start to present obstacles and more severe trade-offs with performance, then a further shift toward a higher efficiency technology or an alternative source of energy is necessary.

Given the motivation provided by the arguments put forward to support the energy–agriculture linkages, Figure 16.1 shows how shifts in the supply–demand equilibrium relate to changes in agricultural resource usage and the role that agricultural productivity plays in moderating those effects. Underlying resources can be land or even water, which would have the same basic trade-off in availability as that shown above.

From this treatment, a shift in the demand curve for the agricultural good (due to an increase in feedstock demand for biofuel production, for example)—from Q_D^0 to Q_D^1—causes the price of the good to rise from P_0 to P_1, holding the supply curve fixed (which may be the case in the short run). The increased demand for the agricultural good and the increased production level needed to meet the demand cause the agricultural resource usage to expand from A_0 to A_1 for a given yield level at y_0. This increase in agricultural land usage decreases the resource's availability for the other sectors and shifts that quantity from R_0 to R_1. Therefore, a correspondence between the price of the agricultural good and the availability of resources for nonagricultural uses can be drawn as the line $G_0(\bullet)$, which depicts a functional relationship such as $R_{\text{non-ag}} = G_0(p_{\text{ag}})$.

In the case where productivity is able to respond to price increase (in the longer term), then the line that describes yield can pivot to y_1 in order to give the same level of agricultural production with less resource usage (Figure 16.2).

The supply curve in such a case would be expected to shift from Q_S^0 to Q_S^1 and the price levels would be reduced from P_1 to P_1'. This resulting shift in yield and production would lead to a new

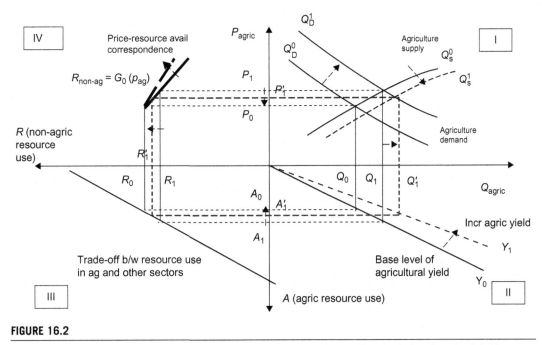

FIGURE 16.2

Linkages between demand-driven market shifts and agricultural land use with price-induced productivity increase.

level of agricultural resource usage at A'_1, which is lower than before (A_1) and entails a higher level of resource availability for nonagricultural uses (R'_1). By following the correspondence between prices and nonagricultural resource availability in this updated case, we obtained a new correspondence $G_1(\bullet)$, whose slope suggests that the decrease in the resource availability for nonagricultural uses is less for the same increase in prices, compared with that in $G_0(\bullet)$.

If availability of land for nonagricultural land uses is viewed as closely connected with land for natural cover (forest, shrubland, grassland, etc.), wildlife habitat, and other nonagricultural human uses, then an increase in agricultural productivity has a "land-saving" effect, allowing production expansion with less area expansion. If this resource is water, increased efficiency in its use allows more to be freed up for the other sectors. In the discussion of the environmental impacts of future agricultural expansion, the issue of yield and productivity gains has been considered a key and decisive element in the avoidance or mitigation of adverse environmental impacts.

The key linkage with energy becomes apparent when the process of resource extraction is considered, as in the case with water. Since energy is needed to pump water out of the ground for irrigation, savings in water withdrawals (and consumption) would mean savings in energy. In the case where an energy feedstock is sourced from agriculture, as in the case with biofuels, then an increase in demand (as seen in Figure 16.1) induced by higher energy prices will mean that more land would be required for that crop (and the agricultural sector) and less would be available for the other sectors. In this case, an improvement in the feedstock crop yield or in the conversion efficiency of feedstock into biofuels would constitute a land savings for the other sectors.

16.3 KEY EXAMPLES OF ENERGY—AGRICULTURE LINKAGES

This section discusses the important energy—agriculture linkages within a richer conceptual framework and within the context of three examples: groundwater, fertilizer, and biofuels. These examples illustrate empirically where the trade-offs and interactions with energy and resource scarcity are most apparent.

16.3.1 ENERGY AND AGRICULTURE LINKAGES IN THE CASE OF GROUNDWATER

The economics of water usage is typically based on the behavioral economics of a profit-maximizing agent, who seeks to maximize the net revenue accruing from irrigated agricultural production and faces a trade-off in terms of costs of inputs (including water) or constraints in water or land use.

Even without considering the dynamics of groundwater usage and how the stock of water held in an underground aquifer evolves over time, the individual profit-seeking agricultural producer can be hypothesized to behave according to the following simple maximization problem:

$$\max_{A,w} \pi(A,w) = p \cdot A \cdot y(w) - c_A A^{\theta} = p \cdot A \cdot kw^{\alpha} - c_A A^{\theta} - c_w w \quad s.t. \ A \le A^{\max} \tag{16.1}$$

where the decision-maker's problem is defined in terms of choosing the optimal level of input w (i.e., water) that enters into the agricultural yield function $y(w)$, while also choosing the optimal land area over which to farm, A, which is available up to a particular limit $A^{\max}$. The price of the agricultural output is denoted by p, whereas the cost of the productive input is given as c_w. The nonlinear cost of land is captured by the constant c_A and the parameter $\theta(>1)$, which represents the decreasing returns to adding land area—due to limited management and labor, as well as variable land quality over the available area—but which are not directly modeled here. The maximization problem in Eq. (16.1) can be stated in terms of the full Lagrangian function, shown in Eq. (16.2), which has the shadow value of the constraint included as a choice variable:

$$\max_{A,w,\lambda} L(w,A,\lambda) = p \cdot A \cdot kw^{\alpha} - c_A A^{\theta} - c_w w - \lambda[A - \overline{A}] \tag{16.2}$$

By following through all of the mathematical derivations necessary to characterize the behavior of the agricultural producer with respect to the key parameters (price, cost) and constraints (land), we could identify the key points of sensitivity in the decision-maker's equilibrium outcome. The detailed derivations are contained in the Technical Appendix; the implications are summarized here. Since our key interest is the behavior of the producer with respect to water use, we focus on those results.

From the analytical derivations (shown in detail in the Technical Appendix), the impact of changes in the volumetric cost of water (c_w) on water usage is given as follows: $(\partial w / \partial c_w) < 0$ in which the marginal effect of increasing the cost serves to decrease the level of optimal water usage, as expected.

Energy enters the groundwater user's problem when the cost of pumping water to the land surface from the groundwater table is considered. The unit cost of water usage, which is described by the single parameter c_w in Eqs. (16.1) and (16.2), can be conceptualized as a function of several variables within the context of groundwater usage. Typically, the marginal cost of groundwater

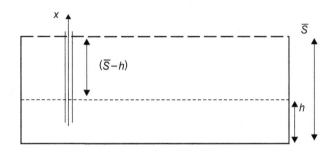

FIGURE 16.3

Simplified representation of pumping from an aquifer.

pumping is thought to vary according to the hydrological conditions under which water is withdrawn from the aquifer. In particular, the distance over which water must be lifted from the groundwater table to the surface (i.e., the "lift") is a key determinant of the marginal cost of pumping a single volumetric unit of water, as well as the energy costs associated with the action of the pump. Therefore, the marginal cost of water can be described as a function of the "state" of the system, which, in turn, can be described by the state variable h, which denotes the height of the groundwater table with respect to a reference level and the distance it lies below the ground surface $\overline{S}$ (Figure 16.3).

Combining the "lift" $(\overline{S} - h)$ with the energy cost of pumping (e), we can express the marginal cost of water usage as:

$$c_w(h, e) = \gamma e \cdot (\overline{S} - h) \tag{16.3}$$

where γ is a conversion factor. Combining this expression for marginal cost with the expression of the farmer's first-order condition for water usage leads to:

$$pA^{\max} f'(w) = \gamma e \cdot (\overline{S} - h(x_{-t})) \tag{16.4}$$

in which the groundwater table is expressed also as a function of water usage decisions in the past x_{-t} and which relates the present state of the natural resource to a history of resource usage behavior which, over time, results in increasing levels of cost to the farmer, both in the immediate period and in the longer term.

Based on this treatment, we get an analogous result to that obtained before if we were to examine the effect $(\partial w/\partial e) = (\partial w/\partial c_w)(\partial c_w/\partial e) < 0$, since the effect of the energy cost on c_w is linear, according to the relationship in Eq. (16.3).

This is a very relevant case to regions in Asia where agriculture is highly dependent on groundwater and where the problem of groundwater overdraft is particularly acute, such as in South Asia and China. For instance, the problems of electricity subsidies (among many other subsidies to agriculture) have been much researched and discussed in India, and a number of authors have noted the implications of continued provision of free or heavily subsidized electricity to farmers on the extraction and future sustainability of groundwater (Monari, 2002; Reddy, 2005). The political economy of subsidies to the agriculture sector in India presents a major stumbling block to reforming the price of irrigation water (IWMI-TATA, 2002; Dubash, 2005; Lal, 2006).

In China, the steady depletion of groundwater on the North China Plain (Lin et al., 2000) is likewise connected to past political motivations for expanding tubewell access to boost food production for staple grains and to the gradual shift in management regimes as the nature of rural governance has changed over time (Wang et al., 2005). Although Chinese farmers are far less subsidized compared with Indian farmers, reversing the trend of depletion in the face of a complex system of village, district, and provincial governance remains to be a challenge in China; it has also been the focus of much policy debate (Lohmar et al., 2003; Wang et al., 2007).

16.3.2 ENERGY AND AGRICULTURE LINKAGES IN THE CASE OF FERTILIZER

By borrowing from the analytical derivations shown in the previous section, we can also consider the effects of fertilizer price changes on production and market prices for agriculture. If input w is fertilizer (instead of water) and if fertilizer price is tied to energy cost (e), then an analogous result would be obtained, where the decision to use input w is determined by the energy cost, such that:

$$pA^{\max}f'(w) = c_w(e) \tag{16.5}$$

hence, the input usage w would be expected to decline with energy cost, i.e., $(\partial w/\partial e) < 0$.

If the problem is modified such that the price of the agricultural good p is endogenized, then the market equilibrium that gives the solution for this price would have to be accounted for. By adding a market-level balance, where the product's production, demand, and trade are accounted for, the following equation would be considered:

$$Q_{\text{Supply}} = Q_{\text{Demand}} + Q_{\text{Export}}$$
$$p \cdot A \cdot kw^\alpha = d_0 \cdot p^\varepsilon + e(\bar{p}_w - p)^\xi \tag{16.6}$$

where exports (Q_{Exports}) are modeled simply as the difference between the (exogenous) world price ($\bar{p}_w$) and the (now-endogenous) country-level price (p), with a proportionality constant (e_0) and an elasticity (ξ)[i]. Treating this as a macro-type of problem provides further justification in imposing the assumption that made agricultural land usage nonbinding ($A < A^{\max}$). This is in contrast to the case where land allocation would be binding, which would be more appropriate to a more microlevel, household problem. (The details of how this expanded problem is solved are in the derivations shown in the Technical Appendix.) Moreover, the key effect of the fertilizer input price on the use of inputs is $(dw/dc_w) < 0$.

Since the output price is now endogenous, the effect of the input price on it can also be derived as $(dp/dc_w) > 0$.

If the energy costs of the input (fertilizer) are linked to its price in a linear way, by the same reasoning that led to the result in Eq. (16.5), energy prices would have a positive impact on the price of agricultural output. This result will be confirmed via an empirical case later in the chapter.

16.3.3 ENERGY AND AGRICULTURE LINKAGES IN THE CASE OF BIOFUELS

As regards biofuels and agricultural markets, if the same kind of producer and profit-maximization problem as above is considered, the same set of first-order conditions would be obtained as:

$$\pi_A(A,w) = p \cdot kw^\alpha - \theta c_A A^{\theta-1} = 0$$
$$\pi_w(A,w) = p \cdot A \cdot \alpha kw^{\alpha-1} - c_w = 0 \tag{16.7}$$

which applies to a case where land is not binding. Next, the demand side of the agricultural market for this single good is expanded and the nonbiofuels use of the good is characterized as responsive only to its price such that $Q_D^{\text{non-biofuel}} = d_0 \cdot p^\varepsilon$—where d_0 is a constant and ε a price elasticity of the demand response (as before). The feedstock demand for the agricultural good in the production of biofuel itself is then added, giving an exogenous price (p_b). The feedstock demand for the agricultural good is expected to be responsive to the price of both the agricultural good (as an input) and the biofuel (p_b).

The demand for the biofuel feedstock can be derived by conceptualizing the biofuel producer's profit-maximization problem as:

$$\max_{Q_d^{\text{fdstk}}} \pi^{\text{biofuels}}\left(Q_d^{\text{fdstk}}\right) = p_b \cdot Q_s^{\text{bioF}} - pQ_d^{\text{fdstk}} = p_b \cdot \left[b_0(Q_d^{\text{fdstk}})^\gamma\right] - pQ_d^{\text{fdstk}} \tag{16.8}$$

where the production level of biofuels (Q_s^{bioF}) is linked to the feedstock usage (Q_d^{fdstk}) of the agricultural good, such that $Q_s^{\text{bioF}} = b_0(Q_d^{\text{fdstk}})^\gamma$, and b_0 and $\gamma(<1)$ are positive constants of the biofuels production function. By taking the first-order conditions of the profit-maximization problem:

$$\pi_{Q_d^{\text{bioF}}}\left(Q_d^{\text{fdstk}}\right) = p_b \cdot \gamma\left[b_0(Q_d^{\text{fdstk}})^{\gamma-1}\right] - p \leq 0, \quad Q_d^{\text{fdstk}} \geq 0 \tag{16.9}$$

we can derive the reduced-form feedstock demand for biofuels production as:

$$Q_d^{\text{fdstk}} = \left[\frac{p}{p_b b_0 \gamma}\left(\frac{p}{p_b}\right)^\gamma\right]^{(1/(\gamma-1))} = \left[\frac{p}{p_b b_0 \gamma}\right]^{(1/(\gamma-1))}\left(\frac{p}{p_b}\right)^{(1/(\gamma-1))} = \hat{b}_0\left(\frac{p_b}{p}\right)^{(1/(1-\gamma))} = \hat{b}_0\left(\frac{p_b}{p}\right)\hat{\gamma} \tag{16.10}$$

which is then added to the nonbiofuels demand for the agricultural good in order to obtain the total demand for the agricultural good as a function of both prices, as follows:

$$Q_D^{\text{total}} = Q_D^{\text{non-biofuel}} + Q_D^{\text{fdstk}} = d_0 \cdot p^\varepsilon + \hat{b}_0\left(\frac{p_b}{p}\right)\hat{\gamma} \tag{16.11}$$

If we consider the market-level balance of the agricultural good, where production balances with demand and exports, an expression is obtained, which is augmented to the relationship (16.6), as shown below:

$$Q_S = Q_D^{\text{non-bioF}} + Q_D^{\text{fdstk}} + Q_{\text{Export}}$$
$$A \cdot kw^\alpha = d_0 \cdot p^\varepsilon + \hat{b}_0\left(\frac{p_b}{p}\right)\hat{\gamma} + e\left(\bar{p}_w - p\right)^\xi \tag{16.12}$$

This expanded problem has three endogenous variables (p, A, w), which will be shifted around due to changes in key parameters, such as those affecting productivity of agriculture (k), cost of productive inputs (c_w), price of biofuels (p_b), and world price of the agricultural good (p_w).

The details of the mathematical derivations are provided in the Technical Appendix; however, the solutions for the price changes of the agricultural good and usage of agricultural land as a function of increased biofuel prices (and feedstock demand for biofuels production) can be shown here. By looking at the effect of biofuel price on agricultural price and land usage, we obtain:

$$\frac{dp}{dp_b} > 0 \qquad \frac{dA}{dp_b} > 0$$

suggesting that a higher price (and incentive to produce biofuels) will lead to upward price pressures on agricultural markets and on the usage of land in agriculture. This result is both intuitive and borne out of a number of empirically based analyses that examined the impacts of biofuels on agricultural markets and land usage.

Since the international price of biofuels is affected by energy prices, the latter's impact on the price of the agricultural good can be derived. While such analysis will not be done here, this chapter argues that the higher the crude oil price, the greater is the incentive to blend crude oil with biofuels, which then increases the international price of biofuels (p_b). The following section demonstrates this effect within an empirical context.

16.4 EMPIRICAL ILLUSTRATIONS OF ENERGY—AGRICULTURE LINKAGES

An empirical and computational framework (i.e., IFPRI's [International Food Policy Research Institute] global, agricultural multimarket model, IMPACT) was used to illustrate the market dynamics that are implicit in energy—agricultural interactions. This framework enables an understanding of the market dynamics between energy and agriculture, beyond what can be provided by the simple, conceptual construct discussed earlier.

16.4.1 THE IMPACT MODEL

The IMPACT model was initially developed at IFPRI to project global food supply, food demand, and food security to 2020 and beyond (Rosegrant et al., 2001). It is a partial equilibrium agricultural model that includes over 40 crop and livestock commodities, including cereals, roots and tubers, meats, milk, eggs, oilseeds, oilcake, vegetable oils, sugar, fruits, and vegetables. IMPACT has 115 country (or, in a few cases, country aggregate) regions, within each of which supply, demand, and prices for agricultural commodities are determined. Large countries are divided into major river basins, which give a total of 281 subnational units that form the basis for livestock and irrigated/rainfed agricultural crop production. Crop production in IMPACT is divided between area and yield response functions and, on the demand side, by relationships for usage as food, feed, oilseed crush, and the demand for biofuels production.[2] Demand and supply are both price responsive and are balanced with levels of exports and imports to the world market that are made to balance at the global level.

A number of exogenous drivers are incorporated into the IMPACT model, both on the supply and demand sides. On the supply side, the critical drivers of growth are those for yield growth and area expansion. Water availability (simulated by a separate but linked model) is also a determinant of future yield growth. On the demand side, the key drivers are those for population and income growth, which affect both size and composition of future agricultural demand. For rapidly growing and emerging economies in Asia, Latin America, and Africa, these drivers are critical in shaping

---

[2]There is also demand for "other" uses that is not price responsive and accounts for the residual utilization given in FAO data, which we did not directly model. These account for usage in industrial or other value-added sectors that are outside the agricultural focus of the IMPACT model.

the future agricultural landscape through the influences they exert in closely linked agricultural markets and patterns of trade.

16.4.2 LINKAGES BETWEEN ENERGY, FERTILIZERS, AND AGRICULTURE

Fertilizer is an example of an energy-intensive, productivity-enhancing input in agricultural production. In many ways, its role is analogous to that of pumped groundwater for irrigation. This is especially so in the way the input price responds to the price of energy products like crude oil (in the case of groundwater) or natural gas, which is a major input for urea-based fertilizers. In the case of phosphate-based fertilizers, the limited availability of key inputs like phosphate rock—whose high quality deposits are found in very few regions of the world, like Morocco—presents an additional constraint to future expansion in the use of this input. When energy prices increase, the expectation is that a shift would be induced toward alternatives with lower levels of energy intensity.

Agricultural intensification is important in ensuring future food security, even as land resources have become limited. Early studies have shown that fertilizer application had increased cereal production by 15% and yield by 56% in developing countries (Pinstrup-Andersen, 1976). In Sub-Saharan Africa (SSA), where soil fertility is relatively low in most areas, fertilizer applications have significantly increased crop productivity. The Food and Agriculture Organization (FAO) of the United Nations had conducted a large number of experiments to demonstrate crop yield response to fertilizer. The final study showed that fertilizer application (i.e., medium amounts of nitrogen, phosphorus, and potassium or NPK) increased maize grain yield by 750 kg/ha (FAO, 1989).

The use of fertilizer (as is the case with pesticides) has environmental benefits and costs. One benefit is that it reduces or eliminates soil nutrient depletion and cultivation of marginal and fragile land (Dorward and Poulton, 2008). Borlaug (2000) estimated that if the cereal yield levels in the 1950s had remained unchanged in 2000, a total of 1.8 billion ha of land would have been converted to cropland to meet cereal demand. Assuming that agricultural productivity grows by 1% per year, Goklany (2001) estimated that by 2050, with the world population projected to reach 8.9 billion people, an additional 325 million ha of forests or other terrestrial habitat would have to be converted to cropland to meet food, fiber, and feed demand. Borlaug and Dowsell (2004) also estimated that about 1 billion mt of cereal will be required annually by 2030, a level that is 50% higher than the demand in 2000. In their analysis, 80% of the production needed to meet this increased demand can be met by intensification and only 20% by area expansion.

On the other hand, about 36 million tons of mineral fertilizers are lost to the environment annually, resulting in polluted water resources, eutrophication, biodiversity loss, and other impacts (Roy et al., 2002; Phipps and Park, 2002). The negative impacts of fertilizers and pesticides have prompted technological and policy options to reduce their use. The recent increases in fertilizer and oil prices have furthered calls to reduce fertilizer use by adopting integrated soil fertility management (ISFM) and promoting nutrient use efficiency (NUE). ISFM and NUE options and the policies and strategies required to increase their adoption will be discussed later.

16.4.2.1 Long-term trends of fertilizer and fossil fuel prices

Fertilizer prices have increased dramatically since 2005 (Figure 16.4). The price of triple super phosphate (TSP) increased by about 51% between 1970 and 1980, and fivefold between 2001 and 2008. Potassium chloride prices also increased by 31% from 1970 to 1980 and 2.5-fold between

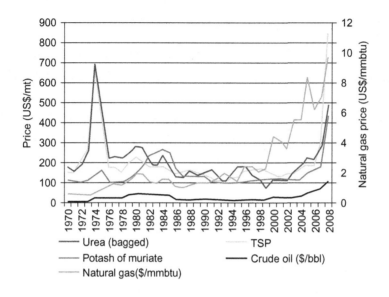

FIGURE 16.4

Trend of fertilizer prices since 1970.

World Bank (2014).

2000 and 2008. The 1970−1980 price increases were due to the oil price crisis in the mid-1970s and the green revolution in Asia, which significantly increased fertilizer demand (Socolow, 1999). Urea and phosphorus prices peaked in 1974 and declined until 2000, when the prices started to rise again (Figure 16.4). As noted above, the upward trend spiked abruptly in 2005/2006.

Increasing oil prices have contributed to high fertilizer prices. Brent crude oil prices increased dramatically from about USD 65/barrel in 2006 to USD 115/barrel in 2008 (Figure 16.4). Since fertilizers are bulky, increasing oil prices contribute to the escalating fertilizer prices due to higher transportation costs. In fact, transportation costs account for as much as 40−60% of fertilizer prices in SSA land-locked countries (Table 16.2). Moreover, natural gas is used to convert atmospheric nitrogen to nitrogenous fertilizer. As such, the increase in natural gas price by almost 50% from USD 8.47/million BTU in 2006 to USD 12.63/million BTU in 2008 also increased the cost of manufacturing nitrogenous fertilizers (Figure 16.4). Notably, phosphorus fertilizer prices increased more dramatically over the period 2004−2006, primarily because of increasing demand and high transportation costs (Morris et al., 2007). Dwindling deposits of nonrenewable phosphorus raw materials have also contributed to the increasing price (IFDC, 2008). In the case of potassium, other than oil prices and other factors, its price trend has been driven by the growing demand for fruits and vegetables, whose cultivation accounts for 19% of potassium consumption in the world (IFA, 2008). Demand for fruits and vegetables has been increasing due to rising incomes in developing countries (IFA, 2008; FAO, 2008).

Food prices have increased by more than 60% since 2006 (Trostle, 2008). Such an increase could have been a windfall to farmers if energy prices had not risen significantly during the same period. The relative change in food and energy prices was determined by analyzing the trend in food/fertilizer price ratios. Figures 16.5−16.7 show a downward trend in food/fertilizer price ratios across all three crops

Table 16.2 Cost of Transportation of Fertilizer in Malawi and Rwanda

Description	Malawi	Rwanda
Farm gate price (USD/ton)	482	513
Fertilizer FOB price (USD/ton)	289	207
Share of fertilizer costs (% of farm gate price)		
Fertilizer FOB (%)	60	40
Sea shipping (%)	12	10
Stevedoring, shore handling, ad-valorem, bagging, warehousing, customs (on value), insurance, and clearing (%)	5	8
Trucking from port to in-country warehouse (+ border fees) (%)	15	34
In country trucking and warehousing (%)	8	8
Total (%)	100	100

Source: *Chianu et al., 2008*

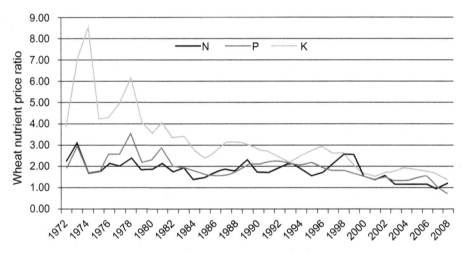

FIGURE 16.5

Trend of wheat and fertilizer price ratio.

(wheat, rice and maize) and major types of nutrients (especially potassium and phosphorus). This suggests that fertilizer prices have been increasing faster than food prices. The dramatic increase in fertilizer prices reflects the marked growth in fertilizer demand, which has been driven by the rapidly growing demand for biofuel (Heffer and Prud'homme, 2008) and rising oil prices. Eastern Asia and North America accounted for about 67% of the total change in fertilizer demand, while Southern Asia and Africa accounted for about 17% and 5% of the change in demand, respectively.

The declining crop/fertilizer price ratios will negatively affect crop production if alternative technologies are not used to enhance soil fertility or NUE.

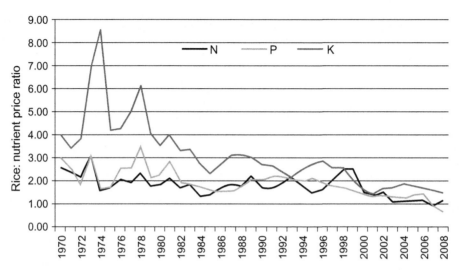

FIGURE 16.6

Trend of rice and fertilizer price ratio.

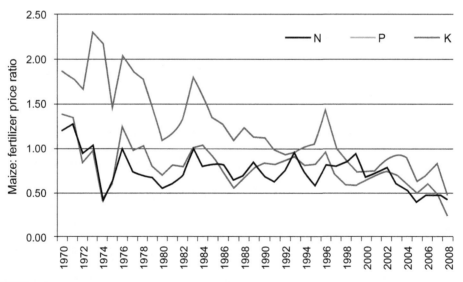

FIGURE 16.7

Trend of maize and fertilizer price ratio.

16.4.2.2 Impact of fertilizer prices on cereal production

The impact of fertilizer prices on the production of wheat, maize, and rice was estimated using the IMPACT model (Rosegrant et al., 2001, 2002). The model disaggregates supply growth into area and yield changes, allowing the comparison of the contribution of agricultural area extensification to production growth versus changes along the intensive margin through yield and productivity growth. Since the yield response functions contain input price responses to fertilizer and labor inputs, the impact of fertilizer prices on agricultural production levels can be directly simulated. The growth of food production was simulated while allowing fertilizer prices to change at 0.9% per annum, which is the average annual rate of NPK price increases from 1990 to 2000. This "reference" case was then compared with the results of a simulation where the rate of change in fertilizer prices was increasing, so that a threefold increase in the composite price of NPK was realized over the period 2000−2030. This is equivalent to an average rate of growth of 3.7% per annum. While a threefold increase may seem extreme, it is actually modest compared with the actual rates of change realized over the decade starting from 2000 for the three major fertilizer components (N, P, and K).[3]

Figures 16.8−16.10 show the changes in yield and crop area of the major cereals due to increase in fertilizer prices. Yields declined in all major producing regions and countries; the

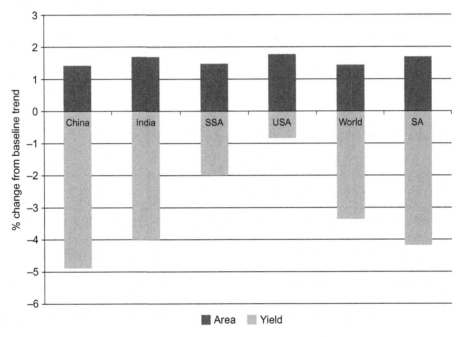

FIGURE 16.8

Wheat yield and area changes in 2030 under threefold fertilizer price increase (% change).

---

[3]The percent change was computed as follows: $[(Y_f - Y_b)/Y_b] \times 100$, where Y_f is production when fertilizer prices increase threefold and Y_b is production when fertilizer prices stay on their original growth trajectory over the 1990−2000 level.

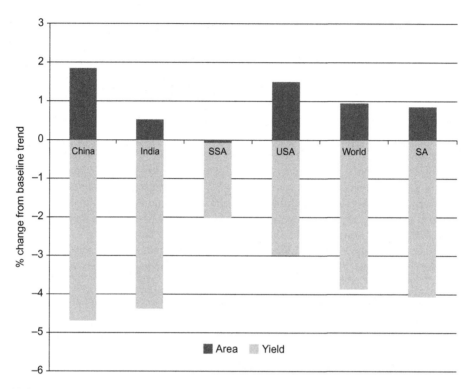

FIGURE 16.9

Rice yield and area changes in 2030 under threefold fertilizer price increase (% change).

decline is highest in China for all three crops. China has a relatively intensive use of yield-boosting inputs, which serves as a land-saving strategy given the country's high-density agricultural landscape. On the other hand, the United States registered the lowest yield decline for wheat and maize. One reason could be that the use of other capital inputs in cereal cultivation is more important than just fertilizer, making the responsiveness of yield to fertilizer prices less than that observed in other capital inputs. The decline of rice yields in SSA is the lowest, mainly due to the relatively low levels of fertilizer use compared with the other regions of the world (Figure 16.9).

In terms of rice area, China is projected to have the largest increase (2%) while that in SSA will not change (Figure 16.9). This underscores SSA's low fertilizer use in rice production and the historically low levels of responsiveness to prices that have been observed there. These differences in response also illustrate the strong contrast existing between relatively input-intensive farming systems in Asia and the Americas and the relatively low-intensity, small-holder dominated farming systems in SSA. All regions and countries considered are projected to increase maize area by about 3% (Figure 16.10).

Figure 16.11 presents the combined impact of yield declines and area expansion on production levels of selected regions and the globe. The global production of wheat and rice will decline while maize production will increase by about 1%, largely due to the US increase of about 3.2%. Maize production in SSA will increase by about 1.8%. Many of these changes are price driven and

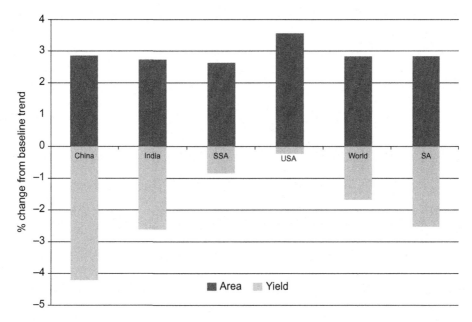

FIGURE 16.10

Maize yield and area changes in 2030 under threefold fertilizer price increase (% change).

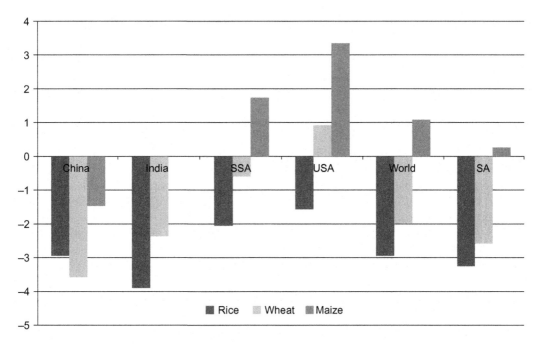

FIGURE 16.11

Production changes in major cereal crops in 2030 due to threefold fertilizer price increase (% change). Note, the production changes of maize in India do not appear because of the very low responsiveness of maize production to fertilizer price changes that are captured in our model.

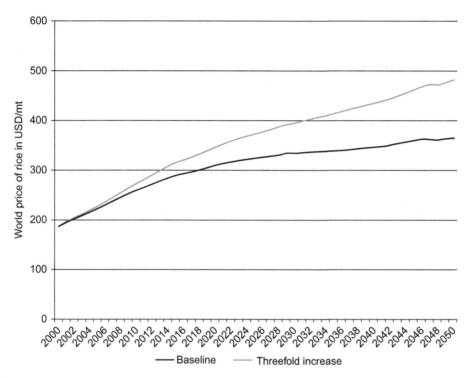

FIGURE 16.12

World price of rice under baseline and a threefold fertilizer price increase scenario.

respond to the overall higher crop prices that result when production is (initially) pushed down by fertilizer price increases. Some positive changes in production reflect the heavier reliance of some regions, such as SSA, on area extensification rather than on intensification and yield growth. Therefore, despite the negative effects of higher fertilizer prices on yield, there are strong area growth trends that push maize and wheat production upward, which are further accelerated by higher food prices (Figures 16.12–16.14). The sizable increases in food prices underscore the overall effect that higher fertilizer prices have on agricultural markets.

These results indicate that the relative price responsiveness of crop area to output price changes and that of crop yield to input price changes drive the overall production response in an environment of high-energy prices. The inevitable increase in food prices that higher energy prices cause, due to increases in fertilizer costs, poses an opportunity for supply-side response to manifest itself through increased output. The above results show that some regions of the world are better poised than others to take advantage of higher output prices and boost their productivity levels, given their resource and capital endowments. These differentials in output response have important implications on how quickly the world food balance can reequilibrate itself in the face of price shocks; it also underscores the importance of maintaining the ability of agricultural production systems to structurally adapt to such changes through continued reforms and investments.

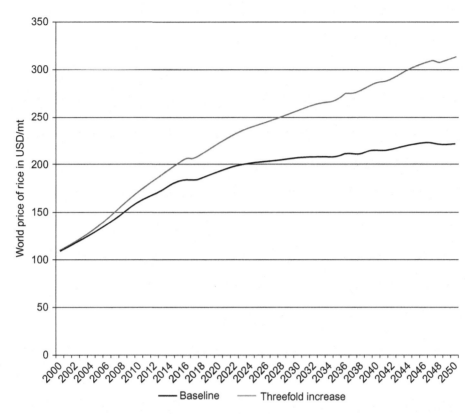

FIGURE 16.13

World price of wheat under baseline and a threefold fertilizer price increase scenario.

16.5 POLICY IMPLICATIONS OF FOOD–ENERGY INTERACTIONS

Given the above results, higher energy prices clearly affect the production and prices of agricultural goods in the same direction as was anticipated from the results of the theoretical model. Even though the IMPACT model deals with a much more diverse array of interactions between various commodities and regions, the basic intuition of the results still comes through.

A key implication from the results of both theoretical and empirical treatments is that agricultural markets will become increasingly tighter as energy prices continue to rise, given the latter's effects on production. While using higher energy inputs results in increased agricultural productivity, over and above the case where very low input usage is observed, the price effects of energy are decreased use of those inputs, eventually necessitating a shift toward other more energy-efficient alternatives. In the case of groundwater, this could mean adopting more energy-efficient methods of irrigation or changing the crops mix to replace those that consume more water with ones that require less water for growth. An analogous effect could be imagined in the case of fertilizer, where application techniques that improve its use efficiency could be employed as the price of its

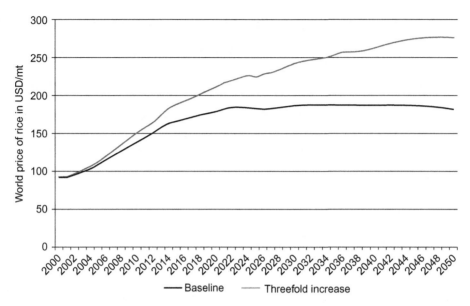

FIGURE 16.14

World price of maize under baseline and a threefold fertilizer price increase scenario.

energy-rich components starts to go up. This could be followed by a switch in the mix of fertilizers toward those types with less of the costly components and eventually by a switch in the crop itself toward those that either require less fertilizer or are nutrient fixing and thus can be alternated with crops that require such nutrients from the soil.

The other dimension of adjustments that agriculture can pursue is moving toward products with higher value. This would better justify the application of increasingly costly inputs due to a higher value marginal product—as suggested by the basic relationship in Eq. (16.5) (i.e., $pAf'(w) = c_w(e)$). Examples of this strategy are seen in many places, where high-value orchard and market crops (e.g., wine grapes) replace lower value field crops in water scarce regions like California and Australia. This trend is expected also in regions facing higher levels of factor scarcity (and prices), as they are able to import the lower value products from more resource-abundant regions while moving higher up the value ladder with respect to their own agricultural production.

In the case of biofuels, the impact on agricultural markets has been explored, both within the context of its effects on food markets (von Braun, 2008; Abbot et al., 2008; Rosegrant et al., 2008; Runge and Senauer, 2007, 2008; de Hoyos and Medvedev, 2009) and its impacts on land use change and the environment (Searchinger et al., 2008; Edwards et al., 2010; Beckman et al., 2011). The continued rise of oil prices enhances the competitiveness of biofuels as a fuel supplement and substitute (especially in regions that are more easily able to produce them, like Brazil). The increased demand for biofuels, and the resulting price increase, will encourage production up to the point where demand can be satisfied, although there would be a corresponding expansion of feed-stock consumption (and production) that would necessitate increasing its area at the expense of other crops, or increasing its use of other scarce factors. Improvements in transportation fuel use efficiency would reduce both fossil-based and bio-based fuels; considerable research efforts are already underway to look for alternative feedstock sources that require less from agricultural starch, sugar, and oil crops, and instead more from either dedicated energy grasses or nonland-based sources such as algae. Some of these innovations have yet to realize economically feasible levels of cost and scale, but many see these as holding considerable promise within the next 10–15 years. The pace of these reforms is expected to accelerate as the price of oil continues to rise.

Therefore, an increasing trend toward efficiency improvement is expected in the future, as well as a shift in cropping patterns across regions, reflecting the relative levels of factor intensity and scarcity, with increasing reliance on trade to balance availability with demand. Two issues would need to be addressed from a policy perspective so that the future of energy and agriculture would unfold in a way that enhances efficient technical change. First is the natural resource management and environmental policy. As long as the prices of resources reflect their true value of extraction and use, rather than an artificially subsidized one (as is often the case with groundwater in India), then agricultural input use and product value will adjust accordingly. Where value distortions exist, adoption of efficiency-enhancing measures will not occur at a pace that matches that of the real market value of the inputs. Second is trade policy. Free movement of goods between regions in ways that reflect their comparative costs and efficiencies of production should be allowed. Experience tells that these conditions are seldom met and are undermined by politically driven desires to protect (often less efficient) producers, support local prices, or extract revenues.

It is, therefore, argued that the path toward an energy-efficient future—which requires a range of policy innovations and incentives—needs to be accompanied by reforms in other sectors so that the evolving dynamics between energy and agricultural markets would occur smoothly. If done in a conscious and coordinated way, efficiency innovations in one sector could encourage and enable technology innovations in another sector, given that the basic linkages to underlying resources discussed in Section 16.2 will require constant innovation and improvement.

At present, a number of countries are pushing for a comprehensive pricing policy reform in energy—a highly contentious and politically sensitive issue. The experience in Nigeria has proved politically difficult, due to public mistrust of how savings from reduced subsidies to fossil fuels would be spent by the government (Adenikinju, 2009). The implications of not fully recovering the costs of power generation, however, are serious and are at the root of much of Africa's poor maintenance and limited expansion of energy generation infrastructure (Foster and Briseño-Garmendia, 2010). As these authors pointed out in the infrastructure diagnostic that the World Bank did for Africa, and reinforced by others (e.g., Alleyne, 2013), improving cost recovery is a key ingredient in rehabilitating the power sector in SSA and enabling the badly needed expansion of capacity. These reforms are most effective when coupled with effective engagement of the private sector, maintaining the financial viability of the entities charged with managing the power system and political will (Alleyne, 2013).

In the case of Brazil, the fossil fuel subsidy has undercut the competitiveness of the country's ethanol sector in recent years, limiting growth in sectoral investments and productivity (de Oliveira and Laan, 2010). This illustrates one of the main reasons why energy pricing reform is a key element in building a favorable environment for renewable energy investments and the importance of incentives—both for energy consumers and for those commercial interests that need to be incentivized to deepen investments in new, resource-saving technologies. The impact of high fossil fuel prices can be a powerful incentive for investment by itself. Babcock (2011) showed that the high oil prices of 2005−2007 likely had more impact on biofuel capacity expansion in the United States than the actual biofuel subsidies and incentive policies that are in place.

These examples underscore the importance of energy prices as triggers for innovation; this is why costly subsidies delay needed transitions to better technologies. Undoubtedly, high-energy prices are often problematic from a political perspective. From a policy perspective, however, they are a useful mechanism for inducing innovation, behavioral change, and efficiency improvements, all of which are key ingredients for transitioning to a "greener" economy that is more environmentally stable and climate smart.

16.6 CONCLUSION

This chapter explored various linkages between energy and agriculture in both theoretical and empirical contexts. The theoretical treatment of energy−agriculture interactions traced the important linkages between output prices and prices of the factors, some of which are closely connected to the underlying energy prices. The cases of groundwater and fertilizer are closely analogous in the conceptual framework, given the very similar way in which they enter into the producer side of the problem. In the case of biofuels, the feedstock demand enters into the demand side of the market, but is also closely tied to the energy price, which induces different levels of biofuel demand and, therefore, production.

Besides the pressure on prices, pressures on other resources such as land and water can also be looked at. The pressure on land is a key concern in the debate over biofuels' environmental sustainability and is under intense research and policy scrutiny. The empirical results of fertilizer prices indicate an expansion in crop area as energy prices went up; that is, there is a possible trade-off in environmental quality as more land is converted to compensate for lower yields. Other dimensions of adjustment, not captured in the model, could occur, such as substitution with crops that are less input intensive or are higher in value. The same can be expected in the case of the more expensive groundwater, which has been long discussed as an important way to induce changes in groundwater pumping behavior in India and reduce groundwater overdraft, although this is very difficult politically.

A connection between energy, agriculture, and the use and availability of certain key natural resources has been clearly seen. Greater resource use efficiency is a key message from the above discussion, since it constitutes a basic response to higher levels of scarcity, but is dependent on the correct price signals reaching the decision makers. This places, therefore, a high priority on policy reform to remove distortions that might prevent this type of response from happening in a timely and progressive manner.

From the perspective of biofuels, food, and agriculture, the future demand of energy for transport will continue to place pressure on land and other resources if starch sugar or oil-based biofuels remain a large part of the renewable energy mix. This will most certainly continue, given the relatively slow growth of cellulose-based biofuels and their difficulty in achieving cost competitiveness, as well as relative to the current "first-generation" biofuels. Given the subsidies provided to the fossil fuel prices in some countries, even first-generation biofuels face challenges, as has been the case in Brazil in recent years.

In view of the important linkages between agriculture and energy, the evaluation of future food outlooks will have to constantly consider energy prices and their effects on both input costs to agriculture and the induced demand for agricultural products in the form of biofuel feedstocks. Improving the understanding and quantification of these effects will require a greater dialogue between the energy and agricultural modeling communities and a common understanding of the technological possibilities and key policy issues relevant to the interaction of food and energy markets.

REFERENCES

Abbot, P.C., Hurt, C., Tyner, W.E., 2008. What's driving food prices? Issue Report. Farm Foundation, Oak Brook, IL.

Adenikinju, A., 2009. Energy pricing and subsidy reforms in Nigeria. Presentation, Environment and Trade Division, Organization for Economic Cooperation and Development (OECD), Paris. Available from: <http://www.oecd.org/tad/events/42987402.pdf> (accessed 25.08.14.).

Alleyne, T., 2013. Energy subsidy reform in Sub-Saharan Africa: experiences and lessons. African Department Discussion Paper 13/2, International Monetary Fund, Washington, DC.

Babcock, B.A., 2011. The impact of US biofuels policy on agricultural price levels and volatility. Issue Paper 35, ICTSD Programme on Agricultural Trade and Sustainable Development, International Center for Trade and Sustainable Development, Geneva.

Beckman, J., Jones, C.A., Sands, R., 2011. A global general equilibrium analysis of biofuel mandates and greenhouse gas emissions. Am. J. Agric. Econ. 93 (2), 334−341.

Borlaug, N., 2000. Ending world hunger: the promise of biotechnology and the threat of antiscience zealotry. Plant. Physiol. 124 (2), 487−490.

Borlaug, N., Dowsell, C., 2004. The green revolution: an unfinished agenda. Lecture at the UN Food and Agriculture Organization, committee on food security. Available from: <http://www.fao.org/docrep/meeting/008/J3205e/j3205e00.htm>.

Chianu, J., Adesina, A., Sanginga, P., Bationo, A., Chianu, J., Sanginga, N., 2008. Structural change in fertilizer procurement method: assessment of impact in Sub-Saharan Africa. Afr. J. Bus. Manage. 2 (3), 65−71.

de Hoyos, R.E., Medvedev, D., 2009. Poverty effects of higher food prices: a global perspective. World Bank Policy Research Working Paper 4887, Development Research Group, World Bank, Washington, DC.

de Oliveira, A., Laan, T., 2010. Lessons learned from Brazil's experience with fossil-fuel subsidies and their reform. Trade, Investment and Climate Change Series Discussion Paper, International Institute for Sustainable Development (IISD), Geneva.

Dorward, A., Poulton, C., 2008. The global fertilizer crisis and Africa, Briefing. Future Agricultures. Brighton. <http://www.future-agricultures.org/pdf%20files/brieffertilisercrisis.pdf>.

Dubash, N.K., 2005. The electricity-groundwater conundrum: a case for a political solution to a political problem. Paper presented at the Fourth IMWI-Tata Annual Partners' Meet Session 5: Groundwater Markets and Livelihoods, February 24−26, Colombo, Sri Lanka.

Edwards, R., Mulligan, D., Marelli, L., 2010. Indirect land use change from increased biofuels demand: comparison of models and results for marginal biofuels production from different feedstocks. JRC Scientific and Technical Reports, Joint Research Center of the European Commission, Ispra, Italy.

FAO (Food and Agriculture Organization), 1989. Fertilizers and Food Production. The FAO Fertilizer Programme 1961−1986, Rome.

FAC, 2008. The global fertiliser crisis and Africa. Policy brief 025. Future Agricultures Consortium (FAC), University of Sussex, Brighton, UK.

Foster, V., Briceño-Garmendia, C., 2010. Africa's Infrastructure: A Time for Transformation. World Bank, Washington, DC.

Goklany, I.M., 2001. The future of food. Forum. Appl. Res. Public Policy 16 (2), 59−65.

Heffer, P., Prud'homme, M., 2008. Outlook for world fertilizer demand, supply and supply/demand balance. Turk. J. Agric. Forestry 32, 159−164.

IFA (International Fertilizer Industry Association), 2008. Assessment of fertilizer use by crop at the global level, Paris.

IFDC (International Fertilizer Development Center), 2008. World fertilizer prices continue to soar as scientists stress the need to increase fertilizer efficiency. IFDC Report 22 (2), 1−7.

IWMI-Tata, 2002. The socio-ecology of groundwater in India. IMWI-Tata Water Policy Briefing, Issue 4. International Water Management Institute, Colombo, Sri Lanka.

Lal, S., 2006. Can good economics ever be good politics? Case Study of India's Power Sector. World Bank Working Paper No. 83, World Bank, Washington, DC.

Lin, Y., Ren, H., Yu, J., 2000. Study of water resource balance in North China Plain. Nat. Resour. 15, 252−258.

Lohmar, B., Wang, J., Rozelle, S., Huang, J., Dawe, D., 2003. China's agricultural water policy reforms: increasing investment, resolving conflicts, and revising incentives. Agriculture Information Bulletin No. 782, Market and Trade Economics Division, Economic Research Service, US Department of Agriculture, Washington, DC.

Monari, L., 2002. Power subsidies: a reality check on subsidizing power for irrigation in India. The World Bank Private Sector and Infrastructure Network. Viewpoint, Number 24477. Available from: <http://www-wds.worldbank.org/external/default/WDSContentServer/WDSP/IB/2002/06/24/000094946_02060704032946/Rendered/PDF/multi0page.pdf> (accessed 10 Feb. 14).

Morris, M., Kelly, V., Kopicki, R., Byerlee, D., 2007. Fertilizer Use in African Agriculture: Lessons Learned and Good Practice Guidelines. World Bank, Washington, DC.

Phipps, R., Park, J., 2002. Environmental benefits of genetically modified crops: global and European perspectives on their ability to reduce pesticide use. J. Anim. Feed Sci. 11, 1−18.

Pinstrup-Andersen, P., 1976. Preliminary estimates of the contribution of fertilizer to cereal production in developing market economies. J. Econ. 1, 169−172.

Reddy, V.R., 2005. Cost of resource depletion externalities: a study of groundwater overexploitation in Andhra Pradesh, India. Environ. Dev. Econ. 10, 533−556.

Rosegrant, M.W., Paisner, M., Meijer, S., Witcover, J., 2001. Global Food Projections to 2020: Emerging Trends and Alternative Futures. International Food Policy Research Institute, Washington, DC.

Rosegrant, M.W., Cai, X., Cline, S., 2002. World Water and Food to 2025: Dealing with Scarcity. International Food Policy Research Institute, Washington, DC.

Rosegrant, M.W., Zhu, T., Msangi, S., Sulser, T., 2008. Global scenarios for biofuels: impacts and implications. Rev. Agric. Econ. 30 (3), 495−505.

Roy, R., Misra, R., Montanez, A., 2002. Decreasing reliance on mineral nitrogen—yet more food. Ambio 31 (2), 177−183.

Runge, C.F., Senauer, B., 2007. How biofuels could starve the poor. Foreign Aff. 86 (3), 41−53.

Runge, C.F., Senauer, B., 2008. How ethanol fuels the food crisis. Author update, May 28, 2008. Foreign Affairs. <http://www.foreignaffairs.com/articles/64915/c-ford-runge-and-benjamin-senauer/how-ethanol-fuels-the-food-crisis> (accessed 25.08.14.).

Searchinger, T., Heimlich, R., Houghton, R.A., Dong, F., Elobeid, A., Fabiosa, J., et al., 2008. Use of U.S. croplands for biofuels increases greenhouse gases through emissions from land-use change. Sci. Expr. 319, 1238−1240.

Socolow, R., 1999. Nitrogen management and the future of food: lessons from the management of energy and carbon. Proc. Natl. Acad. Sci. 96 (11), 6001−6008.

Trostle, R., 2008. Global agricultural supply and demand: factors contributing to the recent increase in food commodity prices. Working Paper #WRS-0801, USDA Economic Research Services, Washington, DC.

von Braun, J., 2008. The World Food Situation: New Driving Forces and Required Actions. Food Policy Report, International Food Policy Research Institute, Washington, DC.

Wang, J., Huang, J., Rozelle, S., 2005a. Evolution of tubewell ownership and production in the North China Plain. Aust. J. Agric. Resour. Econ. 49, 177−195.

Wang, J., Huang, J., Blanke, A., Huang, Q., Rozelle, S., 2007. The development, challenges and management of groundwater in rural China. In: Giordano, M., Villholth, K.G. (Eds.), The Agricultural Groundwater Revolution: Opportunities and Threats to Development. CAB International, pp. 37−62.

World Bank, 2014. World Bank Commodity Price Data (The Pink Sheet): Monthly prices in nominal US dollars, 1960 to present. Updated 04 Agust 2014. Available at: < http://siteresources.worldbank.org/INTPROSPECTS/Resources/334934-1304428586133/pink_data_m.xlsx > (accessed 25.08.2014.).

TECHNICAL ANNEX A: DERIVATIONS OF CONCEPTUAL MODELS

GROUNDWATER EXAMPLE

Starting from the maximization problem given in Eq. (16.2), we proceeded with calculating the optimal behavior under constraints. The necessary conditions for profit-maximizing choice over variables w and A led to the following set of equations:

$$
\begin{array}{lll}
pAf'(w) - c_w \leq 0 & w \geq 0 & w[pAf'(w) - c_w] = 0 \\
pf(w) - \theta c_A A^{\theta-1} - \lambda \leq 0 & A \geq 0 & A[pf(w) - \theta c_A A^{\theta-1} - \lambda] = 0 \\
A - A^{\max} \leq 0 & \lambda \geq 0 & \lambda[A - A^{\max}] = 0
\end{array} \tag{A.1}
$$

which can be used to define the optimal choice of water and land usage by the representative farmer. Assuming that the decision maker always chooses to use as much land as is available, then we reduced the set of conditions in Eq. (A.1) to the following pair of equations:

$$
\begin{aligned}
pA^{\max}f'(w) - c_w &= 0 \\
pf(w) - \theta(A^{\max})^{\theta-1} - \lambda &= 0
\end{aligned} \tag{A.2}
$$

which can be used to examine the sensitivity of the optimal choices to the key economic parameter values. By totally differentiating the pair of first-order conditions in Eq. (A.2), with respect to each variable and parameter, we obtained the following linear system:

$$
\begin{bmatrix} pA^{\max}f''(w) & 0 \\ pf'(w) & -1 \end{bmatrix} \begin{bmatrix} dx \\ d\lambda \end{bmatrix} = \begin{bmatrix} -A^{\max}f'(w) & -pf'(w) & +1 \\ -f(w) & +\theta(\theta-1)(A^{\max})^{\theta-2} & 0 \end{bmatrix} \begin{bmatrix} dp \\ dA^{\max} \\ dc_w \end{bmatrix} \tag{A.3}
$$

which allows us to relate changes in the vector of decision variables, on the left-hand side, to the vector of the key parameters, on the right-hand side. The sign of the principle determinant, D, as shown below:

$$
D = \begin{vmatrix} pA^{\max}f''(w) & 0 \\ pf'(w) & -1 \end{vmatrix} > 0 \tag{A.4}
$$

conforms to expectations of a well-behaved maximization problem, in which the set of production possibilities can be circumscribed by a convex hull. Applying Cramer's rule to the linear system in Eq. (A.3) allowed us to perform the following comparative static calculations, in which we examined the impact of parameter changes on the key decision variable of interest, i.e., water use (w). To see the impact of changes in the volumetric cost of water (c_w) on water usage, we computed the following differential:

$$
\frac{\partial w}{\partial c_w} = \frac{\begin{vmatrix} +1 & 0 \\ 0 & -1 \end{vmatrix}}{D} = \frac{-1}{D} < 0 \tag{A.5}
$$

in which the marginal effect of increasing the cost serves to decrease the water usage level, as expected. Similarly, we can examine the impact of changing the constraint on land, such that we allow for a marginal change in the binding quantity $\bar{A}$.

$$
\frac{\partial w}{\partial A^{\max}} = \frac{\begin{vmatrix} -pf'(w) & 0 \\ +\theta(\theta-1)(A^{\max})^{\theta-2} & -1 \end{vmatrix}}{D} = \frac{pf'(w)}{D} > 0 \tag{A.6}
$$

While these effects are opposite in signs, we can see that their relative magnitude depends on the magnitude of the value marginal product of water $pf'(w)$ relative to unity. If it turns out that the value marginal product exceeds unity, then a change in allowable land area might be more effective in reducing water usage than a change in the unit variable cost of water. The converse would be implied in the case where $pf'(w) < 1$.

FERTILIZER EXAMPLE

From the maximization problem given in Eq. (16.2), we derived a similar result showing the direct effect of fertilizer cost on input use.

We would obtain this result even in the case where we take the original first-order conditions of the production problem in Eq. (A.1), and take the case where land use is not binding ($A < A^{\max}$), in which case $\lambda = 0$, and we have the case where:

$$pAf'(w) - c_w = 0$$
$$pf(w) - \theta A^{\theta-1} = 0 \tag{A.7}$$

If we were to substitute for $f(w) = kw^\alpha$ and take the total derivatives, we would obtain the system:

$$\begin{bmatrix} \alpha pkw^{\alpha-1} & -\theta(\theta-1)c_A A^{\theta-2} \\ \alpha(\alpha-1)Apkw^{\alpha-2} & \alpha pkw^{\alpha-1} \end{bmatrix} \begin{bmatrix} dw \\ dA \end{bmatrix} = \begin{bmatrix} -kw^\alpha & -pw^\alpha & 0 \\ -\alpha Akw^{\alpha-1} & -\alpha pAw^{\alpha-1} & +1 \end{bmatrix} \begin{bmatrix} dp \\ dk \\ dc_w \end{bmatrix} \tag{A.8}$$

which has the characteristics of a "square" linear system $\mathbf{Ax} = \mathbf{By}$, which we can solve by inverting the matrix A and taking the product with the RHS of the system, such as:

$$\mathbf{x} = \mathbf{A}^{-1}\mathbf{By} \tag{A.9}$$

If we consider the principal determinant $\mathbf{D}$ of this linear system, which is:

$$\mathbf{D} = \begin{vmatrix} \alpha pkw^{\alpha-1} & -\theta(\theta-1)c_A A^{\theta-2} \\ \alpha(\alpha-1)Apkw^{\alpha-2} & \alpha pkw^{\alpha-1} \end{vmatrix} < 0 \tag{A.10}$$

the comparative statics result showing the impact of the input (fertilizer) price (c_w) on the usage of the input (w) would be given by:

$$\frac{dw}{dc_w} = \frac{\begin{vmatrix} 0 & -\theta(\theta-1)c_A A^{\theta-2} \\ +1 & \alpha pkw^{\alpha-1} \end{vmatrix}}{\mathbf{D}} = \frac{\theta(\theta-1)c_A A^{\theta-2}}{\mathbf{D}} < 0 \tag{A.11}$$

since $\mathbf{D} < 0$ and the numerator is positive (since $\theta > 1$).

After accounting for the market–balance relationship in Eq. (16.17), we now have an expanded problem, which can be solved by treating it as a linear system of equations.

We now have increased the system of equations that describes the first-order conditions of the market equilibrium and producer problem, as follows:

$$\begin{aligned} p \cdot kw^\alpha && -\theta c_A A^{\theta-1} = 0 \\ p \cdot A \cdot \alpha kw^{\alpha-1} && -c_w = 0 \\ p \cdot A \cdot kw^\alpha - d_0 \cdot p^\varepsilon - e(\bar{p}_w - p)^\xi && = 0 \end{aligned} \tag{A.12}$$

By taking the total derivative of the relationship in Eq. (16.6), we can derive a new linear system that characterizes the equilibrium, as follows:

$$
\begin{bmatrix}
kw^\alpha & \alpha pkw^{\alpha-1} & -\theta(\theta-1)c_A A^{\theta-2} \\
\alpha Akw^{\alpha-1} & \alpha(\alpha-1)pAkw^{\alpha-2} & \alpha pAkw^{\alpha-1} \\
e_0\xi(\bar{p}_w-p)^{\xi-1}-d_0\varepsilon p^{\varepsilon-1} & \alpha Akw^{\alpha-1} & kw^\alpha
\end{bmatrix}
\begin{bmatrix} dp \\ dw \\ dA \end{bmatrix}
$$

$$
=
\begin{bmatrix}
-pw^\alpha & 0 & 0 \\
-\alpha pw^{\alpha-1} & +1 & 0 \\
Aw^\alpha & 0 & e_0\xi(\bar{p}_w-p)^{\xi-1}
\end{bmatrix}
\begin{bmatrix} dk \\ dc_w \\ d\bar{p}_w \end{bmatrix}
$$

(A.13)

The principal determinant of this system

$$
\mathbf{D} =
\begin{vmatrix}
kw^\alpha & \alpha pkw^{\alpha-1} & -\theta(\theta-1)c_A A^{\theta-2} \\
\alpha Akw^{\alpha-1} & \alpha(\alpha-1)pAkw^{\alpha-2} & \alpha pAkw^{\alpha-1} \\
e_0\xi(\bar{p}_w-p)^{\xi-1}-d_0\varepsilon p^{\varepsilon-1} & \alpha Akw^{\alpha-1} & kw^\alpha
\end{vmatrix}
$$

can be shown to be negative, such that when we consider the comparative static result that gives the impact of the input price on the input use, we obtain:

$$
\frac{dw}{dc_w} = \frac{
\begin{vmatrix}
kw^\alpha & 0 & -\theta(\theta-1)c_A A^{\theta-2} \\
\alpha Akw^{\alpha-1} & 1 & \alpha pAkw^{\alpha-1} \\
e_0\xi(\bar{p}_w-p)^{\xi-1}-d_0\varepsilon p^{\varepsilon-1} & 0 & kw^\alpha
\end{vmatrix}
}{\mathbf{D}}
$$

(A.14)

$$
= \frac{k^2 w^{2\alpha} + \theta(\theta-1)c_A A^{\theta-2}[e_0\xi(\bar{p}_w-p)^{\xi-1}-d_0\varepsilon p^{\varepsilon-1}]}{\mathbf{D}} < 0
$$

which comes from the fact that $\theta > 1$ and $\varepsilon < 0$, making the numerator positive. This is analogous to the result given in Eq. (A.11) and conforms to our intuition.

Since the output price is now endogenous, we can examine the effects of the input price on it, which can be derived from the comparative statics result:

$$
\frac{dp}{dc_w} = \frac{
\begin{vmatrix}
0 & \alpha pkw^{\alpha-1} & -\theta(\theta-1)c_A A^{\theta-2} \\
1 & \alpha(\alpha-1)pAkw^{\alpha-2} & \alpha pAkw^{\alpha-1} \\
0 & \alpha Akw^{\alpha-1} & kw^\alpha
\end{vmatrix}
}{\mathbf{D}}
= \frac{-\alpha pk^2 w^{2\alpha}-\theta(\theta-1)c_A A^{\theta-2}\cdot\alpha Akw^{\alpha-1}}{\mathbf{D}} > 0 \quad \text{(A.15)}
$$

which gives a positive impact, since $\theta > 1$ makes the numerator negative.

BIOFUELS EXAMPLE

From the expanded maximization problem, where the market balance for goods was augmented with the demand for biofuels feedstock, as given in Eq. (16.12), we obtained an augmented linear system.

By putting the supply-side response of agriculture together with the market-level balance of supply, demand, and exports[4]—as done in Eq. (A.12)—we obtained the following system of equations:

$$p \cdot kw^{\alpha} - \theta c_A A^{\theta-1} = 0$$

$$p \cdot A \cdot \alpha kw^{\alpha-1} - c_w = 0$$

$$A \cdot kw^{\alpha} - d_0 \cdot p^{\varepsilon} + \hat{b}_0 \left(\frac{p_b}{p}\right) \hat{\gamma} - e(\bar{p}_w - p)^{\xi} = 0$$

(A.16)

which has three endogenous variables (p, A, w) and a number of key exogenous variables (p_w, p_b, k, c_w). Since we are not modeling the demand and trade of biofuels, the price of biofuels (p_b) remains exogenous in this simple analysis.

The solution of this system defines the market equilibrium, which will be shifted around due to changes in key parameters such as those affecting productivity of agriculture (k), cost of productive inputs (c_w), price of biofuels (p_b), and world price of the agricultural good (p_w). If we take the total derivatives of these equations, we obtain a linear system that can help describe the shifts that will happen around the market equilibrium if we were to change the levels of these key variables. With this linear system, given in Eq. (A.17), we can find the principal determinant and solve the comparative statics.

$$
\begin{bmatrix}
kw^{\alpha} & \alpha p kw^{\alpha-1} & -\theta(\theta-1)c_A A^{\theta-2} \\
\alpha A kw^{\alpha-1} & \alpha(\alpha-1)A p kw^{\alpha-2} & 0 \\
e_0 \xi(\bar{p}_w - p)^{\xi-1} - \dfrac{\hat{b}_0 \hat{\gamma}}{p^2}\left(\dfrac{p_b}{p}\right)^{\hat{\gamma}-1} - d_0 \varepsilon p^{\varepsilon-1} & \alpha A kw^{\alpha-1} & kw^{\alpha}
\end{bmatrix}
\begin{bmatrix} dp \\ dw \\ dA \end{bmatrix}
$$

(A.17)

$$
=
\begin{bmatrix}
-pw^{\alpha} & 0 & 0 & 0 \\
-\alpha A p w^{\alpha-1} & 1 & 0 & 0 \\
-Aw^{\alpha} & 0 & d_0 \varepsilon p^{\varepsilon-1} - \dfrac{\hat{b}_0 \hat{\gamma}}{p}\left(\dfrac{p_b}{p}\right)^{\hat{\gamma}-1} & e_0 \xi(\bar{p}_w - p)^{\xi-1}
\end{bmatrix}
\begin{bmatrix} dk \\ dc_w \\ dp_b \\ dp_w \end{bmatrix}
$$

By taking the determinant of the matrix on the LHS, we get[5]:

$$
\mathbf{D} =
\begin{vmatrix}
kw^{\alpha} & \alpha p kw^{\alpha-1} & -\theta(\theta-1)c_A A^{\theta-2} \\
\alpha A kw^{\alpha-1} & \alpha(\alpha-1)A p kw^{\alpha-2} & 0 \\
e_0 \xi(\bar{p}_w - p)^{\xi-1} - \dfrac{\hat{b}_0 \hat{\gamma}}{p^2}\left(\dfrac{p_b}{p}\right)^{\hat{\gamma}-1} - d_0 \varepsilon p^{\varepsilon-1} & \alpha A kw^{\alpha-1} & kw^{\alpha}
\end{vmatrix}
$$

$$= kw^{\alpha} \cdot [\alpha(\alpha-1)pAkw^{\alpha-2} \cdot kw^{\alpha}] - \alpha p kw^{\alpha-1}[\alpha A kw^{\alpha-1} \cdot kw^{\alpha}]$$

---

[4]In this simple treatment, we omitted the possibility of importing, which can be added without loss of generality (as it can also be treated as a positive quantity which is a function of world and domestic prices) but at the expense of additional parameters in the equations.

[5]For simplicity we used the "shorthand," where $[\bullet] = \left[e_0 \xi(\bar{p}_w - p)^{\xi-1} - \dfrac{\hat{b}_0 \hat{\gamma}}{p^2}\left(\dfrac{p_b}{p}\right)^{\hat{\gamma}-1} - d_0 \varepsilon p^{\varepsilon-1} \right]$

$$-\theta(\theta-1)c_A A^{\theta-2}\begin{bmatrix}\alpha Akw^{\alpha-1}\cdot\alpha Akw^{\alpha-1}\\[4pt]-\alpha(\alpha-1)Apkw^{\alpha-2}\left(e_0\xi(\bar{p}_w-p)^{\xi-1}-\dfrac{\hat{b}_0\hat{\gamma}}{p^2}\left(\dfrac{p_b}{p}\right)^{\hat{\gamma}-1}-d_0\varepsilon p^{\varepsilon-1}\right)\end{bmatrix}$$

$$=\alpha(\underbrace{\alpha-1}_{<0})pAk^3w^{3\alpha-2}-\alpha^2pk^3w^{3\alpha-2}+\theta\left(\underbrace{\theta-1}_{>0}\right)c_A A^{\theta-2}\left[\alpha^2k^2A^2w^{2\alpha-2}-\alpha(\underbrace{\alpha-1}_{<0})pAkw^{\alpha-2}(\bullet)\right]$$

which is negative in sign ($\mathbf{D}<0$), given the assumptions on the curvature parameters for yield ($\alpha<1$) and land-related production costs ($\theta>1$) and the quantity $\left[\alpha^2k^2A^2w^{2\alpha-2}-\alpha\underbrace{(\alpha-1)}_{<0}pkw^{\alpha-2}(\bullet)\right]$

which is assumed to be negative. Using this determinant in the implementation of Cramer's rule, we can then derive the following comparative statics results, where:

$$\frac{dp}{dp_b}=\frac{\begin{vmatrix}0 & \alpha pkw^{\alpha-1} & -\theta(\theta-1)c_A A^{\theta-2}\\[4pt]0 & \alpha(\alpha-1)Apkw^{\alpha-2} & 0\\[4pt]d_0\varepsilon p^{\varepsilon-1}-\dfrac{\hat{b}_0\hat{\gamma}}{p}\left(\dfrac{p_b}{p}\right)^{\hat{\gamma}-1} & \alpha Akw^{\alpha-1} & kw^{\alpha}\end{vmatrix}}{\mathbf{D}}$$

$$=\frac{\theta\left(\overbrace{\theta-1}^{(+)}\right)c_A A^{\theta-2}\cdot\alpha(\overbrace{\alpha-1}^{(-)})pAkw^{\alpha-2}\cdot\left(d_0\varepsilon p^{\varepsilon-1}-\dfrac{\hat{b}_0\hat{\gamma}}{p}\left(\dfrac{p_b}{p}\right)^{\hat{\gamma}-1}\right)}{\underset{(-)}{\mathbf{D}}}>0$$

which conforms to our intuition, since we expected that the increase in biofuel prices would encourage the production of biofuels, and hence the demand for the agricultural good as feedstock, which would increase its price in the market equilibrium. In terms of the effect of biofuel prices (and feedstock demand for the agricultural good), we can derive that impact from the following:

$$\frac{dA}{dp_b}=\frac{\begin{vmatrix}kw^{\alpha} & \alpha pkw^{\alpha-1} & 0\\[4pt]\alpha Akw^{\alpha-1} & \alpha(\alpha-1)Apkw^{\alpha-2} & 0\\[4pt]e_0\xi(\bar{p}_w-p)^{\xi-1}-\dfrac{\hat{b}_0\hat{\gamma}}{p^2}\left(\dfrac{p_b}{p}\right)^{\hat{\gamma}-1}-d_0\varepsilon p^{\varepsilon-1} & \alpha Akw^{\alpha-1} & d_0\varepsilon p^{\varepsilon-1}-\dfrac{\hat{b}_0\hat{\gamma}}{p}\left(\dfrac{p_b}{p}\right)^{\hat{\gamma}-1}\end{vmatrix}}{\mathbf{D}}$$

$$=\frac{\left[\alpha Ak^2w^{2\alpha-1}-\alpha^2pAk^2w^{2\alpha-2}\right]\cdot\left[d_0\varepsilon p^{\varepsilon-1}-\dfrac{\hat{b}_0\hat{\gamma}}{p}\left(\dfrac{p_b}{p}\right)^{\hat{\gamma}-1}\right]}{\underset{(-)}{\mathbf{D}}}$$

$$= \frac{\alpha A k^2 w^{2\alpha-1} \cdot \left[1 - \alpha p w^{-1}\right] \cdot \left[d_0 \varepsilon p^{\varepsilon-1} - \frac{\hat{b}_0 \hat{\gamma}}{p}\left(\frac{p_b}{p}\right)^{\hat{\gamma}-1}\right]}{\underset{(-)}{D}} > 0$$

which also conforms to our intuition, since an increase in feedstock demand for the agricultural good, and the resulting increase in price, would likely incentivize the expansion of agricultural area in response to higher market prices.

The effect of the various exogenous parameters on the other endogenous variables of the conceptual model can be derived in a similar fashion.

TRENDS AND FLUCTUATIONS IN AGRICULTURAL PRICE DISTORTIONS

17

Kym Anderson

School of Economics, University of Adelaide, Adelaide SA, Australia;
Australian National University, Canberra, Australia

CHAPTER OUTLINE

JEL: F13; F14; Q17; Q18

Recent spikes in international food prices, and government responses to them, have brought agricultural price and trade policies back into the spotlight. Food-importing developing countries have accused agricultural-exporting countries of exacerbating food security concerns by restricting exports, while other exporters fear such restrictions will lead to a retreat from reliance on international markets as food-deficit countries seek greater self-sufficiency when prices return to trend. Meanwhile, food-importing countries have reduced their import restrictions and a few have even subsidized imports of their staple food, adding to the international food spike.

These recent short-run responses and their possible long-run protectionist consequences beg the question of how governments had responded in the past to international food price trends and fluctuations. Those prices traced a long downward trend from the late nineteenth century until the mid-1980s, apart from a spike around 1974 (Pfaffenzeller et al., 2007; Williamson, 2012; Jacks, 2013)—but, following a flat period of nearly 20 years, an upward trend is observed, with spikes in 2008, 2010/11, and mid-2012.

This chapter reviews past policies by drawing on the World Bank's global database on agricultural price distortions since 1955 to see how governments have dealt with past price trends and spikes.[1] The empirical indicators provided in that database are first outlined, before summarizing trends in national distortions, followed by a review of government responses to fluctuations and their effects on international food prices. The chapter concludes by arguing that new domestic social protection policy options are now available to reduce food insecurity in developing countries, making it more feasible for multilateral action to phase out remaining distortionary farm trade policies.

17.1 BACKGROUND

Agricultural protection and subsidies in high income (and some middle income) countries have been depressing international prices of farm products for many decades, thereby lowering the earnings of farmers and associated rural businesses in developing countries (Johnson, 1991). These policies almost certainly have added to global inequality and poverty, since historically at least three-quarters of the world's poorest people have depended directly or indirectly on agriculture for their main income (Ravallion et al., 2007; World Bank, 2007). In addition to this external adverse influence on incomes of farmers in developing countries, their own governments had taxed them during most of the past half century. This involved both directly taxing farm exports and in some cases production (in-kind), as well as harming farmers indirectly by pursuing an import-substituting industrialization strategy, predominantly by restricting imports of manufactures and overvaluing their exchange rates (Krueger et al., 1988, 1991). These price-distorting policies also reduced the quantity of farm products traded internationally. Such "thinning" of the international market meant that their prices have been more volatile than they otherwise would have been, adding to the influence of the "insulating" feature of farm policies (Tyers and Anderson, 1992, Ch. 6).

Since the mid-1980s, however, many developing country governments have been reforming their agricultural, trade, and exchange rate policies, thereby reducing their anti-agricultural bias. Some high income countries have reduced their farm price supports, too, making it easier for developing countries to compete in the international market. Plenty of diversity in distortions remains across countries and across commodities within each country, so continuing the reform process would still expand farm trade and "thicken" international food markets, which would not only raise the mean but also lower the volatility of prices in these markets. Moreover, the "insulating" feature of farm policies has not diminished in either high income or developing countries, so that it continues to contribute nontrivially to the volatility of international food prices.

---

[1]While Jim Roumasset's recent writings have been in resource and environmental economics, he is also well known for seminal papers from the late 1980s on the economics and political economy of distortions to agricultural incentives (Balisacan and Roumasset, 1987; Roumasset and Setboonsarng, 1988; Clarete and Roumasset, 1990).

17.2 **INDICATORS OF NATIONAL DISTORTIONS TO AGRICULTURAL PRICES**

To gauge how farmer incentives in high income and developing countries have evolved since the 1950s, this chapter draws on time series evidence from a recent World Bank study compiled by Anderson and Valenzuela (2008), summarized in Anderson (2009), and updated to 2010/11 by Anderson and Nelgen (2013). These estimates cover 82 countries, altogether accounting for more than 90% of global agriculture, population, employment, gross domestic product (GDP), and poverty. The key indicator is the nominal rate of assistance (NRA), defined as the percentage by which national government policies raise gross returns to farmers above what they would be without government's intervention—or lower them, if NRA < 0 (see Anderson et al., 2008 for methodological details).

If a trade measure is the sole source of government intervention for a particular product, then the measured NRA will also be the consumer tax equivalent (CTE) rate at that same point in the value chain for that product. But where there are also domestic producer or consumer taxes or subsidies, the NRA and CTE will no longer be equal and at least one of them will be different from the price distortion at the border due to trade measures. Both are expressed as a percentage of the undistorted price. Each industry is classified either as import competing, or a producer of exportables, or as producing a nontradable (with its status sometimes changing over the years), so as to generate for each year the weighted average NRAs for the two different groups of tradables. In calculating the NRA, the methodology also includes the implicit trade tax distortions generated by dual or multiple exchange rates, drawing on Dervis et al. (1981).

In the Anderson and Valenzuela (2008) database, which covers up to 2004 for developing countries and 2007 for high income and transition economies, the NRA and CTE are very highly correlated for most products in all countries. As such, only producer price distortions were estimated in the update to 2010 for developing countries by Anderson and Nelgen (2013). For high income and transition economies and a few of the largest developing countries, the update was up to 2011 and, as with the earlier Anderson/Valenzuela database, was based on the producer and consumer support estimates (PSEs and CSEs) of the Organisation for Economic Cooperation and Development (OECD, 2012).

The coverage of products for NRA estimates averages around 70% of the gross value of farm production in each country. Authors of the country case studies also provided "guesstimates" of the NRAs for noncovered farm products. Weighted averages for all agricultural products were then generated, using the gross values of production at unassisted prices as weights. For countries that also provide nonproduct-specific agricultural subsidies or taxes (assumed to be shared on a pro-rata basis between tradables and nontradables), such net assistance is then added to product-specific assistance to estimate the NRA for total agriculture. Also provided, but as a separate add-on, are so-called decoupled measures, such as whole-farm payments, that in principle do not distort prices.

Farmers are affected not just by prices of their own outputs but also by the incentives nonagricultural producers face. That is, *relative* prices and hence *relative* rates of government assistance affect producer incentives (Lerner, 1936; Vousden, 1990, pp. 46–47). If one assumes the absence of distortions in the markets for nontradables and that the value shares of agricultural and nonagricultural nontradable products remain constant, then the economy-wide effect of distortions to agricultural incentives can be captured by the extent to which the tradable parts of agricultural production are assisted or taxed relative to producers of nonfarm tradables. By generating estimates of the average NRA for nonagricultural tradables, it is then possible to calculate a relative rate of assistance (RRA), defined in percentage terms as

$$RRA = 100[(1 + NRAag^t/100)/(1 + NRAnonag^t/100) - 1]$$

where $NRAag^t$ and $NRAnonag^t$ are the weighted average percentage NRAs for the tradable parts of the agricultural and nonagricultural sectors, respectively. If both sectors are equally assisted, the RRA is zero; if it is below (above) zero, it provides an internationally comparable indication of the extent to which a country's policy regime has an anti- (pro-) agricultural bias (Anderson et al., 2008).

In summarizing pertinent empirical findings from that World Bank study, it is helpful to begin with NRA estimates for the farm sector and then turn to RRA estimates.

17.3 NATIONAL DISTORTIONS TO FARMER INCENTIVES: TRENDS SINCE THE MID-1950S

In both Japan and the European Community (EC) in the 1950s, domestic prices exceeded international market prices for grains and livestock products by <40%. By the early 1980s, however, the difference was more than 80% for Japan and around 40% for the EC, while still being close to zero for the agricultural-exporting rich countries of Australasia and North America (Anderson and Hayami, 1986, Table 2.5). Virtually all of that assistance to Japanese and European farmers in that period was due to restrictions on imports of farm products.

Assistance rose markedly in the mid-1980s, particularly due to the North Atlantic food export subsidy "war." This prompted the launch of the General Agreement on Tariffs and Trade's (GATT) Uruguay Round, and the OECD began to compute annual PSEs and CSEs for its member countries (OECD, 2012). For this country group as a whole, producer support rose slightly between 1986–88 and 2009–11 in US dollar terms (from USD 239 to 248 billion), but, when expressed as a share of support-inclusive returns to farmers, it declined from 37% to 20%. Because of some changes in support instruments, including switching to measures that are based on noncurrent production or on long-term resource retirement, the share of that assistance provided via market price support measures fell from three-quarters to two-fifths. When the PSE payment was expressed as a percentage of undistorted prices to make it like an NRA, the fall was from 59% to 26% between 1986–88 and 2009–11 (OECD, 2012). This indicator suggests that high income country policies have become considerably less trade distorting, at least in proportional terms, even though farmer support in high income countries has continued to grow in dollar terms because of growth in the value of their farm output.

As for developing countries outside Northeast Asia, the main comprehensive set of pertinent estimates over time was, until recently, for the period just prior to when reforms became widespread. It was generated as part of a major study of 18 developing countries from the 1960s to the mid-1980s by Krueger et al. (1988, 1991). That study by the World Bank, whose estimates are summarized in Schiff and Valdés (1992), shows that the depression of incentives facing farmers has been due only partly to various forms of agricultural price and trade policies, including subsidies to food imports. Much more important in many cases have been those developing countries' nonagricultural policies that *indirectly* hurt their farmers. The two key ones have been manufacturing protectionism (which attracts resources from agriculture to the industrial sector) and overvalued exchange rates (which attract resources to sectors producing nontradables, such as services).

The more recent World Bank database, as updated by Anderson and Nelgen (2013), covers 45 developing countries, 13 European transition economies, and 24 high income countries. The results from this study (compared with the earlier Krueger/Schiff/Valdés ones in Anderson, 2010) do indeed reveal substantial reductions in distortions to agricultural incentives in developing countries over the past two to three decades. The reductions, however, have not been uniform across countries and regions, and the reform process is far from complete. More specifically, many countries still have a wide dispersion in NRAs for different farm industries and have a particularly strong antitrade bias in the structure of assistance within their agricultural sector. Some countries have "overshot" in the sense that they have moved from having a negative RRA to farmers, on the average, to a positive one, rather than stopping at the welfare-maximizing rate of zero. Moreover, the variance in rates of assistance across commodities within each country, and in aggregate rates across countries, remains substantial (Figure 17.1).

The global summary of the new results is provided in Figure 17.2. It reveals that the NRA to farmers in high income countries rose steadily over the post-World War II period through to the end of the 1980s, apart from a dip when international food prices spiked around 1973–74. After peaking at more than 50% in the mid-1980s, when international food prices were at a near-record low, the average NRA for high income countries has fallen substantially. This is so even when the new farm programs that are somewhat "decoupled" from directly influencing production decisions are included. For developing countries, the average NRA for agriculture has been moving toward zero also, but from a level of around −25% between the mid-1950s and early 1980s. Indeed it "overshot" in the 1990s by becoming positive, but it is less than half the average NRA for high income countries.

The developing countries average NRA conceals the fact that the exporting and import-competing subsectors of agriculture have very different NRAs. While the average NRA for exporters in developing countries has been negative throughout (coming back from −50% in the 1960s and 1970s to almost zero in 2000–10), the NRA for import-competing farmers in developing countries has fluctuated around a trend rate that has risen from 10% and 30%, even reaching 40% in the years of low international prices in the mid-1980s. This suggests that export-focused farmers in developing countries are still discriminated against in two respects: by the antitrade structure of assistance within their own agricultural sectors and by the protection still afforded farmers in high income countries. The antitrade bias also reflects the more-general fact that NRAs are not uniform across commodities, which in turn indicates that resources within the farm sector of each country are not being put to their best use (Lloyd, 1974). The extent of that extra inefficiency, over and above that due to too many or too few resources in aggregate in the sector, is indicated by the standard deviation of NRAs among covered products in each focus country. This dispersion index has fluctuated between 43% and 60% throughout the covered period, and has not diminished as NRAs have approached zero over the past 25 years (Anderson, 2009, Table 1.6).

The improvement in farmers' incentives in developing countries is understated by the above NRA estimates, because those countries have also reduced their assistance to producers of nonagricultural tradable goods, most notably manufactures. The decline in the weighted average NRA for the latter, depicted in Figure 17.3, was clearly much greater than the increase in the average NRA

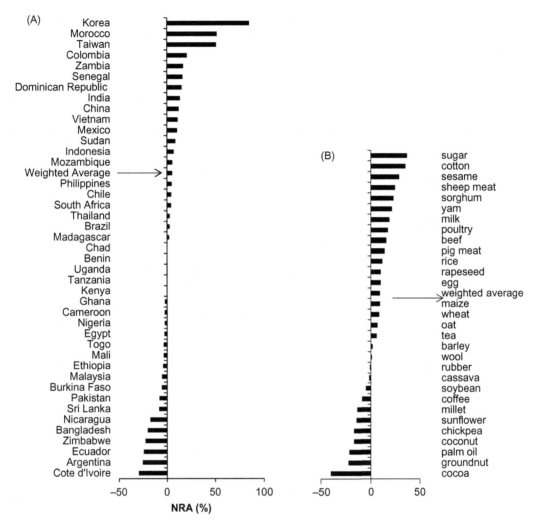

FIGURE 17.1

NRAs across developing countries and across products globally, 2005−10 by (A) country and (B) by-product.

for tradable agricultural sectors for the period to the mid-1980s, consistent with the finding of Krueger et al. (1988, 1991). For the period since the mid-1980s, changes in both sectors' NRAs have contributed almost equally to the improvement in farmer incentives. The RRA for developing countries as a group went from −46% in the second half of the 1970s to just above zero in the first decade of the present century. This increase (from a coefficient of 0.54 to 1.01) is equivalent to an almost doubling in the relative price of farm products—a huge change in the fortunes of developing country farmers in just a generation.

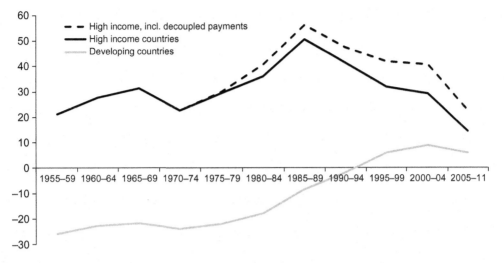

FIGURE 17.2

NRAs to agriculture in high income and developing countries, 1955–2011.

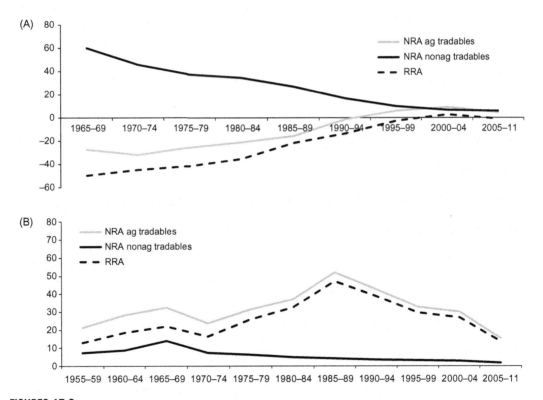

FIGURES 17.3

Developing and high income countries' NRAs to agricultural and nonagricultural tradable sectors, and RRAs, 1955–2011 (A) developing countries and (B) high income countries.

Figures 17.4 and 17.5 illustrate the dramatic changes since the 1980s in the proportions of the global farm population facing various RRAs and in the proportions of the global value (at undistorted prices) of agricultural production facing various NRAs. The data underlying Figure 17.4 involve a transfer from farmers to governments of USD 83 billion and from governments to farmers of USD 182 billion per year in the 1980s, hence an annual net transfer to farmers of USD 99 billion globally; by the 2000s this annual transfer had trebled to USD 298 billion. Associated with those policy reforms are reductions in the distortions to consumer prices of farm products: the altered proportions since the 1980s of the global nonfarm population facing various levels of consumer taxes or subsidies are depicted in Figure 17.6.

It must be kept in mind, however, that within the developing country group the spectrum of national RRA estimates (see Figure 1.8 in Anderson, 2009) remains even wider than the NRA spread in Figure 17.2, indicating a great scope still for global economic welfare gains from further farm trade liberalization. Such reform would not only raise the mean level of real incomes but also reduce the variance of international food prices by "thickening" international food markets: according to global economy-wide modeling results reported in Valenzuela et al. (2009), liberalization of remaining trade barriers as of 2004 would raise the share of farm production exported globally from 8% to 13%. Furthermore, such reform would reduce global income inequality and poverty, according to a follow-up study using numerous global and national economy-wide models all calibrated to 2004 and incorporating the same World Bank estimates of national price distortions as discussed above (Anderson et al., 2010, 2011).

17.4 GOVERNMENT RESPONSES TO FLUCTUATIONS AND SPIKES IN INTERNATIONAL FOOD PRICES

Fluctuations around trend levels of international food prices tend to be transmitted less than fully to national markets. This tendency means the estimated NRA for each product—the percentage by which the domestic price exceeds the border price—also fluctuates from year to year around its long-run trend and in opposite direction to the international price. This propensity has not diminished in either developing or high income countries as part of the trade-related policy reforms that began in the mid-1980s. To estimate the proportion of any international price fluctuation that is transmitted to domestic markets within 12 months, Anderson and Nelgen (2012), following Nerlove (1972) and Tyers and Anderson (1992, pp. 65–75), used a partial-adjustment geometric distributed lag formulation to estimate short-run transmission elasticities for each of nine key traded food products for all focus countries for the period 1985–2010. The elasticity estimates range from 0.73 for soybean to just 0.43 for sugar. The unweighted average across the nine products is 0.56, suggesting that, within one year, barely half the movement in international prices of primary food products was transmitted to domestic markets on average.

When some governments alter the restrictiveness of their food trade measures to insulate their domestic markets somewhat from international price fluctuations (including using specific rather than ad valorem tariffs), the volatility faced by other countries is amplified. That reaction prompts more countries to follow suit. The irony is that when both food-exporting and food-importing countries so respond, each country group undermines the other's attempts to stabilize its domestic

(A) 1980–89[a]

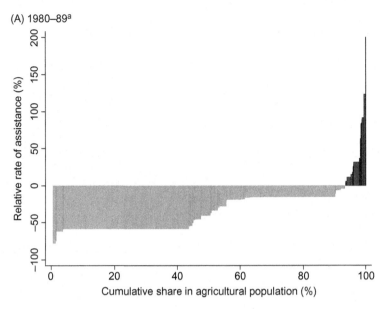

(B) 2000–09

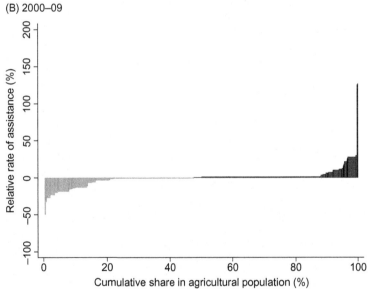

[a]Iceland (299%), Switzerland (290%), and Norway (213%) are shown as just 200%, so as to fit on the page.

FIGURES 17.4

Proportions of global farm population facing various RRAs, (A) 1980–89 and (B) 2000–09.

Generated by Signe Nelgen from estimates in Anderson and Nelgen, 2013.

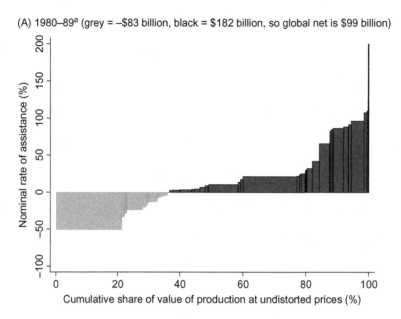

(A) 1980–89[a] (grey = –$83 billion, black = $182 billion, so global net is $99 billion)

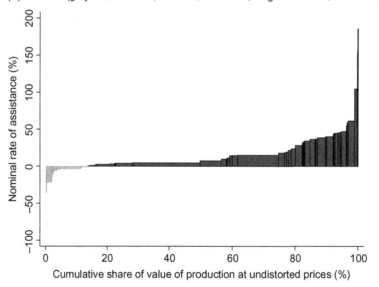

(B) 2000–09 (grey = –$15 billion, black = $313 billion, so global net is $298 billion)

[a]Iceland (299%), Switzerland (290%), and Norway (213%) are shown as just 200%, so as to fit on the page.

FIGURES 17.5

Proportions of the global value (at undistorted prices) of agricultural production facing various NRAs, (A) 1980−89 and (B) 2000−09.

Generated by Signe Nelgen from estimates in Anderson and Nelgen, 2013.

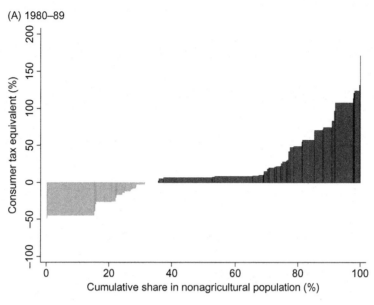

(A) 1980–89

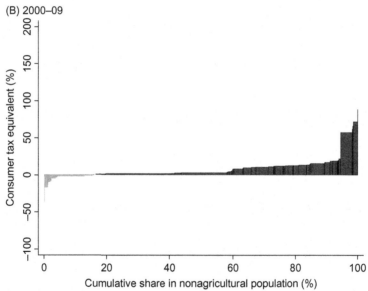

(B) 2000–09

FIGURES 17.6

Proportions of global nonfarm population facing various CTEs on their purchases of farm products, (A) 1980–89 and (B) 2000–09.

Generated by Signe Nelgen from estimates in Anderson and Nelgen, 2013.

Table 17.1 Contributions[a] of High Income and Developing Countries and of Importing and Exporting Countries to the Proportion of the International Price Change That Is Due to Policy-Induced Trade Barrier Changes, 1972−74 and 2006−08

	Total Proportional Contribution	High Income Countries' Contribution	Developing Countries' Contribution	Importing Countries' Contribution	Exporting Countries' Contribution
1972−74					
Rice	0.27	0.04	0.23	0.10	0.17
Wheat	0.23	0.15	0.08	0.18	0.05
Maize	0.18	0.14	0.04	0.06	0.12
2006−08					
Rice	0.40	0.02	0.38	0.18	0.22
Wheat	0.19	0.09	0.10	0.07	0.12
Maize	0.10	0.05	0.05	0.03	0.07

[a]Expressed such that the two numbers in each subsequent pair of columns add to the total proportion shown in column 1 of each row.
Source: Anderson and Nelgen, 2012.

markets. That is to say, what seems like a solution to each importing (or exporting) country's concern *if it were acting alone* turns out to be less effective, the more exporting (or importing) countries—presumably for the same political economy reasons—respond in a similar way.

To see this more clearly, Martin and Anderson (2012) considered the situation in which a severe weather shock at a time of low global stocks causes the international food price to suddenly rise. If national governments wish to avert losses for domestic food consumers, and they do so by altering their food trade restrictions (e.g., by raising export taxes or lowering import tariffs), then only a fraction of that price rise is transmitted to their domestic market. That response raises the consumer subsidy equivalent or lowers the CTE of any such trade measure and does the opposite to producer incentives. However, if such domestic market insulation is practiced by similar proportions of the world's food-exporting and -importing countries, it turns out to be not very effective in keeping domestic price volatility below what it would be in the international marketplace if no governments so respond. Rather, it is like everyone in a crowded stadium standing up to see better: if people are of equal height, no one is better off.

Martin and Anderson (2012) also point out that, with the help of some simplifying assumptions, it is possible to get at least a back-of-the-envelope estimate of the proportional contribution of government trade policy reactions to an international price spike such as in 2006−08. Updated estimates for the key grains are 0.40 for rice, 0.19 for wheat, and 0.10 for maize (Anderson and Nelgen, 2012). Those policy contributions could be apportioned between country groups. Table 17.1 reports the contributions of high income versus developing countries and of exporting versus importing countries. During 2006−08, developing countries were responsible for the majority of the policy contribution to the price spikes of the three grains, whereas in 1972−74 the opposite was the case, except for rice. As for exporters versus importers, exporters' policies appear to

Table 17.2 Comparison of the Domestic Price Rise with the Rise in International Grain Prices Net of the Contribution of Changed Trade Restrictions, Rice, Wheat, and Maize, 2006–08 (Percent, Unweighted Averages)

	International Price Rise		Domestic Price Rise		
	Including Contribution of Changed Trade Restrictions	Net of Contribution of Changed Trade Restrictions	All Countries	Developing Countries	High Income Countries
Rice	113	68	56	48	74
Wheat	70	56	77	65	81
Maize	83	75	73	62	82

Source: *Anderson and Nelgen, 2012.*

have the majority of the influence, except for wheat in the 1970s, but importers made a very sizeable contribution as well. It is also possible, in the light of these estimates, to get a sense of how effective changes in trade restrictions were in limiting the rise in domestic prices. The proportional rise in the international price *net of* the contribution of changed trade restrictions, when multiplied by the international price rise, is reported in the second column of Table 17.2, where it is compared with the proportional rises in the domestic price in the sample countries. The numbers for 2006–08 suggest that, on average for all countries in the sample, domestic prices rose slightly more than the adjusted international price change for wheat, only slightly less for maize, and just one-sixth less for rice. These results suggest that the combined responses by governments of all countries have been sufficiently offsetting as to do very little to insulate domestic markets from this recent international food price spike. A new study shows that those policy responses do not even reduce global poverty when account is taken of the combined effect of all countries' actions in exacerbating the international price spikes (Anderson et al., 2014).

17.5 POLICY IMPLICATIONS AND CONCLUDING REMARKS

The rapid growth of the developing economies' share of global industrial production and exports, led by China, appears to be continuing (Anderson and Strutt, 2014; Hanson, 2012). Industrialization in these emerging economies also continues to drive the strong demand for farm products. Food production variability is expected to increase, too, thanks to climate change and biofuel mandates. One might therefore expect the recent trend and fluctuations of rates of assistance to agriculture to go on. In particular, people and governments in emerging/industrializing economies—especially large ones such as China, India, and Indonesia—may well feel more food insecure as their farm sectors become less competitive while their food and feed demands grow. As such, continuing growth in their agricultural protection cannot be ruled out, even if international food prices remain high (Anderson and Nelgen, 2011; Anderson et al., 2013). This will raise their domestic prices of foods increasingly above those at their borders, thereby undermining food security for all their households except those that are net sellers of food. Such a rise is already evident in China and Indonesia, whose agricultural NRAs averaged -3% in 1999; by 2010 the average had risen to 21% in China and 27% in Indonesia (Anderson and Nelgen, 2013).

As for fluctuations in NRAs around trend, past behavior leads one to expect both high income and developing country governments to continue to alter their food trade restrictions so as to somehow insulate their domestic markets from international food price volatility. For reasons laid out in the previous section, this behavior will continue to amplify price fluctuations in the international market and, if both exporting and importing countries continue to respond similarly, such interventions will keep being rather ineffective in preventing fluctuations in domestic food prices. How severe such volatility might be will depend on the size of any unanticipated exogenous shocks to world food markets and the global stocks-to-use ratios of the affected products at the time of any such exogenous shocks. If stocks would be very low when harvests fail in significant regions, food price spikes of the magnitude experienced in mid-2008, early 2011, and mid-2012 could well be repeated under current policies.

The above empirical evidence supports the view that national trade restrictions add nontrivially to international food price volatility in at least two ways: through "thinning" international food markets and through "insulating" domestic food markets from international price fluctuations. Both policy attributes magnify the effect on international prices of any shock to global food supply or demand.

The solution to the first problem ("thinning") is simply for countries to open further their markets to food trade. The political difficulty and the adjustment costs associated with doing that would be minimized if countries can agree to liberalize their food and agricultural markets multilaterally, and to do so at the same time as nonagricultural markets are liberalized. That was what happened in the Uruguay Round, and the same is aspired by members of the World Trade Organization (WTO) via their Doha Development Agenda (DDA). After more than a decade of negotiating, however, the DDA has come to a standstill. While prospects look dim (Bureau and Jean, 2013), there is still some hope that the talks will be revived. Meanwhile, various plurilateral negotiations on options for regional integration and free-trade areas are under discussion, but the benefits from them are always far smaller than those from a multilateral agreement—and often agriculture is the sector least liberalized.

The optimal solution to the second problem (insulating) also involves the WTO. In a many-country world, it is clear from the above analysis that the trade policy actions of individual countries can be offset by those of other countries to the point that the interventions become ineffective in achieving their stated aim of reducing domestic food price volatility. This is a classic international public good problem that could be solved by a multilateral agreement to restrain the variability of trade restrictions (e.g., by converting specific tariffs into ad valorem ones).

In the current Doha round of WTO negotiations, there are proposals to phase out agricultural export subsidies and to bring down import tariff bindings, both of which would contribute to having more stable international food prices. However, proposals to broaden the Doha agenda to also introduce disciplines on export restraints have struggled to gain traction.

Whether or not WTO member countries liberalize their food trade and bind their trade taxes on exports as well as imports at low or zero levels, there would still be occasions when international food prices spike. This raises the question as to what alternative instruments governments could use to avert losses for significant groups in their societies. A standard answer from economists is that food security for consumers, most notably food affordability for the poor, is best dealt with using generic social safety net measures that offset the adverse impacts of a wide range of different shocks on poor people—net sellers as well as net buyers of food—without imposing the costly by-product distortions that necessarily accompany the use of nth-best trade policy instruments for social protection. That might take the form of targeted income supplements to only the most vulnerable households and only while the price spike lasts.

This standard answer has far greater power now than just a few years ago, thanks to the digital information and communication technology (ICT) revolution. In the past, the claim has often been that such payments are unaffordable in poor countries because of the fiscal outlay involved and the high cost of administering such handouts. However, recall that in half the cases considered above, governments *reduced* their trade tax rates, so even that intervention may require a drain on the budget of many finance ministries. In any case, the option of using value-added taxes in place of trade taxes to raise government revenue has become common practice in even low income countries over the past decade or two. The ICT revolution has made it possible for conditional cash transfers to be provided electronically as direct assistance to even remote households.

What if countries are still unsatisfied with the contribution of their farmers to national food security, as reflected in food self-sufficiency ratios, or feel their farmers are missing out on the benefits of rapid economic growth and industrialization? Again, agricultural import protection measures are far from first-best ways of dealing with these socio-political concerns. Alternative measures include subsidizing investments in agricultural research and development, in rural education and health, and in roads and other rural infrastructure improvements. If the social rates of return from those investments are currently high and above the private rates of returns, as is typically the case in developing countries, expanding such investments will boost economic growth (the same for improvements in land and water institutions that determine property rights and prices for those key farm inputs). Such investments almost certainly would reduce poverty and boost food security by raising net farm incomes while lowering consumer prices of food in towns and cities.

The political challenge of encouraging countries to switch from trade to domestic policy instruments to address nontrade domestic concerns is evidently nontrivial, although the evidence summarized above shows some reform has been possible during the past three decades. The emergence of new, lower cost social protection mechanisms involving conditional cash e-transfers might edge governments one more step away from the use of beggar-thy-neighbor trade measures. Researchers could contribute to this process of policy reform by analyzing the prospective economic effects of such new measures as compared with price-distorting measures to achieve societal objectives. They could also use political economy theory to try to explain why some governments have gone down this reform path more than others and use insights from those findings to suggest politically feasible ways to move toward more-optimal policies.

ACKNOWLEDGMENTS

This chapter draws on the work of many participants in a large World Bank project (www.worldbank.org/agdistortions). Thanks are due especially to Ernesto Valenzuela, Johanna Croser, and Signe Nelgen for assistance in compiling the global agricultural distortions database, and to the Australian Research Council, the Rural Industries Research and Development Corporation, and the World Bank for financial assistance. Views expressed are the author's alone.

REFERENCES

Anderson, K. (Ed.), 2009. Distortions to Agricultural Incentives: A Global Perspective, 1955–2007. Palgrave Macmillan/World Bank, London/Washington, DC.

Anderson, K., 2010. Krueger/Schiff/Valdés revisited: agricultural price and trade policy reform in developing countries since the 1980s. Appl. Econ. Perspect. Policy 32 (2), 195–231.

Anderson, K., Hayami, Y., 1986. The Political Economy of Agricultural Protection: East Asia in International Perspective. Allen and Unwin, London.

Anderson, K., Nelgen, S., 2011. What's the appropriate agricultural protection counterfactual for trade analysis? In: Martin, W., Mattoo, A. (Eds.), The Doha Development Agenda: An Assessment. Centre for Economic Policy Research and the World Bank, London.

Anderson, K., Nelgen, S., 2012. Agricultural trade distortions during the global financial crisis. Oxford Rev. Econ. Policy 28 (2), 235–260.

Anderson, K., Nelgen, S., 2013. Updated National and Global Estimates of Distortions to Agricultural Incentives, 1955 to 2011. World Bank, Washington, DC. Available from: <www.worldbank.org/agdistortions>.

Anderson, K., Strutt, A., 2014. Emerging economies, productivity growth, and trade with resource-rich economies by 2030. Aust. J. Agric. Resour. Econ. 58. Available from: http://dx.doi.org/doi:10.1111/1467-8489.12039.

Anderson, K., Valenzuela, E., 2008. Estimates of Global Distortions to Agricultural Incentives, 1955 to 2007. World Bank, Washington, DC. Available from: <www.worldbank.org/agdistortions>.

Anderson, K., Kurzweil, M., Martin, W., Sandri, D., Valenzuela, E., 2008. Measuring distortions to agricultural incentives, revisited. World Trade Rev. 7 (4), 1–30.

Anderson, K., Cockburn, J., Martin, W. (Eds.), 2010. Agricultural Price Distortions, Inequality and Poverty. World Bank, Washington, DC.

Anderson, K., Cockburn, J., Martin, W., 2011. Would freeing up world trade reduce poverty and inequality? The vexed role of agricultural distortions. World Econ. 34 (4), 487–515.

Anderson, K., Rausser, G., Swinnen, J.F.M., 2013. Political economy of public policies: insights from distortions to agricultural and food markets. J. Econ. Lit. 51 (2), 423–477.

Anderson, K., Ivanic, M., Martin, W., 2014. Food price spikes, price insulation and poverty. In: Chavas, J.-P., Hummels, D., Wright, B. (Eds.), The Economics of Food Price Volatility. University of Chicago Press for NBER, Chicago, IL and London.

Balisacan, A., Roumasset, J., 1987. Public choice of economic policy: the growth of agricultural protection. Weltwirtsch. Arch. 123 (2), 232–248.

Bureau, J.C., Jean, S., 2013. Trade liberalization in the bio-economy: coping with a new landscape. Agric. Econ. 44 (S): 173–182, November.

Clarete, R., Roumasset, J., 1990. The relative welfare cost of industrial and agricultural policy distortions: a Philippine illustration. Oxford Econ. Papers 42 (2), 462–472.

Dervis, K., de Melo, J., Robinson, S., 1981. A general equilibrium analysis of foreign exchange shortages in a developing country. Econ. J. 91, 891–906.

Hanson, G.H., 2012. The rise of middle kingdoms: emerging economies in global trade. J. Econ. Perspect. 26 (2), 41–63.

Jacks, D.S., 2013. From boom to bust: a typology of real commodity prices in the long run. NBER Working Paper 18874, Cambridge, MA.

Johnson, D.G., 1991. World Agriculture in Disarray, revised edition. St. Martin's Press, London.

Krueger, A.O., Schiff, M., Valdés, A., 1988. Agricultural incentives in developing countries: measuring the effect of sectoral and economy-wide policies. World Bank Econ. Rev. 2 (3), 255–272.

Krueger, A.O., Schiff, M., Valdés, A., 1991. The Political Economy of Agricultural Pricing Policy, Volume 1: Latin America, Volume 2: Asia, and Volume 3: Africa and the Mediterranean. Johns Hopkins University Press for World Bank, Baltimore, MD.

Lerner, A., 1936. The symmetry between import and export taxes. Economica 3 (11), 306−313.

Lloyd, P.J., 1974. A more general theory of price distortions in an open economy. J. Int. Econ. 4 (4), 365−386.

Martin, W., Anderson, K., 2012. Export restrictions and price insulation during commodity price booms. Am. J. Agric. Econ. 94 (2), 422−427.

Nerlove, M., 1972. Lags in economic behavior. Econometrica 40 (2), 221−252.

OECD (Organisation for Economic Co-operation and Development), 2012. Producer and Consumer Support Estimates. Online database. Available from: <www.oecd.org>.

Pfaffenzeller, S., Newbolt, P., Rayner, A., 2007. A short note on updating the Grilli and Yang commodity price index. World Bank Econ. Rev. 21 (1), 151−163.

Ravallion, M., Chen, S., Sangraula, P., 2007. New evidence on the urbanization of poverty. Popul. Dev. Rev. 33 (4), 667−702.

Roumasset, J., Setboonsarng, S., 1988. Second-best agricultural policy: getting the price of Thai rice right. J. Dev. Econ. 28 (3), 323−340.

Schiff, M., Valdés, A., 1992. The Political Economy of Agricultural Pricing Policy, Vol. 4: A Synthesis of the Economics in Developing Countries. Johns Hopkins University Press for the World Bank, Baltimore, MD.

Tyers, R., Anderson, K., 1992. Disarray in World Food Markets: A Quantitative Assessment. Cambridge University Press, Cambridge and New York, NY.

Valenzuela, E., van der Mensbrugghe, D., Anderson, K., 2009. General equilibrium effects of price distortions on global markets, farm incomes and welfare. In: Anderson, K. (Ed.), Distortions to Agricultural Incentives: A Global Perspective, 1955−2007. Palgrave Macmillan/World Bank, London/Washington, DC.

Vousden, N., 1990. The Economics of Trade Protection. Cambridge University Press, Cambridge.

Williamson, J.G., 2012. Commodity prices over two centuries: trends, volatility, and impact. Annu. Rev. Resour. Econ. 4 (6), 1−22.

World Bank, 2007. World Development Report 2008: Agriculture for Development. World Bank, Washington, DC.

GETTING THE PRICE OF THAI RICE RIGHT: EPISODE II

18

Suthad Setboonsarng

Former Thailand Trade Representative, Prime Minister Office, Government House,
Dusit District, Bangkok, Thailand

CHAPTER OUTLINE

JEL: Q17; Q18; N55

18.1 INTRODUCTION

Thailand is one of the world's largest rice exporters, making rice its major source of income. In addition to greatly contributing to the growth of the country's agriculture sector and the economy, rice is also a source of national pride. As such, a centerpiece Thai government policy typically involves rice.

After the 1960s, the Thai government experimented with various agriculture policies, each with different objectives and levels of success. Roumasset and Setboonsarng (1988) explained that the Thai export tax on rice performed at least three functions: an optimum tax given the market power in the international market, generating revenue at a relatively low collection cost, and suppressing the domestic price of rice to help the urban consumer (which supported the lower wages in the manufacturing sector). To compensate the farmer, the government advanced many programs to subsidize the rice farmer, including the investment in irrigation and providing loans.

While the export tax on rice was abolished in 1986, the price of rice in the domestic market was still kept at a minimum level to protect the urban consumer and many programs were advanced to compensate the rice farmers. The most recent government rice policy is the Paddy Mortgage Scheme (PMS),[1] which was implemented in 2011.

The PMS guaranteed Thai farmers a purchase price of THB 15,000 (US$492) per ton of regular rice paddy and THB 20,000 (US$656) for jasmine (*hom mali*) rice. In 2011, the government purchased through the PMS 53% of rice production, which was milled and auctioned. It hoped to hold a large enough rice stockpile to sell it at a profit. However, the increase in rice export from India, compounded with leakage, corruption, and inefficient management, disrupted the scheme.

This chapter explains the motivation for rice policy given the context of the global rice market and analyzes its successes and failures. It also suggests a possible exit strategy and gives policy recommendations for the future of Thailand's rice and agriculture industry.

18.2 ORIGINS OF THE PMS

18.2.1 INTERNATIONAL RICE MARKETS

Before discussing the details and analyzing the outcomes of the PMS, it is important to understand that as a major rice exporter, Thailand has an agriculture policy that caters to a domestic constituency and is influenced by international market conditions.

The global rice market shares many characteristics with that of other grains such as wheat, corn, and maize. Most nations keep the consumer price of rice low and provide subsidies to compensate the farmer. They tried to be self-sufficient in production and rely on the international markets for the deficiency or absorbing the surplus. Such a strategy results in transmitting domestic fluctuation to the world market.

---

[1] It is also called the rice pledging scheme. However, the scheme involves acquiring paddy from the rice farmer.

As a staple grain in many countries, however, rice is subject to more government interventions and subsidies than other agriculture products. Together, the domestic price control and production subsidies create a downward price bias that results in low rice prices.[2] This comes at the cost of rice exporting countries and taxing the livelihood of farmers. The low price of rice also discourages investment in research and development and innovation throughout its value chain, from the production to consumer market. Rice exporting countries often pursue policies to combat these negative price biases.

18.2.2 PRECURSOR OF THE PMS

From 1998 to 2000, the world price of rice declined steadily from about US$302 to 202 per ton. Its severe depression became a pressing issue for the Thai government, given that rice production in the country involves about 3.73 million farming households (Office of Agricultural Economics, 2012). Thus, rice is not only a source of national income, but of the welfare of a large number of Thai farmers as well.

Understanding this imperative of assisting the agriculture sector, the Thaksin administration, on its assumption of office in 2001, introduced a Paddy Price Guarantee Scheme, a minimum support price scheme. The government announced the buying price of paddy at a fair market price and farmers had the option of pledging their paddy with the Government Warehouse Organization and Marketing Organization for Farmers as collateral for a loan of up to 80% of the crop value. The loan had a 3- to 4-month payback period and carried a 3% interest rate. The scheme attracted little participation because farmers preferred receiving their payment in full. Furthermore, they found the documentation process too cumbersome.

From 2002 to 2009, the government adjusted the program to make it more flexible and to increase the transfer to farmers.[3] When the Democrat Party took office in late 2009, it replaced the Paddy Price Guarantee Scheme with the Income Guarantee Scheme (IGS), wherein the government covered the difference between the market price and the announced price.[4]

18.2.3 OVERVIEW OF THE PMS

After winning the general election in 2011, the Yingluck administration implemented a full-fledged PMS, which was part of its campaign promise. It cited four key objectives of the PMS: (i) increase farmers' income, (ii) stimulate domestic consumption, (iii) control the rice market system, and (iv) increase export price.[5] It set the buying price at THB 15,000 (US$492) per ton of long grain paddy and THB 20,000 (US$656) per ton of aromatic jasmine (*hom mali*) paddy.

In crop year 2011−12, over 21 million tons of paddy (52% of total production)[6] were brought under the scheme, giving the government control of the domestic and export supply of rice. Rice prices increased between August and December 2011, especially for the high quality jasmine rice.

---

[2]Section 18.9.2 discusses the impact of this policy.
[3]Isvilanonda (2012) described the development of the minimum price support programs during this period.
[4]Poapongsakorn and Siamwalla (2009).
[5]National Rice Policy Committee, October, 2012, p. 5.
[6]Further details will be discussed in the following section.

However, the increase was not enough to make the PMS profitable, hence the government did not release its rice inventory. The rice stockpile increased from 5.6 million tons in 2010 to 9 million tons in 2011 and 11 million tons by the end of 2012. Unable to sell rice at the desired value, the government was forced to stockpile its rice purchase. Over time, the quality of rice in stockpiling deteriorated. Moreover, leakages decreased the value of the rice inventory.

The government continued implementing the PMS in crop year 2012–13 at the same guaranteed price despite criticism and protest from exporters and academics. Mounting losses, inefficient administration, and corruption within the PMS eventually led to the ousting of the Commerce Minister, contributing to the fall of the Yingluck administration.

18.2.4 ANALYSIS OF THE PMS

A simple model (Figure 18.1) was devised to serve as a framework in estimating the impact of the PMS. A liner model was used in the estimation.

The quantity that will maximize revenue is where marginal revenue (MR) from export quantity equals marginal cost. The optimum price is determined by the price on the demand curve for that quantity. The marginal cost of export or the supply of export (S_w) in the international market is determined by the excess of domestic supply (S_D) over the domestic demand (D_D). The world demand curve for Thai rice (D_w) is the sum of all the excess demand over domestic supply from rice importing countries. It was found to be elastic because the export quantity of Thailand, being a major exporter, has an impact on the world price. The MR of export, reflecting the incremental revenue from each unit of export, is derived from the world demand D_w.

Leaving it to competition, the quantity of rice export would be at Q_w, where S_w intersects D_w and the world price is at P_w. However, this quantity and price will not maximize the revenue for Thailand. The quantity of export that maximizes revenue for exporting Thai rice is Q_1, where the MR curve intersects S_w (area B). The optimum price of rice is P_1.

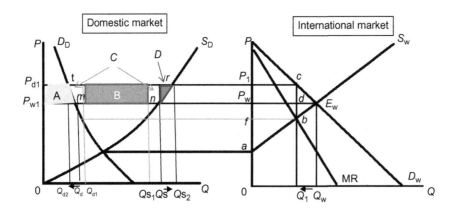

FIGURE 18.1

Optimum tariff and international rice price.

If world price increases from P_{w1} to P_{d1}, the producer surplus will increase from OnP_{w1} to OrP_{d1}, which is equal to area $P_{w1}nrP_{d1}$. This area, consisting of $P_{w1}mtP_{d1}$, comes from the reduction of domestic consumer surplus and $mnrt$, which is the reduction of the consumer surplus in the foreign market. This will raise the farmer's income at the expense of consumers, with the government bearing the administrative cost.

To attain this target, the government can either increase the export **price**, for example, by imposing an export tax equal to 1/elasticity (Corden, 1974, p. 176) or sell the marginal **quantity** of rice until Q_1 is reached and the price will be at P_1.

The PMS employed the export quantity management measure. It acquired paddy from the farmer at a price higher than the market to have enough stock in the government's hand. Then, the government auctioned limited quantities for export to pressure an increase in prices. The government also auctioned the rice in the inventory for the domestic market at current market price to keep domestic prices stable. However, a combination of internal and external factors prevented the government from fulfilling its objectives for the PMS. Its own inability to manage leakages and curb corruption led to heavy financial losses. Externally, good harvest yields and changes in the export law of other rice producing countries dramatically increased the world supply, forcing prices down.

18.3 **PRICE ELASTICITY OF DEMAND FOR THE EXPORT OF THAI RICE**

The fact that Thai rice is facing a downward sloping demand is well established in the past literature.[7] A downward sloping demand suggests that Thailand can earn optimum revenue if the quantity of export is reduced.

Table 18.1 presents the elasticities of demand for the five categories of rice export—jasmine rice (*hom mali*, high quality, aromatic), *pratum* (an aromatic rice), 100% white rice, 5% white rice, and others—which were calculated using data in 2011 and 2012.

The results show that

- The demand for the higher quality categories is less elastic than that for the normal quality.
- The average rice price increased by 15.9%, ranging from 8% (100% white rice) to 20% (*pratum*).
- The export quantity decreased by 37%, mostly from second quality categories: 5% white rice (−66%) and *pratum* (−59.6%). Jasmine rice was the least affected.

Given the demand elasticity for import of each type of rice, the optimum tax varies from 60.3% (*hom mali* rice) to 23.7% (5% white rice). On average, the optimum export tax is 42.9%. The actual price increase, however, was only 11.4% for jasmine rice and 15.7% for 5% white rice. This suggests that although the PMS increased the price of rice, it did not reach the optimum level. This could be because of the competition from Viet Nam and India, as noted by Chulaphan et al. (2013).

---

[7]Choeun et al. (2006) summarized the various estimation of elasticity of demand for Thai rice export during 1978–1995.

Table 18.1 Calculation of Elasticity of Demand for Thai Rice Export by Types of Rice

| | 2011 | | | 2012 | | | Changes | | | | | |
	Q (mill. ton)	V (mill. baht)	P (baht/ton)	Q (mill. ton)	V (mill. baht)	P (baht/ton)	dQ (mill. ton)	dV (mill. baht)	dP (baht/ton)	%dQ	%dP	Ed
Hom mali	2.36	63,584	26,954	1.91	57,434	30,028	−0.4463	−6,150	3074	−18.9%	11.4%	−1.6589
Pratum	0.21	4965	23,416	0.09	2410	28,148	−0.1264	−2554	4732	−59.6%	20.2%	−2.9499
100%	0.82	13,673	16,639	0.59	10,564	17,990	−0.2346	−3110	1351	−28.5%	8.1%	−3.5159
5%	2.23	33,309	14,920	0.76	13,051	17,258	−1.4763	−20,257	2338	−66.1%	15.7%	−4.2190
Others	5.08	80,587	15,860	3.39	59,517	17,543	−1.6883	−21,069	1682	−33.2%	10.6%	−3.1328
Total	10.71	1,96,117	18,318	6.73	1,42,976	21,231	−3.9718	−53,141	2913	−37.1%	15.9%	−2.3332

Note: Q, quantity; V, value; P, price; dQ, dV, dP are changes in Q, V, and P, respectively; Ed = elasticity of demand = (%dQ/%dP).
Source: Rice Export Statistics, Office of Agricultural Economics, Ministry of Agriculture and Cooperatives, 2013.

A drawback of this model (Figure 18.1) is that it does not consider the long-run dynamic impact of the PMS. It does not address the issue of leftover stock. For instance, the higher price of paddy would induce further production in the subsequent year. The inventory of leftover rice will accumulate over the years. Therefore, due consideration has to be paid to the long-run implications.

As Table 18.1 shows, the demand for *hom mali* and *pratum* is less elastic. As such, the PMS could be more effective in enhancing prices of products over which Thailand has market power. On the other hand, most of the reductions in quantity came from the lower quality white rice market. This indicates that the PMS is not appropriate for low quality rice because Thailand has limited market power over it and has high rice production costs due to the relatively higher labor costs. With some modifications, the PMS could be made to focus on appropriate products (i.e., high quality rice) and use other schemes to deliver the intended income transfers.

18.4 PMS OUTCOMES

The shortcomings of the PMS have been well documented by experts in and outside Thailand. Still, the scheme had achieved three notable positive effects desired by the Thai government: (i) increased farmers' income, (ii) increased world price of rice, and (iii) Stabilized Farm Gate Price. There are caveats, however, as to how these achievements were reached.

18.4.1 INCREASED FAMERS' INCOME

In 2011, the Thai government, like other governments in Asia, implemented economic stimulus packages to increase domestic consumption to compensate for the slowdown in global demand. As farmers have higher marginal propensity to consume (MPC), transfer of income to them was expected to generate relatively higher multiplier effects. The rice price offered under the PMS was 30–40% higher than the prevailing market price, resulting in increased income of farmers who participated in the scheme. As market price rose, the income of nonparticipating farmers also increased.

Figure 18.1 shows P_1Q_1 as the optimum price and quantity in the international market. Export reduction is Q_w to Q_1 and export price increase is P_{w1} to P_{d1}. Domestic production increase is Q_s to Q_{s1} and domestic consumption reduction is Q_d to Q_{d1}.

The increase in producer surplus is area $P_{w1}nrP_{d1}$, which consists of four components:

- Area A—Reduction of consumer surplus because of increase in rice price;
- Area B—Reduction of consumer surplus in the export market, which is equivalent to $P_w dcP_1$ in the export market;
- Area C—Producer surplus from the additional production in response to higher price; and
- Area D—Loss from unsold rice stock.

In 2012, the total export revenue from rice increased to THB 192 billion (about US$6.3 billion), from THB 142 billion (about US$4.65 billion) in 2011. This was despite the decline in export quantity from 10.7 million tons in 2011 to only 6.7 million tons in 2012. The average export price increased from THB 18,313 (US$600) per ton in 2011 to THB 21,231 (US$696) per ton in 2012.

Table 18.2 Calculation of Welfare Transfer

Area	Description	Welfare Transfer (billion baht)	Proportion (%)
A	Reduction of domestic consumer surplus	31,194	45.2
B	Reduction of consumer surplus in the export market	19,615	28.4
C	Producer surplus from the additional output	1762	2.6
D	Remaining unsold stock carries by the government	16,473	23.9
Total		69,043	100.0%

Note: *Based on Figure 18.1.*

Domestic retail price did not change and consumption grew slightly from 10.3 million to 10.4 million tons. Rice output increased from 21.89 million tons in 2011 to 23.1 million tons in 2012.[8]

Given this information, the total transfer of income to rice farmers was calculated at THB 65.51 (US$2.15) billion. The sources of funding for this transfer are given in Table 18.2.

Since the government did not increase the selling price in the domestic market, it thus absorbed the cost of area A. The government also incurred costs due to additional production induced by the higher rice price and the remaining inventory. Consumers abroad absorbed about THB 19.6 (US$0.64) billion (28% of the total transfer). In short, it cost the government THB 49.428 (US$1.62) billion to transfer THB 69.043 (US$2.07) billion to Thai rice farmers.

It should be noted that if the government increased the price of rice in the domestic market to the level implied by the world price, the government's cost would be reduced drastically to only THB 18.23 (US$0.59) billion.

The effectiveness of income distribution hinges on transferring income to poorer people. However, transferring income by increasing the price of paddy is innately biased in favor of larger farmers. Poapongsakorn (2013) found that among the PMS participants, 52.8% of farm households were in the low income brackets (less than THB 100,000) and they received only 20.4% of the total transfer for the main crop. The transfer went to the medium sized farmers (Table 18.3).

However, just because medium size farmers are getting the higher proportion of the transfer is not sufficient reason to write off the potential benefit of the transfer. The effectiveness of using the PMS as a stimulus package depended on the relative MPC of these farm households. Its assessment needs to be further studied to compare with alternative options.

18.4.2 INCREASED WORLD PRICE OF RICE

The PMS indirectly increased farmers' income by increasing world prices. As the largest exporter of rice for many years, Thailand has been the benchmark for the world market price. As such, for Thailand to maximize the revenue from export, the optimum price and quantity should be determined at the quantity where the MR equals marginal cost or the supply for export. This is no easy task because in the real world market imports are usually carried out by government agencies that

---

[8]Based on the assumption that 0.05% of paddy is used for seeding and paddy to rice conversion is 65%.

Table 18.3 Number of Farm Household Participating in the PMS and Amount of Money Received, Classified by Income Levels

Income Level of Recipient (THB)	Main Crop		Second Crop	
	Number of Households	Amount of Transfer (billion THB)	Number of Households	Amount of Transfer (billion THB)
<100,000	4,39,919	24,122	1,96,804	11,108
(% share)	52.28%	20.36%	32.03%	7.83%
100,000−600,000	3,93,590	88,352	3,84,286	1,03,690
(% share)	46.78%	74.56%	62.55%	73.11%
More than 600,000	7882	6030	33,309	27,034
(% share)	0.94%	5.09%	5.42%	19.06%
Total	8,41,391	1,18,504	6,14,399	1,41,832

Source: *Poapongsakorn, 2013, Table 7.1*

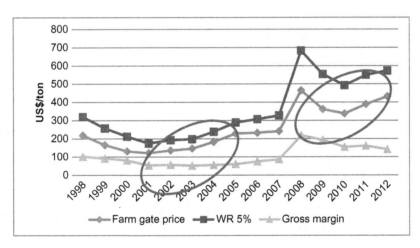

FIGURE 18.2

Farm gate price and export price of rice (US$per ton).

Thai Rice Exporter Association, 2013.

manage their respective countries' food security, and they have their own legal requirements and conditions.

The exporters (countries and companies), on the other hand, compete among themselves. By managing the export of rice through auctioning out stock, the Thai government controlled the quantity and price of rice, but placed pressure on rice exporters.

The world price of rice increased from July to December 2011 (Figure 18.2). Without the PMS, it is unlikely that the world price would have increased during this period because of the overwhelming amount of rice exported from India. Still, as a consequence, Thai rice exports declined in 2011, leaving a large stock of rice for the subsequent years.

18.4.3 STABILIZED FARM GATE PRICE

At harvest time, farmers sell their paddy to pay off their loans and spend on household needs. Storage facilities for paddy at the farm level are usually inadequate. Moreover, holding the stock risks damage from pests, theft, and quality deterioration. Thus it is favorable to sell the paddy as early as possible.

On the other hand, farmers are not equipped with appropriate market information and skills to negotiate with middlemen or rice millers. For instance, the paddy's moisture level is one of the parameters used to determine the value of paddy, and this is usually determined by the middlemen or rice miller. The PMS thus helped farmers achieve stable farm gate price and make appropriate decisions on investment at planting.

The PMS is one tool with many objectives. There could be other more efficient measures that can achieve these objectives.

18.5 CHALLENGES TO THE PMS

18.5.1 EXTERNAL FACTORS

The PMS had created an impact on the international rice market and vice versa. The attempt to manage the supply of rice in the international market sought to stabilize the market. In 2011, two important events affected the international rice market.

- In July 2011, India lifted the control on rice export and subsequently released its surplus into the international market.[9] Its export thus increased from 4 million (2010) to 10.7 million tons, making India the world's largest rice exporter (Figure 18.3).
- Weather conditions in 2011 negatively affected Chinese production so that the following year its grain import increased by 294%, from about 0.58 million tons in 2011 to 2.2 million tons in 2012. China became the world's second largest rice importer after Nigeria.[10]

When India increased its exports toward the end of 2011, world prices took a dive, especially for the lower quality market segment. This segment was the source of the PMS' main loss. The increase of imports from China helped absorb some of this surplus in the international market, thus preventing the price from dropping further. However, the transition of its leadership constrained China from participating more actively in the international market.

---

[9]http://www.dawn.com/news/643820/india-lifts-ban-on-export-rice-prices-crash-in-local-world-markets.

[10]The official statistics of grain production in China between 2011 and 2012 increased slightly. The years 2011−12 saw the transition of leadership from President Hu Jintao to President Xi Jinping. Uncertainty in food security during the delicate transition period was not acceptable to Chinese leadership, thus China's official rice market information was not transparent. After the political transition was complete, the rice and grain market became clearer. China said that the increase in the rice import will not affect global food security. (http://www.fao.org/docrep/017/ap047e/ap047e00.pdf).

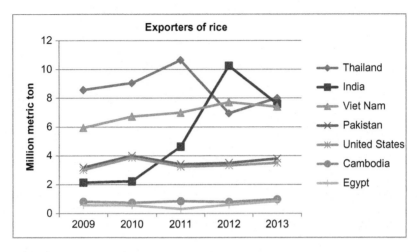

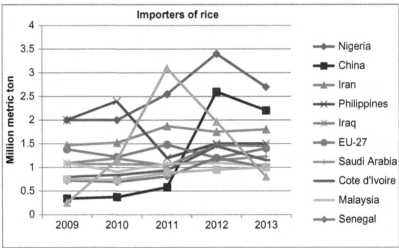

FIGURE 18.3

Major importing and exporting countries.

Grain: World Markets and Trade, USDA, 2013.

By 2013, Thailand had regained its leading position in the world rice market. It also has sufficient stock at hand to prevent exporters from undercutting prices. As a consequence, in the next few years, all else being equal, the price of rice is expected to continue to increase.

18.5.2 INTERNAL FACTORS: CORRUPTION AND INEFFICIENCY

Poapongsakorn (2013) reported that corruption occurred throughout the supply chain and many methods of embezzling money were employed. Khoman (2014) noted that corruption already existed in the implementation of a similar scheme in 2009 (e.g., falsification of warehouse receipts to claim subsidy).

Democrat Party (2013) pointed out five main areas of corruption in the PMS: (i) falsification of warehouse receipts in terms of paddy weight, purity, and moisture level; (ii) surrogating name of non-rice farmers or using smuggled paddy; (iii) rotation of paddy for multiple claims; (iv) false surveyor reports; and (v) nontransparent selling of rice, especially the government-to-government transactions.

In other countries, such as the Philippines, the rice self-sufficiency policy is also crammed with corruption throughout the supply chain (Clarete, 2014). Thang et al. (2013) also questioned the allocation of rice export from Viet Nam. However, the scale may be different from the case of PMS in Thailand.

These corruptions are crimes which are punishable by law, but the cost of enforcement is high. Violations have been continuously added to the growing pile of cases for investigation, prosecution, and conviction. Because it takes a long time to reach a verdict and deliver punishment, the law has lost its effectiveness. Rampant violations in the PMS have rendered it almost impossible to investigate and prosecute these crimes.

It is difficult to measure the cost of inefficiency created by the PMS. As some aspects are of important concern, more research is needed to quantify them.

18.6 NEGATIVE OUTCOMES OF THE PMS

The Thai government had been amply criticized and warned against the dangers and corruption of the PMS by both Thai and foreign analysts (over 3500 articles in Thailand and 356 articles by foreign analysts).[11] This section discusses the common critiques of the PMS and examines other areas that were overlooked.

18.6.1 LOSSES FROM SELLING RICE BELOW COST

The government incurred losses when it acquired paddy through the PMS at above-market prices and then sold it at regular prices. This loss was borne by the government's budget. In 2011, the world price of rice increased, but it was not sufficient to cover the cost of paddy plus milling costs and other expenses, resulting in a sizeable loss by the government. Poapongsakorn (2013) estimated the revenue earned from 2011/12 was THB 218.1 (US$7.15) billion, while the cost was THB 388.4 (US$12.73) billion, or a loss of THB 170 (US$5.57) billion (Table 18.4).

The government objected to the above estimates because it did not consider the value of the rice in the stockpile. The government estimated the cost at about THB 100 (US$3.27) billion. It should be noted, however, that the size of stock had never been made clear because it was recorded by several agencies and done so inconsistently. Moreover, the opaque government-to-government transactions did not disclose the number of buyers, prices, and tonnage.

Aside from the accounting cost, the quality of services—milling, storage, and transportation—performed by the rice mill and the owner of storage facilities was different from when they were

---

[11]Democrat Party, 2013, p. 10.

Table 18.4 Estimated Revenue and Cost of PMS, 2011/2012			
	Main crop	**Second crop**	**Total**
Paddy October 31, 2012 (mmt)[a]	6.95	14.81	21.76
Rice equivalent (mmt)[a]	4.21	9.15	13.36
Estimated revenue received[b]	83,889	1,34,855	2,18,745
Loan (cost of paddy)[b]	1,18,594	2,08,627	3,27,221
Interest[b]	6616	9819	16,434
Operating costs[b]:	11,888	20,280	32,169
— Gov. agencies expenses	4800	6945	11,745
— Rice miller costs	7088	13,335	20,424
Quality deterioration[b]	5527	7063	12,590
Estimated cost[b]	1,42,624	2,45,789	3,88,414
Loss[b]	−58,735	−1,10,934	−1,69,669

Note:
[a]mmt = million metric ton.
[b]billion THB.
Source: Derived from Poapongsakorn, 2013, Table 7.2

doing these for their own paddy and rice. Added to this, government agencies supervising these operations did not have sufficient time or sufficient expertise to monitor the activities daily. When they did, their level of commitment and expertise did not match that of traditional rice millers.

18.6.2 DEPLETION OF RICE DIVERSITY

The more long-term damage of the PMS is the reduction in diversity of rice. Only *hom mali* and *pratum* paddy were given different prices; all other varieties of white rice were priced the same. Thus, farmers used the variety with the highest yield, regardless of its quality (e.g., taste and cooking quality). This went against the long-term strength of the Thai rice market, which is based on quality,[12] as evidenced by the prominence of *hom mali* rice in the world market.

The PMS signaled the domestic market to focus only on a few strains of high output rice, which reduced the genetic diversity of rice in Thailand. It is difficult to assess the size of damage created by this development.

18.6.3 EROSION OF TRADITIONAL MARKET MECHANISMS

Under the PMS, farmers brought their paddy to the designated government warehouses and the government hired millers to process and store the rice until the government would auction it. This

---

[12]The higher cost of farm labor in Thailand prevents Thai rice from competing with lower wage countries. The niche market for Thai rice is higher quality rice (Setboonsarng, 1990).

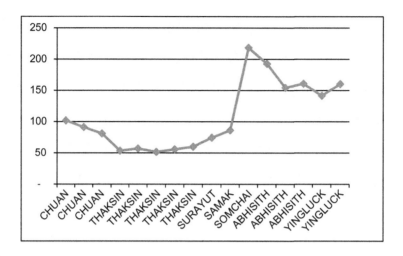

FIGURE 18.4

Gross margin for export (US$ per ton).

Calculated from data of the Thai Rice Exporter Association, 2013.

procedure diminished the role of middlemen as well as the local and central paddy markets. Isvilanonda (2012) noted that the central paddy market in the central plain was at risk of bankruptcy as middlemen who held paddy stock to speculate for higher prices no longer had a role in the new scheme. The futures market for paddy also became irrelevant.

The income of rice millers and exporters was also affected. This is because the gross margin of export, which is the difference between the cost of rice (calculated from the farm gate price of paddy) and the export price, reflects the revenue accruing to the exporter. The storage cost of rice, on the other hand, adds to the income of the rice miller.

Figure 18.4 shows that the gross margin for export exhibits had an increasing trend from 2001 to 2007. Export prices also increased in 2008 during the global food crisis.[13] During the PMS' implementation, however, the margin declined. Middlemen, millers, and exporters made less for each ton of rice.

18.6.4 LOSS OF EXPORTER LEADERSHIP

Thailand is equipped with the appropriate infrastructure to be the hub for international rice trading. Some academics and the international media were of the view that the PMS had reduced Thai export and pushed Thailand from the number one rank among rice exporting countries. The concern

---

[13]Gross margin is revenue minus cost of goods sold. In this case, it is the price of rice minus the cost of paddy, assuming a milling rate (i.e., percent of rice from 1 kg of paddy) of 65%.

is that in the long run Thailand may lose more market share to the other countries. The reduction of Thai export can be attributed to three short-term factors:

- Export decreased because Thailand used the PMS to push up world price.
- Blocking paddy from Cambodia and Lao PDR diverted these paddies to Viet Nam. Viet Nam has now become an exporter of aromatic rice.[14] (It is known, although not officially sanctioned, that some portion of Thai rice exports originate from neighboring countries.)
- The increase in India's export put pressure on the global rice price and market.

The determination of price of rice in the international market is carried out not by country, but by the international rice traders.[15]

18.7 DOMESTIC POLITICS

The PMS took place against the backdrop of intense political competition in Thailand between the incumbent Pheu Thai Party and the opposition, Democrat Party. During the Democrat's government in 2009—11, an IGS was implemented to stabilize farmers' income. However, this scheme did not become popular and had implementation problems (e.g., cumbersome documentation requirements and corruption charges).

The IGS implementation suggests that both political parties share the same objective of transferring income to rice farmers. Since PMS and IGS had been problematic, a more effective and efficient scheme should be considered, especially one designed to minimize corruption.

In fact, to say that PMS or IGS is a transfer of income to the farmer is **incorrect**. As shown in Figure 18.1, if the Thai farmers received the market price, they should have been given the full amount of producer surplus. But in order to keep down the consumer price, the government compensated THB 31 (US$1) billion to the farmer.

This brings to fore the issue of consumption subsidies, which is the root cause of the farm subsidy. Although there is no explicit measure to control the retail price of rice but as a product in the "Control List" of the Control of Price of Goods and Service Act, B.E.2542, retailers, wholesalers, and rice millers keep a check on the price of rice.

The lower price of rice also invites wasteful consumption. However, such subsidies are politically popular because they are beneficial to the urban population, which is more politically vocal, and they enhance the development of the manufacturing sector by keeping wages down.[16] Rice consumption subsidy should be carried out through a more efficient program that directly targets poor consumers, not by keeping the price of rice low.

Taking away consumption subsidies will not be politically popular, but it would automatically reduce the bias against the farming sector, which shoulders most of the burden.

---

[14]http://oryza.com/news/rice-news/fao-2013-aromatic-rice-price-sub-index-averages-21-higher-2012.

[15]Calpe (2006) provides a description of the role and function of the international rice traders and the list of these entities.

[16]Roumasset and Setboonsarng (1988) estimated that this objective is equivalent to an export tax of 6.25—7.32%.

18.8 POLICY RECOMMENDATIONS

18.8.1 AN EXIT STRATEGY FOR THE PMS

- Abolish consumer subsidies through price control

    If the government sells rice in stock at the world market price for the domestic rice market, the government will earn more revenue to pay the farmer. Domestic price of rice will increase and reduce wasteful rice consumption.
- Utilize direct subsidies for lower income households

    Trairatvorakul (1984) found that the price increase will harm the poor consumer more than benefit the farmer. To compensate for this weakness, the government may consider having a program that would directly subsidize rice consumption for the poor household. For example, a rice voucher program for poor households can be introduced.

18.8.2 REGIONAL RICE MARKET ARRANGEMENT

Given economic growth in Asia where most of the rice is produced and consumed, the international rice market will be changing in the coming decade. The surge of export from India and the increase in the import from China during the past two years were the prelude of the change in the international rice market. The stability of rice supply will become a bigger issue, affecting the economic welfare of millions of consumers and producers, if governments were to continue lower rice price to benefit consumers and subsidize the production cost. These policies have systematically reduced the world rice price and increased its fluctuation. As Figure 18.5 shows, production subsidies in both importing and exporting countries push out the supply curve of rice, respectively, reducing the world price of rice from P_w to $P_{w'}$ and $P_{w''}$.

The price of rice has been kept lower than that of other staple grains even during the global food crisis in 2008 (Figure 18.6).

China and India, the two largest rice producers and consumers, should take the lead in forging a regional arrangement in Asia to improve security in supply, put in place a price stabilizing mechanism, standardize support for poor consumers, and respect the market mechanism. There is already a platform

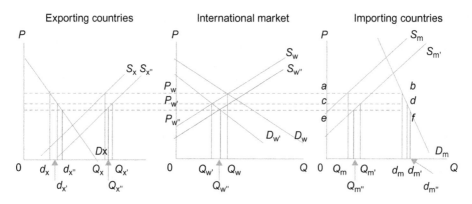

FIGURE 18.5

Impacts of production subsidies on world rice price.

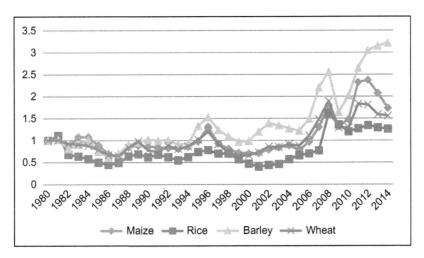

FIGURE 18.6

Price index for major grains.

IMF, World Economic Outlook, October 2013.

that could be developed to perform this function—the ASEAN Plus Three Emergency Rice Reserve (APTERR) Agreement, which involves the 10 ASEAN countries, China, Japan, and Korea. Article 2 of the Agreement indicates the ASEAN Plus Three's intention to deal with the stability of supply.

These 13 countries accounted for 57% of global production, 42% of world export, and 24% of world import in 2012. With India, the coverage increases to about 75% of global production and 68% of world export. Therefore, India should join the APTERR arrangement.

18.9 CONCLUSION

There are three broad issues that the government of Thailand has been addressing: (i) increase farmer income, (ii) increase export price, and (iii) ensure affordable rice prices.

Agriculture and trade policy of rice importing countries artificially bring down rice prices, causing the government of rice exporting countries to react. Thailand's PMS was intended to raise world rice price by stockpiling rice purchased at above-market rates. The government purchases paddy from farmers at higher than market price, but as global prices increase, each ton of rice in inventory can be auctioned off for a profit.

An important consideration for the government is to purchase only the type of rice over which Thailand can exercise market power. Estimations show that the demand for jasmine (*hom mali*) rice is more inelastic. This suggests that the Thai government would have more success in pursuing a policy that would increase the world price of *hom mali*, rather than the low quality white rice, for which demand is elastic.

The Thai government should rethink consumer subsidies innate in the PMS. Direct subsidy programs can be directed to low income households, while domestic prices should be allowed to increase and match world prices. The cost of political popularity, however, has prevented politicians from subscribing to an economically sound and efficient policy.

REFERENCES

Calpe, C., 2006. Rice International Commodity Profile, Food and Agriculture Organization of the United Nations, Market and Trade Division, December.

Choeun, H., Godo, Y., Hayami, Y., 2006. The economics and politics of rice export taxation in Thailand: a historical simulation analysis, 1950–1985. J. Asian Econ. 17, 103–125.

Chulaphan, W., Chen, S.E., Jatuporn, C., Wen, C.H., 2013. Different causal relationships of export rice prices in the international rice market. Am. Eurasian J. Agric. Environ. Sci. 13 (2), 185–190.

Clarete, R., 2014. Philippines rice self-sufficiency program: pitfalls and remedies. In: Balisacan, A., Chakravorty, U., Ravago, M.-L. (Eds.), Sustainable Economic Development: Resources, Environment and Institutions. Elsevier Inc, San Diego, CA.

Corden, W.M., 1974. Trade Policy and Economic Welfare. Clarendon Press, Oxford.

Democrat Party, 2013. In-depth knowledge, Really knowing the Cheating of Paddy Mortgage Scheme. (In Thai). Democrat Party, Bangkok.

Grain: World Market and Trade, USDA, 2013. Available from: < http://apps.fas.usda.gov/psdonline/circulars/grain.pdf > .

IMF, World Economic Outlook, October 2013. Available from: < http://www.imf.org/external/pubs/ft/weo/2013/02/weodata/download.aspx > .

Isvilanonda, S., 2012. 1 Year and Government Economic Policy: Agricultural Policy. Paper presented at the Economics Conference, National Research Council, Bangkok.

Khoman, S., 2014. Corruption, transactions costs and network relationships: governance challenges for Thailand. In: Balisacan, A., Chakravorty, U., Ravago, M.-L. (Eds.), Sustainable Economic Development: Resources, Environment and Institution. Elsevier Inc, San Diego, CA.

National Rice Policy Committee, 2012. In-depth knowledge, Really Knowing the Government's Paddy Mortgage Scheme. (In Thai). Department of Internal Trade, Nonthaburi, Thailand.

Office of Agricultural Economics, 2012. Basic Information on Thai Agricultural Sector. (In Thai). Ministry of Agriculture and Cooperatives, Bangkok.

Poapongsakorn, N., 2013. Impacts of government rice price intervention. (In Thai). In the Report on Thai Rice Strategy: Research and Looking Forward. Thailand Development Research Institute, Bangkok.

Poapongsakorn, N., Siamwalla, A., 2009. New approach for rice price intervention. (In Thai). TDRI's Report No. 75, Thailand Development Research Institute, Bangkok.

Rice Export Statistics, Office of Agricultural Economics, Ministry of Agriculture and Cooperatives, 2013. Available from: < http://www.oae.go.th/more_news.php?cid=95 > .

Roumasset, J., Setboonsarng, S., 1988. Second-best agricultural policy: getting the price of Thai rice right. J. Dev. Econ. 28, 323–340.

Setboonsarng, S., 1990. Objectives of Public Agricultural Research in Thailand, Issue 28 of ACIAR/ISNAR Project Papers, Australia.

Thai Rice Exporter Association, 2013. Price database. < http://www.thairiceexporters.or.th/database/AVG%20prices%202013.html > .

Thang, T.C., Huong, D.L., Minh, L.N., 2013. Who Has Benefited from High Rice Price in Viet Nam? Oxfam International, Hanoi, Viet Nam. Available from: < http://blogs.oxfam.org/sites/blogs.oxfam.org/files/who-benefits-rice-prices-vietnam-full-20131017.pdf > .

Trairatvorakul, P., 1984. The effects on income distribution and nutrition of alternative rice price policies in Thailand. Report No. 46, International Food Policy Research Institute (IFPRI), Washington, DC.

PHILIPPINE RICE SELF-SUFFICIENCY PROGRAM: PITFALLS AND REMEDIES

19

Ramon L. Clarete

School of Economics, University of the Philippines, Diliman, Quezon City, Philippines

CHAPTER OUTLINE

JEL: Q18; H54; F13

The Philippines is implementing its Food Self-Sufficiency Program, which, among other objectives, aims to make the country self-sufficient in rice by 2013 or 2016. The country is past the first deadline and failed to become so. There remains the second deadline in 2016. To realize it, the Philippines has restricted rice imports, supported palay prices, and has invested billions of pesos largely in irrigation facilities. This chapter discusses the possible rationale for it, describes the program, identifies its pitfalls, and suggests a few fixes.

19.1 RICE SELF-SUFFICIENCY AND FOOD SECURITY

Food self-sufficiency, as understood in the Philippines, stems from a rather early concept of food security. At the World Food Summit in 1974, the delegates considered the need for an adequate supply of food to support a steady expansion of food use and ensure stable food supplies and prices. The predominant thinking then proceeded from inadequate food supplies in 1972 and 1974 (Maxwell and Slater, 2003).

The concept evolved through the years. Sen's work (1981) on poverty and famines shifted policy makers' grasp of food security from availability of to economic access to food, particularly of the poor (FAO, 1983). World Bank (1986) reported that food price fluctuations explain significantly transitory food insecurity. But alongside with that is the recognition that changing levels of foreign exchange earnings, local food production, and household incomes of the country are important determinants of food access. Keeping food prices stable and affordable as well as alleviating poverty have become paramount tasks to secure economic access to food since the 1980s.

It is not that policy makers had abandoned earlier ideas. Rather, they expanded the meaning of food security. In the 1990s, the safety and nutritional dimensions of food security came to the forefront. At the World Food Summit in 1996, policy makers articulated the prevailing meaning of food security as the situation where "all people, at all times, have physical, social, and economic access to sufficient, safe and nutritious food that meets their dietary needs and food preferences for an active and healthy life" (FAO, 1996). Embedded in it are three dimensions of food security, namely food availability, access, and safe use.[1]

Despite the added dimension of safe use, the uncertain and weak access to food of the world's poor has remained the paramount concern of policy makers. In its Food Insecurity Report 2001, FAO (2002) had emphasized the intertwined problems of poverty and food insecurity. People are vulnerable to food insecurity because their economic status has reduced their entitlements to adequate, safe, and nutritious food. The improved economic status of the population enables households to better cope with transitory food price fluctuations. With low levels of income, households find adjusting their respective budgets to cope with any unexpected increases of food prices difficult; for those in the lowest income levels, access to food may altogether be economically barred. Even those that ordinarily find food affordable become food insecure in times of sharp and unexpected increases of food prices.

A renewed effort to avoid extreme swings of food prices surfaced once again in the aftermath of the food crisis in 2007 and 2008.[2] In these years, exporting countries like Vietnam and India restricted their exports of rice in order to keep their respective local rice prices stable and affordable. In response, importing countries like the Philippines accelerated their respective rice imports to insulate their national markets from the extreme fluctuations of global food prices. Others subsidized rice imports. Inadvertently, these actions fulfilled the crisis they all wanted to avoid.

The 2008 rice crisis encouraged ASEAN member states to increase their local rice outputs. In Malaysia, the government targeted an increase in yield from 2.47 to 4.48 tons per hectare with public support. Sabah and Sarawak are identified as the new frontiers for production. In April 2008, the Philippine government launched its FIELDS program, targeting the country to be capable of producing at least 98% of its rice consumption in two years' time. It failed to attain its objective. Instead, it heavily imported rice from 2008 to 2010. Indonesia has been working for full self-sufficiency, devoting public resources to increasing rice production. Even Brunei Darussalam, which easily obtains its rice requirement from trade, launched in September 2009 its rice hybrid development program, targeting 1344 ha to help attain 26% self-sufficiency in rice.

---

[1]The International Conference on Nutrition organized by the World Health Organization and FAO in Rome in 1992 is responsible for the added dimension of food safety.
[2]For example, see Timmer (2009). Headey (2011) and G20 (2011) report on extreme food price volatility.

19.2 **WHY COUNTRIES PURSUE SELF-SUFFICIENCY IN RICE**

Being able to produce at least all of a country's rice use may be regarded as insurance against three food insecurity risks: (i) abnormal rice price volatility in the rest of the world spilling into the home market, (ii) the relatively thin rice trade of the world, and (iii) exporting countries restricting rice exports. In the following, the validity of each is assessed.

19.2.1 **RICE PRICE VOLATILITY**

World rice prices have fluctuated through the years, with a few of these episodes involving price spikes and plunges. In Figure 19.1, there appears to be at least one rice price surge of Thai white rice, 5% broken, in each decade since the 1960s.[3] The price spikes in the 1970s and 1980s were larger than those in the 1960s and 1990s. After declining in the late 1990s and early 2000s, rice prices sharply increased in 2008. Such extreme price fluctuations may compel policy makers to insulate their domestic rice markets from extreme price volatility in the world market by becoming self-sufficient.

Table 19.1 compares the average annual rice price volatilities[4] of Philippine wholesale and Thai, 5% broken rice prices, which represent world prices. Philippine rice prices are relatively more stable than world rice prices from the 1970s to 2000s.

Has rice self-sufficiency contributed to relatively more stable rice prices in the Philippines? It appears from the data that the Philippines has not been self-sufficient at all since the 1960s. Although there had been significant rice exports in the late 1970s and the early 1980s, this performance was more than offset by the rice imports in the early part of the 1970s and the second half

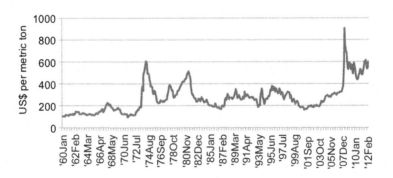

FIGURE 19.1

Average monthly world rice prices (Thai, 5%): 1960–2012.

World Bank Commodity Price Data.

---

[3]Timmer (2009) claims that on average there are three food crises per century.
[4]Price volatility is measured as the standard deviation of the natural logarithm of yearly changes in prices, multiplied by the square root of the total number of observations.

Table 19.1 Average Volatility of Annual Rice Prices and Average Annual Philippine Rice Trade

Period	Volatility (%)		Rice Trade (in metric tons)	
	Philippines Wholesale	Thai, 5% Broken	Imports	Exports
1961–1969	43.87	38.25	188,986	4287
1970–1979	53.81	128.83	154,993	23,788
1980–1989	56.59	74.06	104,435	51,448
1990–1999	24.11	39.91	589,753	4545
2000–2010	37.76	83.26	1,501,277	230

Source: Author's computation using data from BAS on Philippine wholesale prices and from World Bank on Thai 5% rice prices; FAOStat on Philippine Rice Trade Data.

of the 1980s.[5] From this, the country need not be self-sufficient in rice to keep rice price fluctuations elsewhere in the world out of the country.

Since the late 1990s, the government's attitude toward rice imports has changed. The country became increasingly reliant on rice imports. Since 2001, rice imports have steadily climbed, depicting the change of attitude of the government with respect to rice imports. In 2005, the Philippines became the world's largest rice importer but slid down to second in 2006. Starting in 2007, it consistently became the top rice importer with a share of nearly 12% of the total rice imports of the world in 2008. The data shows that the country can be a significant rice importer and at the same time stabilize local rice prices.

19.2.2 THIN RICE TRADE

A second argument for rice self-sufficiency is that it addresses the risk that rice suppliers may not be able to meet significantly increased demands in the event of simultaneous crop failures in large rice consuming countries. Global rice trade is thin, presently at 6–7% of world output. In contrast, the corresponding figures for wheat and maize are at least 13%. Thus, simultaneous rice crop failures in China and Indonesia can increase global prices to levels the country may not be able to afford.

The ASEAN region contributes nearly half of the marketable surplus of rice in the world. Its rice imports, however, are relatively low, about a fourth of global rice imports, as Figure 19.2 shows. In the 2000s, ASEAN rice imports hardly increased, unlike its rice exports, which increased sharply. A greater part of this surplus went to markets outside the region.

A stochastic multi-year simulation was conducted using the Arkansas Global Rice Model (AGRM) in order to determine the likelihood that these countries may run out of exportable surplus. To undertake the simulation, data on historical rice yields in the region's four largest rice trading countries were gathered and their respective probability distributions were

---

[5]Presenting annual data would show that the Philippines was a net exporter of rice from 1978 to 1982. This performance has not been repeated after those years.

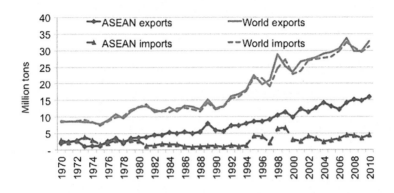

FIGURE 19.2

Rice trade in ASEAN and the world: 1970−2010.

UN FAOStat.

estimated.[6] The exporting countries were Thailand and Vietnam, while Indonesia and the Philippines were selected as the importing countries. The AGRM was subjected randomly to plausible national rice yields, the levels of which were consistent with their respective probability distributions. After each counterfactual yield was introduced as a shock to the model, the model's economic equilibrium was computed. The probability distribution of the equilibrium net exports corresponding to the random feeds of rice yields into the AGRM were then estimated. Using data in the distribution, the 90% and 10% percentiles of net exports, i.e. those occurring with the likelihood of 10% in the upper and lower tails of the distribution as shown in Figure 19.3, were obtained.

Figure 19.3 shows the projected 10th and 90th percentiles and mean rice net exports of Vietnam and Thailand. Both countries have adequately positive rice surpluses from 2012 to 2022. In the case of Vietnam, its surplus is projected to decline. However, even at its lowest level in 2022, the country has 90% chance of exporting at least 3.3 million metric tons a year and the 10% likelihood that it will have at least 4.6 million tons of exportable surplus. The mean surplus is estimated at about 3.9 million tons in 2022.

Thailand's net rice exports are larger than Vietnam's and are projected to increase. At its peak in 2022, the country's net exports will reach at least 12.8 million tons with the likelihood of 10%. On the other hand, the country has a 90% chance of exporting at least 10.3 million tons, significantly exceeding the 7.6 million tons in net rice exports of the country in 2012.

Based on Figure 19.3, extreme price volatility of rice is likely not due to a lack of rice to buy, but rather to other factors like lack of good quality market information, plus the plausible situation that market players have varying assessments of the market, with hardly a mechanism for converging into a more accurate and more widely shared depiction of the market. The rice price crisis in 2008 illustrates how the lack of good quality market information had driven buyers to excessively stock up rice and suppliers to withdraw stocks in anticipation of higher prices.

---

[6]University of Arkansas Distinguished Professor Eric Wailes developed the AGRM. This global rice economic model comprises 40 rice importing and exporting countries, grouped under one of its six geographical regions of the world (Wailes et al., 1997). The author of this chapter collaborated with Prof. Wailes in conducting the stochastic simulations. Both were consultants of the ADB under its Support for the ASEAN Plus Three Integrated Food Security Framework project (TA 7495).

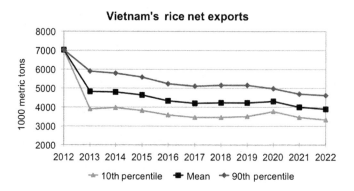

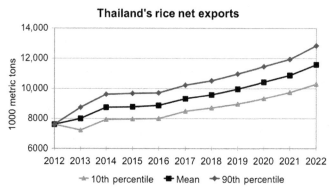

FIGURE 19.3

Projected 10th and 90th percentiles and mean rice net exports of Vietnam and Thailand, 2012–2022.

Stochastic simulations by E. Wailes using the AGRM (Wailes et al., 1997).

Disseminating accurate information on market trends mitigates the risk of unproductive price speculation. Using the definition of extreme rice price volatility of Martins-Filho, Torero, and Yao (Martins-Filho et al., 2010), Labao and Clarete (2013) identified the rice price crisis days using daily rice future prices from July 1991 to March 2013. They identified a total of eight such rice price crisis periods with an average length of 2.41 months since 1991. The price crisis in 2008 was the latest to occur. The authors estimated an 8.4% chance that a rice trading day qualifies to be part of a rice price crisis period.

Thin rice trade is self-perpetuating. Because regional rice trade is shallow, rice-deficit countries tend to pursue self-sufficiency, which in turn reduces rice trade levels. Rice trade in proportion to world production has been the smallest among the top three cereals (Table 19.2). From 1962 to 2010, the export to output ratio (XOR) in the world was 4.98% for rice, while wheat and maize had 18.63% and 13.57%, respectively. The ratio for rice has grown through time but apparently not at a rate as to catch up with those of maize and wheat.

In Table 19.2, annual price volatility appears inversely related to tradability. For the entire period from 1961 to 2010, rice had the largest price volatility at 152.28% compared to wheat (139.08%) and maize (133.7%). This may be explained by an important role of trade of relieving temporary

Table 19.2 Average Cereal Price Volatility and XORs: 1961–2010 (%)	Maize		Rice		Wheat	
	Price Volatility	XOR	Price Volatility	XOR	Price Volatility	XOR
1961–1969	32.64	9.99	44.97	4.46	29.40	16.95
1970–1979	62.72	14.88	89.45	3.94	80.86	17.07
1980–1989	64.78	16.20	56.02	4.21	40.73	19.97
1990–1999	56.19	13.29	70.69	5.20	61.46	18.64
2000–2010	74.52	13.18	71.41	6.86	80.22	20.18
1961–2010	133.72	13.57	152.28	4.98	139.08	18.63

Source of basic data: *Food and Agriculture Organization of the United Nations (FAOStat) for trade data; World Bank Commodity Price Data (Pink Sheet) for prices.*

production shortfalls in some parts of the world. Had rice trade been as large as those of wheat and corn, fluctuations of rice production due to natural causes would not have ended up in extreme rice price volatility in the world market. The deeper the trade the more stable its market price.

With adequate rice demand and investments in the supply chain in and out of exporting countries, the rice export supply capacity of the region can increase further. In Myanmar and Cambodia, investments to modernize their road infrastructure, logistics, and rice mills have the potential of increasing their respective exportable surpluses. However, if rice self-sufficiency remains a key objective of rice-deficit countries, as it has been, then they only prove the obvious result that rice trade is thin and unreliable, which in turn requires them to pursue rice self-sufficiency programs to insure against the risk of not finding enough rice to buy from the world market.

The actions of these countries illustrate an off–on participation in the global rice trade, which fails to support a sustained expansion of rice trade. Exporting countries, particularly India, regard the world market as where they can dispose of their rice surpluses (Timmer and Falcon, 1975). When they determine that they may run out of supply, they slow down export flows or stop export to ensure their national food security. In the case of importing countries like the Philippines, the world market is regarded as its source of last resort. When these countries bet on an impending high price, they start to draw all the rice they want in the shortest time possible, causing prices to shoot up. When their self-sufficiency programs manage to deliver a good harvest, they stop importing.

19.2.3 EXPORT RESTRICTIONS

A recently encountered risk of rice importing countries is export restrictions. In 2007 and 2008, Vietnam and India, two of the largest rice suppliers of the world, prohibited or restricted rice exports. The spike of rice prices in 2008 could have been avoided as the rice market fundamentals indicated there were adequate supplies of rice (Timmer, 2009). Instead, the uneven level and low quality of information about the true state of the market among the millions of players in the international rice market triggered panic buying.

Slayton (2009) identified three factors that caused the rice price hike in 2008. The first was that India in 2007 restricted its exports of rice for political reasons. Drive for profits explained Vietnam's contribution to the rice price crisis in 2008. Initially anticipating rice shortages due to a prolonged cold spell threatening the rice harvest at the Red River Delta in January 2007, Vietnam restricted its rice exports. Only Vinafood 2 and a few provincial exporters participated to supply the NFA imports of that year, according to Slayton. As rice prices in the world climbed in the first half of 2008, Vietnam banned rice exports. Local traders held on to their rice inventories in anticipation of higher export prices. The Philippines added to the growing price crisis by issuing in the regional market unprecedentedly large rice tenders. In that year, the Philippines imported about 2.4 million tons, most of which came from Vietnam.

The crisis in 2008 which exporting countries triggered could have been eased. Two regional cooperation measures could have been used to help dampen the rice price increase in 2008. One, the information about the world rice market was largely uneven. With inadequate information, rice market players act on the basis of the information about the market situation that they may have as rice prices start to climb. Had the correct information on the rice supply and demand balance been widely disseminated, rice traders in Vietnam would likely not have sat on rice inventories, and the NFA would not have issued the amount of rice tenders the way it did in 2008, i.e., the country's annual imports in just the first four months of the year. Households, who heard about the developing rice price crisis in 2008, could not have seen any need for stocking up.

Two, the dynamic interaction between a sizable number of buyers and sellers results in the convergence of price offers and requests into the equilibrium price that truly reflects the underlying scarcity of rice. Thus, reducing the market shares of government trading companies like the NFA and Vinafood 2 in favor of the private sector combined with requiring transparency in rice trades has the potential of moderating any unwarranted increases of rice prices. Private market players are out to maximize profits. If market players can view the rice transactions that go on, importers seek out and transact with the least cost exporters; the latter tend to move their rice supply to buyers willing to pay the higher prices.

The NFA and Vinafood 2 are two of the largest players in the regional rice market. The two largely accounted for at least half of the regional rice trade in 2008. It is certain that if they committed mistakes in assessing the market, these errors were incorporated into the government-to-government rice procurement agreements they made in 2008. For example, the NFA, acting on a mistaken assessment that a full-blown rice shortage in 2008 had transpired, issued unprecedentedly large tenders of rice in the world market. Errors like these can be made less likely to happen or completely avoided with more private sector importers bringing in the volume of rice that the country needs.

19.3 PHILIPPINE RICE SELF-SUFFICIENCY PLAN

In 2010, the Philippine government articulated its plan of making the country self-sufficient in rice in about three years (Figure 19.4). To attain the objective, the country had to produce the palay equivalent of its total use of rice. For example, the Philippines required 20.25 million tons of palay in 2010 and 21.12 million in 2013. The baseline rice paddy output in 2010 was 16.24 million tons.

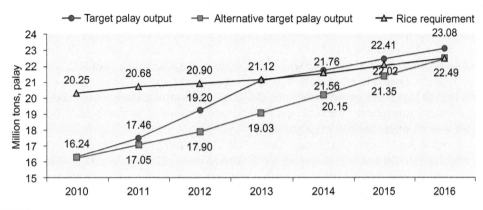

FIGURE 19.4

The Philippine rice self-sufficiency plan, 2010−2016.

Department of Agriculture.

Table 19.3 Sources of Incremental Production (in thousands metric tons of palay)

Program Intervention	2011	2012	2013	2014	2015	2016
More area harvested	411.66	264.04	318.45	320.54	346.79	339.56
Improved water management	232.95	179.08	217.25	226.71	225.29	175.52
R, D, and extension service	139.73	259.95	389.10	352.88	360.23	360.23
Farm mechanization	19.20	36.80	37.12	30.74	31.66	32.61
Reduced postharvest loss	7.11	84.67	93.14	30.74	31.66	32.61
Organic fertilizer application	–	27.50	77.50	155.00	200.00	200.00
Target incremental production	810.64	852.03	1132.55	1116.59	1195.62	1140.52

Source: *Department of Agriculture.*

Two target production plans are considered. The more optimistic plan makes this country become self-sufficient in 2013. Under the alternative conservative plan, the country meets its self-sufficiency target in 2016. The difference between the two is in the added response of the rice farmers to other interventions such as credit assistance and marketing support, and the private businesses, which can help in disseminating the hybrid rice as well as in increasing the use of agro-chemicals and fertilizers.

Table 19.3 sums up the relative contributions of the selected program interventions to the 2016 plan. These numbers indicate the reliance of the Department of Agriculture (DA) on expansion of irrigated areas and improved management of water, particularly in the initial years. Starting in 2013, incremental outputs would come from research, development, and extension services and organic fertilizer application.

The largest contributor to generating the additional production is expanding the area harvested. This is made possible with added investments in irrigation, either by developing new service areas or rehabilitating existing ones. About 60% of the total budget goes into irrigation, with the National Irrigation Administration (NIA) taking the bigger share of the funds compared to the Bureau of Soils and Water Management (BSWM). The program plans to establish new small-scale irrigation projects, diversion dams, and shallow tube wells; restore and rehabilitate national and communal irrigation systems; expand rice-based farming systems in upland and indigenous peoples' areas; and to proactively engage local government units (LGUs) on use of idle lands and extension services. Besides irrigation investments, the NIA and BSWM will implement programs aimed at improving rice yields with more dependable water sources.

Higher yields due to improved management of rice farms using the PhilRice's PalayCheck program is the second contributor to increasing palay outputs under this plan. The PalayCheck program comprises several steps that the farmer goes through from farm preparation to postharvest management to ensure the proper application of the technology and realize its potential in increasing yields. This is training intensive and makes use of pilot demonstration farms to showcase the benefit of the proper use of the technology. The use of certified seeds, for example, is one of the checks in this system. The DA plans to field monitors who will advise farmers and help them carry out the palay checks. Following these measures are three program interventions with relatively smaller contributions. These are farm mechanization, organic fertilizer use, and postharvest loss reduction.

The self-sufficiency plan introduces programs aimed at shifting some of the food requirement from rice to white corn. With this feature of the plan, the DA reduces the requirement for rice and makes the country more likely to be self-sufficient in rice. The per capita rice consumption is presently estimated at 120 kg per year. By shifting some of the rice consumers to white corn (presumably these are residents in traditionally corn consuming areas), the DA aims to reduce this to 114 kg per year in 2013. Due to lack of attention to improving white corn availability, traditional corn users, such as those in Cebu and Western Mindanao, have shifted to rice. Presently, the per capita use of this grain is 7 kg per year. The DA plans to increase this to 13 kg in 2013. Consumption surveys will be undertaken every three years starting in 2012.[7]

The plan will cost the government about PhP 141.98 billion over the next six years, or approximately PhP 23.67 billion pesos a year, as shown in Table 19.4. These numbers are higher compared to the cost of the program when this was unveiled in 2011 by about PhP 4 billion each year. This plan focuses on public services such as irrigation, as well as research, development, and extension. Of the total planned spending, the two items account for 83%. The rest is budgeted for organic fertilizer subsidies (4%) as well as support for the acquisition and operation of farm machineries and postharvest equipment (13%).

Table 19.4 adjusts target incremental outputs for the rice self-sufficiency plan compared to what the DA declared at the start of the program in 2011. The adjusted numbers frontloaded the incremental outputs of the plan and increased the total target incremental output by 0.71 million metric tons. In the original plan, the DA was targeting the total of 6.25 million tons of palay.

The total cost of this 6-year banner program in agriculture is PhP 141.98 billion. The program is designed to deliver about 6.96 million tons of palay, equivalent to 4.524 million tons of rice. At the average annual world price of rice from Vietnam during the period 2010−2012 of PhP 18,362 per metric ton (assuming PhP 40:$1), plus 10% shipping/insurance cost, the equivalent import bill

---

[7]The plan likewise talks about encouraging the mix of sweet potato in the diet, but this is not formally taken up in terms of budgetary allocation.

Table 19.4 Planned Output and Budget of the Rice Self-Sufficiency Program: 2011–2016 (Output in million metric tons and Budget in billion pesos)

	2011	2012	2013	2014	2015	2016	Total
Incremental output (palay)[a]	1.19	1.5	1.58	1.46	0.63	0.6	6.96
Budgetary cost by intervention[a]							
Improved water management							
NIA	13.59	30	20	15	10	5	93.59
BSWM	0.5	0.72	0.62	0.65	0.68	0.71	3.88
R, D, and extension service							
PalayCheck field schools	1.17	1.86	2.55	2.58	2.86	3.07	14.09
Bureau of plant industry	0.16	0.25	0.18	0.18	0.19	0.19	1.15
PhilRice	0.32	0.52	0.47	0.43	0.38	0.35	2.47
Farm mechanization and postharvest loss reduction							
PhilMech	0.2	5.46	5.99	2.58	2.12	2.18	18.53
Organic farming							
BSWM/NOAB	0.5	0.75	0.75	0.75	0.75	0.75	4.25
Others	1.17	0.57	0.57	0.57	0.57	0.57	4.02
Total budgetary cost	17.61	40.13	31.13	22.74	17.55	12.82	141.98

[a]*The author obtained the data from the National Economic Development Authority (NEDA) based on the DA's submission in the 2011-2016 Public Investment Plan. According to the 2011 version of the program, as shown in a DA's PowerPoint presentation on Food Self-Sufficiency Roadmap, 2010-2016, the budgetary cost was originally estimated at PhP 177.44 billion.*

would have been PhP 91.38 billion. It implies that the country is spending an extra PhP 50.6 billion for the expected output of rice of the program.

19.4 PITFALLS OF THE PROGRAM

The self-sufficiency program is widely recognized by policy makers and stakeholders as necessary for food security. However, three concerns are raised, which hopefully contribute to a more informed assessment of the economic desirability of the program, namely the likelihood of success, governance issues, and the economic cost of the program.

19.4.1 LIKELIHOOD OF SUCCESS

Even before climate change, as reflected by the increased frequency of more severe rainfall and flooding affecting the productivity of rice farms, the Philippines has had routine bouts of typhoons, and, occasionally, droughts, as the El Niño weather phenomenon in 1997. These conditions reduce the productivity of the country's rice farms. For the success of the self-sufficiency program, the target production ought to exceed the consumption requirements of the country by the expected damage to rice output from these adverse weather conditions.

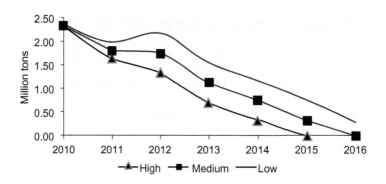

FIGURE 19.5

Rice import requirements under alternative production scenarios, 2010–2016.

Department of Agriculture.

The DA considered in its plan the possibility that typhoons or drought may occur during the plan's implementation. Accordingly, it considered two scenarios. The plan is set back if such conditions moderately reduce the country's planned production. Crop damage such as what occurred during Tropical Storms Ondoy and Pepeng in 2009 is considered moderate. At worst, the rice output of the country fell by 650,000 tons in that year. The other scenario places the damage at 1.3 million tons.

In either case, the DA plans to offset shortfalls with rice imports. Figure 19.5 shows the level of rice imports needed to make up for the shortfall of production under the two scenarios. The high production level scenario is when weather is as expected. In 2013, the DA plans to import about 0.7 million tons of rice, or significantly lower than the 2.33 million tons it planned to import in 2010. The medium and low production scenarios correspond to the moderate and high damage scenarios due to typhoons, respectively. The country may need to import more compared to the expected level of imports in the plan (high production scenario). The plan should have programmed to produce the consumption requirement of the country plus the expected annual damage to the rice crop from typhoons or other adverse weather conditions. The NFA imported 0.215 million metric tons. In 2013, or about two-thirds lower compared to the expected rice imports in that year under the high production scenario. But with super typhoon Yolanda in 2013, import levels should have reached 1.5 million metric tons. Since actual rice import was low, rice stocks in the country were depleted causing 2013 rice prices to increase by at least 20%.

How likely is it that the DA can attain its target? De la Peña (2011) forecasted the country's rice supply and utilization to 2019. She estimated the parameters of an annual forecasting model for the country's rice industry, the equations of which were obtained from a quarterly version of the partial equilibrium rice model (Clarete et al., 2000). De la Peña's forecasts indicate that the Philippines may not become fully self-sufficient within the next three years ending 2016. Imports are projected to be at 4.57 million tons (in palay terms) by the end of 2016 (Table 19.5).

While De la Peña's forcast is deterministic, a multi-year stochastic projection of the Philippine supply and utilization of rice until 2022 was done using the AGRM. The simulation results indicate that in 2013 the Philippines had only a 5% probability of becoming self-sufficient in rice. Truly enough, the country did not attain the target. However, if current spending and policies are sustained, there is up to a 10% chance that the Philippines will develop positive net exports given

Table 19.5 Rice Supply and Utilization Forecast, Philippines: 2010–2019 (in million metric tons of palay)

Year	Food	Feed, Seed, Proc'sing, Waste	Ending Stock	Rice Output	Imports	Beg Stocks	Total Supply/ Use
2010	19.00	2.12	3.22	17.04	3.26	4.04	24.34
2011	19.63	2.20	3.38	17.74	4.25	3.22	25.21
2012	20.23	2.26	3.66	18.35	4.43	3.38	26.16
2013	20.79	2.36	4.30	19.24	4.55	3.66	27.45
2014	21.32	2.41	4.78	19.68	4.53	4.30	28.51
2015	21.84	2.46	5.25	20.22	4.55	4.78	29.55
2016	22.34	2.52	5.69	20.73	4.57	5.25	30.55
2017	22.77	2.54	5.96	20.98	4.59	5.69	31.26
2018	23.18	2.55	6.03	21.16	4.64	5.96	31.76
2019	23.49	2.52	5.79	21.00	4.76	6.03	31.79

Source: De la Peña, 2011.

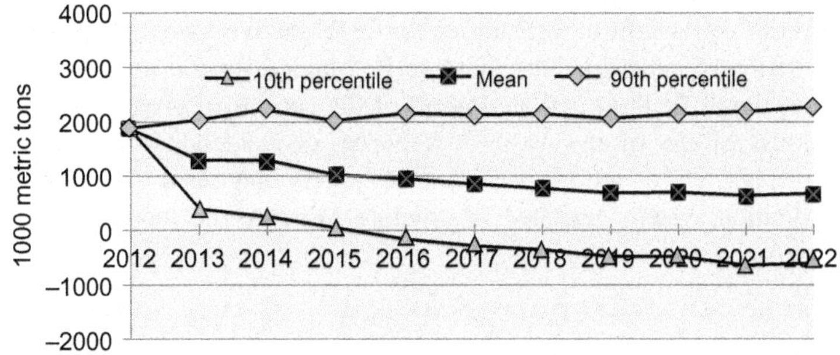

FIGURE 19.6

Philippines' rice net imports (10-year stochastic projections).

Stochastic simulations by E. Wailes using the AGRM (Wailes et al., 1997).

continued high price policy for rice due to import restrictions starting in 2016 (Figure 19.6). However, it must be pointed out that sustaining this success beyond 2016 is not assured given that there is at most a 90% chance that the country will have to import rice.

Cuerdo and Maines (2014) updated the forecast of the supply and use of rice in the Philippines that De la Peña did. The authors innovated by doing a stochastic projection of rice output, something not done by De la Peña and by Barbero and Lastimosa (2010, p. 54). The study noted that demand consistently surpassed the deterministic projection, which simply uses trend values through time of the exogenous variables of the rice market model. Moreover, the stochastic forecast indicated that the nation's expected mean production would be 12% lower than the demand in 2014 and 7% in 2023. They concluded that "given the projections made in this study, such desired production outcome in 2016 seems to be unrealistic."

19.4.2 **GOVERNANCE**

There are individuals who regard a lack of resources as critically preventing the attainment of the rice self-sufficiency objective or the modernization of the agriculture sector. Undoubtedly, the lack of resources for the agriculture sector has always been a challenge. However, the effectiveness of project administration has turned out to be more critical. In 1995, when the country acceded to the WTO and Congress passed a law providing for the modernization of the agricultural and fisheries sector, the government invested billions of pesos in the sector to make it more competitive. After approximately 19 years transpired, however, the added budget did not translate into a modern agricultural sector; quite the contrary occurred. The sector continues to be populated with farmers living below the poverty line, and worse, public spending for the sector has been contaminated with corruption.

Asymmetric information, particularly the principal agent problem, had turned out to be the more important constraining factor to attaining self-sufficiency in rice and modernizing agriculture. Program managers may act, rather than in accordance with the public interest, in pursuit of their own private interest or that of their private principals. It is difficult to monitor the actions of officials and staff in government to ascertain if they continue to act in the public interest.

The practice followed in setting performance/production targets is rather mechanistic and borders on being overly optimistic. Coming up with a plan that proposes to spend not a trivial amount of resources but without the benefit of whether or not its objectives can be realistically attained can only be possible if there are natural events that can be blamed for the nonattainment of the plan's objectives. Moreover, the difficulty of observing all of the actions of program managers and implementers implies accountabilities of any waste of resources or incidence of graft and corruption hard to establish. Thus, the practice of setting performance targets may need to be improved and a good monitoring and evaluation system designed in order to mitigate the moral hazard problem in the rice self-sufficiency plan.

19.4.3 **HIGH COST OF THE PROGRAM**

The economic cost of the plan and policies in support of it nets out the cost to rice consumers from the benefit to rice producers of these policies, particularly the import restrictions and the subsidies that the NFA provides to rice farmers and to poor consumers. Earlier work in estimating the economic cost of rice policies in the Philippines (Roumasset, 1999) pointed out that the government's rice policies had not prevented the general public from paying relatively high rice prices and farmers from receiving less than the trade protection implied by current policies.

Five components of rice market inefficiencies were defined and estimated using a partial equilibrium model of the rice market: foregone tariff revenues, consumer surcharges, producer losses, excess burden for consumers, and excess burden for producers. From 1995−1998, on a yearly basis, the total of all these estimated rice market inefficiencies ranged from PhP 11.95 billion to PhP 31.24 billion. On average, the total loss amounted to PhP 31.24 billion a year.

Clarete et al. (2000) documented that the government lost revenues amounting to PhP 3.72 billion a year by restricting rice importation to the NFA. If permits to import rice were auctioned to the private sector, the government would have earned revenues. Alternatively, if rice imports were liberalized subject to import taxes, tariff revenues would accrue to the government. Currently, the

Table 19.6 Welfare Cost of Rice Policies in the Philippines, 2000–2005 (in billion pesos)

	2000	2001	2002	2003	2004	2005	Average 2000–2005
Changes in producer surplus	31.1	20	23.9	12.7	0.2	1.7	16.8
Changes in consumer surplus	−100.1	−79.7	−84.7	−66.9	−38.9	−61.1	−72.8
Changes in budget surplus	0	0	0	0	0	0	0
Collected and paid duties	0	0	0	0	3.9	12	2.7
Reimbursement via tax subsidy	0	0	0	0	3.9	12	2.7
Net change	−69	−59.7	−60.8	−54.9	−38.7	−59.4	−56

Source: World Bank, 2007.

NFA, the sole importer of rice in the country, receives a tax subsidy to pay for the customs duties when it imports rice.

Consumers lost about PhP 6.67 billion each year due to the segmented rice marketing system and the less transparent and predictable import regime—conditions giving rise to imperfect competition in the Philippine rice market. On a per unit basis, the penalty is measured by the excess of the actual price of rice resulting from the pricing and import policies over the equilibrium price of rice under an integrated rice marketing system and a tariffs-only import regime. In addition, consumers suffered losses due to the policy of protecting rice producers, amounting to PhP 4.24 billion each year.

World Bank (2007) also estimated the welfare cost of rice policies using a partial equilibrium model but involving the following items: (i) the consumer income losses due to implicit trade protection, (ii) the producer gains due to the same policies of restricting imports, and (iii) public spending (net of taxes collected on rice imports, if any) by the government for the rice self-sufficiency program and the pricing interventions by the NFA. Table 19.6 shows the estimates of the World Bank (2007) of the welfare cost of rice policies in the Philippines.

Because they pay rice prices higher than border prices, consumers reduce their use of rice. Their loss ranged from PhP 100.1 billion to PhP 38.9 billion. Their loss is partly gained by producers. Due to higher take home prices, producers increase their production relative to if they had to operate on the basis of the lower border price. The gains of producers ranged from PhP 31.1 billion to as low as PhP 0.2 billion. If the government collected the import duties, which were at 50% from 2000 to 2005, then the budget surplus would have risen. The government provides a tax subsidy to the NFA, who is the sole importer of rice in the Philippines. Adding up all the changes in agent surpluses, the Philippine economy suffered welfare efficiency effects ranging from PhP 38.7 billion to PhP 60.8 billion. The average annual welfare cost was PhP 56 billion. The bulk of this amount is accounted for by the average annual loss of rice consumers of PhP 72.8 billion. Gains of producers were 23% of the losses of consumers.

These estimates are updated in this study, but only focusing on the import rice restrictions (Table 19.7). The analysis identifies four stakeholders: rice farmers, consumers, importers, and the government. Rice farmers gain with high rice prices due to import restrictions. In 2008, however,

Table 19.7 Economic Welfare Cost of Rice Policies in the Philippines, 2006–2012 (in billion pesos)

	Consumer Surplus[1]	Producer Surplus[1]	Govt Surplus	Importers' Surplus	Tariff Revenue	Tax Subsidy[2]	Net Surplus
2006	− 65.39	51.21	–	7.68	–	–	− 6.50
2007	− 74.05	58.03	–	8.62	–	–	− 7.40
2008	43.75	− 39.43	–	− 7.05	–	–	− 2.73
2009	− 89.52	69.54	–	9.82	–	–	− 10.17
2010	− 130.10	91.81	0.00	18.54	27.29	− 27.29	− 19.76
2011	− 81.51	69.40	0.00	3.62	6.69	− 6.69	− 8.50
2012	− 165.99	128.76	0.00	4.98	5.39	− 5.39	− 32.25
Average							
2006–09	− 46.30	34.84	0.00	4.76	–	–	− 6.70
2010–12	− 125.87	96.65	0.00	9.05	13.12	− 13.12	− 20.17

[1]*The demand and supply equations used in computing the surpluses are respectively: $Qt = A_t − 350^* Pt$ and $Qt = A_t + 300^* Pt$. The intercepts (A_t) are computed each year to account for demand and supply shifters.*
[2]*This subsidy is given to the NFA (including the private sector importers that it gave permits to import rice) to be used to pay customs duties on imported rice.*
Source: *Author's computation.*

world prices of rice were extremely high but local wholesale prices were lower. In that year, the rice farmers would have obtained more revenues had the local price moved up with world prices. The loss of rice farmers, which amounted to PhP 39.43 billion in 2008, was gained by rice consumers.

In the years other than 2008, the rice consumers were worse off as they paid a higher price for rice as a result of the policies of restricting imports and the import monopoly of the NFA. Based on the average figures in Table 19.7, rice consumers lost about PhP 46.3 billion each year from 2006 to 2009 and PhP 125.87 billion from 2010 to 2012. Roughly three-fourths of these average annual consumer losses went to rice producers, who as a group gained PhP 34.44 billion and PhP 96.65 billion in the two time periods.

The NFA and the private sector rice importers stood to gain as they purchased Vietnamese rice at a lower price than what they sold that rice at locally. As a group, these importers gained PhP 4.76 billion in 2006–2009 and PhP 9.05 billion in 2010–2012. The NFA sells the imported rice at lower than market to stabilize rice prices during the lean months of the year. It makes use of accredited rice retailers, to whom it sells the rice at a peso lower than the official release price, with the instruction that they sell that rice at the official price. Monitoring of the actual sales of these retailers of the NFA-supplied rice is imperfectly conducted, if not absent. Leakages occur, benefitting these retailers. These retailers get the import rents and not the NFA.

The government's cost is zero. While in theory it should receive tariff revenues from rice imports, it provides a tax subsidy of equal amount to NFA and the private sector rice importers that it delegated half of the country's rice imports to by waiving those taxes in favor of the NFA.

Summing up, the country lost on average about PhP 20.17 billion each year from 2010 to 2012. The amount is higher than what transpired in the period from 2006 to 2009, in which the economy's loss averaged PhP 6.7 billion each year. The amount may be regarded as the economic cost of the program arising from the policies in support of rice self-sufficiency, which distort rice prices.

As Table 19.7 shows, the high price of rice due to import restrictions forces consumers to give up incomes in favor of rice producers and importers. Rice farmers gained on average about PhP 96.65 billion in 2010–2012. But how much do rice farmers, particularly the poorer rice farm households (RFHs), benefit from this high price of rice? Dawe (2006) estimated that only the richer 40% of RFHs enjoy two-thirds of the benefit of import restrictions, while the remaining 60% divide among themselves the remaining third.

Table 19.8 shows these numbers and how each RFH could have benefited from restricting imports. The RFHs with larger farms, i.e., the top two quintiles, may have benefited from import restrictions. The RFH belonging to the top quintile received an average net gain equal to PhP 94,478 each year in 2010–2012, while the corresponding number for the second quintile was PhP 47,277. Dawe (2006) attributes this to the relatively larger marketable surplus of the top two quintiles of RFH.

In contrast, the RFH belonging to the bottom three quintiles receive from PhP 36,039 to PhP 9067 annually. Interestingly, Dawe likewise estimates that in any year about 7% of all RFHs, which fall under the bottom quintile, may not have any marketable surplus, and thus are penalized equal to PhP 4419 per year due to the high rice price, i.e. these farmers are practically rice consumers rather than producers.

The pattern is similar to what transpired in Thailand's paddy mortgage system (PMS). Setboonsarng (2014, p. 318) documented the uneven distribution of benefits of PMS. Citing Poapongsakorn (2013), the author observed that among the farmer beneficiaries of Thailand's PMS, only "52.8% of farm households were in the low-income brackets (<THB 100,000) and they received only 20.4% of the total transfer for the main crop. The transfer went to the medium sized farmers."

While there is room to improve rice yields, increase outputs, and reduce postharvest losses, this chapter suggests that aiming to attain full self-sufficiency at whatever cost is not in the best interest of the country or even the majority of the rice farmers. One important fix of the current plan is to remove the import restrictions. The charter of the NFA needs to be changed and the agency's monopoly of rice imports ended. Instead, the import restriction has to be converted initially at its going equivalent tariff, and subsequently gradually reduced to low levels as is ordinarily expected under the single market ASEAN economic community.

There are performance indicators that could indicate progress toward self-sufficiency in rice at reasonable cost. The DA need not declare that it plans to attain self-sufficiency in three years or so. It may work for improvements in yields, expanding the area harvested and reducing post harvest losses. Provided the indicators are moving toward the goal of higher rice output at just about world prices, then there is evidence that public resources are used wisely. In the meantime, while the country is not self-sufficient in rice, the DA continues to import rice in order to fill in the gap between demand and supply and stabilize rice prices.

19.5 **CONCLUSION**

This chapter looked into the current self-sufficiency program in rice of the Department of Agriculture. The plan, based on the multi-year stochastic simulations of rice yields using the AGRM, has up to a 10% chance of succeeding in 2016. This study, however, does not have any

Table 19.8 Net Benefit of RFHs Due to Rice Import Restrictions

RFHs Income Quintiles[a]	Share of Output of Rice Marketable Surplus[a]	Benefit from Import Restrictions (billion pesos)[b]	No. of RFHs[c]	Average Benefit from Rice Import Restrictions (PhP per year per RFH)	Penalty on RFH Due to Import Restrictions[d](PhP per year per RFH)	Net Benefit (PhP per year per RFH)
Top	44	43	430,000	98,898	4419	94,478
Second	23	22	430,000	51,697	4419	47,277
Third	18	17	430,000	40,458	4419	36,039
Fourth	9	9	430,000	20,229	4419	15,810
Bottom	6	6	430,000	13,486	4419	9067
Bottom 7%	–	–	150,500	–	4419	4419
	100	96.65	2,150,000			

[a]The distribution of marketable surplus shares came from Dawe (2006). The share estimates for the third and fourth quintiles are from the author of this memorandum.
[b]Total producer gain from import restriction is PhP 54.75 billion.
[c]Total number of RFH is 2.15 million.
[d]Assumptions: RFH size is 5; per capita rice consumption per year per RFH is.547 kg.

claim on the chances of the government sustaining the country to be self-sufficient. In the late 1970s and early 1980s, the Philippines had the chance of becoming a net exporter but unfortunately was not able to sustain it. A country does not need to be self sufficient in rice to be food-secure. Other countries, e.g., Malaysia, aimed to be only 80% self-sufficient. The country may well adjust its target for self-sufficiency and allow the private sector with the NFA to import the remaining gap. This study shows that the risk of food insecurity that policy makers fear the country is vulnerable to under trade is low and with regional cooperation in ASEAN, some of these risks, as in 2008, can be avoided or reduced.

Possible future research on this topic may proceed along the following two strands. First, while the projected supply and use of rice in the Philippines was stochastically modeled using the AGRM, it may be useful to validate the result of the analysis this time by conducting a stochastic projection of rice supply and use in the Philippines using a national instead of a global model of the rice market. Cuerdo and Maines (2014) have initially worked on it, but their work needs improvement by doing the stochastic projection to all of the rice producing provinces of the country. Second, the data on Table 19.8 can be reassembled from another survey similar to what Dawe (2006) did, with a more varied average per household rice use, including household rice inventories.

REFERENCES

Barbero, P., Lastimosa, P., 2010. Philippine rice self-sufficiency: issues and prospects. (Unpublished undergraduate thesis). The University of the Philippines, School of Economics.

Clarete, R., et al., 2000. AGILE NFA Report: Strategic Reorganization of the NFA for the New Millenium. USAID AGILE Project.

Cuerdo, B., Maines, R., 2014. Stochastic Projections of Philippine Rice Self-Sufficiency Amid Climate Variability. (Unpublished undergraduate thesis). The University of the Philippines, School of Economics.

Dawe, D., 2006. Rice trade liberalization will benefit poor. In: Dawe, D., Moya, P., Casiwan, C. (Eds.), Why Does the Philippines Import Rice? Meeting the Challenge of Trade Liberalization. Joint Publication of International Rice Research Institute (IRRI) and PhilRice.

De la Peña, B., 2011. Situation and Outlook of the Rice Market in the Philippines. A Country Project Report Prepared in Support for the Association of Southeast Asian Nations Integrated Food Security Framework. Asian Development Bank TA 7495.

Food and Agricultural Organization of the United Nations (FAO), 1983. World Food Security: a Reappraisal of the Concepts and Approaches. Director General's Report. FAO, Rome.

Food and Agricultural Organization of the United Nations (FAO), 1996. Rome Declaration on World Food Security. FAO, Rome.

Food and Agricultural Organization of the United Nations (FAO), 2002. The State of Food Insecurity in the World 2001. FAO, Rome.

G20, 2011. Price Volatility in Food and Agricultural Markets: Policy Responses. Policy Report including contributions by FAO, IFAD, IMF, OECD, UNCTAD, WFP, the World Bank, the WTO, IFPRI and the UN HLTF.

Headey, D., 2011. Rethinking the global food crisis: the role of trade shocks. Food Policy 36, 136–146.

Labao, A., Clarete, R., 2013. Predicting time periods of excessive price volatility: the case of rice. ReSAKSS-Asia Conference in Siem Reap, Cambodia, September 25–26.

Martins-Filho, C., Yao, F., Torero, M., 2010. Two-step Conditional-Quantile Estimation via Additive Models of Location and Scale. International Food Policy Research Institute (IFPRI). Available from: <http://www.foodsecurityportal.org/sites/default/files/Martins-FilhoToreroYao2010.pdf>.

Maxwell, S., Slater, R., 2003. Food Policy Old and New. Dev. Policy Rev. 21 (5−6), 531−553.

Poapongsakorn, N., 2013. Impacts of government rice price intervention. The Report on Thai Rice Strategy: Research and Looking Forward. Thailand Development Research Institute, Bangkok (In Thai).

Roumasset, J., 1999. Market Friendly Food Security: Alternatives for Restructuring NFA. Unpublished report commissioned by the USAID-supported AGILE project for the NFA.

Sen, A., 1981. Poverty and Famines. Clarendon Press, Oxford.

Setboonsarng, S., 2014. Getting the price of thai rice right: episode II. In: Balisacan, A., Chakravorty, U., Ravago, M.-L. (Eds.), Sustainable Economic Development: Resources, Environment, and Institutions, p. 318. Elsevier, San Diego, CA.

Slayton, T., 2009. Rice Crisis Forensics: How Asian Governments Carelessly Set the World Rice Market on Fire. Working Paper Number 163. Center for Global Development, Washington, DC.

Timmer, C.P., 2009. Did Speculation Affect World Rice Prices? ESA Working Paper No. 09-07. Agricultural Development Economics Division, Food and Agriculture Organization of the United Nations, Rome.

Timmer, C.P., Falcon, W.P., 1975. The Political Economy of Rice Production and Trade in Asia. In: Reynolds, L. (Ed.), Agriculture in Development Theory. Yale University Press, New Haven, CT, pp. 373−408.

Wailes, E.J., Cramer, G.L., Hansen, J.M., Chavez, E.C., 1997. Structure of the Arkansas Global Rice Model. Department of Agricultural Economics, University of Arkansas, Fayetteville, AR.

World Bank, 1986. Poverty and Hunger: Issues and Options for Food Security in Developing Countries. World Bank, Washington, DC.

World Bank, 2007. Philippines: Agriculture Public Expenditure Review. Technical Working Paper No. 40493. World Bank, Washington, DC.

PRODUCTION SPECIALIZATION AND MARKET PARTICIPATION OF SMALLHOLDER AGRICULTURAL HOUSEHOLDS IN DEVELOPING COUNTRIES

20

Upali Wickramasinghe

Centre for the Alleviation of Poverty through Sustainable Agriculture (CAPSA) of the United Nations Economic and Social Commission for Asia and the Pacific (UNESCAP), Bogor, Indonesia

CHAPTER OUTLINE

JEL: C34; Q12; Q13; Q18

20.1 INTRODUCTION

Integrating smallholder agricultural households into local, national, and international agricultural markets has recently been promoted as a strategy toward realizing the goal of sustainable development (United Nations, 2012). The premise behind the proposal is that market participation of smallholders is likely to contribute toward agricultural growth, thereby triggering the much-needed structural transformation in the agricultural sector and a movement toward the alleviation of poverty and food insecurity of agrarian households in the developing countries.

Apart from the potential contribution of smallholder market participation on poverty reduction and food security, a number of compelling reasons for such a strategy can be identified. One such argument is that market participation facilitates better use of resources according to their comparative advantages (Timmer, 2005) and contributes to large-scale production and dynamic technological change (Romer, 1994), which, in turn, could facilitate structural transformation in agriculture, thus allowing rural farming communities an opportunity to move from subsistence agriculture into more specialized, market-oriented systems (Lewis, 1954; Kuznets, 1973; Chenery et al., 1986), potentially increasing farmer incomes and reducing poverty (IFAD, 2011). Mazumdar (1987) has shown that enhancing opportunities in the agricultural sector through commercialization can encourage farmers to move into productive agriculture rather than migrate out of the agricultural sector. Lack of opportunities in the agricultural sector to enhance living standards, on the other hand, is known to push farmers to cultivate marginal lands such as swamps and watersheds that are unsuitable for agriculture, leading to flooding, loss of top soil, and damage to biodiversity (Falkenmark et al., 2007).

While the need to promote smallholders' market participation is widely accepted, there is little consensus on a suitable strategy to integrate smallholder households with markets. This chapter attempts to contribute to strengthening our understanding of smallholder agriculture sectors in a specific country context and the link between production specialization and smallholder market participation, including factors that might explain the observed behavior of agrarian households toward commercialization. The chapter is organized as follows: Section 20.2 reviews literature on production specialization and market participation with a view to inform the theoretical framework to be used in the empirical analysis. Section 20.3 presents the conceptual framework and empirical model. Section 20.5 presents the data and a description of the production and marketing environment of Tanzania used in the empirical model. Section 20.5 presents the econometric results, and Section 20.6 concludes.

20.2 LITERATURE REVIEW

Adam Smith is credited with observing the link between markets and specialization. He noted that "the greatest improvements in the productive powers of labor ... seem to have been the effects of the division of labor" (Smith, 1776, Book I, Chapter 1), and that "it is the power of exchanging that gives rise to the division of labor, so the extent of this division must always be limited by the extent of that power, or, in other words, by the extent of the market" (Smith, 1776, Book I, Chapter 3). According to the classical view, a larger market allows a greater division of labor by generating adequate demand for specialized products and skills. Specialization over skills improves labor productivity, leading to greater production and supply, effectively enlarging the size of the market. This process continues until all exchange possibilities are exhausted and as long as there are no external restrictions on the functioning of markets.

Young (1928) extended the theory to include increasing returns to scale as another key factor in the process of specialization and market exchange. Accordingly, it is the capacity of an economy to utilize increasing returns to scale, not larger operations per se, which determines a country's growth and development. Stigler (1951) extended the theory further by including the functional operation of firms including purchasing and storing materials, transforming materials into semi-finished products and semi-finished products into finished products, storing and selling the output, and extending credit to buyers. Borland and Yang (1992) and Yang (2003), in their formalization of the theory of specialization, considered the possibility for the division of labor to evolve along product specialization and through expansion of the number of final and intermediate products. The theory suggests that an increase in the size of the market along with lower average fixed costs of new intermediate products induces firms to become specialized, increases the number of products, and opens up the possibility of increasing the number of transactions.

The theory of specialization, as explained above, mostly referred to manufacturing industries. Smith, as well as Marshall (1920), viewed agriculture as a sector with limited opportunities for specialization, too little potential for exploiting economies of scale given the small size of agricultural markets, sharp seasonality, and tasks that are not amenable to specialization, and thus with little possibility for greater division of labor over cropping tasks (Yang et al., 2013). In Smith's words, "[t]he nature of agriculture, indeed, does not admit of so many subdivisions of labor, nor of so complete a separation of one business from another, as manufactures" (Smith, 1776, Book I, Chapter 1). This impossibility of the division of labor over tasks was identified as a reason for the inability of agricultural laborers to keep pace with improvements as in the manufacturing sector. Stigler (1951) noted that the common perception that farmers were unable to keep up with improvements in technology as in the manufacturing sector was a result of empirical judgment; it was not that division of labor in agriculture was impossible, but less likely compared to the manufacturing sector. In the manufacturing sector, technology improvement due to division of labor received prominence, whereas in agriculture, increasingly intensive use of a relatively fixed supply of land was widely observed to yield diminishing returns.

Research conducted by economists on the agriculture sector in recent times, however, has changed this perception significantly. This strand of research not only recognized that the agricultural sector has the capacity to specialize over products but also that the division of labor is a central driving force of agrarian transformation from subsistence agriculture to incomplete agricultural contracts and toward commercialized agriculture. Some of the pioneering research in this stand includes Roumasset and Uy (1980), Roumasset and Smith (1981), Roumasset (1995), and Roumasset et al. (1995). This stand of literature has made two important contributions to our understanding on specialization in agriculture: (i) labor contracts and other organizational arrangements are the institutions that facilitate specialization in agriculture and (ii) specialization of production and institutions in agriculture coevolve. The fundamental tenets of this stand of literature are summarized below with a view to inform the conceptual framework to be used in this study.

At pure subsistence level, the family unit is the source of everything it consumes. It uses family labor to carry out all activities, and no possibility of specialization or market participation. As an agrarian economy expands, the need to engage outside labor for cultivation increases, allowing family members to specialize in certain tasks, especially of supervisory nature. At this level, exchange labor appears to be the preferred form, wherein neighbors and friends provide assistance, effectively limiting the need to monitor for shirking, dishonesty, and theft as family bonds ensure that workers keep to a certain moral conduct. However, exchange of labor only partly substitutes for family labor, being limited to tasks such as harvest collection. With agricultural development, however, it becomes increasingly difficult to organize a mutual exchange of labor and thus households begin to use hired labor, but family members work with hired laborers to reduce the need for monitoring and supervision. Hired labor tends to specialize in tasks that are routine, standardized, and arduous (Roumasset and Smith, 1981), whereas farmers themselves specialize in tasks involving joint labor and management, such as supervising hired labor and applying fertilizer (Eswaran and Kotwal, 1985; Kikuchi and Hayami, 1999). Variations have been observed within the system of hired labor: daily wage, piece rates, and time rates (Roumasset and Uy, 1980). Time rates are dominant in activities that are difficult to monitor (e.g., chemical application) whereas piece rates dominate activities that are easily monitored and have a more heterogeneous workforce (Roumasset and Uy, 1980). Further transformation of agrarian economies reduces the incidence of piece and time rates, but emerges piece rates where teams begin to play a larger role compared to individuals.

As opportunities for the division of labor further expand, attempts by farmers and farm workers to specialize in tasks are likely to be associated with changes in marginal productivity over tasks. It has been noted that marginal rate of human capital accumulation involving routine tasks diminishes more rapidly than those involving managerial tasks, which in turn leads to a widening differential between wages of hired labor and owner-operators. The increased commercialization and modernization of agriculture also contributes to changes in the composition of family labor through fertility choices (Evenson and Roumasset, 1986), which in turn changes the opportunity costs of family labor. This leads to an increased demand for hired workers to perform a variety of tasks in agriculture and hence greater opportunities for them to specialize in specific tasks, such as operating farm machinery and equipment. Human capital accumulation can be considered as a key contributor to the specialization of agricultural tasks and the overall process of agricultural modernization.

The other significant contribution of this stand of literature is to recognize the role of transaction costs, especially wedges between buying and selling prices, in explaining the emergence and evolution of agrarian institutions (Roumasset, 1981; de Janvry et al., 1991). The theory posits that changes in relative prices allow for increased intensification of production and productivity, which in turn induces greater specialization of agricultural production (Roumasset and Lee, 2007). Therefore, it can be argued that the initial thrust for a movement toward production specialization among smallholders comes from changes in relative prices induced by reduction in transaction cost wedges. Intensification of agricultural production gives further impetus to the process.

Recent models have given prominence to the wedges between buying and selling prices as a key contributor to the emergence of specialization in household decisions regarding sale of their labor and labor-leisure choices (de Janvry et al., 1991). The impact of transaction costs on smallholder market participation, as identified in the evolutionary theory of institutions, has been a subject of recent academic inquiry using a variety of econometric techniques. Goetz (1992), in a pioneering study, shows that transaction costs affect farmer decisions on whether or not to participate in markets and, when in the market, how much to produce and sell. This study confirms that fixed transaction costs hinder market participation while better information stimulates it. The same study identifies that changes in grain price have a differential impact on new sellers compared to those already in the market. Key et al. (2000) show that proportional transaction costs affect the quantity sold whereas fixed transaction costs affect decisions on whether to participate in markets. Fixed transaction costs affect a household's decision to participate in markets but they have no impact on the sales (Goetz, 1992; Skoufias, 1994; Key et al., 2000). Henning and Henningsen (2007) show that transaction costs and labor heterogeneity influence household behavior.

Several other factors and behavioral attributes have been shown to affect smallholder market participation and specialization. Heltberg and Tarp (2001) show that farm size per household worker, animal traction, mean yield, age of household head, and climate risk significantly affect market participation, particularly on cash crop sales. Kurosaki (2003) observed that cropping patterns of subsistence agriculture changed significantly with the concentration of crop acreage in districts with higher and growing productivity. Omamo (2007), using a numerical simulation model with Kenyan data, shows that it is possible for households located far away from markets to allocate a larger share of productive resources to intercropping and less to more specialized activities. He observed the need for empirical analyses to determine whether high farm-to-market transaction costs are associated with low participation in markets and high degree of farm diversification. Based on evidence provided by Njehia (1994) as cited in Omamo (2007), the study suggests that the average area devoted to intercropping rises as market access falls.

In recent work, the impact of traditional, personal connections and relationships, known as "guanxi" in China, on which an individual can build to secure resources or to draw benefits in business as well as in social life has been examined (Hualiang et al., 2008). Their findings suggest that traditional connections reduce market transaction costs, which in turn helps farmers to get access to modern markets such as supermarkets and international marketing channels. Trust is also an important element that determines the capacity of a farm family to enter into such contracts (Zhang and Hu, 2011). Even with limited access to technical and financial support, those who entered into contracts have been able to develop better opportunities to negotiate with buyers to get better prices and flexible conditions. Armed with their experiences, farmers are trying to move away from traditional "arm-length" business relationships and establish strong relationships with preferred buyers to reduce costs, increase efficiency, and enhance competitive advantage.

Land quality differentials have also been attributed an important role in determining the process as well as the pace of agricultural modernization (Benjamin, 1995). Access to productive technologies and adequate private and public goods to produce marketable surplus plays a significant positive role in smallholder market participation (Barrett, 1997). A significant amount of public investment in institutional and physical infrastructure is necessary to ensure broad-based, low cost access to competitive, well-functioning markets. Public investment alone is not sufficient to get smallholders to enter into markets. They must be complemented by investment by farmers themselves on their lands and other assets. Without such complementary investments, smallholder farmers may not benefit from public infrastructure provided by the state. However, investing in their lands and productive technology will require smallholders to save enough, which depends on their ability to generate a marketable surplus.

20.3 CONCEPTUAL FRAMEWORK AND EMPIRICAL MODEL

A simple model of household choice can capture the core issue of specialization and market participation of smallholder agrarian households. Consider a household that maximizes its utility of consumption. In the case of agrarian households, the bundle of consumption consists of home-produced staples, market-bought staples, market-purchased other commodities, and services (c_i). The household may earn income by selling staple food and cash crops it produces (q_i), labor, and other services it owns, such as renting out draft animals (e_i). Production of staple and cash crops is a function of assets held by the household (e.g., land, draft animals), flows of services provided by resources held by the household (e.g., labor), the public sector (e.g., irrigation and extension services), and the private sector (e.g., transport services, land tilling, and harvest collection).

The household is assumed to maximize utility (U) by choosing how much of each product or service to consume c_i, produce q_i, buy b_i, and sell s_i, subject to a standard set of constraints: cash constraint, resources availability, and the production function. The cash constraint shows that the total value of purchases of the household must be equal to or less than its income earned by selling staple or cash crops it produces and the revenue it generates by supplying labor and other services. This effectively assumes away the possibility of borrowing or lending, but the endowment can technically incorporate household savings from previous years or seasons. Given market determined prices (p_i^*), the household maximizes the utility function (20.1) subject to Eqs. (20.2)–(20.4)

$$Max\ U(c_i; z_u) \tag{20.1}$$

$$\sum_{i=1}^{N} p_i^* c_i \leq \sum_{i=1}^{N} p_i^*(q_i + e_i) \tag{20.2}$$

$$c_i \leq q_i - s_i + b_i + e_i, \quad i = 1, \ldots, N \tag{20.3}$$

$$f(q_i; z_q) \geq 0. \tag{20.4}$$

where z_u and z_q, respectively, are exogenously determined consumption and production shifters and f is the production technology.

Transaction costs, as discussed above, can influence the household decision on whether or not to participate in markets for goods or services. Three possibilities can be identified as below recognizing that agrarian households face different prices when they are sellers or buyers, which can be represented by

$$p_i^* = p_i - \tau_i(Z, A, G, Y) \quad \text{if } s_i > b_i (\text{net seller}) \tag{20.5}$$

$$p_i^* = p_i + \tau_i(Z, A, G, Y) \quad \text{if } s_i < b_i (\text{net buyer}) \tag{20.6}$$

$$p_i^* = p_i^a \quad \text{if } s_i = b_i = 0 \text{ (self sufficient)} \tag{20.7}$$

where p_i are market prices; τ_i are commodity-specific transaction costs, determined by: (i) household characteristics (e.g., distance to markets, number of family members, age and education of household head, social connections) Z; (ii) assets owned by the household (land, draft animals) A; (iii) infrastructure provided by the government (e.g., irrigation and extension services), G; and (iv) liquidity position of the household Y.

As shown by de Janvry et al. (1991), market participation in the presence of transaction costs gives rise to a kinked price schedule within the price band defined between the market price plus or minus the transaction cost margin. The optimization problem defined above would therefore require evaluating the utility function for each choice variable under the three possible market prices, namely at autarky, buyer, and seller prices. Guided by net returns to market participation (Omamo, 2007; Boughton et al., 2007), households are modeled using the two-step decision process wherein they first decide whether or not to participate in the market as a seller, and second, how much to sell. The structural model above suggests that a farmer faces a unique price for each crop depending on a unique set of transaction costs, conditional on idiosyncratic factors unique to the household. Often these transaction costs are unobservable but can be inferred from household and other conditions identified above. Based on the structural model (followed by Goetz, 1992; Key et al., 2000), the first order conditions of the maximization of the utility function will yield the reduced form of the output marketed supply, conditional on market participation.

Two questions are of particular interest to this exercise: (i) how market participation in crop sales varies with the distance to markets and (ii) whether one can reasonably expect transaction costs to have an influence on the choice of farm diversification. If the answer to the first issue is affirmative, then transaction costs can be considered as a critical variable that determines market participation, conditional on household and other production characteristics. If distance to markets affects farm decisions on whether to participate in markets and how much to sell, then policy options with respect to promoting market participation among smallholders will need to include policy tools that directly address those transaction costs, including enhanced public good provisions, such as extension services. This question is analyzed using a data set from Tanzania in the next section.

20.4 DATA AND DESCRIPTION OF PRODUCTION AND MARKETING ENVIRONMENT

This chapter used data from the 2010/11 Tanzania National Panel Survey, which is a nationally representative household survey, including a separate module on agriculture and fisheries, administered to 3265 households in 409 areas (National Bureau of Statistics, 2011). The data set provides information on agricultural production at the plot level and inputs usage, agricultural sales, and trading costs. Using this rich set of data, this chapter reviews the production and trading environment of agricultural households both at plot and household levels. The objective is to identify broad patterns to inform the nature of production specialization and market participation and test if production specialization is linked to their market entry and participation.

Out of 8206 farm plots, 83% were cultivated, 12% left to fallow, and 0.5% rented out. Out of those cultivated plots, 65% were intercropped. At the outset, it is important to note that 84% of the households indicated that the motivation for intercropping was to fully exploit soil fertility, and only 4% indicated that it was to ensure generating income from at least one crop if others were to fail. Thus, risk is not a motive for intercropping, removing a major reason that had been a subject of debate for intercropping. The fact that only 5% of the farmers had said that their crop failed in that season is further proof to conclude that risk is not a reason for intercropping.

Maize was planted in 41% of the plots, cassava 15%, paddy 12%, banana 6%, sorghum 4%, beans 3%, and all other crops numbering close to 50 cover the rest. On market participation, 39% of the households have sold agricultural commodities. Maize is sold by 28% of the households, followed by paddy (13%), beans (11%), and groundnuts (8%). To generate policy relevant conclusions, crop sales data were aggregated at main commodity groups and clusters of farm-to-market distance. Figure 20.1 shows the incidence of sales by key commodity group and clusters of distance. It reveals some interesting features of market participation by agrarian households. First, it is clear that agricultural households across all different distances participate in commodity markets. As expected, a larger proportion of sellers can be found near markets, and the incidence of sales, at least of those who sell larger volumes, declines with distance. Second, only a limited number of agricultural households participate in the market for vegetables and legumes, and they are mostly

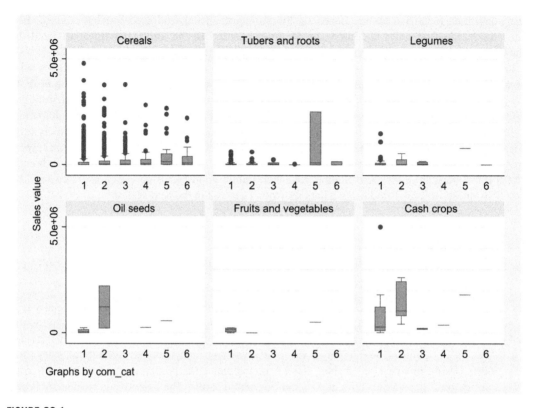

FIGURE 20.1

Sales classified by commodity group and distance to market.

located closer to markets. Those who are located away from market centers are found to be relatively large producers, who own relatively large land parcels.

To further investigate the potential association between far-to-market distance and commodity structure of sales, cumulative density functions (CDFs) of sales, classified by clusters of distance, were calculated and drawn (Figure 20.2). This shows that the sellers with a distance of less than 10 km from markets constitute the largest group, but a general stochastic dominance over distance categories does not exist. If distance can consistently explain market participation, we would see nonoverlapping CDFs. As observed, CDFs of several clusters overlap and there is no particular pattern for their locations. This might mean that distance does matter, but there are other factors that may be dominant in certain environments. It may be that farmers located closer to markets have some advantage over those located away from the market, but those farmers having their farms away from markets may enjoy other benefits that outweigh transaction cost disadvantages.

The household cultivation choices may explain part of this apparent paradox. At a superficial level, one could observe that certain products are primarily cultivated for the market. In Tanzania, it was observed that almost 100% of the farmers who cultivated some crops sold their products.

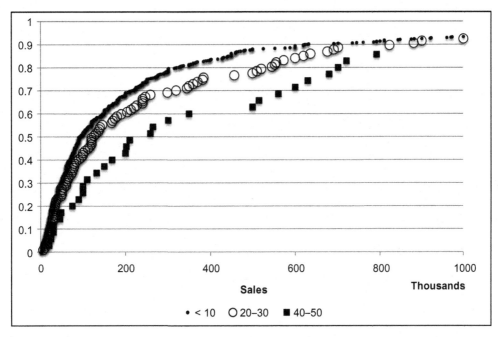

FIGURE 20.2

CDF of sales by distance categories. Note: To avoid clutter, distance categories 10–20, 30-40 and over 50 have been dropped. Except for the over 50 category, others fall somewhat between the respective categories.

These crops included cotton, chrysanthemum, carrot, soya beans, cucumber, spinach, eggplants, cauliflower, tobacco, chickpeas, onions, tomatoes, cabbage, and sesame. To investigate this further, CDFs of sales, classified by key commodity groups identified above, were drawn (Figure 20.3). The most striking feature of this graph is the clear stochastic dominance of commodity groups, with the exception of vegetables and oil seeds. The second key message of the graph is the stochastic dominance of cash crops over all other categories. At the other end of the spectrum is the roots and tubers category. What it means is that those who cultivate some categories, such as roots and tubers, generate less income than those who sell cash crops.[1]

20.5 ECONOMETRIC RESULTS

The analysis below is restricted to maize production and sales. This allows us to avoid the complex issue of commodity aggregation bias[2] and focus on the issue of interest of this chapter: to statistically test if

---

[1]Consider 80% of households selling roots and tubers. Note that it corresponds to Shilling 200,000 on the horizontal axis. Reading on the percentage of households with respect to cash crops, one could observe that only 30% of them fall below Shilling 200,000.

[2]A discussion on the issue can be found in Azam et al. (2012).

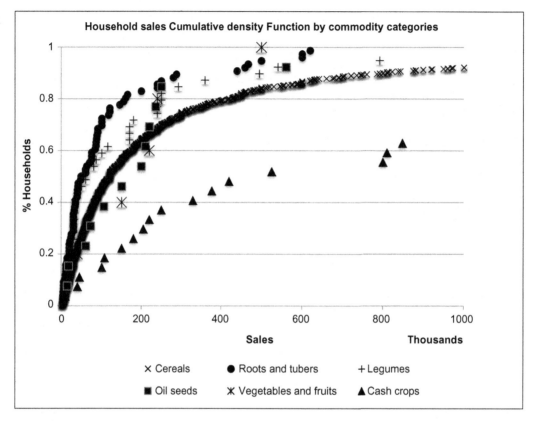

FIGURE 20.3

CDF of sales classified by commodity groups.

transaction costs influence farmers' decisions on their production arrangements, market entry, and participation. This however does not preclude us from using the rich data set available on Tanzania given that maize is cultivated and traded by farmers living in different environments with so vastly different endowments.

Before turning into econometric results, it may be useful to describe the maize production and trading environment in Tanzania. To do so, data were aggregated at two levels: pure stands and mixed stands. An additional group combining all maize producers was also constructed. Table 20.1 provides a summary of statistics. On land endowment, a typical household owns 5.2 acres of land, and a majority of households in both pure and mixed stands owns smaller prices of land (Figures 20.4 and 20.5). Farms are located 2.2 km away from main roads and 12 km away from main markets. The weighted unit transport cost is 4.33 Shillings per kg, and transport costs are observed to escalate with rising farm-to-market distances. A typical head of household is 48 years of age, has approximately four years of schooling, and in 69% of cases s/he can read. Of these households, 75% are headed by a male and have on average six family members. Rain-fed agriculture dominates the agricultural landscape, and access to irrigation is limited to a mere 4% of

Table 20.1 Descriptive Statistics of Maize Farm Households in Tanzania, 2010/11

| | Maize | | | | All Farmers | |
| | Pure-Stand | | Mixed Stand | | Mean | STD |
	Mean	STD	Mean	STD	Mean	STD
Maize sales (TS)	143,771	19,694	252,462	25,646	188,353	15,912
1 = HH head is male	0.713	0.022	0.790	0.017	0.750	0.120
HH head age (years)	47.400	0.750	49.500	0.650	48.500	0.440
1 = HH head can read	0.693	0.022	0.718	0.020	0.690	0.130
Years of schooling of HH head	4.668	0.164	4.675	0.140	4.600	0.100
No of family members	5.278	0.122	5.667	0.127	5.600	0.070
Land owned (acres)	3.453	0.231	7.595	0.517	5.160	0.260
1 = soil is considered good	0.583	0.252	0.601	0.022	0.560	0.130
1 = HH used irrigation	0.043	0.012	0.046	0.011	0.040	0.007
Value of agricultural equipment (TS)	487,549	115,276	963,131	178,740	1,004,057	203,401
1 = HH used hired labor	0.376	0.025	0.377	0.024	0.380	0.014
1 = HH used organic fertilizer	0.219	0.024	0.248	0.026	0.208	0.016
1 = HH used chemical fertilizer	0.186	0.024	0.266	0.028	0.180	0.016
1 = HH received extension services	0.097	0.014	0.112	0.014	0.090	0.008
1 = HH received seed voucher	0.133	0.016	0.111	0.015	0.117	0.009
1 = HH used improved seed	0.778	0.022	0.725	0.017	0.728	0.011
1 = HH received price information	0.363	0.025	0.477	0.024	0.445	0.014
No of mobile phones	0.753	0.052	0.709	0.042	0.757	0.027
Distance to road (km)	2.100	0.248	2.065	0.141	2.210	0.130
Distance to market (km)	13.910	1.270	13.134	0.946	12.800	0.778
1 = HH transported maize to market	0.129	0.029	0.253	0.062	0.190	0.033
Distance transported for selling maize (kg)	3.674	1.123	3.379	0.477	3.433	0.460
No of times maize transported	0.438	3.100	0.955	3.300	0.605	0.073
Cost of transportation (TS)	1741	691	2239	626	2281	512
Quantity of maize stored (kg)	82.68	13.00	135.90	19.57	108.00	12.00
1 = HH used pesticides	0.128	0.019	0.164	0.020	0.127	0.012
1 = HH harvested less area than cultivated	0.437	0.025	0.506	0.019	0.433	0.013
1 = HH experienced crop damages	0.392	0.026	0.506	0.021	0.437	0.015
1 = HH received non-farm income	0.578	0.025	0.674	0.021	0.650	0.013
1 = HH has wage employment	0.111	0.016	0.128	0.015	0.130	0.008
1 = HH has secondary employment	0.019	0.006	0.039	0.008	0.036	0.005
1 = HH has self employment	0.071	0.011	0.064	0.011	0.064	0.006
1 = HH has livestock	0.708	0.022	0.816	0.019	0.740	0.013
1 = HH sold livestock	0.342	0.021	0.444	0.023	0.369	0.013
1 = HH has livestock income	0.119	0.015	0.147	0.016	0.120	0.009
No of observations	548		700		2154	
Estimated population	1.6 million		2.0 million		5.5 million	

Notes: HH refers household, TS refers to Tanzanian Shillings.
Source: Author's calculations based on Tanzania National Household Panel Survey 2010/11.

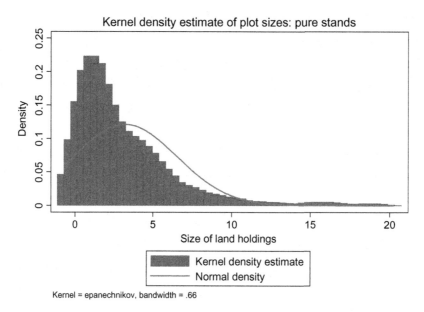

FIGURE 20.4

Distribution of plot sizes operating pure stands.

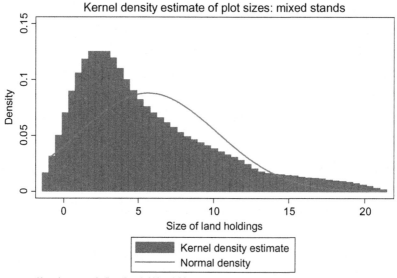

FIGURE 20.5

Distribution of plot sizes operating mixed stands.

households. Fertilizer is used by 18%, pest control practices by 13%, and improved crop varieties by 73%. The use of mobile phones is significantly high, which may explain the fact that 44% of farmers had received price information before sales were made.

Seven broad categories of variables were used in the model: household characteristics (age of household head and education) which may explain part of idiosyncratic nature factors related to transaction costs; assets (land and agricultural equipment); access to nontradable services (access to labor and the equipment rental market); publicly provided services (e.g., irrigation and extension services); transaction costs (e.g., distance to road and markets); risks (e.g., crop damages); and exclusion variables (e.g., availability of nonfarm income and livestock).

The approach to measuring transaction costs may need elaboration. Transaction costs are further classified into fixed and proportional components (Key et al., 2000). Fixed transaction costs may include the costs of searching buyers and markets, negotiation and bargaining, screening, enforcement, and supervision. Variable transaction costs include the cost of transporting commodities, time spent in delivering products to markets, and others involving accessing marketing, including bribing. Proportional transaction costs thus lower the price received for output and raise the price paid for inputs, effectively creating a price band within which farmers may find it unprofitable to operate. Although useful in explaining household behavior in rural settings, measuring them is extremely difficult, if not impossible. In particular, it is possible for households facing high transaction costs not to participate in markets and therefore we may not observe them at all. In general, the approach many researchers adopt is to use observable exogenous variables that are likely to explain transaction costs. In this chapter, we use distance from plot to markets, ownership of transportation equipment, ownership of equipment that is likely to contribute to getting better information and market intelligence, and household characteristics, such as age and education level, gender of the head of household, and the number of family members.

An additional area that may need elaboration is the approach to estimating the model. A method that recent empirical work has extensively used is to model market participation decisions using the two-step estimation procedure (Goetz, 1992; Key et al., 2000).[3] This gives us the estimate for the

---

[3]Following Goetz (1992) and Key et al. (2000), households first make their decisions on market participation, while the supply decision, conditional on market participation, depends on transaction costs and production characteristics. The econometric model of interest is posited as follows: $Q_i^s = \beta X_i + \varepsilon_i$, where the dependent variable is net sales of the household, X are observed variables, and ε is the standard error term that measures all unobserved determinants of market participation. As only some households participate in markets, we do not observe Q for all the households, leading to the "selection bias." To avoid the selection bias, as many other papers on this subject, this chapter uses a two-step procedure to estimate the model. In the first step a Probit model of market participation, defined below, is estimated: $Q_i^{s*} = Z_i \gamma + \mu_i$, where Q_i^{s*} is the choice of households on market participation taken on the basis of their evaluation of the underlying costs and benefits of market participation, and Z are all the independent variables. This Probit model yields an estimate of γ allowing us to construct consistent estimates of the inverse Mills ratio term, $\hat{\lambda}_i(-Z_i\hat{\gamma}) = (\phi(Z_i\hat{\gamma}))/(\Phi(Z_i\hat{\gamma}))$, where $\phi(\bullet)$ and $\Phi(\bullet)$ are the standard normal probability distribution function and standard normal cumulative probability function, respectively. In the second stage, the outcome equation is estimated by ordinary least squares, for the uncensored observations, where the outcome equation includes both the original X whose coefficients are the parameters of the overall market participation equation and the constructed value of the inverse Mills ratio, which is $Q_i^s = \beta X_i + \theta \hat{\lambda}_i(-Z_i\hat{\gamma}) + \varepsilon_i$. This procedure is expected to generate unbiased and consistent results. The standard t-test of the null that $\theta = 0$ is a test of the null that there is no selection bias, conditional on the assumptions of model. Values however should be interpreted carefully given the known problems, particularly the correlation problem which arises when the variables used in estimating predictors are correlated with independent variables. The model was implemented on STATA version 11.

Heckman estimation procedure and thus becomes the most obvious choice for estimating the model.[4] This however requires additional information, especially a set of variables known as exclusion variables. This need arises from the fact that only some farmers participate in markets as sellers, and thus, regression results using only those who decided to sell can lead to a selection bias purely because of the particular sample framework. Thus, regression results may yield inconsistent estimates. Following exclusion variables used in similar models (e.g., Boughton et al., 2007 and Azam et al., 2012), this chapter uses access to non-crop income, wage earnings, and livestock revenue as exclusion variables.

The chapter attempts to answer if there is a significant difference of maize market entry, and of sales once entered the market, between those who operate pure stands and those who operate mixed stands. We expect households with pure stands to be in lower transaction cost environments and have relatively large endowment compared to those operating mixed stands. Transaction costs are, therefore, expected to be of less significance to farmers holding pure stands. Farmers operating mixed stands on the other hand may have chosen to intercrop and forgo potential benefits from specialization primarily because of the underlying transaction cost environment and their lower endowments. We could thus expect transaction costs to have a relatively bigger impact on their decisions on market entry and how much to sell.

The first important observation of the econometric results is that the Mills ratio is significant at 99% probability for the selection model of pure stand sellers (Table 20.2), thus the use of Heckman procedure is justified. Second, it could be noted that transaction related variables (distance to road, distance to market, distance transported maize and no of time maize transported and cost of transporting maize) are not significant at any probability in both pure- and mixed-stand farmer groups (Table 20.2). This suggests that transaction costs have not affected farmers' market entry decisions. However, several other factors have influenced market entry decisions (selection model) of both farmer groups. Whether these variables have influenced market entry positively or negatively have also been consistent in both cases. Age of the household head, fertilizer usage and seed vouchers are significant at the 0.001 probability of rejection in both farmer groups. Three additional variables (access to irrigation, availability of a labor market and mobile phones) for market entry of pure-stand farmers, and one additional variable (land area) for mixed stand farmers, are significant at the same level of probability. Out of the several selection variables used in the model, only livestock ownership is significant in explaining market entry of pure-stand farmers at the 0.001 probability of rejection. Livestock sales and livestock income are similarly significant in explaining market entry in case of pure-stand farmers, but respectively at 0.05 and 0.01 probabilities of rejection.

What would explain farmers' sales decisions once they have entered the market? The OLS regression models explain the volume of maize sales as a function of some exogenous variables. The model on mixed stand farmers is considered first. Some variables that measure market transaction costs in the model viz. distance to markets and cost of transporting commodities are significant in explaining market sales, but they do not have the theoretically expected negative sign. The two variables that represent farmers' access to assets (extent of land owned and agricultural equipment) are significant with the probability of rejection only at 0.001. Similarly, the variables that represent

---

[4]An exception is the variable measuring access to mobile phones, which negatively affects market entry decisions of farmers operating pure stands.

Table 20.2 Regression Results of Maize Selling in Tanzania, 2010/11

Variables	Pure-stand						Mixed-Stand						Overall					
	Selection			OLS			Selection			OLS			Selection			OLS		
	1 = HH sold maize			Maize sales (value)			1 = HH sold maize			Maize sales (value)			1 = HH sold maize			Maize sales (value)		
	Coef.	sig.	z	Coef.	sig.	z	Coef.	sig.	z	Coef.	sig.	z	Coef.	sig.	z	Coef.	sig.	z
1 = HH head is male	−0.133	c	−0.9	0.203		1.21	0.315	b	2.1	−0.179	b	−1.1	0.077		0.8	0.039		0.3
Age of HH head (years)	−0.976	c	−4.4	−0.150		−0.58	−0.685	c	−3.2	−0.470	b	−2.4	−0.713	c	−5.2	−0.348	b	−2.3
1 = HH head can read	−0.007		0.0	−0.245		−0.88	0.023		0.1	0.086		0.5	0.031		0.2	−0.036		−0.2
Years of schooling of HH head	0.004		0.1	0.049		1.28	−0.021		−0.7	−0.006		−0.2	−0.007		−0.4	0.018		0.8
Family members (number)	−0.122		−0.9	0.255	b	1.7	−0.172		−1.5	0.106		1.0	−0.196	b	−2.5	0.130		1.5
Land area (acres)	0.121	b	1.9	0.328	c	4.88	0.201	c	3.0	0.405	c	6.5	0.150	c	3.9	0.309	c	7.4
1 = soil quality good	0.236	a	1.7	0.340	b	2.43	0.223	a	1.8	0.052		0.5	0.184	b	2.2	0.186	b	2.1
1 = HH has access to irrigation	1.340	c	3.2	0.746	b	2.4	0.550		1.5	0.785	c	3.2	0.589	c	2.5	0.743	c	3.9
Agricultural equipment (TS)	0.087	b	2.2	0.086	b	2.01	−0.062		−1.6	0.165	c	5.2	0.033		1.4	0.140	c	5.9
1 = HH use hired labor	0.361	c	2.4	0.328	b	2.15	0.167		1.3	0.234	b	2.0	0.270	c	3.0	0.331	c	3.6
1 = HH use organic fertilizer	−0.154		−0.9	−0.098		−0.58	0.135		0.8	−0.292	b	−2.0	−0.055		−0.5	−0.208	a	−1.9
1 = HH use fertilizer	0.619	c	3.4	0.537	c	3.06	0.558	c	3.4	0.381	c	2.8	0.525	c	4.8	0.440	c	4.2
1 = HH use pesticides	0.145		0.7	0.645	c	3.43	−0.091		−0.5	0.375	b	2.4	0.016		0.1	0.407	c	3.4
1 = HH use extension service	−0.032		−0.1	−0.518	b	−2.32	−0.337	a	−1.7	−0.040		−0.2	−0.139		−1.0	−0.171		−1.2
1 = HH received seed voucher	−0.559	c	−2.7	−0.247		−1.07	−0.646	c	−3.2	−0.183		−0.9	−0.535	c	−4.2	−0.175	c	−1.2
1 = HH used improved seed	−0.034		−0.2	0.063		0.4	0.090		0.6	−0.087		−0.7	−0.031		−0.3	−0.020		−0.2
1 = HH received price information	0.230		1.6	0.121		0.84	0.093		0.8	0.269	b	2.4	0.166	a	1.9	0.184	b	2.1
1 = HH use mobile phone	−0.431	c	−2.8	0.159		1.04	−0.107		−0.8	0.089		0.7	−0.214	b	−2.3	0.146		1.6
Distance to road (km)	0.026		0.4	−0.001		−0.01	−0.009		−0.2	−0.008		−0.1	−0.003	a	−0.1	0.019		0.5
Distance to market (km)	0.081		1.3	0.120	a	1.83	0.058		1.0	0.100	b	2.0	0.065	a	1.7	0.138	c	3.5

Table 20.2 Regression Results of Maize Selling in Tanzania, 2010/11 Continued

Variables	Pure-stand Selection 1 = HH sold maize			Pure-stand OLS Maize sales (value)			Mixed-Stand Selection 1 = HH sold maize			Mixed-Stand OLS Maize sales (value)			Overall Selection 1 = HH sold maize			Overall OLS Maize sales (value)		
	Coef.	sig.	z	Coef.	sig.	z	Coef.	sig.	z	Coef.	sig.	z	Coef.	sig.	z	Coef.	sig.	z
Distance transported maize (km)				−0.133		−1.39				−0.090		−1.2				−0.091		−1.6
No. of times maize transported				0.287		1.66				0.339	c	3.2				0.350	c	4.0
Cost of transporting maize (TS)				0.128	c	4.18				0.097	c	4.6				0.110	c	6.4
Maize storage (kg)	0.062	b	2.0	−0.014		−0.45	0.032		1.3	0.008		0.4	0.052	c	3.0	0.003		0.2
1 = HH had plot damages	−0.040		−0.3	−0.360	b	−2.56	−0.180		−1.5	−0.120		−1.1	−0.045		−0.5	−0.195	b	−2.3
1 = Hh had crop damages	0.065		0.5	0.064		0.45	0.087		0.7	−0.052		−0.5	0.103		1.2	0.005		0.1
1 = HH has non-farm income	−0.065		−0.5				0.064		0.5				0.008		0.1			
1 = HH has wage employment	−0.017		−0.1				0.253		1.3				0.173		1.3			
1 = HH has second employment	0.200		0.4				0.707		1.5				0.341		1.1			
1 = HH is self employed	0.211		0.9				−0.108		−0.4				0.028		0.2			
1 = HH has livestock	0.559	c	3.3				−0.125		−0.8				0.190	a	1.8			
1 = HH sold livestock	0.111		0.7				0.254	a	1.8				0.158		1.6			
1 = HH received livestock income	−0.465		−2.1				0.372	b	1.9				0.007		0.1			
Mills Ratio	−0.166	b	−0.3	0.398	b	2.15				0.2826		1.14				0.398	b	2.15
Constant	0.275	c	0.8	8.845	c	14.8	1.280	b	2.2	9.115	c	19.6	0.275	c	0.8	8.888	c	24.5
Wald chi2(49)	231.41	c					352.21	c					505.91					
Number of obs	542						699						1351					
Censored obs	251						219						526					

Notes
a = p < 0.05.
b = p < 0.01.
c = p < 0.001; HH refers to household and TS refers to Tanzanian Shillings.
Source: Author's estimates.

the agricultural inputs environment (access to irrigation, fertilizer use, and receipt of seed vouchers) are significant in maize market sales. These results in combination suggest that farmers with better endowment and capacity enter the maize market regardless of the transaction cost environment, but access to assets and agricultural inputs greatly determine the volume of maize sold. The significance of livestock and maize storage in market entry decisions along with the factors that explain the volume of maize sales implies that farm households enter into markets once they reach a certain income and food security threshold, perhaps with the objective of generating additional income for purchasing other essential commodities such as fuel, clothing etc.

In case of pure-stand farmers too, variables that represent transaction costs (distance to market and cost of transportation) are significant but not of the expected negative sign. Several variables that represent household endowment viz. land area, the number of family members good soil quality, access to irrigation and agricultural equipment are significant in the volume of maize sales with the probability of rejecting the null hypothesis at 0.001 in case land area and at 0.01 in all others. Variables related to agricultural inputs (access to labor market and irrigation, and fertilizer and pesticide use) are significant with fertilizer and pesticides use having a higher probability of acceptance. Crop damages have negatively affected sales volume. The overall results indicate that pure stand farmers practice intensive agriculture with intensive usage of hired labor, fertilizer, pesticides, irrigated water and extension services. They all indicate signs of enhanced commercialization.

Broadly, these results are consistent with the production environment described in the previous section where we found that distance is not as significant a variable for market entry and sales in Tanzania. It may be that farms located away from markets enjoy other pecuniary benefits, such as access to relatively less costly labor, large land parcels, and local inputs.

20.6 CONCLUSION

The findings of the chapter are in broad agreement with the theory that transaction costs affect the probability of market entry as well as how much to sell once in the market. Given the high-transaction cost environments, farmers first make their cultivation decisions, followed by market entry, and how much to sell once entered into the market. Results suggest that the choice of farm specialization over mixed cultivation is a result of their relatively large capacity to effectively use input and output markets. The very fact that a farmer has opted to specialize in cultivating certain crops in pure stands might be a result of a favorable transaction cost environment, in particular relatively easy access to markets. Therefore, transaction costs may not be as binding a constraint on their market entry. This may explain why transaction cost-related variables are not significant for farmers with pure stands, yet variables representing endowment and agricultural services, such as access to land and irrigation, availability of a labor market and agricultural equipment, and the use of fertilizer, are significant for market entry.

Significant behavioral differences are observed between farmers with pure and mixed stands on their sales volumes once in the market. Farmers with pure stands appear to raise productivity by enhancing the use of agricultural inputs such as fertilizer, pesticides, irrigated water, and agricultural equipment. These variables are significant for farmers with mixed stands as well, but in addition, information and distance continue to play roles in their decisions on sales. Results of this chapter provide evidence to suggest that distance is an important variable for market participation, but not a factor of overwhelming significance. Access to assets, mainly land and soil quality with access to other agricultural input, is equally or more important in some instances.

In broad terms, the results indicate that policies for promoting market participation are likely to yield results only so far as they are able to address the underlying transaction cost environments. Policy interventions targeting farmers holding pure stands need to focus more on enhancing agricultural productive capacity, including through providing better access to input markets and agricultural mechanization, whereas farmers with a larger share of mixed stands are likely to be induced to enter agricultural markets by policies that address both transaction costs and agricultural productivity.

REFERENCES

Azam, S., Imai, K.S., Guiha, R., 2012. Agricultural supply response and smallholders market participation: the case of Cambodia. Discussion paper, Research Institute for Economics and Business Administration, Kobe University, Japan.

Barrett, C.B., 1997. How credible are estimates of peasant allocative scale, or scope efficiency? A commentary. J. Int. Dev. 9 (2), 221−229.

Benjamin, D., 1995. Can observed land quality explain the inverse productivity relationship? J. Dev. Econ. 46 (1), 51−84.

Borland, J., Yang, X., 1992. Specialization and a new approach to economic organization and growth. Papers and Proceedings of American Economic Association. Am. Econ. Rev. 82 (2), 386−391.

Boughton, D., Mather, D., Barrett, C.B., Benfica, R., Abdula, D., Tschirley, D., et al., 2007. Market participation by rural households in a low-income country: an asset-based approach applied to Mozambique. Faith Econ. 50 (Fall), 64−101.

Chenery, H.B., Sherman, R., Moises, S., 1986. Industrialization and Growth: A Comparative Study. Oxford University Press, New York, NY.

de Janvry, A., Fafchamps, M., Sadoulet, E., 1991. Peasant household behaviour with missing markets: some paradoxes explained. Econ. J. 101, 1400−1417.

Eswaran, M., Kotwal, A., 1985. A theory of two-tier labor markets in agrarian economies. Am. Econ. Rev. 75 (1), 162−177.

Evenson, R.E., Roumasset, J., 1986. Markets, institutions and family size in rural Philippine households. J. Philipp. Dev. 23, 141−162.

Falkenmark, M., Finlayason, C.M., Gorgen, L.J., 2007. Agriculture, water, and ecosystems: avoiding the costs of going too far. In: Molden, D. (Ed.), Water for Food Water for Life: A Comprehensive Assessment of Water Management in Agriculture. International Water Management Institute. Earthscan, Sterling, VA, pp. 233−277.

Goetz, S.J., 1992. A selectivity model of household food marketing behaviour in Sub-Saharan Africa. Am. J. Agric. Econ. 74 (2), 444−452.

Heltberg, R., Tarp, F., 2001. Agricultural supply response and poverty in Mozambique. Discussion Paper # 2001/114, World Institute for Development Economics Research (WIDER). United Nations University, Japan.

Henning, C.H.C.A., Henningsen, A., 2007. Modelling farm households' price responses in the presence of transaction costs and heterogeneity in labor markets. Am. J. Agric. Econ. 89 (3), 665−681.

Hualiang, L., Trienekens, J.H., Omta, S.W.F., Shuyi, F., 2008. The value of guanxi for small vegetable farmers in China. Br. Food J. 110 (4−5), 412−419.

International Fund for Agricultural Development, 2011. Rural Poverty Report 2011: New Realities, New Challenges: New Opportunities for Tomorrow's Generation. IFAD, Rome.

Key, N., Sadoulet, E., de Janvry, A., 2000. Transaction costs and agricultural household supply response. Am. J. Agric. Econ. 82, 245−259.

Kikuchi, M., Hayami, Y., 1999. Technology, Market, and Community in Contract Choice: Rice Harvesting in the Philippines. Econ. Dev. Cult. Change 47 (2), 371−386.

Kurosaki, T., 2003. Specialisation and diversification in agricultural transformation: The case of West Punjab 1903−92. Am. J. Agric. Econ. 85 (2), 372−386.

Kuznets, S., 1973. Modern economic growth: findings and reflections. Am. Econ. Rev. 63 (3), 247−258.

Lewis, A.W., 1954. Economic development with unlimited supplies of labor. Manchester Sch. 22 (2), 139−191.

Marshal, A., 1920. Principles of Economics. Library of Economics and Liberty. Available from: < http://www.econlib.org/library/Marshall/marP.html > .

Mazumdar, D., 1987. Rural-urban migration in developing countries. In: first ed. Mills, E.S. (Ed.), Handbook of Regional and Urban Economics, vol. 2 (Chapter 28). Elsevier, North-Holland, pp. 1097−1128.

National Bureau of Statistics, 2011. Basic Information Document: National Panel Survey (NPS 2010-11), United Republic of Tanzania.

Njehia, B.K., 1994. The impact of market access on agriculture productivity: A case study of Nakuru district, Kenya. Ph.D. Dissertation, Hohenheim University, Germany.

Omamo, S.W., 2007. Farm-to-market transaction costs and specialization in small-scale agriculture: explorations with a non-separable household model. J. Dev. Stud. 35 (2), 152−163.

Romer, P.M., 1994. The origins of endogenous growth. J. Econ. Perspect. 8 (1), 3−22.

Roumasset, J., 1981. Population, technological change and the evolution of labor markets. Popul. Dev. Rev. 7, 401−419.

Roumasset, J., 1995. The nature of the agricultural firm. J. Econ. Behav. Organ. 26, 171−177.

Roumasset, J., Lee, S., 2007. Labor: Decisions, Contracts and Organization. In: Evenson, R., Pingali, P. (Eds.), Handbook of Agricultural Economics, Agricultural Development: Farmers, Farm Production and Farm Markets, Volume 3. Elsevier, Amsterdam.

Roumasset, J. Setboonsarng, S., Wickramasinghe, U., Estudillo, J., Evenson, R., 1995. Specialization and coevolution of agricultural markets. Centre for Institutional Reform and the Informal Sector, University of Maryland at College Park, Working Paper No. 165.

Roumasset, J., Smith, J., 1981. Population, technological change and the evolution of labor markets. Popul. Dev. Rev. 7, 401−419.

Roumasset, J., Uy, M., 1980. Piece rates, time rates and teams: explaining patterns in the employment relation. J. Econ. Behav. Organ. 1 (4), 343−360.

Skoufias, E., 1994. Using shadow wages to estimate labor supply of agricultural households. Am. J. Agric. Econ. 76, 215−227.

Smith, A., 1776. An inquiry into the nature and causes of the wealth of nations. Available from: < http://www.econlib.org/library/Smith/smWN.html > .

Stigler, G.J., 1951. The division of labor is limited by the extent of the market. J. Polit. Econ. 59 (3), 185−193.

Timmer, P., 2005. Agriculture and pro-poor growth: An Asian perspective. Centre for Global Development, Working paper No. 63, July 2005.

United Nations, 2012, Report of the United Nations Conference on sustainable development—Rio de Janeiro, Brazil, 20−22 June, New York (A/CONF.216/16).

Yang, X., 2003. Economic Development and the Division of Labor. Blackwell, New York, NY.

Yang, J., Huang, Z., Zhang, X., Reardon, T., 2013. The rapid rise of cross-regional agricultural mechanization services in China. Am. J. Agric. Econ. 95 (5), 1245−1251, October.

Young, A.A., 1928. Increasing returns and economic progress. Econ. J. 38 (152), 527−542.

Zhang, X., Hu, D., 2011. Farmer-buyer relationships in China: the effects of contracts, trust and market environment. China Agric. Econ. Rev. 3 (1), 42−53.

DEVELOPMENT, VULNERABILITY, AND POVERTY REDUCTION

DEVIANT BEHAVIOR: A CENTURY OF PHILIPPINE INDUSTRIALIZATION

21

Emmanuel S. de Dios* and **Jeffrey G. Williamson**[†]

University of the Philippines, Diliman, Quezon City, Philippines
[†]*Harvard University, Cambridge, MA and University of Wisconsin, Madison, WI*

CHAPTER OUTLINE

JEL: F1; N7; O2

21.1 **INTRODUCTION**

Recent research has now documented industrial output growth around the poor periphery since 1870, finding unconditional convergence on the leaders by countries including those in Asia (Bénétrix et al., 2012). Industrial growth accelerated over the century between the 1870s and the 1970s, especially during the interwar 1920–1938 and import-substitution-industrialization (ISI) 1950–1972 years when the precocious poor periphery leaders underwent a surge (intensive growth) and more poor countries joined the modern industrial growth club (extensive growth). Furthermore, by the 1920s and 1930s, the majority were even catching up with the three core industrial leaders—Germany, the United States, and the United Kingdom, a process that accelerated during 1950–1972. In short, there was unconditional industrial convergence long before the modern BRIC and even before the Asian Tigers.

The Philippines was very much part of that industrial catching-up. After decades of nineteenth-century deindustrialization in the face of American and European competition (Legarda, 1999; Williamson, 2011: Chapter 5), Philippine industrial growth quickened in the early twentieth century. Like every other emerging industrial nation, it was led by small-scale, labor-intensive manufacturing—without much inanimate power—that first specialized in commodity processing. Still, in the decade or so up to 1913, Philippine industrial output grew at 6.3% per annum, way above that achieved by the three leaders, thus catching up. Indeed, the Philippines was a regional leader, since it was the third Asian country to enter the 5% industrial growth club. The Asian leaders were Japan 1899, China 1900, the Philippines 1913, Taiwan 1914, Korea 1921, and India 1929.[1] The Philippines continued its industrial catch-up during the interwar years 1920–1938. This impressive industrial performance also contained during the ISI years 1950–1972, when Philippine industry grew at 7% per annum, 1.8% faster than the three leaders even though the latter were undergoing a postwar growth miracle.[2]

While the Philippines conformed to the industrial convergence pattern, it began to deviate sharply from the pack in the 1980s. Indeed, it left the industrial catching-up club in 1982 following the country's worst post-World War II economic and political crisis. While per capita incomes eventually recovered in the mid-1990s, the Philippines never reentered the industrial growth club. Instead, services have served as the platform of growth for more than a quarter century.

This premature transition from manufacturing to services is a significant puzzle. Several factors will be examined to help explain this deviant behavior after almost a century of impressive industrial growth. Was it some institutional weakness absent (or at least benign) in the pre-1982 past but made more powerful since? Was it the liberal policy package introduced at a time which penalized manufacturing when it was least able to defend itself? Was it the labor emigration surge starting in the 1980s that stripped the work force of industrial skills? Was it some massive and subsequent Dutch Disease created by huge emigrant remittances? Was it some other force making for an

---

[1]No other Asian nation joined the industrial growth club until after the ISI period.
[2]These figures are reported in Bénétrix et al. (2012), and they are based on the following sources: 1902–1951: gross value added in manufacturing in 1985 pesos (Hooley, 2005: Table A.1: 480–481); 1951–1960: industrial production (Mitchell, 2007: Table D1: 368); and 1960–2007: manufacturing in constant pesos (World Bank, 2011).

overvalued peso, thus reducing domestic manufacturing's competitiveness? We think the answers only emerge when the Philippine deviant performance is placed in a comparative context, and when the successful 1900−1972 experience is compared with the past four decades or so of failure.

21.2 THE PHILIPPINES AND THE COMPETITION: CATCHING UP SINCE 1870

World economic history since 1800 has largely been one of how the international economic system adjusted to the dramatic asymmetric shock that was the Industrial Revolution.[3] The transition to modern economic growth created a system that was lopsided in the extreme. The new energy-intensive manufacturing technologies originated in Britain and spread with a short lag to western continental Europe and North America. As a consequence, per capita income gaps between the West and the rest widened. The Industrial Revolution also gave rise to a "new world trade order" (Lewis, 1978). The new technologies gave Britain, Germany, the United States, and other industrial leaders a powerful comparative advantage in manufacturing relative to the economies of the periphery, like Asia. This comparative advantage was increasingly realized across the nineteenth century, as ocean freight rates declined, the Suez Canal opened, railroads linked port to interior, world peace prevailed, and trade-fostering gold standard regimes flourished. The result was an exchange of manufactures from what we will call the industrial core for commodities from what we will call the poor periphery. This exchange posed both challenges and opportunities for the periphery. It allowed the periphery to expand its commodity exports greatly and to enjoy steeply rising terms of trade. The same forces led to deindustrialization. If modern industry provides the route to modern growth as endogenous growth theory suggests (Krugman, 1981, 1991; Romer, 1986, 1990; Lucas, 1988; Helpman, 2004), then the static benefits of trade were potentially offset, or even outweighed, by the dynamic consequences of deindustrialization. Up to 1870, this potential became a fact (Williamson, 2011).

Although most poor countries became a little richer from their commodity export trade, the key question was how to join the faster-growing industrial club. Falling transport costs cut both ways. On the one hand, domestic industries were increasingly exposed to European competition. On the other hand, transport costs eventually fell to the point where having thick coal seams, large iron ore deposits, extensive oil fields, and land suitable for producing fibers was losing its importance: increasingly, poorly endowed industrial laggards (like the Philippines) could purchase these inputs on world markets at competitive prices, and well-endowed leaders lost some of their industrial competitive edge (Wright, 1990). In the years following 1870, poor industrial followers interacted with a world economic system that went through several radically different phases: the globalization of the late nineteenth century; its disintegration during the world wars and the interwar between; the reintegration of the Atlantic economy following World War II; decolonization and state-led import-substitution (ISI) in Asia; and the second wave of globalization which embraced more and more of the world from the 1970s onward. Through all these phases, there was overall twentieth-century industrial catch-up, and the Philippines was part of it.

---

[3]This section draws heavily on Bénétrix et al. (2012).

21.2.1 **THE INDUSTRIAL OUTPUT DATA**

We focus on six periods. The years before World War I are divided into two parts, before and after 1890. There is then the interwar period from 1920 to 1938; the postwar reconstruction years from 1950 to 1972; the period following the oil crises from 1973 to 1989; and the years of rapid globalization between 1990 and 2007. There are 175 countries in the 1990–2007 sample, but the farther back into the past we go, the fewer are the countries whose manufacturing growth we can document, and the smaller are the samples. Thus, our sample falls to 141 countries in 1973–1989; 93 in 1950–1972; 54 in the interwar period; 41 in 1890–1913 (the Philippines entering the sample in 1902); and 31 in 1870–1889.

These countries are divided into nine groups in what follows. First, there are the three traditional industrial leaders: the United Kingdom, Germany, and the United States. Next, there are other rich industrial countries in the European core: Belgium, France, Luxembourg, the Netherlands, and Switzerland. A third, intermediate group lying between the European core and periphery contains the three Scandinavian countries, while the fourth, the European periphery, includes all other European countries in the south and east. The settler economies of Australia, Canada, and New Zealand form a fifth group. The remaining four groups are the Middle East and North Africa (MENA), Asia, sub-Saharan Africa, and Latin America plus the Caribbean (hereafter simply Latin America). We often refer to these last four regions, together with the European periphery, as "the periphery," or as "followers," contrasting the experience of these five regions with what we call the "core" or the "leaders."

21.2.2 **REGIONAL GROWTH RATES: WHEN AND WHERE DID INDUSTRIAL GROWTH BEGIN?**

When did the poor periphery start recording rapid manufacturing output growth? When did it begin to experience higher growth than the rich industrial core, thus catching up? Table 21.1 reports average annual growth rates of industrial output in our nine regions and six time periods between 1870 and 2007. In each case, the regional growth rate is a simple unweighted average of individual country growth rates. Table 21.2 presents the growth rates in each region relative to the core, the latter a GDP-weighted average of the three leaders.

Since the country samples change over time, use of Tables 21.1 and 21.2 should be limited to growth rate comparisons between regions in any given period. Of course, we can only compute growth rates where output data are available, and one can surmise that where output data are missing for earlier periods, there was probably not much modern manufacturing to measure. For example, according to Table 21.1, there was an unweighted average manufacturing growth rate of 4.2% per annum in Asia between 1890 and 1913. This figure represents an average of Japan, China, British India, Indonesia, Korea, Burma, the Philippines, Taiwan, and Thailand. These nine account for a very large share of the late nineteenth-century Asian economy, but it might be reasonable to assume that the average Asian industrial growth rate was in fact a bit lower than 4.2% during this period, reflecting lower rates in those (much smaller) countries for which we do not have data (like Malaya, Indo-China, and Nepal).

Tables 21.1 and 21.2 tell us that growth among the leaders was fairly steady between 1870 and 1913, averaging 3–3.4% per annum, followed by a decline to 1.9% during the troubled interwar years. The rates for the core reached 5.2% per annum during the post-World War II growth miracle era (Panel A); if the United Kingdom is replaced by Japan, the core growth rates reached 7.9% per annum (Panel B). These were, of course, the years of the German *Wirtschaftswunder* and the

Table 21.1 Industrial Growth Rates

Panel A: Leaders Always United States, Germany, and United Kingdom

Groups	1870–1889	1890–1913	1920–1938	1950–1972	1973–1989	1990–2007
Leaders	3.0	3.4	1.9	5.2	1.0	2.1
European Core	2.5	2.8	2.9	4.0	1.4	2.0
Scandinavia	2.8	4.8	3.9	4.9	1.1	3.1
European Periphery	4.7	5.0	4.7	8.6	3.5	2.8
Newly Settled	4.9	4.6	2.3	5.2	2.0	2.3
Asia	1.5	4.2	4.2	8.1	5.5	4.2
Latin America	6.3	4.4	2.8	5.2	2.9	2.2
MENA	1.2	1.2	4.9	7.6	6.4	4.5
Sub-Saharan Africa			4.6	5.0	3.5	3.8
Countries	31	41	54	93	141	175

Panel B: Leaders are United States and Germany, plus United Kingdom before 1939, Japan after

Groups	1870–1889	1890–1913	1920–1938	1950–1972	1973–1989	1990–2007
Leaders	3.0	3.4	1.9	7.9	2.3	2.2
European Core	2.5	2.8	2.9	4.0	1.1	1.8
Scandinavia	2.8	4.8	3.9	4.9	1.1	3.1
European Periphery	4.7	5.0	4.7	8.6	3.5	2.8
Newly Settled	4.9	4.6	2.3	5.2	2.0	2.3
Asia	1.5	4.2	4.2	7.8	5.5	4.3
Latin America	6.3	4.4	2.8	5.2	2.9	2.2
MENA	1.2	1.2	4.9	7.6	6.4	4.5
Sub-Saharan Africa			4.6	5.0	3.5	3.8
Countries	31	41	54	93	141	175

Note: *The table reports unweighted average industrial growth rates by region. Individual country growth rates are computed as the β coefficient of the following regression:* $Y = \alpha + \beta t$, *where* Y *is the natural logarithm of industrial production and* t *is a linear time trend. Regressions are performed only where at least four observations are present.*
Source: *Bénétrix et al. (2012), Table 1.*

Japanese growth miracle. Thus, this postwar recovery set the bar very high for any poor region to surpass, although Asia did (Table 21.2, Panel B). While the Philippines could not match Japan's miraculous 12.4% per annum, its 7% per annum beat out the other leading three—5.2% per annum between 1950 and 1972 (Table 21.3)—by quite a margin. There was a leader slow down after 1973, partly due to the fact that war reconstruction forces were exhausted and partly due to the poor macroeconomic conditions following the oil crises, but secular deindustrialization forces in the rich core were probably playing the bigger role, as suggested by the continued slow leader growth between 1990 and 2007 (Table 21.1). One of the biggest—and premature—slowdowns was in the poor but previously catching up Philippines, its manufacturing growth falling from 7% per annum in the ISI period to 2.5% in the quarter century thereafter.

Table 21.2 Catching-Up: Industrial Growth Rates Relative to the Leaders

Panel A: Leaders Are Always United States, Germany, and United Kingdom

Groups	1870–1889	1890–1913	1920–1938	1950–1972	1973–1989	1990–2007
European Core	−0.4	−0.6	1.1	−1.0	0.0	−1.1
Scandinavia	−0.1	1.3	2.1	0.0	−0.2	0.0
European Periphery	1.8	1.5	3.0	3.6	2.1	−0.3
Newly Settled	2.0	1.1	0.6	0.2	0.7	−0.8
Asia	−1.4	0.8	2.5	3.1	4.1	1.1
Latin America	3.4	0.9	1.1	0.2	1.5	−0.9
MENA	−1.7	−2.3	3.1	2.7	5.0	1.3
Sub-Saharan Africa			2.8	0.0	2.1	0.7

Panel B: Leaders are United States and Germany, plus United Kingdom before 1939, Japan after

Groups	1870–1889	1890–1913	1920–1938	1950–1972	1973–1989	1990–2007
European Core	−0.4	−0.6	1.1	−2.4	−1.1	−1.0
Scandinavia	−0.1	1.3	2.1	−1.5	−1.1	0.3
European Periphery	1.8	1.5	3.0	2.1	1.2	0.0
Newly Settled	2.0	1.1	0.6	−1.3	−0.2	−0.5
Asia	−1.4	0.8	2.5	1.3	3.3	1.5
Latin America	3.4	0.9	1.1	−1.3	0.7	−0.6
MENA	−1.7	−2.3	3.1	1.2	4.1	1.6
Sub-Saharan Africa			2.8	−1.5	1.2	1.0

Note: *Average industrial growth rates by region relative to the leaders are computed in two steps. First, we compute the average growth rates for each region as in Table 21.1. Second, we subtract the GDP-weighted average of the three leaders' growth rates. Note that the leader averages in Table 21.1 are unweighted.*
Source: *Bénétrix et al. (2012), Table 2.*

21.2.3 WHEN DID RAPID INDUSTRIAL GROWTH BECOME WIDESPREAD?

The average regional growth rates presented above have their limitations. After all, we are interested not only in when modern industrial growth began in each region but also when it began to be widespread. Figure 21.1 addresses this issue. It exploits information on the first year each country joined the modern industrial growth club, where membership is defined as recording 5% per annum or more over the previous decade. The share of the countries in each region which had joined the "modern industrial growth club" is calculated for each year and then plotted in Figure 21.1. The shares are monotonically increasing, since we are not concerned with the industrially mature as they permanently exit from the club late in the postwar period: after all, every successful economy eventually starts to deindustrialize as it moves on to high-tech services.

Figure 21.1 shows the spread of rapid manufacturing growth across the poor periphery: first Scandinavia, then the European periphery, then Latin America, then Asia, then MENA, and finally

Table 21.3 Industrial Growth of Early Members in the Modern Growth Club									
Group	Country	In	1870–1889	1890–1913	1920–1938	1950–1972	1973–1989	1990–2007
European Periphery	Finland	1880	3.7	5.0	6.7	5.9	3.5	6.4
	Russia	1880	5.3	4.6	15.3	8.3	4.2	−0.5
	Austria	1883	4.9	3.3	2.3	5.8	2.5	2.8
	Hungary	1883	4.9	3.3	4.0	7.3	2.3	5.9
	Spain	1884	3.4	1.3	−0.5	8.8	1.2	2.9
Asia	Japan	1899	3.0	5.3	6.7	12.4	3.9	1.0
	China	1900		7.8	5.3	9.2	8.4	9.8
	Philippines	1913		6.3	3.4	7.0	1.7	3.3
	Taiwan	1914		5.1	4.4	11.5	9.0	4.9
	Korea	1921		8.0	7.1	13.2	11.8	7.4
Latin America and Caribbean	Chile	1881	7.5	3.9	2.6	5.2	2.0	3.5
	Brazil	1884	7.2	0.0	3.2	7.8	2.9	2.1
	Argentina	1886	6.4	8.8	4.2	4.9	−0.9	1.7
	Uruguay	1886	4.2	3.9	3.2	1.4	1.5	0.1
	Mexico	1902		6.0	3.7	7.1	3.1	3.2
MENA	Turkey	1931	1.2	1.2	8.1	7.6	5.0	4.1
	Morocco	1949				4.8	4.2	2.9
	Tunisia	1950				3.5	7.7	4.6
	Algeria	1959				9.7	7.9	0.1
	Egypt	1962			1.6	6.9	7.9	5.6
Sub-Saharan Africa	South Africa	1924			6.7	6.9	2.8	2.6
	Congo, Dem. Rep. of	1940			2.4	−4.2	−0.4	−3.9
	Zimbabwe	1951				−0.3	2.7	−3.7
	Kenya	1964				8.5	5.4	1.7
	Zambia	1966				8.3	2.1	2.8

Note: *"In" indicates the first year that a country experienced a 10-year backward looking average growth rate greater than 5%.*
Source: *Bénétrix et al. (2012), Table 4.*

sub-Saharan Africa. By 1913, 31% of the European periphery had joined the club, 18% of Latin America had, and 10% of Asia had. By 1938, club membership had been attained by half of the European periphery, 24% of Latin America, and 15% of Asia. By 1973 and the end of the ISI period, the threshold had been attained by 31% of Asia.

The percentages plotted in Figure 21.1 are conservative for two reasons. The first is that where we cannot document industrial performance, we are forced to exclude the country in question from the club. The second is that these percentages are based on a denominator which includes a large number of modern-day countries, several of which are very small, some of which did not exist in previous periods, and for many of which we do not have data for

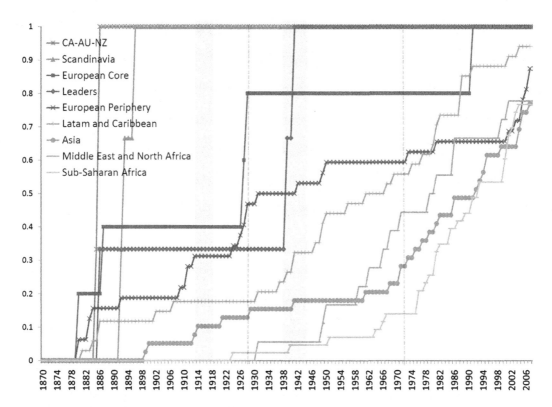

Note: The figure shows the proportion of countries for which the 10-year backward looking average industrial growth rate exceeded a 5% threshold. Countries for which data are missing are assumed not to have exceeded this threshold.

FIGURE 21.1

Regional diffusion curves: reaching the 5% threshold.

these earlier periods. Figure 21.2 provides an alternative perspective which deals at least to some extent with the second of these problems, since it weights the different country experiences by their populations in 2007. More precisely, it asks: what proportion of a region's 2007 population was living in countries which had attained the 5% growth threshold by any given year?

By giving more weight to China than Bhutan, we increase the measured diffusion rates in the periphery dramatically. By World War I, the modern growth threshold had been attained in countries accounting for 42% of Asia's 2007 population, already a very large number. By 1938, it had been achieved by three-quarters of the Asian 2007 population. By 1973, the club had been attained by countries accounting for 94% of the Asian population. Asia saw its greatest diffusion in the 1890–1938 years, and the Philippines was part of it.

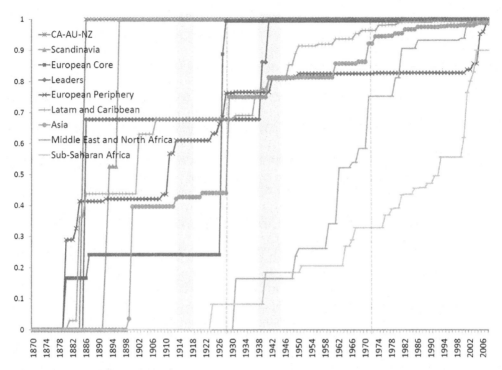

Note: The figure shows the proportion of the region's population in 2007 living in countries for which the 10-year backward looking average industrial growth rate exceeded a 5% threshold. Countries for which data are missing are assumed not to have exceeded this threshold.

FIGURE 21.2

Regional population-weighted diffusion curves: reaching the 5% threshold.

21.3 **UNCONDITIONAL INDUSTRIAL CONVERGENCE**

21.3.1 **UNCONDITIONAL CONVERGENCE**

There is a vast empirical literature that asks whether poorer countries grow more rapidly than richer ones (Abramovitz, 1986; Barro, 1997; Bourguignon and Morrisson, 2002), and, when measured by aggregates like GDP per capita, it finds that they do not.[4] More recently, however, Rodrik (2013) has found evidence of unconditional convergence in labor productivity for individual *manufacturing* sectors. Since we do not have comparable data on manufacturing employment, we cannot repeat Rodrik's exercise. However, we can answer a related question: did less industrialized economies experience more rapid industrial growth than more industrialized ones? More precisely, did countries with lower manufacturing output per capita experience more rapid growth in

[4]Economists have only found evidence of conditional convergence (Durlauf et al., 2005).

manufacturing output than countries with more manufacturing output per capita? In order to answer this question, we need to be able to compare levels of manufacturing output across countries. Comparable manufacturing output *levels* are much more difficult to measure over long time periods than are rates of growth. Here we use two approximations. First, the World Bank's World Development Indicators report comparable manufacturing output levels for 2001, expressed in US dollars. We extrapolate these 2001 output levels back in time using our output indices and then divide these by population taken from the World Development Indicators (2011) and Maddison (2010). This procedure yields estimates of manufacturing output per capita back to 1870, for 179 countries during the most recent 1990–2007 period, 145 for 1973–1989, 101 for 1950–1972, 54 for 1920–1938, 42 for 1890–1913, and 29 for 1870–1889. Second, we take Bairoch's (1982) data on cross-country industrial output per capita for two benchmark years (1913, 1928), and then use our annual output indices (and population data) to generate comparable absolute levels of per capita output for each year within the periods 1870–1913 and 1920–1938. Similarly, we use UN data for 1967 to generate comparable absolute levels of per capita output for 1950–1972 and World Bank data to generate comparable absolute levels for 1973–2007.

Table 21.4 provides the slope coefficients from regressions of manufacturing output growth rates against initial levels of per capita output. The first column presents our preferred estimates, using the data on output per capita generated from period-specific benchmarks (i.e., the Bairoch data for 1913 and 1928 and the UN data for 1967). One problem with these results is that the number of observations is not constant across time periods, making the coefficients difficult to compare.

Table 21.4 Unconditional Industrial Convergence

Period	Using Period-Specific Benchmarks	Country Sample					
		1870–1889	1890–1913	1920–1938	1950–1972	1973–1989	1990–2007
1870–1889	− 0.384	− 0.106					
	(0.493)	(0.275)					
1890–1913	− 0.589	− 0.049	− 0.271				
	(0.388)	(0.118)	(0.225)				
1920–1938	− 0.766**	− 0.464*	− 0.380*	− 0.646***			
	(0.329)	(0.256)	(0.189)	(0.207)			
1950–1972	− 3.095***	− 1.066*	− 1.067**	− 1.091***	− 1.004***		
	(0.387)	(0.516)	(0.395)	(0.287)	(0.222)		
1973–1989	− 0.523***	− 0.584**	− 1.178***	− 0.937**	− 0.804***	− 0.540***	
	(0.168)	(0.233)	(0.397)	(0.386)	(0.282)	(0.169)	
1990–2007	− 0.175	− 0.363	− 0.908**	− 0.471	− 0.115	− 0.106	− 0.175
	(0.166)	(0.346)	(0.382)	(0.293)	(0.262)	(0.227)	(0.166)
No. of countries		23	28	44	56	87	134

Note: *Coefficients are obtained by regressing average growth rates per annum on the log level at the beginning of the period. The first column reports coefficients using period-specific benchmarks; subsequent columns use backward extrapolation from a 2001 benchmark. Statistical significance at the 10%, 5%, and 1% levels is indicated by *, **, ***, respectively.*
Source: *Bénétrix et al. (2012), Table 5.*

Subsequent columns address this issue, using the data on levels constructed by extrapolating backward from the 2001 World Bank data. In this manner, the coefficients in any given column are comparable with each other, based as they are on the same country samples.

Table 21.4 tells a consistent story. While there is evidence of unconditional convergence between 1870 and 1913, it only became statistically significant at conventional levels after World War I. Clearly, the highpoint of unconditional industrial convergence in the periphery was the ISI period between 1950 and 1972: while strong unconditional convergence persisted after the first oil shock, it was less pronounced than before, and it fizzled out entirely after 1990.

21.3.2 WAS THERE PERSISTENCE?

How persistent were high growth rates over time? More precisely, were high growth countries in one period also the high growth countries in the following period? For each region and time period, Table 21.5 provides a list of the top 10 performers, ranked by their average growth performance over the period as a whole.[5] Some countries appear consistently in the table: Russia, China, Japan, India, and Brazil being perhaps the most prominent.[6] The table also documents a good deal of churning over time, with many countries entering, exiting, and reentering the top 10 leader board over time (like Burma, Indonesia, and Thailand). What is surprising about the Philippines is that it didn't churn: rather, it dropped off the Asia leader board entirely and forever.

21.4 UNDERSTANDING THE PHILIPPINES' DEVIANT BEHAVIOR

The Philippines' early inclusion in and then sudden disappearance from league tables of leading industrial performers in otherwise economically backward countries warrants explanation. Since manufacturing output per employed person is simply the product of manufacturing productivity and the share of manufacturing in total employment[7], our search will focus on changing manufacturing employment shares and manufacturing productivity growth. The evidence makes it clear that dynamism has been absent from Philippine industry and its manufacturing in particular. The share of industry value added in GDP remained constant at around 25% between 1970 and 1990, then fell to 20% in the next decade (Figure 21.3). Similarly, the manufacturing employment share has stagnated at some 10% for more than five decades, and the industry employment share fared no better, staying essentially at around 15%. Thus, the structural shift since the 1950s has not been the classic one from agriculture to industry but rather from agriculture to services. Without a dynamic industrial sector, the relatively slow transformation has meant too many poor farmers for

---

[5]Table 21.3, in contrast, ranked countries according to how early they joined the modern growth club, which was defined in terms of growth performance over just 10 years.
[6]It looks like the BRIC's rapid industrialization is a phenomenon with deep historical roots.
[7]This is a close measure of manufacturing output per capita, which was used in the first half of this chapter. Labor force participation and dependency rates for the population at large will account for the difference.

Table 21.5 The Top Ten Performers by Region and Period

1870–1889	1890–1913	1920–1938	1950–1972	1973–1989	1990–2007
European Periphery					
Bosnia	Bosnia	Russia	Albania	Cyprus	Ireland
Russia	Romania	Latvia	Bulgaria	Malta	Lithuania
Austria	Serbia	Romania	Romania	Ireland	Slovak Republic
Hungary	Finland	Finland	Yugoslavia	Bulgaria	Poland
Finland	Russia	Bulgaria	Poland	Portugal	Finland
Spain	Bulgaria	Ireland	Cyprus	Russia	Hungary
Bulgaria	Italy	Estonia	Spain	Yugoslavia	Bosnia
Italy	Austria	Hungary	Italy	Latvia	Czech Rep.
Portugal	Hungary	Greece	Russia	Italy	Belarus
	Portugal	Poland	Greece	Finland	Estonia
Asia					
Japan	Korea	Korea	Singapore	Indonesia	Cambodia
Indonesia	China	Japan	Korea	Korea	Burma
Thailand	Philippines	China	Japan	Bhutan	Afghanistan
India	Japan	Taiwan	Malaysia	Tonga	Vietnam
	Taiwan	Philippines	Taiwan	Taiwan	China
	India	India	Pakistan	Hong Kong	Kazakhstan
	Thailand	Indonesia	Mongolia	China	Bhutan
	Indonesia	Burma	China	Maldives	Korea
	Burma	Thailand	Vietnam	Malaysia	Malaysia
			India	Thailand	Laos
Latin America and Caribbean					
Chile	Argentina	Colombia	Panama	St. Lucia	Trinidad and Tobago
Brazil	Peru	Peru	Puerto Rico	Grenada	Costa Rica
Argentina	Mexico	Argentina	Nicaragua	Dominica	Dominican Rep.
Uruguay	Chile	Costa Rica	Costa Rica	Paraguay	Honduras
	Uruguay	Mexico	Brazil	St. Vincent and Grenadines	Belize
	Colombia	Guatemala	Venezuela	Antigua and Barbuda	Nicaragua
	Brazil	Brazil	Mexico	Belize	El Salvador
		Uruguay	El Salvador	Puerto Rico	St. Kitts and Nevis
		Chile	Honduras	Cuba	Peru
			Peru	Ecuador	Suriname

Table 21.5 The Top Ten Performers by Region and Period *Continued*					
1870–1889	1890–1913	1920–1938	1950–1972	1973–1989	1990–2007
Middle East and North Africa					
Turkey	Turkey	Turkey Egypt	Iran Israel Saudi Arabia Algeria Turkey Egypt Morocco Tunisia Syria	UAE Algeria Egypt Tunisia Saudi Arabia Syria Sudan Turkey Jordan Morocco	UAE Oman Jordan Iran Syria Yemen Egypt Saudi Arabia Sudan Tunisia
Sub-Saharan Africa					
		South Africa Congo, Dem. Rep.	Mozambique Central African Rep. Kenya Zambia Cameroon South Africa Botswana Ghana Senegal Gambia	Cameroon Cape Verde Swaziland Lesotho Botswana Mauritius Mali Central African Rep. Gambia Congo, Rep.	Equatorial Guinea Mozambique Namibia Uganda Lesotho Sierra Leone Angola São Tomé & Principe Burkina Faso Benin

too long and thus too much inequality and poverty for too long: the agricultural employment share only fell below 50% in the early 1980s.

The link between the growth of overall labor productivity and that of manufacturing can be seen more clearly if we divide the growth in aggregate labor productivity into two components: the growth of labor productivity *within* each sector and the growth that reflects *structural change*, as labor is pulled toward sectors where productivity growth is fastest and productivity levels are highest (following McMillan and Rodrik, 2011; see Appendix 1 for details). When these components are computed for 1956–2009, we find that *within*-sector manufacturing productivity has grown fairly steadily, except for two periods: 1980–1985 and 1990–1995 (Figure 21.3). The first of these periods coincides with the largest postwar recession the economy experienced, a combined financial and political crisis, while the second relates to a less prolonged but severe power-sector crisis (1991–1992). The steady growth of within-sector productivity in Philippine manufacturing is

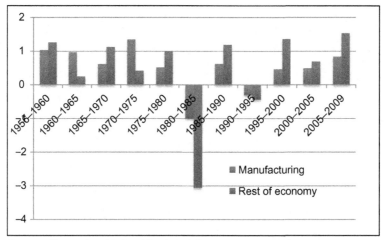

Source of basic data: National Statistical Coordination Board (output measured in constant 1984 prices).

FIGURE 21.3

Within-sector productivity growth by sector 1956−2009 (annual rates, in percent).

consistent with Rodrik's (2013) documentation of an unconditional global convergence in manufacturing productivity. This is the good news.

The bad news is that manufacturing's productivity contribution due to structural shift has been weak at best and a drag on growth at worst (Figure 21.4). For most subperiods, productivity gains due to structural shift were typically negative, reflecting the secular fall in the manufacturing employment share. Thus, the Philippines illustrates how convergence in manufacturing productivity by subsector can coexist with divergence in aggregate manufacturing productivity, economy-wide productivity, and industrial output per capita. Briefly put, although manufacturing productivity within sectors has more or less kept abreast with global trends, aggregate manufacturing productivity has not. Between 1970 and 1985, the *structural contribution* to manufacturing productivity growth was negative and thus failed to reinforce the effects of the 1962 devaluation. A fact relevant to any explanation for Philippine deviant industrialization behavior is that the period in question was also characterized by persistent current-account deficits that were financed by heavy government borrowings from private sources. These loans were used in part to finance industrial projects of Marcos cronies. The capital-intensity and inefficiency of many of these projects are a likely explanation for their weak impact on manufacturing employment and productivity. This is also reflected in the behavior of *total factor* productivity which Hooley (1985) estimates fell throughout most of the 1970s, a trend he attributes to the inefficiency of the industries set up at the time.

There have been brief episodes in which manufacturing did contribute positively to productivity via structural shift. These occurred during incipient recoveries from preceding crises. The 1965−1970 years, for example, coincide with a revival of manufacturing following the dismantling

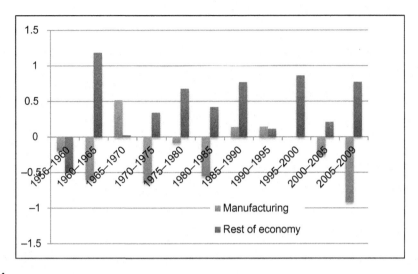

FIGURE 21.4

Productivity growth from structural shift by sector 1956–2009 (annual rates, in percent).

of the system of quantitative restrictions (decontrol) in 1962. This was not sustained, however, and gave way to structural productivity losses between 1970 and 1985. The same pattern is evident after the most acute postwar crisis 1980–1985, when a positive structural shift into manufacturing was subsequently wiped out.

Why weren't these promising recoveries sustained? What caused the 1970–1984 and post-1995 periods to be net negatives for industrialization even as within-sector productivity appeared to grow? Four important narratives have been advanced to explain parts or all of the Philippine deviant industrial behavior. We denote these as the institutional story; the liberalization story; the real exchange-rate story; and the overseas migration story. We take up each in what follows.

21.4.1 INSTITUTIONS: POLITICAL INSTABILITY AND THREATENED PROPERTY RIGHTS?

The recent literature has trained attention on the developmental role of political regimes and the distribution of economic power (North, 1990; Acemoglu et al., 2002; Acemoglu and Robinson, 2012; Engerman and Sokoloff, 2012). Moreover, the development discussion in the Philippines has stressed the legitimacy of political institutions and the control of corruption (NEDA, 2011). The main institutional hypothesis is that perennial political instability and legitimacy crises have been a major hindrance to investment and growth (see, for example, de Dios, 2011). This hypothesis finds its strongest support in the turbulent years of 1983–1986, when the debt–repayments crisis combined with political instability to produce the worst postindependence recession. The 1983–1986 crisis began as a debt-repayments problem as the over-leveraged

Philippine economy—like some Latin American countries but unlike most of East Asia—became caught in the pincers forged in America's Volcker recession: rising interest rates on the country's foreign loans and slumping exports (as major markets slipped into recession). Heavy debt servicing commitments necessitated a unilateral debt-payment moratorium by early 1984. The moratorium cut off the supply of imports, while the implementation of IMF conditionalities (mostly through a radical monetary contraction) depressed domestic demand. Both factors precipitated huge declines in total output and total employment, but industry—the most import-dependent part of the economy—was hardest hit. This can be seen in the large productivity declines in manufacturing during the 1980—1985 period, both for within-sector and structural terms. Industrial output fell by 19% between 1980 and 1985, while investment fell by 48%. The automotive, electronics, garment, and textile industries were affected especially severely as trade credits dried up, and both home-demand and exports collapsed.

Feeding into the financial crisis was the political instability that culminated in a popular revolt leading ultimately to the overthrow of the authoritarian Marcos regime. Political uncertainty did not immediately subside even with the restoration of democratic rule, however, with major putsch attempts (i.e., 1987, 1989, 1990) and farmer and worker strikes preoccupying the new government. Aside from these severe political threats, the post-Marcos government was also confronted with the problem of sorting out (and sometimes squabbling with Marcos cronies over) the ownership and operations of several dominant firms, notably food-processing conglomerates, iron and steel, drugs and chemicals, power distribution and generation, and telecommunications. Throughout the period, therefore, not only was political stability in doubt, but so too were property rights.

Further political instability occurred in 2000—2001, following the corruption scandal and aborted impeachment process involving the Estrada administration. This led to a second popular revolt that installed the Arroyo administration. The latter, however, became embroiled in scandals involving corruption and electoral anomalies that undermined its legitimacy and gave rise to mass demonstrations and attempted putsches (2003, 2006, and 2007). On the whole, the country fared poorly on political stability and property rights assurances after the early 1980s: investor-services cited the Philippines as a "high political risk" for the entire period 1984—1991.[8] Econometric evidence suggests that the Philippine investment ratio has been suppressed by political instability and corruption. In a previous work, one of this chapter's authors has shown that the borrowing rate is raised by political instability, corruption, and internal conflict. In addition, corruption diminished investment demand directly (de Dios, 2011).

The timing of political crises and institutional failure mattered. The period of deepest political crisis 1984—1991—years spanning the downfall of the Marcos dictatorship and the continuing political challenges to the Aquino government that replaced it—was also a period of large-scale relocation to Southeast Asia of Japanese manufacturing industries in response to the yen revaluation following the Plaza-Louvre Accords. This wave of foreign direct investments benefited Malaysia, Thailand, and Indonesia and led to the buildup of a significantly export-oriented manufacturing in those countries. Owing to the Philippines' political instability, however, Japanese (as well as

---

[8]The ICRG measure of "political risk" is consistently below 50 (described as "very high risk") for this entire period.

Taiwanese and Hong Kong[9]) FDI largely bypassed the country. The volume of this FDI entering Thailand during the 1987–1991 period has been estimated at $24 billion, as compared with $1.6 billion entering the Philippines (Yoshihara, 1994: 49). Philippine political instability relative to its neighbors explains about half of the differential in per capita direct foreign investment for the period 1985–1992 (de Dios, 2011: 89; Table 21.5). This fact probably accounts for much of Coxhead's (2013) finding that the Philippines missed most of the powerful regional spillovers generated by fast growth first of Japan and then China. The Philippines' failure to benefit from such spillovers—which effectively jump-started the industrialization of Malaysia, Thailand, and Indonesia—may in turn ultimately be attributed to the political instability that plagued the country at just the worst possible historical juncture.

Investment exceeded 25% of GDP only during 1975–1983 (Figure 21.5), the most stable years of the Marcos regime. This coincided with the global recycling of petrodollars from the Mideast oil bonanza, at which time the Philippines obtained access to large amounts of foreign loans. Since then, however, the investment ratio has never exceeded 25%. While there was a predictable drop following the Asian crisis, a further decline took place after 2004, when the investment share fell and remained at 20% of GDP or even less. These shares are very low by Asian standards.

To the extent that political uncertainty and a dysfunctional government affected investment negatively; that import-dependent manufacturing was hit especially hard by these crises; and that political instability and disputed property rights caused the country to miss out on the massive relocation of Japanese, Taiwanese, and Hong Kong manufacturing—then institutional factors must be considered a fundamental explanation for the failure of industrialization to continue after the ISI period. However, what the institutions narrative is unable to explain is why the structural

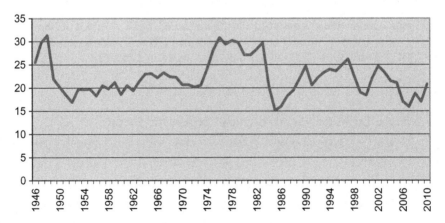

FIGURE 21.5

Gross investment 1946–2010 (percent of GDP).

---

[9]The inclusion of Taiwan and Hong Kong recognizes the pattern of Japanese investment that relies on ancillary industries.

contribution to productivity growth was already negative even before the crisis, the missed FDI opportunities, and unexploited regional spillovers. It also cannot explain why fast industrial growth did not resume after political stability returned. In any event, the ability of Southeast Asian latecomers like Vietnam, Cambodia, and Laos to adapt and increase industrial output per capita undermines the sufficiency of the explanation. Although political instability and governance issues may explain the poor industrial performance from the 1970s to the 1990s, they cannot fully account for the poor industrial performance after the 1990s.

21.4.2 **TRADE LIBERALIZATION?**

The notion that trade liberalization may have caused the failure of Philippine industrialization is a long and widely held view (e.g., Bello et al., 2004). Before the Philippine industrial slowdown, the country had maintained a protectionist stance toward industry for some time. It began with import-controls and exchange-restrictions during 1949–1961 and continued with the replacement of import-controls by high tariffs in 1962 (Power and Sicat, 1971; Bautista et al., 1979; Medalla et al., 1995). The cascading tariff structure was largely maintained throughout the extended Marcos regime, modified only by modest concessions to new exports in the form of tax incentives and the creation of a few export-processing enclaves. A tariff reform program that the Marcos regime finally undertook in 1981 initially cut nominal tariffs to the 10–50% range (from as high as 100%) and eased quantitative restrictions. This reform was quickly undone by the debt crisis when strict quantitative restrictions were reimposed to ration foreign exchange as all trade financing dried up. After import-controls were removed, the effective rate of protection for manufacturing was still a high 64.7% (Bautista, 2003: 19).

The decisive liberalization move was taken in 1991, when, in its last year in office and after the restoration of macroeconomic stability, the Aquino administration reduced tariffs to a 3–30% range. The Ramos administration continued the liberalization trend, acceded to the World Trade Organization in 1994, and made further tariff reductions in 1998 with a trajectory of reaching a uniform 5% by 2004.[10] These 1981–2003 trends are documented in Table 21.6.

Table 21.6 Average Tariffs for Various Economic Sectors (in percent)									
Sector	**1981**	**1985**	**1990**	**1991**	**1995**	**1998**	**2000**	**2001**	**2003**
Agriculture	43.23	34.61	34.77	35.95	27.99	18.91	14.40	14.21	11.04
Mining	16.46	15.34	13.97	11.46	6.31	3.58	3.27	3.25	2.84
Manufacturing	33.74	27.09	27.49	24.61	13.96	9.36	6.91	6.68	5.43
Overall	34.60	27.60	27.84	25.94	15.87	10.69	7.95	7.70	6.19
Source: *Philippine Tariff Commission.*									

---

[10]This measure was also an attempt to avoid possible trade diversion with the advent of the low tariffs on intra-ASEAN trade under the ASEAN Free Trade Area.

Even conceding a role for trade liberalization in accounting for the failure of industrialization after the early 1990s, the liberalization thesis fails to provide an explanation for the dismal industrial performance after the early 1970s, by which time the country had already dropped out from the league of high growth performers.

21.4.3 REAL CURRENCY OVERVALUATION?

Another prominent explanation for Philippine deviant industrial development has been currency real overvaluation. Policy and academic debates have periodically[11] focused on the possible role of the exchange rate as a developmental tool rather than simply a lever for attaining macroeconomic stability. The debate has been further stoked by difficulties currently faced by the authorities in stemming the nominal appreciation of the peso, as well as by a recent academic literature (e.g., Rodrik, 2007; McMillan and Rodrik, 2011) which highlights the salutary growth effects of systematic real currency undervaluation in emerging economies.

Until 1970, currency overvaluation under fixed exchange rates was generally associated with the import-substitution strategy in place at the time, with the system being supported by foreign-exchange controls, import quotas, and tariffs (Bautista et al., 1979). A second source of overvaluation treated in the literature has been the accumulation of debt, first involving heavy public borrowing during the authoritarian Marcos episode 1974–1981 (Fabella, 1996), then the debt accumulation by the private sector and portfolio capital inflows in the period preceding the Asian financial crisis 1992–1997 (de Dios et al., 1997). Finally, overseas workers' remittances have recently emerged as a major influence on the current account and the exchange rate. Together with monetary expansion in the United States and other countries, overseas remittances have been associated with a historically unprecedented secular appreciation of the peso in nominal terms (by some 33% relative to the dollar between 2004 and 2012).

To what extent does currency overvaluation contribute to an explanation of Philippine industrial failure after the early 1970s? Mustering evidence can be thorny, beginning with the definition of "real overvaluation" itself, a notion that depends heavily on the selection of a reference point. Although Rodrik's (2007) evidence lends support to undervaluation as a successful industrialization policy tool, its applicability to the Philippines is somewhat awkward, since his own data suggest that the Philippines—together with other countries in Southeast Asia—had consistently *undervalued* its currency for most of the postwar period. Figure 21.6 uses Rodrik's (2007) data to show that the only episode of overvaluation for the Philippines light-colored line) was 1950–1961 which preceded the 1962 devaluation. This result emerges partly because Rodrik applies the Balassa-Samuelson adjustment, which shifts the observed real exchange rate by an amount depending on a country's rate of growth. (The darker line in Figure 21.6 displays the trend in the real exchange rate without that adjustment.) Rodrik's adjusted series seems inconsistent with the fact that the Philippines has run current-account deficits throughout most of the entire postwar period[12] until continuous surpluses began to appear after 2003. It may be more constructive, therefore, to speak only of the trends of real appreciation and depreciation rather than levels.

---

[11]These policy debates have occurred at least in the late 1970s, the early 1990s, and more recently the present decade. (On these, see Bautista et al., 1979; de Dios et al., 1997; Fabella, 2013).
[12]The exceptions are the years 1962–1966, 1986, and 1998.

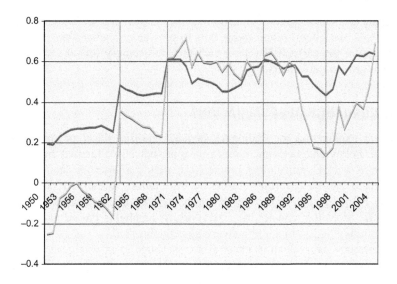

FIGURE 21.6

Real exchange rates and undervaluation index 1950–2004 (in natural logarithms).

Table 21.7 Indices of Real Effective Exchange Rates of the Peso (1980–2010)								
	1980	**1985**	**1990**	**1995**	**2000**	**2003**	**2005**	**2010**
Major	99.44	89.28	66.20	79.19	71.92	59.94	61.98	84.08
Broad	101.44	100.51	86.90	100.78	109.12	99.91	101.51	137.65
Narrow	99.52	101.98	124.41	146.35	169.40	142.67	149.52	173.16

Notes: *Index is based on the dollar price of a peso; a higher value of the index signifies peso appreciation;* major index weights: *United States, Japan, Eurozone;* broad index weights: *Singapore, South Korea, Taiwan, Malaysia, Thailand, Indonesia, Hong Kong;* narrow index weights: *Indonesia, Malaysia, Thailand.*
Source: *Bangko Sentral ng Pilipinas.*

Whether one uses the undervaluation index or the (natural logarithm of the) real exchange rate, some general trends are common: large real depreciations resulting from the 1962 devaluation and the adoption of a floating rate in 1970; a real appreciation between 1973 and 1979; another real appreciation between about 1991 and the Asian financial crisis; and significant real depreciation thereafter. Not plotted in the above series, but evident in others, is the real appreciation of the peso from around 2003 onward (Table 21.7). Focusing on Rodrik's undervaluation index, the most favorable conditions for industrial development in terms of the real exchange rate should have been the years bracketed by the 1970 peso float and the debt crisis from 1983. Yet, these are precisely the years when industrial growth rates dropped off so markedly. The real appreciation prior to the Asian financial crisis (1990–1995) is well understood as the result of the renewed access to foreign credit and heavy borrowing by the private sector, fueling a real-estate boom as it did in other countries. That this was also a period of positive structural contributions from manufacturing is probably due to the recovery of industrial employment after the brief but crippling power-related recession of 1990–1992.[13] In contrast, between 2003 and 2010, the peso appreciated in real terms (Table 21.7) by as much as 38%.[14] This phase coincides with the emergence of regular current-account surpluses owing to overseas workers' remittances, rapid decline in manufacturing competitiveness, and a further loss in manufacturing jobs.

In short, currency overvaluation does not by itself offer a consistent explanation for a quarter century of deviant industrial growth in the Philippines.

21.4.4 OVERSEAS LABOR MIGRATION?

As a candidate for explaining the failure of Philippine industrialization, Dutch Disease caused by overseas migrant remittances is only relevant beginning with the early 1990s. Data on remittances are deficient before the mid-1980s, but overseas deployment, with a lag, can serve as a proxy for remittance trends. Prior to the early 1980s, overseas migration was a minor phenomenon. In relation to the domestic labor force, overseas migration became significant only from 1983, when registered annual deployment shot up to more than 2% of the labor force. A further acceleration occurred in 1998, when registered deployment rose 56% and exceeded 3% of the labor force (Table 21.8, line 1). These growth spurts in overseas deployment coincided with or occurred shortly after major economic crises at home, when domestic urban employment opportunities were shrinking significantly[15] and overseas markets were relatively open.

Increasing foreign deployment is mirrored, with a lag, by rising inward remittances by workers based overseas. The leap in the importance of remittances occurred in the late 1990s (Table 21.8, line 2),

Table 21.8 Deployment of Overseas Workers and Remittances (1975–2009) (percentage of domestic labor force; annual averages by period)

	1975– 1979	1980– 1984	1985– 1989	1990– 1994	1995– 1999	2000– 2004	2005– 2009
Deployment	0.48	1.70	2.02	2.65	2.75	2.96	3.44
Remittances	5.26	7.86	7.75	10.48	13.56	13.95	19.78

Notes: *Annual overseas deployment is a percent of the domestic labor force. Remittances (current transfers) are a percent of total current-account receipts. The labor force figure for 1979 is a between-year interpolation.*
Sources: *Philippine Overseas Employment Authority; National Statistical Coordination Board.*

---

[13]The country's strained power situation and resulting high electricity costs relative to the rest of the region since 1992 are also perennially cited locally as a major reason for manufacturing's lack of competitiveness. This is an oversimplification, however, since it glosses over the fact that various manufacturing subsectors are not uniformly reliant on electricity and that the share of electricity in production cost in manufacturing is only 4.63% on average (Usui, 2012).

[14]That is, 137.65 from 99.91 in the second row of Table 21.7. The real appreciation is 38 (respectively 21) percent if the values of the major (respectively Narrow) indices are used for the same years.

[15]The slight slowdown in overseas deployment immediately after the 1986 revolution is also evident and may indicate the goodwill and optimism early in the post-Marcos period. This soon faded, however, and the gradual rise in deployment resumed and shot up in 1998.

when they came to represent 14% of all current-account receipts (from only 5% in the late 1970s). This figure rose further to 20% by the late 2000s, but the country had already begun to run current-account surpluses on a regular basis as early as 2003. The upshot is that overseas remittances had only modest impact on the current-account and exchange rates before the early 1990s. If they have generated a significant industrial Dutch Disease, they can only have done so from that time onward.[16]

Furthermore, it is unlikely that the surge in emigration was exogenous to the poor industrial performance. Rather, it seems more probable that the poor industrial performance after 1972 pushed out emigrants in the 1980s, thus raising remittances and making the exchange rate less competitive for manufacturing subsequently. Of course, it might also be argued that once the labor force was stripped of potential new industrial workers by emigration, firms were faced with higher per unit labor costs when other conditions improved.

21.4.5 REAL EXCHANGE RATE AND TRADE REGIME INTERACTIONS?

Any cross-country real exchange-rate assessment should also take into account the trade regime. While real exchange rates deal with the prices of tradables relative to nontradables, they do not deal with the relative prices between exportables and importables. Thus, it is critical to sort out the net incentive effect of real over- or undervaluation of the peso in conjunction with the trade regime. Policy debates in the 1960s over the trade regime were superseded after the 1990s by controversies over the exchange rate, but an explicit consideration of the two together is rarely made. Bautista (2003) is the rare exception. Its most relevant point, which we schematize in Table 21.9, is that a failure to coordinate exchange-rate management and trade policy could lead to paradoxical or even perverse results.

Episodes of currency undervaluation or real depreciation are best accompanied or preceded by a liberalization of imports (Quadrant I). The other extreme (Quadrant IV) is a restrictive import regime combined with an overvalued currency. This latter case may roughly characterize the entire import-substitution period 1949−1961 before the 1962 devaluation. This was a period when manufacturing growth ran up against limited market size, which consisted almost exclusively of the protected domestic market. Indeed, an expansion of manufacturing exports was stifled by the implied penalty of currency overvaluation. The 1990s, before the Asian financial crisis hit the

Table 21.9 Trade Liberalization and Real Depreciation

	Liberalized Imports	Restricted Imports and Export Penalties
Real depreciation	I. Demand- and supply-constraints relaxed	II. Supply-side constraints binding
Real appreciation	III. Demand-side constraints binding	IV. Demand- and supply-constraints binding

Source: *Bautista (2003).*

---

[16]Another hypothesis related to migration relates to the depletion of skilled workers needed in manufacturing. An attempt to represent Philippine events is given in de Dios (2013). Like the Dutch Disease phenomenon, however, the timing issue rules this out as a primary cause, since it is conditional on the fact of migration itself, although it cannot be ruled out as an explanation of the difficulty of industrialization in the 2000s.

Philippines in 1998, were years commonly characterized by import liberalization, when tariffs were lowered and import-licensing requirements (introduced in the crisis years 1980–1988) were removed.[17] However, this was also a period of real currency appreciation if not outright overvaluation. Private access to foreign borrowing resumed, fueling a real-estate and equity market boom, and overseas workers' remittances began to contribute a significant share (more than 10%) of current-account receipts. This regime falls in Quadrant III of Table 21.9. Import liberalization eased the input-supply constraints to manufacturing, but real appreciation undercut its competitiveness in both domestic and foreign markets. From this perspective, the observed falloff in the structural contribution of manufacturing growth is hardly surprising.

The most enigmatic part of the historical record, however, is the 1971–1981 decade. These years marked the first stage of the secular collapse of Philippine manufacturing growth. As already noted, it was also a period of relative political stability (if through repression) and included years when investment was at an unequaled high. Dutch Disease from overseas migration could not have played a role during the decade, since that phenomenon attained a significant magnitude only in the 1990s. Also, the adoption of the managed float in 1970 resulted in a real depreciation that should have stimulated manufacturing and its exports. In short, many conditions were favorable for industrialization. So, why the poor industrial performance?

Bautista (2003) argues that this period's potential was seriously diluted by the continuing protection of importables, which raised the cost of imported inputs. His thesis finds support when one considers the effective exchange rates confronted by various tradables during the period. Reading off the entries for 1970–1975 and 1976–1980 in Table 21.10, it is evident that export industries were still constrained on the supply side, given the exchange-rate penalty on producer goods. Over that decade, a nontraditional exporting firm would have confronted a 7.6–11.4% penalty on foreign exchange as between its final product and its essential production input. The penalty is even larger if one considers the import of "semiessential" and "nonessential" producer goods. To be sure, a few export processing zones (Bataan, Cavite, and Mactan) and some export incentives relieved

Table 21.10 Effective Exchange Rates for Various Categories of Goods (1949–1980; period averages; pesos per dollar)

	Nominal	NEC	SEC	EC	NEP	SEP	EP	TX	NX	*Ratio*: EP/NX
1949	2.00	2.05	2.05	2.00	2.05	2.00	2.00	2.00	2.24	0.89
1950–1959	2.00	3.65	2.46	2.06	2.43	2.44	2.44	2.00	2.29	1.065
1960–1969	3.90	10.56	5.27	3.91	6.91	4.22	4.61	3.46	3.70	1.245
1970–1975	6.86	21.19	9.16	7.56	12.46	8.08	8.24	6.17	7.66	1.076
1976–1980	7.42	25.49	10.17	8.82	13.46	9.34	9.40	7.12	8.44	1.114

Notes: *Categories refer to nonessential consumer goods (NEC); semiessential consumer goods (SEC); essential consumer goods (EC); nonessential producer goods (NEP); semiessential producer goods (SEP); essential producer goods (EP); traditional exports (TX); and nontraditional exports (NX).*
Source: *Senga (1983) for 1972–1980 and Baldwin (1975) for 1949–1971.*

---

[17]Aldaba (2005) provides a brief chronology.

these constraints somewhat, but these efforts were limited. The export boom in 1970−1971 was short-lived, the drive by the Marcos regime to establish "major industrial projects," including a number of heavy (capital-intensive) industries, was aborted (though not before sucking away large investible resources from potentially more efficient industries), and the supply-constrained effort failed to generate a manufactured-export takeoff. The resulting current-account deficits were covered by massive government foreign borrowings, which in turn laid the foundation for the debt crisis beginning in 1983 when global conditions became adverse. The failure of the new export industries to expand and the capital-intensity of the regime's favored projects explain the weak structural impact of manufacturing during the decade (especially on employment), in spite of the higher investment ratios.

21.5 DEVIANT BEHAVIOR AND PATH DEPENDENCE

We conclude our assessment of the sources of the Philippine industrialization collapse by stressing the role of path dependence. The 1983−1986 political crisis and recession shunted the economy off on a debt-driven trajectory. By the 1990s, helped along by a popular restoration of democracy, the protectionist regime, which had been the principal obstacle to industrial growth in the 1970s, was eliminated. If the global crisis had not occurred and political instability had been averted, would the regime's debt-financed industrial effort—crony-led and corrupt as it was—have ultimately transitioned into a more typical East Asian growth pattern? Or alternatively, if the post-Marcos Aquino government had not been beleaguered by successive putsch attempts, would the Southeast Asian flood of Japanese FDI in the 1990s have given the Philippines a second chance at an industrial future? Path dependence made it unlikely.

The widespread joblessness occasioned by poor manufacturing growth in the 1970s and 1980s gave birth to a new phenomenon that would further stifle industrial growth in the future: large-scale overseas workers' migration. The size and growth of this migration, and its resulting foreign remittances, would by the early 2000s resolve the foreign-exchange constraint that had been Philippine industry's other perennial nemesis. Indeed, it did more: increasing remittance inflows would be responsible for a Dutch Disease phenomenon by the late 2000s, causing a sustained real appreciation and imposing a penalty on tradable manufacturing. If the outmigration had not occurred or had been much more modest (like the rest of Southeast Asia), would Philippine manufacturing have fared better if liberalization had been combined with a currency that was competitively depreciated after the Asian crisis?

The path followed has led to a new stable equilibrium where a largely liberalized trade in goods coexists with a recurrent current-account surplus built on remittances and strong (skill-intensive) service-sector exports. The peso is under steady pressure to rise in real terms, which leaves little room for (lower-skill) manufacturing to compete and expand. A considerable rise in the investment rate—still low by East Asian standards—would relieve the current-account pressure for real appreciation and create more jobs. The low investment rate may itself be part of an equilibrium where capital requirements are low simply because a significant share of the urban labor force is already abroad.

It appears that the deviant behavior of Philippine manufacturing since the early 1970s was produced by a "perfect storm" of protectionist policy, political instability, missed FDI opportunities, foreign capital dependency, and financial crisis. The new equilibrium which has emerged since the 1990s suggests that the Philippines has deviated from the well-trodden industrial path to modern economic growth and is unlikely ever to find it again.

21.6 **FUTURE RESEARCH**

Old World maps are of little help in *terra nova*, and the Philippines' deviation from the well-beaten East Asian industrialization path[18] means the past can offer few clues to guide future policy. Nonetheless somewhat guilelessly, the Philippine government has defined a manufacturing renaissance as being one of its main growth objectives (Updated Philippine Development Plan 2014), creating the expectation that a first-generation Asian Tiger story might still be in the cards. Whether this will be borne out remains to be seen.

An important question to ask is whether the conditions that led to the economy's current position will persist or ultimately reverse themselves. To what extent, for example, can current trends be treated simply as a Philippine edition of Dutch Disease, where booming sectors (in this case, overseas employment and outsourced business-process) penalize the tradable manufacturing sector? If the problem is essentially transitory and self-correcting, one accepted response is simple benign neglect,[19] and the only prescription would be the prudent disposition of foreign-exchange windfalls while they last (say, through investment in infrastructure or the creation of sovereign-wealth funds). Against benign neglect, it could be argued—through an appeal to path dependence—that the remittances-phenomenon is really long-term in nature. (It has after all persisted for more than a decade.) If so, an extended period of real currency appreciation could permanently impair the chances of any future manufacturing expansion. This would imply a need for corrective positive fiscal or monetary policy, combined with the appropriate exchange-rate regime, to moderate or even reverse the trend (see, for example, Corden, 2004: 102−104).

Suppose this latter path is taken and some proactive policy was implemented? Regardless of how deliberate currency undervaluation has historically worked for high growth East Asia (as Rodrik suggests), one must still ask to what extent it remains a feasible strategy to stimulate manufactured exports and industry more generally. Several complicating factors exist, such as the lower global tolerance for "currency manipulation" in the twenty-first century and the fact that such an action is proposed for a country already in perennial current-account surplus. Undervaluation also imposes an additional policy objective for monetary authorities, giving rise to some form of the well-known "impossible trinity." More palpable and immediate are the financial losses of central banks that must absorb the "negative carry" of accumulating low-yielding foreign-exchange denominated assets in order to tamp down real appreciation. What about fiscal expansion then? Especially if financed by domestic borrowing, fiscal expansion could help reverse the trend in principle by reducing the current surplus through greater absorption and encouraging a real

---

[18]Fabella (2013) uses the term "development progeria" to refer to the same idiosyncratic Philippine pattern where services overtake manufacturing at an early rather than an advanced stage of development.

[19]Under a fixed exchange-rate system, the penalty to tradables is resolved by current account deficits. Under flexible exchange rates, the contraction of the tradables sector ultimately outweighs the booming-sector inflows to cause a nominal depreciation.

depreciation. The exchange-rate effect can be partly undone if fiscal expansion raises home interest rates and attracts inward capital flows. The latter is exacerbated for the Philippines by narrowing sovereign-risk premiums and improved credit ratings.

In sum, even if deliberate undervaluation were a desired objective, the tools to engineer it would be far from obvious. While the main directions of many policy choices can therefore be outlined in theory, their applicability or even relevance in the specific conditions of the Philippines depend crucially on real historical circumstances and the magnitudes of empirical parameters.

Apart from the exchange-rate issue, a larger research question for the Philippines is just how much of a stimulus exports can still provide to manufactures in the current historical period. The "rebalancing" of the Chinese economy away from exports and investment and toward domestic consumption—alongside the rise in wages and other costs—presents opportunities for lower-wage countries to step into the breach as a new source of manufactured exports and destination for manufacturing foreign investments in the by now familiar "flying-geese" pattern.[20] But while the industrial-relocation phenomenon itself is well understood, it is still unknown how far the Philippines will actually benefit from it, given its high-wage status relative to countries further down the flying-geese chain. Also to be considered is the reverse trend of "onshoring" in response to both tax incentives (or penalties) from governments of rich investor-countries and the falling costs of digital and machine replacements of labor in some industries (e.g., robotics and 3D printing). Such trends erode the edge of mass manufacture based on cheap labor, but few studies, if any, exist that define the long-term horizon for the Philippines in these terms.

Besides macroeconomic tools, it would be useful to ask whether other policy-levers exist for late-industrializers such as the Philippines to expand manufactures. Put differently, are there peculiar internal obstacles or penalties that keep manufacturing industry from attaining its potential? It is useful to know, for example, to what extent formal labor protection in the Philippines has indeed been "overdeveloped," e.g., through wage legislation and rules on hiring and firing (see Esguerra, 2010). Other "structural" impediments floated in policy-opinion, but little-researched, include low productivity and incomes in agriculture, which might conceivably have had indirect effects through supply channels (e.g., through the impact of high food prices on wages and supply of inputs) or through demand. Organized private sector opinion has also suggested that constitutional ownership restrictions have been important (though these do not explicitly cover manufacturing), as well as physical-infrastructure bottlenecks, energy, and labor productivity itself. The direct impact of overseas migration in draining the pool of skilled labor that can be tapped for manufacturing also needs to be pinned down.

Most writing about the sluggishness of Philippine growth—especially industry growth—has hitherto focused on external demand stimuli or general supply constraints (hence the focus on exchange rates, tariffs and tax incentives, wage laws, infrastructure, and energy). This has also tended to guide the design of policy. By comparison, few studies of the Philippines have been

---

[20]K. Akamatsu first used this term to describe the shifting locational assignment of dynamic comparative advantage, a prediction vindicated in the subsequent massive Japanese relocation following the Plaza-Louvre accords. More recently, Chandra et al. (2013) argue that the same pattern is being repeated in the case of China in the "leading-dragon phenomenon."

made to assess the differing levels of productivity across firms themselves[21] and their real potential for exports, along the lines suggested by the "New New Trade Theory" (Mélitz, 2008).

At the very least, all of this should help form more realistic expectations about the government's current thrusts. For example, while expanded infrastructure spending may well be indicated generally, this will not remove hindrances peculiar to manufacturing. Instead it is services such as tourism that may benefit and even more prominently. The effects of such policy thrusts on the program's goals may therefore be quite different from what is expected.

Finally, it would be helpful to study not only what has failed (e.g., manufacturing) but also what has succeeded and why. In particular, the growth of a highly competitive modern services sector—starting from simple call-center services but now moving into back-office support operations for foreign-based firms, accounting, and big-data analysis—has been built on a large labor force component with better-than-average educational attainment and English-language skills, together with technological developments that have allowed "trading in tasks" (Grossman and Rossi-Hansberg, 2008; Baldwin and Robert-Nicoud, 2010). The country's competitive new services sector has been little studied, a curious situation considering that sector's significance in the accumulation of foreign exchange, its impact on labor supply (especially skilled labor) to other industries, and the growth of a future middle class and patterns of internal demand more generally.

Where history is a poor guide, research will be less authoritative, although decision making will no doubt be helped by answering partial questions closely related to policy choices, such as those enumerated above. For the rest, one can only rely on attentive pragmatism, a creative use of first principles, and a willingness to adjust as developments unfold.

REFERENCES

Abramovitz, M., 1986. Catching up, forging ahead, and falling behind. J. Econ. Hist. 46 (2), 385–406.

Acemoglu, D., Johnson, S., Robinson, J., 2002. Reversal of fortune: geography and institutions in the making of the modern world income distribution. Q. J. Econ. 117, 231–294.

Acemoglu, D., Robinson, J., 2012. Why Nations Fail. Crown Business, New York, NY.

Aldaba, R., 2005. The Impact of Market Reforms on Competition, Structure, and Performance of the Philippine Economy. Discussion Paper Series No. 2005–24. Philippine Institute for Development Studies, Makati.

Aldaba, R., 2012. Surviving Trade Liberalization in Philippine Manufacturing. Discussion Paper Series No. 2012–10. Philippine Institute for Development Studies, Makati.

Bairoch, P., 1982. International industrialization levels from 1750 to 1980. J. Eur. Econ. Hist. 11, 269–333.

Baldwin, R., 1975. Foreign Trade Regimes and Economic Development: The Philippines. National Bureau of Economic Research, New York, NY.

Baldwin, R., Robert-Nicoud, F., 2010. Trade-in-goods and Trade-in-tasks: An Integrating Framework. Working Paper. Graduate Institute, University of Geneva.

Barro, R.J., 1997. Determinants of Economic Growth: A Cross-Country Empirical Study. MIT Press, Cambridge, MA.

Bautista, R., 2003. Exchange Rate Policy in Philippine Development. Research Paper Series, 2003-01. Philippine Institute for Development Studies, Makati.

---

[21]An initial attempt in presenting such data is made in Aldaba (2012).

Bautista, R., Power, J. and Associates, 1979. Industrial Promotion Policies in the Philippines. Philippine Institute for Development Studies, Makati.

Bello, W., Docena, H., de Guzman, M., Mulig, M.L., 2004. The Anti-Development State: The Political Economy of Permanent Crisis in the Philippines. University of the Philippines Department of Sociology and Focus on the Global South, Quezon City.

Bénétrix, A., O'Rourke, K.H., Williamson, J.G., 2012. The Spread of Manufacturing to the Periphery 1870−2007: Eight Stylized Facts. NBER Working Paper 18221. National Bureau of Economic Research, Cambridge, MA (July).

Bourguignon, F., Morrisson, C., 2002. Inequality among world citizens: 1820−1992. Am. Econ. Rev. 92 (4), 727−744.

Chandra, V., Lin, J., Wang, Y., 2013. Leading-dragon phenomenon: new opportunities for catch-up in low-income countries. Asian Dev. Rev. 30 (1), 52−84.

Corden, W.M., 2004. Too Sensational: On the Choice of Exchange Rate Regimes. MIT Press, Cambridge, MA.

Coxhead, I., 2013. Southeast Asian economics and development: an overview. Handbook of Southeast Asian Economics (forthcoming, Routledge).

de Dios, E., 2011. Institutional constraints on Philippine growth. Philipp. Rev. Econ. 49 (1), 71−124.

de Dios, E., 2013. Skills, migration, and industrial structure in a dual economy. UPSE Discussion Paper 2013-3. University of the Philippines School of Economics, Quezon City.

de Dios, E., Fabella, R., Medalla, F., Monsod, S., 1997. Exchange rate policy: past failures and future tasks. Public Policy [Philippines] 1 (1), 15−41.

Durlauf, S., Johnson, P., Temple, J., 2005. Growth econometrics. In: Aghion, P., Durlauf, S. (Eds.), Handbook of Economic Growth. North-Holland, Amsterdam, PP. 555−677.

Engerman, S.L., Sokoloff, K.L., 2012. Economic Development in the Americas since 1500: Endowments and Institutions. Cambridge University Press, New York, NY.

Esguerra, E., 2010. Job Creation: What's Labor Policy Got to Do with it? AC-UPSE Lecture. Unpublished.

Fabella, R., 1996. The debt-adjusted real effective exchange rate. J. Int. Money Finan. 15 (3), 475−484.

Fabella, R., 2013. Development Progeria: Genesis and Healing. Professorial Chair Lecture, Bangko Sentral ng Pilipinas.

Grossman, G., Rossi-Hansberg, E., 2008. Trading tasks: a simple theory of offshoring. Am. Econ. Rev. 98 (5), 1978−1997.

Helpman, E., 2004. The Mystery of Economic Growth. Harvard University Press, Cambridge, MA.

Hooley, R., 1985. Productivity Growth in Philippine Manufacturing: Retrospect and Future Prospects. PIDS Monograph Series No. 9. Philippine Institute for Development Studies, Makati.

Hooley, R., 2005. American economic policy in the Philippines, 1902−1940: exploring a statistical dark age in colonial statistics. J. Asian Economics. 16, 464−488.

Krugman, P., 1981. Trade, accumulation, and uneven development. J. Dev. Econ. 8, 149−161.

Krugman, P., 1991. Increasing returns and economic geography. J. Polit. Econ. 99, 483−499.

Legarda, B.J., 1999. After the Galleons: Foreign Trade, Economic Change and Entrepreneurship in the Nineteenth-Century Philippines. University of Wisconsin Press, Madison, WI.

Lewis, W.A., 1978. The Evolution of the International Economic Order. Princeton University Press, Princeton, NJ.

Lucas, R.E., 1988. On the mechanics of economic development. J. Monet. Econ. 22 (1), 3−42.

McMillan, M., Rodrik, D., 2011. Globalization, Structural Change, and Productivity Growth. NBER Working Paper 17143. National Bureau of Economic Research, Cambridge, MA (June).

Maddison, A., 2010. Statistics on World Population, GDP and Per Capita GDP, 1-2008 AD. <http://www.ggdc.net/MADDISON/oriindex.htm>.

Medalla, E., Tecson, G., Bautista, R., Power, J., and Associates, 1995. Philippine Trade and Industrial Policies: Catching Up with Asia's Tigers, vol. I. Philippine Institute for Development Studies, Makati.

Mélitz, M., 2008. The impact of trade on intra-industry reallocation and aggregate industry productivity. Econometrica 71, 1695−1725.

Mitchell, B.R., 2007. International Historical Statistics: Africa, Asia & Oceania 1750−2005, sixth ed. Palgrave Macmillan, London.

National Economic and Development Authority [NEDA], 2011. Philippine Development Plan 2011−2016 (Pasig City).

North, D., 1990. Institutions, Institutional Change, and Economic Performance. Cambridge University Press, New York, NY.

Power, J., Sicat, G., 1971. The Philippines: Industrialization and Trade Policies. Oxford University Press, London.

Rodrik, D., 2007. The Real Exchange Rate and Economic Growth: Theory and Evidence. Working Paper 2008-0141. Weatherhead Center for International Affairs, Harvard University.

Rodrik, D., 2013. Unconditional convergence. Q. J. Econ. 128, 165−204.

Romer, P.M., 1986. Increasing returns and long-run growth. J. Polit. Econ. 94, 1002−1037.

Romer, P.M., 1990. Endogenous technological change. J. Polit. Econ. 98, S71−S102.

Senga, K., 1983. A note on industrial policies and incentive structures in the Philippines 1949−1980. Philipp. Rev. Econ. Bus. 20 (3−4), 299−305.

Usui, N., 2012. Transforming the Philippine Economy: Industrial Upgrading and Diversification. Presentation before the ADB-AFD-JICA Joint Forum, 28 February. <http://fpi.ph/fpi.cms/News/Norio%20Usui%20-%20Presentation%20During%20the%20Summit.pdf>.

Williamson, J.G., 2011. Trade and Poverty: When the Third World Fell Behind. MIT Press, Cambridge, MA.

World Bank, 2011. World Development Indicators. Washington, DC.

Wright, G., 1990. The origins of American industrial success, 1879−1940. Am. Econ. Rev. 80, 651−668.

Yoshihara, K., 1994. The Nation and Economic Growth. Oxford University Press, Kuala Lumpur; New York, NY.

APPENDIX 1

Let y_t denote labor productivity in the whole economy at time t and Δy_t its change relative to a previous period k. Then:

$$\Delta y_t = \sum_{i=1}^{n} \theta_{i,t-k}(y_{i,t} - y_{i,t-k}) + \sum_{i=1}^{n} y_{i,t}(\theta_{i,t} - \theta_{i,t-k})$$

$$\frac{\Delta y_t}{y_t} = \sum_{i=1}^{n} w_{it} + \sum_{i=1}^{n} s_{it} = \sum_{i=1}^{n}(w_{it} + s_{it})$$

Here, $y_{i,t}$ is labor productivity in sector i at time t, and $\theta_{i,t}$ is the employment share of sector i at time t, with $\sum_i \theta_{i,t} = 1$ for all t. Dividing both sides by y_t (and properly annualizing) gives the annual rate of growth of productivity for the period as the sum of two terms: $\sum_{i=1}^{n} w_{it}$ is the change in productivity "within" the sector, while the second term $\sum_{i=1}^{n} s_{it}$ is a measure of structural change, reflecting the extent to which labor is pulled toward sectors where productivity is growing rapidly.

APPENDIX 2

Rodrik (2007: 6) explains his undervaluation index as:

$$\ln \text{UNDERVAL}_t = \ln \text{RER}_t - \ln \text{RER}_t^*$$

where $\ln \text{RER}t = \ln(\text{XRAT}_t/\text{PPP}_t)$, while $\ln \text{RER}^*_{it}$ is the predicted value from the equation: $\ln \text{RER}^*_t = a + b \ln \text{RGDPCH}_t + f_t + u_{it}$, which is estimated across countries with f_t being a time fixed-effect and a value of $b = -0.24$ is obtained to represent the "Balassa-Samuelson" effect of lower prices of nontraded goods in developing countries (so that currencies strengthen as their economies prosper). In the case of overvaluation, $\text{RER}_t/\text{RER}^*_t < 1$, or $\ln \text{UNDERVAL}_t < 0$. On the other hand, $\ln \text{UNDERVAL}_t > 0$ implies undervaluation.

BUNDLING DROUGHT TOLERANCE AND INDEX INSURANCE TO REDUCE RURAL HOUSEHOLD VULNERABILITY TO DROUGHT

22

Travis J. Lybbert and Michael R. Carter

Department of Agricultural and Resource Economics, University of California, Davis, CA

CHAPTER OUTLINE

JEL: O33; Q12; D80

22.1 INTRODUCTION

Drought is fundamentally a weather event characterized by below normal precipitation, but its human impacts are much more complex and nuanced than this simple climatic definition might suggest. In human welfare terms, drought effects are shaped by the geophysical, agronomic, social, economic, and political features of a given context. Among many of the world's poor, frequent drought combined with an unfortunate confluence of these mediating features makes drought one of their most worrisome perennial concerns. In rural rainfed settings, drought can damage or destroy crops and livestock, causing hunger and illness and reducing household food security. These immediate effects can induce lasting harm by triggering costly coping responses such as liquidating assets, skipping meals, and choosing inferior foods that can have persistent effects and reduce long-run household welfare. Moreover, it is not just the occurrence of drought that carries a hefty burden; for vulnerable households the very threat of drought can prevent them from taking full advantage of their resources and keep them poorer than they otherwise might be.

Since World War II, development and relief efforts have largely aimed to provide frontline responses to weather shocks such as floods and droughts and, more recently, to reduce household vulnerability to these shocks. While the essence of these policy objectives remains the same, climate change projections of more extreme and more frequent weather events in many poor places have added a new urgency to mitigating the weather risks borne by the rural poor. In the case of drought risk, a flurry of innovative interventions to reduce these risks has attracted substantial attention and funding from both the public and private sectors. Drought tolerance (DT) in crops and index insurance (II)—agronomic and financial innovations, respectively—have generated particularly high expectations.

This chapter argues that some of the hype around DT and II should be moderated by an appreciation of their respective limitations. In isolation, DT may protect against crop losses due to moderate drought but leave farmers' exposure to extreme drought risk virtually unchanged. While II might provide good protection against both moderate and severe drought, rural households may be unwilling to pay actuarially fair prices to access II that offers complete protection.

By exploring a simple complementarity between the two, this chapter demonstrates that properly bundling DT and II may restore a good deal of these innovations' promise. In particular, when bundled with DT that protects against moderate drought, II could be redesigned to cover only the kind of severe drought events that might overpower any protection offered by DT. Such an extreme event II product could be offered at substantially lower prices, making it more accessible to poor farmers. Further, an increase in II demand due to the implicit subsidy provided by DT may improve the long-run viability of such emerging financial markets.

22.2 DROUGHT RISK, VULNERABILITY, AND DEVELOPMENT INTERVENTIONS

The United Nations recently estimated that 1.5 billion people are vulnerable to drought (United Nations, 2012). Although difficult to assess rigorously, this vulnerability conceptually begins with drought as purely a precipitation-based measure; the necessary biophysical, infrastructural, social, economic, and political filters are then added to identify those populations for whom low precipitation imposes a serious welfare burden in terms of nutrition, income, assets, and future well-being. Given that each of these layers is spatially heterogeneous, the hundreds of millions of vulnerable households tend to be concentrated in specific regions. Drought vulnerability at the national level is highest in Asia and, especially, Africa, although pockets of it also exist in Central and South America (Eriyagama et al., 2009).

To appreciate the welfare effects of drought among vulnerable households, a few important nuances should be kept in mind. First, although drought often conjures up images of a weather-worn and weary farmer leaning on a hoe and surveying a dusty plot of withering crops that were supposed to produce food for his family, such images can be misleading. Most poor farmers rely heavily on markets for selling their products and even more heavily for *buying* food. This means that for many of the rural poor the direct impact of drought on their own food production can be less worrisome than the indirect effect on food prices and households' access to food.

Second, as mentioned earlier, drought can trigger a series of behavioral responses that imply that the total welfare burden of drought is much higher than what might be observed in the immediate aftermath of a drought event. In addition to the so-called *ex post* risk effect that encompasses these immediate effects, the threat of drought, like a bully, induces households to opt out of higher return livelihoods and store their assets in forms that have low or negative returns but high liquidity. Elbers et al. (2007), using a panel of household data from Zimbabwe to quantify the magnitude of these effects, found that more than half of the drought risk burden is due to this *ex ante* threat of drought rather than its actual occurrence. Similarly, much of the drought burden is essentially hidden from direct observation in many settings.

Lastly, individual drought coping responses can be aggregated across households in ways that magnify the overall drought burden in a region or country. For example, when livestock markets are poorly integrated, a shared drought event that prompts many to sell animals in order to fund food purchases can cause livestock prices to plummet once local markets aggregate this individual drought response. In an even more dramatic fashion, severe and persistent drought can spur mass migrations of displaced populations and lead to social and political tensions that, when mixed with other instabilities, can create serious conflict. As a case in point, several years of extreme drought in Syria after 2005—which forced many rural households to migrate to urban edges—seem to have set the stage for the subsequent socio-political instability (Hoerling et al., 2011). While extended drought was surely not the only cause of this civil conflict, it acted as a "threat multiplier" that interacted in potent ways with other existing threats (Johnstone and Mazo, 2011).

Since it is difficult to overstate the burden that drought can impose in some settings, several different angles have emerged to remedy its impacts.[1] Table 22.1 organizes several common drought-related interventions. While this table is intended primarily to provide a backdrop to the discussion of DT and II in the subsequent section, a few things are worth noting. Aside from infrastructure, information, and organizations, many of the interventions treat individuals or households directly, raising some important targeting issues. These issues arise because drought is less of a shared experience than it may seem; idiosyncratic factors (e.g., spatial variation in rainfall, local soil and topographical differences that affect how much moisture is retained in the soil, differences in cropping and livelihood strategies, and heterogeneous mitigation, coping and recovery capacities) mediate how a particular household bears up in a drought—and how much of an impact a particular intervention might have.

Many of the interventions in Table 22.1 aim to address *ex post* drought effects directly. Whether or not they also reduce *ex ante* effects depends largely on how reliable they are as perceived by key decision makers in households before drought occurs. Farmers are unlikely to put their households' welfare on the line based on interventions that may or may not help in the wake of a drought, and drought is likely to continue to impose an important *ex ante* burden. Agronomic technologies and practices may ultimately reduce households' vulnerability to drought, but such households must gain sufficient familiarity and experience with these innovations before they are likely to feel less bullied by the threat of drought.

---

[1]In an entirely different spirit, drought can bring distinct and important benefits to some interest groups. This can be particularly apparent in political realms where "everybody loves a good drought" (Sainath, 1996). Especially egregious and disturbing are claims that droughts provide opportunities to oppress or punish specific ethnic groups or other factions.

Table 22.1 Different Categories of Drought Interventions with Examples for Each

Type of Drought Intervention	Example
Acute relief	Food aid and humanitarian aid
	Refugee security and support
	Cash transfers and vouchers
Agronomic	Breeding staple crops for better drought resistance and early maturity
	Resource conservation practices such as zero tillage and water harvesting
	Permanent and supplemental irrigation
	Extension to promote adoption of improved inputs and practices
Financial	Income diversification and livelihood support programs
	Microcredit and savings
	Insurance
Organizational	Cooperatives and producer associations to provide access to inputs and higher value markets
	Social organizations to provide informal safety nets
Infrastructural and informational	Improved roads to reduce transport and transaction costs in order to better integrate markets
	Improved weather information and climate models to provide better daily, weekly, and seasonal weather forecasts

22.3 DT AND DROUGHT II: PROSPECTS AND COMPLEMENTARITY

The interventions explored in this chapter, DT and II, aim to directly reduce farmers' vulnerability to drought. While both are quite simple conceptually and closely related to familiar approaches to mitigating drought risk, each has evolved rapidly in recent years, thanks to a flurry of investments and innovations in the past decade. This section describes these recent experiences and the near-term prospects for DT and II to reduce drought vulnerability among the rural poor. It also introduces the complementarity between the two, which is the basis for the argument that a bundled DT—II product may be much more potent than each of them in isolation.

22.3.1 DROUGHT TOLERANCE

As an agronomic intervention, DT has been a long-standing breeding objective. Conventional breeding has often selected for resistance to moisture or temperature stress in order to target specific agroclimatic zones with improved varieties. What is innovative in the current generation of DT is the set of tools and techniques breeders are bringing to bear on the problem. Agricultural biotechnology tools such as marker-assisted selection have rapidly advanced conventional plant breeding. Genetic engineering likewise pushes the frontier of DT innovation. Although there are dissenting views about how much can be expected from this DT innovation stream (Gurian-Sherman, 2012), the promise and potential of these new breeding tools have attracted almost unprecedented investments aimed at DT research from both the private and public sectors. Major private sector players see substantial profit opportunities for these new generation DT technologies in North America, Australia, and elsewhere in the coming decades. For example, Monsanto estimates that the market for a DT trait in maize in the United States alone could be USD 500 million by 2020 (Padgette et al., 2010).

Monsanto, Syngenta, and Pioneer have all recently released new maize hybrids with new generation DT traits. Monsanto's genetically engineered DT products are marketed as Genuity Droughtgard Hybrids. Syngenta and Pioneer released their (nongenetically engineered) DT maize products in 2012 under the names Agrisure Artesian ("Maximize your yield without risking it all on weather") and Optimum AQUAmax, respectively. Although farmer experiences with these new DT maize hybrids are limited by their recent release, the severe drought conditions prevalent across much of the corn belt of the United States in 2012 provided a useful test of the potential DT benefits they confer. As reported by these seed companies (not peer-reviewed), on-farm trials of these hybrids showed average DT yield benefits in the range of 5−17%. Whereas the DT benefit of the Agrisure Artesian maize hybrids is negligible for environments with average yields above 100 bu/acre (6250 kg/hectare), the DT benefits for plots with more pronounced drought pressure are more dramatic (48% for plots with yields below 50 bu/acre [3125 kg/hectare]) (Syngenta, 2012).

The private sector's research and development of DT crops since 2000 has been matched by a similar explosion of public sector investments in DT technologies. Universities, national agricultural research institutes, and international research centers of the Consultative Group for International Agricultural Research (CGIAR) have made important breakthroughs in the area, with financial support from development agencies, research foundations, and other donors. Several public−private partnerships focused on DT crops have emerged. Two important such partnerships are focused on breeding DT crops for Africa. One is the Drought-Tolerant Maize for Africa (DTMA) project, which was started in 2007 and is coordinated by two CGIAR centers: International Maize and Wheat Improvement Center (CIMMYT) and International Institute of Tropical Agriculture (IITA). It involves several national agricultural research centers, public and private seed companies, and farmer groups in 13 countries. The other is the Water Efficient Maize for Africa (WEMA) project, which was begun in 2008 and is led by the African Agricultural Technology Foundation in collaboration with Monsanto, CIMMYT, and several national research centers. Together, these two initiatives have attracted substantial investments in DT research.

Four features of agricultural research in this area facilitate these public−private partnerships. One, DT research is often characterized by large upfront fixed costs and small (near zero) marginal costs of producing DT traits. Two, agroecological variation implies that breeding these DT traits into maize hybrids that are well adapted to local growing conditions can require significant late-stage breeding effort. Three, private seed companies clearly segment profitable markets for their products from lower priority markets. Four, many private and public agricultural research organizations are eager to demonstrate the potential of agricultural biotechnology in ongoing debates about the costs and benefits of new technologies based on these techniques. Thanks to the confluence of these four features, private seed companies have been eager to collaborate with public research organizations, and the public sector is similarly eager to collaborate with the private companies.

While it is premature to quantify the impact of the DTMA and WEMA projects on farmers' maize yields in Africa, the growing body of agronomic and field trial results suggests that the hybrids emerging from these public−private initiatives are likely to confer meaningful DT benefits. The WEMA project aims to release its first maize hybrids in 2014 and expects that "increasing yield under moderate drought could mean an additional two million tons of maize during drought years that could feed 14 to 21 million people" (WEMA, 2012). The DTMA project has started releasing some breeding material to national seed systems with the ultimate objective of "generating maize hybrids with a 1 ton per hectare potential under 'drought stress' conditions and increasing the average productivity of maize under smallholder farmers' conditions by 20−30%" (Abdoulaye et al., 2012). Based on many reports of field trials of these improved DT hybrids, the average yields under "managed drought stress" are often in the 1−3 tons/ha range (Makumbi, 2012). It is presumably based on such results that the main DTMA website quotes a maize farmer, saying, "This is like crop insurance within the seed."

In some settings, DT may indeed function like crop insurance, but in others it most assuredly does not—and knowing why not is important to understanding the inherent limitation of DT. Figure 22.1 provides a stylized explanation. Both panels depict the probability distribution function (pdf) of stochastic drought pressure ($\tilde{D}$) as function $g(\tilde{D})$. For now, think of this drought severity measure as a conventional measure like the Palmer drought index with average conditions indicated by zero and increasing positive values indicating increasing drought severity. This figure also depicts the net benefits associated with DT as function $\Delta y(\tilde{D}) = y_{DT}(\tilde{D}) - y_0(\tilde{D})$. The left panel shows these relationships for a hypothetical location that experiences moderate drought stress but never severe or extreme drought. In such a location, the farmer on the DTMA website is quite right: DT is like crop insurance built in to the seed. The right panel shows these relationships for a second hypothetical location that experiences drought ranging from moderate to severe to extreme. In such a drought-prone setting, DT is far from insurance-like: precisely when protection is needed most—during severe and extreme drought events—the DT benefits fall as drought worsens, eventually providing no benefit whatsoever. In this setting, smallholders will have a difficult time learning the value of DT, and—seemingly paradoxically—the most risk-averse farmers may be the least eager to adopt DT crops (Lybbert and Bell, 2010). Furthermore, in such a setting, DT crops do little to alleviate the *ex ante* drought burden since households remain under threat of extreme drought.

With this conceptual model in mind, it is worth returning to the DTMA results. Crops were subjected to "managed drought stress"—that is, crops were "grown during a rain-free period, with irrigation applied at the beginning of the season to establish a good plant stand, then irrigation was

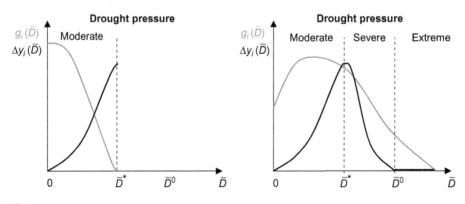

FIGURE 22.1

Stylized differences in drought severity, drought probabilities, and net DT benefits for two hypothetical locations.

withheld so that the crop suffered drought stress during flowering and grain-filling" (Makumbi, 2012). It is not clear whether this treatment simulated drought stress to the right or left of the "optimal drought stress" ($\tilde{D}^*$) at which DT benefits are maximized, but it is quite clear that reduced irrigation at the beginning of the season would quickly erode any DT benefits. While none of the DTMA trials enable one to understand the drought-yield profiles as shown in Figure 22.1, it is almost certainly true that smallholders in Sub-Saharan Africa are more likely to face drought risks like the panel on the right than the one on the left, which has direct implications on the prospects of DT maize in these settings. To underscore this point, consider anew the on-farm trial results reported by Syngenta (Syngenta, 2012). The lowest "yield environment" reported in these results (<50 bu/acre [3125 kg/hectare]) suggests an impressive 48% yield gain, but a maize yield of this level corresponds to the 90th percentile of maize yields in Ethiopia. It seems likely that the median Ethiopian maize farmer would confront much more modest DT benefits in severe drought.

22.3.2 INDEX INSURANCE

Agricultural II works not by insuring the farm household directly against its own specific income or yield losses,[2] but instead by insuring against a direct or predicted measure of the average or typical losses experienced by farms located in the vicinity of the household. An II contract can be represented as an indemnity schedule that links payments to an index that predicts typical losses in the zone covered by the index. To avoid problems of moral hazard and adverse selection, the level of the index cannot be influenced by the actions of the insured, nor can its level depend on which particular individuals choose to purchase the insurance.

Recent technological advances in remote sensing and automated weather measurement that permit estimation of crop losses (as well as the potential of older ideas like area yield insurance[3]) have opened the door to innovative II contracts. To tap the risk transfer potential of these advances one must appreciate both demand- and supply-side constraints, although a number of recent projects have shown that the supply-side challenges can be overcome.

Despite this supply-side progress, contract demand and uptake have been sometimes tepid. As discussed in Carter (2012), a number of demand-side challenges remain, including devising insurance indexes that are highly correlated with individual farmer outcomes (reducing residual uninsured or "basis" risk). An emerging but still small body of research shows that when these problems are resolved, the impacts of II can be substantial (Elabed et al., 2013; Janzen and Carter, 2013; Karlan et al., 2012).

Even though II for small farm agriculture is still a work in progress, this chapter focuses on its complementarities with DT varieties. That is, II can insure extreme event losses where even DT varieties fail, whereas DT is a more cost-effective risk management tool than II for less extreme events.

---

[2]Myriads of experiences show that trying to insure all sources of variation in agricultural outcomes for small farmers is beset by a host of problems rooted in the costs of obtaining information on small farm outcomes that renders such insurance infeasible (Hazell, 1992).

[3]Area yield insurance measures average yields in a defined geographic area (e.g., a valley or administrative district) and makes payments when these average yields fall below a specified "strike point" level.

As a package, they thus offer a relatively low cost but highly effective risk management solution. Demand for such a bundle is likely to be high,[4] creating a sustainable market for both products.

Prior to discussing the analysis that substantiates these points, it is worth mentioning one key difference between DT varieties and II. The development of DT varieties has extremely high fixed, upfront costs, but once the varieties are developed the marginal costs of offering the DT trait to farmers is close to zero. On the other hand, II has modest upfront development costs, but it requires the payment of an annual premium, which is composed of the expected payout plus an additional loading required to cover distribution and other administrative costs. While governments and donors appear willing to subsidize the upfront development costs for both DT varieties and II, they seem less willing to subsidize an annual premium. For this reason, in the analysis to follow, the DT seeds are assumed to cost the farmer no more than non-DT seeds, and it is assumed that the farmer must pay the full annual cost of the insurance premium.

This feature of the cost structure of the II industry provides a central motivation for bundling DT and II. Specifically, it implies that farmers' willingness to pay for these insurance products will critically determine the viability of these financial markets. In other words, poor farmers will only have access to an appropriate selection of II products if their demand for these products is sufficient to sustain and encourage the development of these markets. In some contexts, this reality may be at odds with existing evidence suggesting that farmers can be quite sensitive to price when facing II products.

22.3.3 THE DT−DROUGHT II COMPLEMENTARITY

Given the specific limitations of stand-alone DT and II, this section tackles the complementarity that makes bundling them potentially interesting. This complementarity, which is tied to drought severity, is quite simple. A bundled DT−II product can offer monotonic benefits as drought severity increases because II can incrementally cover the severe drought pressure beyond the point where relative DT benefits begin to fade. On the flip side, with DT covering low to moderate drought events, II can be designed so it only covers more rare and extreme droughts, which can substantially reduce the actuarially fair premium associated with the insurance. In this way, bundling II with DT may offer what governments and donors seem unwilling to provide: a long-term—albeit implicit—subsidy on II premiums.

Figure 22.2, which borrows the format and notation in Figure 22.1, graphically depicts the essence of this complementarity. The net yield gains associated with DT are represented as a triangular distribution centered on $\tilde{D}^*$. As before, these net DT gains begin to decrease in severe drought and are zero in extreme drought when crops fail (i.e., when yield is too meager to justify harvesting). Two II contracts are depicted in Figure 22.2: (a) a "full coverage" contract that pays out in moderate, severe, and extreme drought and (b) a potentially much cheaper "limited coverage" contract that pays out only in severe and extreme drought. Finally, the figure shows what a stylized DT−II bundle would look like in this case. Note that as drought intensity increases in the severe drought range the falling DT net gains are offset by rising II payoffs. In extreme drought, II continues to provide a payout. In this stylized example, the resulting payoff profile for bundled DT−II

---

[4]Work by Karlan et al. (2012) and McIntosh et al. (2013) shows that the demand for index insurance appears to be quite price elastic.

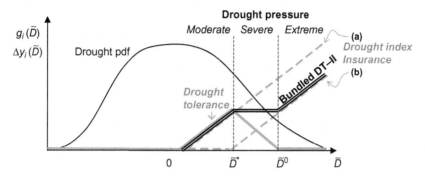

FIGURE 22.2

Stylized depiction of the complementarity between DT and II with two drought II contracts depicted: (a) high coverage, relatively expensive II and (b) low coverage, cheaper II designed to be bundled with DT.

is never decreasing in drought severity, which implies that such a bundled product can reduce both the *ex post* and *ex ante* drought risk burdens.

In practice, bundling a particular DT crop with II raises a few important—if somewhat nuanced—considerations. First, whereas the net payoff profile for a DT crop can only be altered through hard-earned breeding breakthroughs and innovations in agronomic practices or other inputs, the payoff profile for II can easily be changed to fit a given context. Before one can properly construct an II contract that complements a given DT crop, one must understand the net yield profile of the DT crop. Second, although the depiction in Figure 22.2 suggests that modifying the II contract to complement the DT crop is simply a matter of changing the strike point (i.e., the point along the drought index continuum where the contract begins to pay out), in practice optimizing the design of an II contract to fit a particular DT crop is likely to be more complex than this. To the extent that a DT trait changes the relationship between drought and yield at different levels of drought severity, optimizing the bundled II contract may require more than adjusting the strike point.

22.4 CALIBRATING AND EVALUATING A DT—II BUNDLE FOR MAIZE IN ECUADOR

This section presents an analysis of how a DT—II bundle might operate in practice using maize data from a drought-prone region of Ecuador. The Ecuadoran government annually collects yield data from random samples of producers in different regions of the country (Carter et al., 2014).[5] This analysis used maize yield data covering 2001—2011 from the coastal Guayquil department to estimate the underlying probability structure for the traditional maize yields. In Figure 22.3, the below average yields were standardized by the long-term average yields in the region (i.e., the mode of the probability distribution is at 100%). While factors other than drought stress explain some of the yield

---

[5]Data came from the annual Encuesta de Superficie y Producción Agropecuaria Continua (ESPAC) survey collected by the Ministry of Agriculture. More details on the survey can be found in Carter et al. (2014).

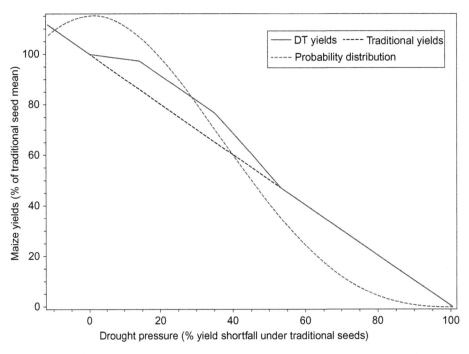

FIGURE 22.3

Probability distribution of drought pressure in Ecuador (measured as yield shortfall from average yield of traditional maize) and the stylized, hypothetical yield advantage of DT maize.

variability, for simplicity all yield declines were assumed to be driven by drought stress, and shortfalls from average yield were used as a proxy for drought pressure on the horizontal axis.

Based on the limited evidence described above, some stylized assumptions were made about DT impacts. The dashed 45° line in Figure 22.3 graphs the yield shortfall under the traditional technology as a function of itself as a benchmark, and the solid line displays the assumptions about DT yields. The assumptions are as follows:

- For moderate drought pressure, DT stabilizes yields at nearly their long-term expected average even as yields under the traditional technology fall up to 15% short of that average (there is a 15% probability that yields will fall in this range).
- As drought pressure increases, and traditional yield shortfall increases from 15% to 35% of the long-term average, DT yields also begin to slowly decline.
- DT maintains a 20% yield advantage, compared with non-DT seeds, over this range (there is a 20% probability that drought pressure and yields will fall in this range).
- As drought pressure further increases the yield shortfall to 55% of the long-term average (a 10% probability), the advantages of DT disappear and DT yields become identical to traditional yields over the lowest 5% tail of the probability distribution.

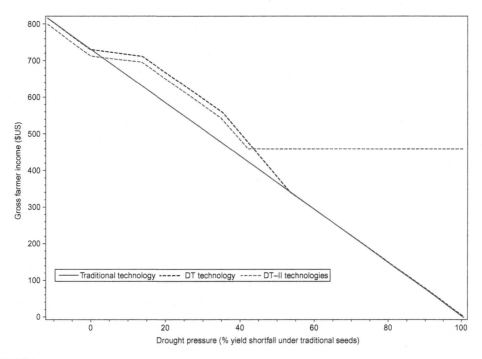

FIGURE 22.4

Performance of DT and bundled DT—II products measured in gross farmer income as a function of drought pressure in Ecuador.

In summary, the assumption is that DT affords modest to strong yield advantages 90% of the time under unfavorable conditions.

Further, two stylized II contracts were considered. The first has a strike point of when the yield shortfall reaches 15% of the long-term average. Under this contract, farmers are compensated dollar-for-dollar for every loss in area yield that occurs beyond this point. Under this high coverage contract, farmers would never receive less than the value of 85% of their long-term mean yield, less the cost of the premium. The other contract considers a lower strike point and has a low coverage; it begins to pay off when the yield shortfall hits 35% of the average, stabilizing farm incomes at levels that would be realized under these conditions. This lower coverage strike point coincides with the point where drought pressure begins to overpower DT and DT benefits begin to fade.

Figure 22.4 compares these alternative agronomic and financial technologies. Using the current market value of maize in Ecuador, the traditional maize returns an average gross income of USD 716 per acre (1790 per hectare), whereas DT varieties under the study's assumptions would return USD 750 (1875 per hectare).[6] While this is a modest increase, DT also has a substantial impact on risk. Figure 22.4 illustrates gross incomes under traditional (solid line) and DT technologies

---

[6]The proportionate increase in net income would, of course, be much larger, assuming that the cost structures of DT and traditional production are identical.

Table 22.2 Consumption and Certainty Equivalent Performance of DT, II, and Bundled DT−II

	Additional Cost Above Traditional Technology (USD/acre)	Mean Gross Income, USD (Net of Insurance Costs)	Certainty Equivalent, USD/acre	% Change Certainty Equivalent
Traditional maize	–	716	675	
DT maize	–	750	715	6.1
II-high coverage (15% yield shortfall strike)	66	710	692	2.6
II-low coverage (35% yield shortfall strike)	20	718	688	1.9
Bundled DT−II with low coverage II "optimized" for DT yield distribution	13	748	723	7.2

(dashed non-horizontal line) as a function of drought pressure. To value this change in risk, the certainty equivalent of the gross income streams implied by the two technologies was calculated, assuming that individuals' preferences can be characterized by a constant relative risk aversion utility function, with a coefficient of relative risk aversion of 1.1. As Table 22.2 shows, the certainty equivalent of the DT technology is 6.1% higher than that of the traditional technology (USD 715 vs. 675 per acre [USD 1787 vs. USD 1687 per hectare]).

Table 22.2 also reports the certainty equivalent value of stand-alone II that is introduced to farmers using traditional technology. The high coverage contract would cost USD 66 per acre (USD 165 per hectare) and modestly reduces expected gross income because the insurance is priced at 20% above the actuarially fair price. However, even with this markup, this insurance would increase the certainty equivalence by almost 3%. The low coverage contract, when introduced with traditional technology, costs USD 20 per acre (50 per hectare) and increases certainty equivalence by about 2%.

As the comparison of the two II contracts makes clear, farmers would value the additional stabilization of yields provided by the higher strike point contract. However, can this additional stabilization be provided more efficiently by DT seeds rather than by a high strike point insurance contract? Similarly, the certainty equivalence of DT technology is reduced by the fact that the technology fails under extreme conditions. Can the additional stabilization provided by a low strike point contract offer substantial additional value to the risk-averse farmer?

The final row in Table 22.2 addresses these questions. The analysis assumed that the low strike contract is calibrated to DT yields. That is, the contract only begins to pay off when the DT (not traditional) yield shortfall is 35% of the long-term average of traditional seeds. Since these lower yields are less likely with DT seeds, the cost of this insurance falls from USD 20 to 13 per acre (USD 50 to 32.50 per hectare).[7] In combination, the DT−II bundle yields a gross income function given by the dotted line with the horizontal "floor" in Figure 22.4. The bundle stabilizes gross farmer income levels under the most extreme conditions. It also reaps the benefits of DT under

---

[7]This insurance price would further incentivize the adoption of DT varieties.

moderate drought pressure. As reported in Table 22.2, the certainty equivalent of the DT−II bundle is USD 723, or 7.2%, higher than the certainty equivalent returns of the traditional technology. While it is doubtlessly possible to design more finely-tuned packages, this simple example does illustrate the complementarity between DT and II technologies, based on their complementary statistical properties.

22.5 CONCLUSION

Many of the world's poor are vulnerable to drought in some way. Climate change adds urgency to alleviating their drought vulnerability, which imposes a serious welfare burden on the poor. While DT and II have attracted substantial attention as stand-alone interventions, this chapter argues that their true potency is likely to emerge only when they are properly bundled. Such bundling overcomes their stand-alone limitations by leveraging a fundamental complementarity between the two. This complementarity was demonstrated conceptually using maize data from Ecuador; the analysis illustrated how a bundled product could look and how it might affect household welfare.

The argument for bundling DT and II rests firmly on its prospective benefits to farm households. This perspective (i.e., benefits to farming households) must remain central to any effort to calibrate and refine a DT−II product in practice. How should an II contract be optimized to properly reflect a given DT benefit profile? What information will farmers need about these DT-optimized contracts to appreciate the synergies associated with bundled DT−II? In this analysis, the case for bundled DT−II was found compelling in contexts where the risk of extreme drought and total crop loss is nontrivial, although several key design and delivery questions remain. With the expected release of many DT crops in the near future, now is the time to begin exploring these questions.

REFERENCES

Abdoulaye, T., Bamire, A.S., Wiredu, A.N., Baco, M.N., Fofana, M., 2012. Characterization of maize-producing communities in Bénin, Ghana, Mali, and Nigeria: West Africa regional synthesis report. Drought Tolerant Maize for Africa (DTMA) Project—Community Surveys. IITA, Ibadan, Nigeria.

Carter, M., 2012. Designed for development impact: next generation approaches to index insurance for small farmers. In: Churchill, G., Matul, M. (Eds.), Protecting the Poor: A Microinsurance Compendium, volume 2. International Labor Organization, Geneva.

Carter, M., Boucher, S., Castillo, M.J., 2014. Index Insurance: Innovative Financial Technology to Break the Cycle of Risk and Rural Poverty in Ecuador. Working Paper.

Elabed, G., Bellemare, M.F., Carter, M.R., Guirkinger, C., 2013. Managing basis risk with multi-scale index insurance contracts. Agric. Econ. 44, 419−431.

Elbers, C., Gunning, J-W., Kinsey, B., 2007. Growth and risk: methodology and micro evidence. World Bank Econ. Rev. 21 (1), 1−20.

Eriyagama, N., Smakhtin, V., Gamage, N., 2009. Mapping drought patterns and impacts: a global perspective. IWMI Research Report. International Water Management Institute, Colombo, Sri Lanka.

Gurian-Sherman, D., 2012. High and Dry: Why Genetic Engineering Is Not Solving Agriculture's Drought Problem in a Thirsty World. Union of Concerned Scientists, Cambridge, MA.

Hazell, P.B., 1992. The appropriate role of agricultural insurance in developing countries. J. Int. Dev. 4 (6), 567−581.

Hoerling, M., Eischeid, J., Perlwitz, J., Quan, X.W., Zhang, T., Pegion, P., 2011. On the increased frequency of Mediterranean drought. J. Clim. 25, 2146−2161.

Janzen, S.A., Carter, M.R., 2013. After the drought: the impact of microinsurance on consumption smoothing and asset protection. NBER Working Paper (No. 19702).

Johnstone, S., Mazo, J., 2011. Global warming and the Arab spring. Survival 53 (2), 11−17.

Karlan, D., Osei, R.D., Osei-Akoto, I., Udry, C., 2012. Agricultural decisions after relaxing credit and risk constraints. NBER Working Paper (No. 18463).

Lybbert, T., Bell, A., 2010. Stochastic benefit streams, learning, and technology diffusion: why drought tolerance is not the new Bt. AgBioForum 13, 1.

Makumbi, D., 2012. Results of the 2011 Regional Trials Coordinated by CIMMYT-Kenya. CIMMYT, Nairobi, Kenya.

McIntosh, C., Sarris, A., Papdopoulos, F., 2013. Productivity, Credit, Risk, and the Demand for Weather Index Insurance in Smallholder Agriculture in Ethiopia. Agric. Econ. 44 (4-5), 399−417.

Padgette, S., Goette, J., Mazour, C., 2010. Drought-tolerant corn. Retrieved from < http://www.monsanto.com/SiteCollectionDocuments/whistlestop-drought-posters.pdf > . (Accessed 13 02 13.)

Sainath, P., 1996. Everybody Loves a Good Drought: Stories from India's Poorest Districts. Penguin Books, New York, NY.

Syngenta, 2012. Genuity droughtgard hybrids: do more with less water - that's hydroefficiency. Retrieved from < http://www.genuity.com/corn/Pages/Genuity-DroughtGard-Hybrids.aspx > . (Accessed 16.07.14.)

United Nations, 2012. Desertification, drought affect one third of planet, world's poorest people, second committee told as it continues debate on sustainable development. Retrieved from < http://www.un.org/News/Press/docs/2012/gaef3352.doc.htm > . (Accessed 16.07.2014.)

WEMA, 2012. Water efficient maize for africa report. may. Retrieved from < http://wema.aatf-africa.org/wema-report-may-2012 > . (Accessed 13.02.2013.)

HAVE NATURAL DISASTERS BECOME DEADLIER?

23

Raghav Gaiha*, **Kenneth Hill**[†], **Ganesh Thapa**[‡], and **Varsha S. Kulkarni**[§]

**Department of Global Health and Population, Harvard School of Public Health, Boston, MA and Faculty of Management Studies, University of Delhi, Delhi, India* [†]*Harvard Centre for Population and Development Studies, Boston, MA* [‡]*International Fund for Agricultural Development, Rome, Italy* [§]*School of Informatics and Computing, Indiana University, Bloomington, IN*

CHAPTER OUTLINE

JEL: H84; Q54; Q58

23.1 **INTRODUCTION**

Recent analyses point to greater vulnerability of developing countries to natural disasters, as reflected in a marked increase in the frequency of their occurrence and in their costliness in recent years (World Bank, 2006, 2010; Birkmann, 2006; Kahn, 2005; Gaiha and Thapa, 2006; *The Economist*, 2012b; Thomas et al., 2012).

The present study sought to build on earlier work by identifying factors associated with the frequency of natural disasters and the resulting mortality. From the main findings, observations were made from a policy perspective, focusing on key elements of a disaster and mortality reduction strategy.

Few would question the rising cost of natural disasters, especially in developing countries. The Indian Ocean tsunami in December 2004 killed over 250,000 people. The earthquake that followed, centering on Kashmir, killed tens of thousands and left over three million homeless. Meanwhile, poor harvests and pests threaten famine in the Sahel and Southern Africa. The overall picture of disaster impacts is one of large-scale human suffering, loss of lives, and a precipitous rise in financial costs. The global costs of natural disasters have risen 15-fold since the 1950s and more sharply in recent years.[1]

Perhaps more serious in the context of developing countries is their vulnerability to a multitude of disasters in short spells. A report in *Time* (Beech, 2009), for example, cited a few countries in the Asia-Pacific region that suffered disasters in a week's time and consequent damage and loss of human lives:

> In late September, tropical storm Ketsana killed more than 160 people in Vietnam and nearly 300 in the Philippines, submerging 80% of Manila. Just hours before Sumatra was jolted, another earthquake triggered a tsunami that inundated the Samoan islands and Tonga, extinguishing some 180 lives. In the latest catastrophe, southern India was ravaged by some of the worst torrential rains in decades, killing around 300 people and leaving some 2 million others homeless.

Hazards are broadly classified into two types: hydrometeorological (e.g., floods, droughts, storms) and geophysical (e.g., earthquakes, volcanic eruptions, and related tsunamis). Their impacts differ,[2] which is partly due to differences in their frequency of occurrence and predictability. Some of these differences are discussed in Annex 2.

23.2 **ISSUES**

The present study was motivated by a broad concern for human well being, in which resilience against natural disasters is a key element. A specific concern is that natural disasters are often far deadlier in

---

[1]Disaster costs in material losses rose from USD 38 billion (at 1998 values) in 1950−1959 to USD 652 billion in 1990−1999 (World Bank, 2006). More recent evidence points to sharply rising economic losses. "... Multi-billion dollar natural disasters are becoming common. Five of the ten costliest, in money rather than lives, were in the past four years... Munich RE, a reinsurer, reckons their economic costs were $378 billion last year, breaking the previous record of $262 billion in 2005 (in constant 2011 dollars)" (*The Economist*, 2012a).

[2]For definitions of natural hazards and classification into hydrometeorological and geophysical, see Table A.1.1.

low-income countries. For example, between 1980 and 2002, India experienced 14 major earthquakes that killed 32,117 people, while the United States, which experienced 18 major earthquakes, had only 143 people dead (Kahn, 2005). Designed to throw light on the underlying geographic, institutional, and development variables, this study specifically addresses the following issues:

- Have natural disasters become more frequent?
- Have the death tolls from these disasters increased in more recent years?
- How important is the role of geophysical and climate-related factors in causing the different types of natural disasters?
- Do relatively affluent countries suffer fewer deaths than low-income countries[3]?
- Do institutions matter? Specifically, are there fewer deaths in a democracy?
- Is the payoff to disaster prevention high? Specifically, does learning from past experience help to save lives?

Some of these issues are addressed in Kahn (2005). However, the specifications and samples used in the present study differ, as do the findings.[4]

Extant literature is not cast in a theoretical framework, but it addresses the larger issues of whether or not frequency of extreme events (e.g., droughts, floods) is linked to climate change and if the latter is largely anthropogenic.[5] In both cases, the links are conjectural. This study was thus essentially an empirical investigation. Its points of departure from extant literature are confined to the specification and method of estimation.

23.3 **DATA**

The above issues were addressed with the help of a database compiled from Emergency Events Database (EM-DAT), World Development Indicators (WDI), Food and Agriculture Organisation Statististical Database (FAOSTAT), and from the website of the Kennedy School of Government at Harvard, among others. The main component is EM-DAT, which covers all countries over the entire twentieth century. It reports the types of disasters, their dates and locations, and the numbers killed, injured, made homeless, and otherwise affected. An event qualifies for inclusion in EM-DAT if it is associated with: (1) 10 or more people reported killed; (2) 100 or more people affected, injured, or homeless; or (3) a declaration of a state of emergency and/or an appeal for international assistance made.

The EM-DAT quality has improved since the 1970s. For a better focus, the present analysis used data for the period 1980−2004, with different subperiods for specific exercises.

A recent review draws attention to the following problems/gaps in the EM-DAT:[6]

- Data coverage is incomplete for several categories. The numerical data categories (e.g., numbers killed, total affected) are unsatisfactorily represented before 1970, with many recorded

---

[3]For a classification of countries into low-, lower middle-, and upper middle-income groups, see Table A.2.1.
[4]When the number of disasters in a country is known, it is intriguing that a probit is used that allows for a dichotomous classification of whether or not a country experienced a natural disaster. More seriously, while occurrence of natural disasters is endogenous, in mortality equations, number of natural shocks/disasters is treated as exogenously determined.
[5]For a clear exposition of difficulties in specifying a theoretical framework, see Kousky (2013).
[6]For details, see Brooks and Adger (2005).

events having no entries for numbers killed or affected. Even after 1970, data are patchy for some countries and event types.

- According to a report by Working Group 3 of the Inter-Agency Task Force of the International Strategy for Disaster Reduction, a comparison between EM-DAT and the DesInventar disaster database[7] for Chile, Jamaica, Panama, and Colombia shows a substantial difference in the reported number of people affected. Differences in numbers killed are, however, much smaller and "generally of the same order of magnitude" (Brooks and Adger, 2005, p.15). In any case, a general consensus is that mortality data are more robust across different data sets.[8]
- The economic losses consist of direct and indirect components (Andersen, 2005). Direct losses refer to the physical destruction of assets, comprising private dwellings, small business properties, industrial facilities, and government assets, including infrastructure (e.g., roads, bridges, ports, telecommunications) and public facilities (e.g., hospitals, schools). Indirect losses, on the other hand, refer to the disruption of economic activities and loss of employment and livelihoods. In addition, business pessimism could dampen investment and consequently growth. As such, the relationship between destruction of capital and loss of income may vary a great deal. Although economic losses have sharply risen—especially in recent years—available estimates are incomplete and unreliable (Andersen, 2005; Kousky, 2013).

A World Bank study (2006) points out the lack of private insurance against natural hazard risk in most developing countries. Specifically, while about half of the costs of natural disasters are covered by insurance in the United States, less than 2% of them are covered in the developing world. Besides, the costs of hedging against natural hazard risks in developing countries often exceed the cost of simply paying for the damages when they occur. Also, rich countries are much more aware of and prepared for such risks. This present study thus restricted its analysis to the sample of countries other than the rich ones (including members of the Organization for Economic Cooperation and Development [OECD] members and non-OECD members).

23.3.1 FREQUENCY AND DEADLINESS OF NATURAL DISASTERS

The first part of this section considers the frequency distributions of natural disasters and their deadliness. Deadliness could be measured as number of deaths per disaster or number of deaths per million population. While this study used both definitions, it prefers the latter as it allows for the fact that the same disaster could be deadlier in more populous countries.[9] These measures are given in Table 23.1.

The aggregate sample indicated floods as the most frequent disaster (more than one-third of the total disasters during 1985–1994), followed by windstorms (more than a quarter of the disasters). With the exception of insect infestations, all natural disasters became more frequent in the next

---

[7]http://www.desinventar.org.

[8]For further validation, see Annex 3, and, from different perspectives, Thomas et al. (2012), Jennings (2011), and Munich (2011). Annex 3 reports results of whether frequency of disasters and mortality varied with exposure to newspapers. The absence of significant relationships is shown in Figures A.3.8 and A.3.9.

[9]For a disaggregated analysis by type of disaster (i.e., whether hydrometeorological or geophysical), see Annex 2 and for an update, see Thomas et al. (2012) and Jennings (2011).

Table 23.1 Types of Natural Disasters and Their Death Tolls

Disaster Type	Frequency of Disaster		Deaths		Deaths per Million		Deaths per Disaster		Disasters per Million	
	1985–1994 (%)	1995–2004 (%)	1985–1994 (%)	1995–2004 (%)	1985–1994	1995–2004	1985–1994	1995–2004	1985–1994	1995–2004
Flood	506 (34.59)	904 (39.86)	51,383 (13.35)	89,975 (12.97)	12	19	102	100	0.12	0.19
Earthquake	200 (13.67)	238 (10.49)	104,124 (27.06)	69,434 (10.01)	25	14	521	292	0.05	0.05
Drought	96 (6.56)	188 (8.29)	2911 (0.76)	1798 (0.26)	1	0.4	30	10	0.02	0.04
Famine	25 (1.71)	32 (1.41)	8640 (2.25)	223,644 (32.24)	2	46	346	6989	0.01	0.01
Extreme temperature	42 (2.87)	112 (4.94)	3897 (1.01)	13,721 (1.98)	1	3	93	123	0.01	0.02
Insect infestation	44 (3.01)	17 (0.75)	0 0.00	0 0.00	0.0	0.0	0.0	0.0	0.01	0.004
Landslide	117 (8.00)	169 (7.45)	7695 (2.00)	8236 (1.19)	2	2	66	49	0.03	0.04
Wave surge	4 (0.27)	21 (0.93)	111 (0.03)	228,943 (33.00)	0.03	48	28	10,902	0.00	0.00
Volcano eruption	35 (2.39)	38 (1.68)	24,425 (6.35)	227 (0.03)	6	0.05	698	6	0.01	0.01
Wildfire	26 (1.78)	73 (3.22)	438 (0.11)	332 (0.05)	0	0	17	5	0.01	0.02
Windstorm	368 (25.15)	476 (20.99)	181,222 (47.09)	57,379 (8.27)	44	12	492	121	0.09	0.10
Total	1463 (100.00)	2268 (100.00)	384,846 (100.00)	693,689 (100.00)	92	144	263	306	0.35	0.47

Calculations based on EM-DAT.

Table 23.2 Frequency Distribution of Natural Disasters by Region

Region	Number of Disasters		Relative Frequency of Disasters (%)	
	1985–1994	1995–2004	1985–1994	1995–2004
Latin America and the Caribbean	326	499	22.28	22.0
South Asia	249	327	17.02	14.42
East Asia and the Pacific	465	659	31.78	29.06
Europe and Central Asia	116	270	7.93	11.90
Middle East and North Africa	95	145	6.49	6.39
Sub-Saharan Africa	212	368	14.49	16.23
Total	1463	2268	100	100

Calculations based on EM-DAT.

period (i.e., 1995–2004). The share of floods and droughts rose, while those of windstorms, famines, and insect infestations declined.

Ratios of disasters to populations (disasters per million population) also increased from 0.35 to 0.47. Corresponding ratios of floods and droughts rose, while the values for other disasters changed little. Total deaths due to disasters rose sharply in the second subperiod, faster than the increase in the number of disasters. Deaths per disaster rose by more than 16%, from 263 to 306. Deaths per million population also increased by about 57%, from 92 to 144.

Among the deadliest (per disaster) during 1985–1994, volcano eruptions and earthquakes became less deadly in the next decade. In terms of deaths per million population, the deadliest were windstorms and earthquakes during 1985–1994; both also became less deadly during the next decade. By contrast, the deadliness of wave surges and famines sharply rose.

In sum, natural disasters had become more frequent and deadlier during the periods studied. However, this observation is subject to the caveat that the averages reported cannot be the basis of trends.[10]

Table 23.2 shows the frequency distribution of natural disasters by region for two subperiods, 1985–1994 and 1995–2004.

During 1985–1994, the largest number of disasters occurred in East Asia and the Pacific, accounting for just under one-third of the total natural disasters. Latin America and the Caribbean was a close second, accounting for over 22%. The Middle East and North Africa as well as Europe and Central Asia recorded the lowest numbers, with relative frequencies between 6% and 8%.

The number of disasters significantly increased across regions during the next decade, from a total of 1463 to 2268. On the other hand, the relative frequencies changed slightly, with lower

[10](i) For confirmation of rising trends in the frequency of natural disasters, especially hydrometeorological, see Thomas et al. (2012) and Jennings (2011). (ii) For an analysis of the more rapid rise in the frequency of hydrometeorological disasters and in mortalities associated with them, see Table A.2.1.

Table 23.3 Number of Natural Disasters by Region

Region	Number of Disasters (per Country)		Number of Disasters (per Million)	
	1985–1994	1995–2004	1985–1994	1995–2004
Latin America and the Caribbean	13.58	20.79	0.77	1.0
South Asia	35.57	46.71	0.23	0.25
East Asia and the Pacific	24.47	34.68	0.29	0.37
Europe and Central Asia	7.25	16.88	0.28	0.65
Middle East and North Africa	8.64	13.18	0.44	0.56
Sub-Saharan Africa	7.57	13.14	0.49	0.66
Total	13.93	21.60	0.35	0.47

Calculations based on EM-DAT.

concentrations in South Asia, East Asia and the Pacific, and Latin America and the Caribbean. By contrast, the shares of Europe and Central Asia and of Sub-Saharan Africa rose.

The vulnerability of different regions to natural disasters may be better understood when the disasters' impacts are expressed per country and per million population in each region. As may be noted from Table 23.3, *ex post* vulnerability of regions to disasters differs depending on the ratio used.[11]

South Asia had the highest number of disasters per country during 1985–1994, while Europe and Central Asia and Sub-Saharan Africa had the lowest. All regions recorded large increases in the number of disasters in the next decade, with South Asia continuing to be the most vulnerable, followed by East Asia and the Pacific. Sub-Saharan Africa remained the least vulnerable, followed by the Middle East and North Africa.

In the aggregate sample, the number of disasters increased substantially over the period in question, implying greater vulnerability. However, the picture strikingly changes when disasters are expressed per million population. This is not surprising, though, as populations of countries *within* a region vary. Specifically, the most vulnerable were Latin America and the Caribbean during 1985–1994, followed by Sub-Saharan Africa and by the Middle East and North Africa. Latin America and the Caribbean remained the most vulnerable in 1995–2004, followed by Sub-Saharan Africa and by Europe and Central Asia. On the other hand, South Asia and East Asia and the Pacific remained the least vulnerable.

All regions witnessed a rising frequency of disasters—using either ratio—but the increase in disasters per million was relatively small in South Asia.

Another classification, this time based on per capita income (based on World Bank's classification), yields somewhat surprising results.[12] In Table 23.4, the highest frequency of disasters occurred

---

[11]Vulnerability to disasters is in principle measured as the probability of occurrence of a disaster. As an approximation, actual frequency of disasters was used in this study. The higher the frequency of disasters in a region or another group of countries in a given period, the greater is their vulnerability to disasters.
[12]For details of the World Bank classification, see Table A.1.2.

Table 23.4 Number of Natural Disasters by Income Groups

Income Group	Number of Disasters		Relative Frequency of Disasters (%)	
	1985–1994	1995–2004	1985–1994	1995–2004
Low income	544	871	37.18	38.40
Lower middle income	774	1141	52.90	50.31
Upper middle income	145	256	9.91	11.29
Total	1463	2268	100	100

Calculations based on EM-DAT.

Table 23.5 Ratio of Natural Disasters by Income Groups

Income Group	Number of Disasters (per country)		Number of Disasters (per million)	
	1985–1994	1995–2004	1985–1994	1995–2004
Low income	11.83	18.93	0.32	0.42
Lower middle income	19.85	29.26	0.35	0.46
Upper middle income	7.25	12.80	0.59	0.90
Total	13.93	21.60	0.35	0.47

Calculations based on EM-DAT.

in lower middle-income countries during 1985–1994, followed by low-income countries. These two groups accounted for 90% of the disasters recorded in the sample. Although (absolute) frequencies rose sharply in both groups in the following decade, their (combined) share decreased only slightly.

In Table 23.5, disaster ratios are presented according to the countries' income group. Using the ratio of disasters per country, lower middle-income countries also had the highest frequency during 1985–1994, followed by low-income countries. In each group, the frequency rose sharply and, as a result, so did the aggregate sample during the following decade. When the ratio of disasters to population was used, upper middle-income countries recorded the largest number of disasters per million during 1985–1994, followed by lower middle-income countries. Vulnerability to natural disasters rose in all groups, especially in the upper middle-income countries.

In sum, the frequency of disasters rose in the period 1995–2004, relative to 1985–1994. However, there was no clear correspondence (1) between the frequency of disasters and level of income and (2) between the relative frequency of disasters and level of income. Changes in the (absolute and relative) frequency of disasters also followed a somewhat complex pattern.

The second part of this section examines the distribution of deaths associated with natural disasters over the period 1985–1994 and its changes in 1995–2004, using regional and income classifications.

As Table 23.6 shows, more than half of the deaths in 1985–1994 occurred in South Asia, with nearly equal but considerably lower shares in Latin America and the Caribbean, East Asia and the

Table 23.6 Regional Distribution of Deaths Due to Natural Disasters

Region	Deaths		Number of Deaths per Disaster		Number of Deaths (per million)	
	1985–1994 (%)	1995–2004 (%)	1985–1994	1995–2004	1985–1994	1995–2004
Latin America and the Caribbean	48,016 (12.48)	66,826 (9.63)	147	134	114	136
South Asia	200,357 (52.06)	132,838 (19.15)	805	406	181	102
East Asia and the Pacific	48,210 (12.53)	436,517 (62.93)	104	662	30	245
Europe and Central Asia	30,231 (7.86)	7804 (1.12)	261	29	74	19
Middle East and North Africa	44,002 (11.43)	36,789 (5.30)	463	254	205	140
Sub-Saharan Africa	14,030 (3.65)	12,915 (1.86)	66	35	32	23
Total	384,846 (100)	693,689 (100)	263	306	92	144

Calculations based on EM-DAT.

Pacific, and the Middle East and North Africa. A dramatic change occurred in 1995–2004: the share of East Asia and the Pacific climbed to about 63%, while that of South Asia dropped to less than one-fifth. The shares of the Middle East and North Africa, Latin America and the Caribbean, Europe and Central Asia, and Sub-Saharan Africa also registered reductions.

When the ratio of deaths to disasters was used as an indicator of their deadliness, South Asia ranked the highest in 1985–1994, followed by the Middle East and North Africa and by Europe and Central Asia. In the following decade, East Asia ranked highest, followed by South Asia and by the Middle East and North Africa. East Asia and the Pacific also registered a large increase in the ratio in question, while South Asia recorded a huge reduction.

Alternatively, deadliness of disasters could be expressed as deaths per million population. In this case, the aggregate sample indicates that the disasters became deadlier between 1985–1994 and 1995–2004. However, the regional ranking changes. In 1985–1994, the Middle East and North Africa were at the top, followed closely by South Asia. In the following decade, deadliness of disasters was highest in East Asia and the Pacific, followed by the Middle East and North Africa and by Latin America and the Caribbean. More than a moderate reduction in the deadliness of disasters was observed in South Asia and in the Middle East and North Africa.

The preceding analysis was repeated for income groups. The main findings are as follows (Table 23.7):

- A vast majority of disaster-related deaths occurred in low-income and lower middle-income countries (about 97% of total deaths in 1985–1994).

Table 23.7 Deadliness of Disasters by Income Group

Income Group	Deaths		Number of Deaths per Disaster		Number of Deaths (per million)	
	1985–1994 (%)	1995–2004 (%)	1985–1994	1995–2004	1985–1994	1995–2004
Low income	222,863 (57.91)	351,214 (50.63)	410	403	132	173
Lower middle income	149,639 (38.88)	307,439 (44.32)	193	269	67	123
Upper middle income	12,344 (3.21)	35,036 (5.05)	85	137	50	124
Total	384,846 (100)	693,689 (100)	263	306	92	144

Calculations based on EM-DAT.

- Although large increases in deaths were recorded in the following decade, the combined share of these two income groups recorded a slight reduction only (it fell to about 95%).
- Ratio of deaths to disasters was highest in low-income countries, followed by lower middle-income countries. Disasters, however, were only about half as deadly in the latter compared with the former. While the deadliness of disasters in low-income countries reduced slightly in the next decade, it remained substantially higher than those of the other income groups.
- There was a marked increase in the deadliness of disasters in the remaining two groups, especially in lower middle-income countries.
- A similar pattern can be seen in the analysis using deaths per million, except that deadliness of disasters rose in all income groups in 1995–2004. It increased 2.5 times in the case of upper middle-income countries.

The preceding analysis indicates an increase in the frequency of natural disasters, as well as higher deaths. Some regional and income group contrasts, however, require more detailed investigation. Again, these averages are subject to the caveat that they cannot be interpreted as trends.

23.3.2 RECENT PATTERNS IN NATURAL DISASTERS

Three studies (Thomas et al., 2012; Jennings, 2011; Munich, 2011) offer updated and insightful accounts of frequency of different types of natural disasters (broadly distinguished into geophysical and hydrometeorological) up to 2010.

The analysis of Thomas et al. (2012) has two interesting aspects. One, it focused on intense disasters (i.e., those killing 100 or more persons or affecting 1000 or more persons) over the period 1971–2010, which were likely to be better recorded and thus less likely subjected to reporting biases. Two, it disaggregated natural disasters into geophysical and hydrometeorological. The related issue of whether or not hydrometeorological disasters are causally linked to climate change, given the available evidence, was examined first globally and then in greater detail for Asia and the Pacific, particularly the Philippines.

Between 1971 and 1980, 539 intense natural disasters were recorded. The frequency of their occurrence continued to increase across decades; the total number increased four times between 2001 and 2010. Such global trends are largely a result of the rise in hydrometeorological disasters: about two-thirds of natural disasters during 1971–2010 were hydrometeorological. Their number increased by 66%, from 1210 in 1991–2000 to 2004 in 2001–2010.

An econometric analysis of the risks of intense hydrometeorological disasters in Asia and the Pacific region suggests that these risks are greater in more populous countries.[13] This is understandable, as disasters are defined in terms of number of people killed or affected. On the other hand, population also serves as a useful proxy for exposure. The higher the vulnerability of a country (defined in terms of high population density and low income), the greater is its disaster risk. Southeast Asia is thus highly prone to such disaster risks. In addition, controlling for the effects of population and vulnerability, climate factors have significant roles in explaining risks from hydrometeorological disasters. Specifically, precipitation anomalies explain risks from hydrometeorological disasters across South Asia and Southeast Asia, while temperature anomalies explain such risks in East Asia.

Jennings (2011), however, takes a more nuanced view, arguing that: "there is insufficient evidence to exclude the possibility that climate change is increasing hazards and hence trends in reported disasters" (p.19). He elaborates that: "the effect is unlikely to be very large, because the magnitude of climate change over the past 20–30 years is relatively small when compared with, for example, the growth in world's population over that time" (p.19).[14]

23.3.3 DETERMINANTS OF DISASTERS AND THEIR DEADLINESS

Frequency of disasters is specified in the reduced form as a function of geophysical features of a country (e.g., degree of elevation, share of coastal land, size in area (km^2), whether landlocked), its population density, income level, and how democratic the regime/polity is, log of lagged deaths (in 1970–1979), and lagged number of disasters (in 1970–1979). The lagged number of disasters is an identifying instrument for disasters in 1980–2004. It is justified on the grounds that it directly influences the frequency of disasters but without affecting deaths in the sample period 1980–2004. Lagged deaths are a measure of severity of disasters initially. As IV estimation is used, (log of) deaths in the structural equation is hypothesized to depend on all exogenous variables in the reduced form equation (except natural disasters in 1970–1979) and predicted frequency of disasters obtained from it.[15]

Following the standard exposition, the structural equation of deaths is first specified:

$$\text{Log of } D_i = \beta_0 + \beta_1 ND_i + \beta_2 Z_{1i} + u_i \tag{23.1}$$

where D_i denotes deaths due to natural disasters in country i in 1980–2004, ND_i refers to natural disasters during the same period, and Z_{1i} denotes a vector of exogenous variables (which vary by country). Exogenous variables include geophysical characteristics of a country (whether landlocked,

---

[13]A random effect logistic regression model was used. For details, see Technical Notes in Thomas et al. (2012).
[14]For a more nuanced but emphatic contrary view, see IPCC (2007).
[15]For convenience of exposition, log of deaths and deaths are used synonymously.

area, elevation, share of coastal land), socioeconomic characteristics (ethnic fractionalization, per capita income level), and how democratic the political regime is (an aspect of institutional quality). Number of disasters, ND_i, is supposed to be endogenous. Hence another equation is needed—a reduced form—with number of disasters as the dependent variable:

$$ND_i = \pi_0 + \pi_1 Z_{1i} + \pi_2 Z_{2i} + v_i \tag{23.2}$$

where, in addition to Z_{1i}, the exogenous variables from the structural equation, there is an instrument variable, Z_2, which is lagged natural disasters (in 1970−1979). There is a caveat, however. If natural disasters are serially correlated, using lagged natural disasters as an identifying instrument for natural disasters (in 1980−1994) does not fully control for unobserved heterogeneity across countries. A key identification condition is $\pi_2 = 0$, among others.[16] Accordingly, IV estimation is used.

Since hardly any country escaped a natural disaster during 1980−2004, use of IV procedure was not problematic, with adjustment of standard errors for heteroscedasticity. Several specifications were experimented with to avoid omitted variable bias.

The fact that lagged natural disasters significantly influenced the occurrence of disasters in 1980−2004 suggests that the former is a relevant instrument. This was corroborated by the F-test of excluded instrument(s). The Cragg-Donald Wald F statistic and Kleibergen-Paap Wald rk F statistic rejected the null of weak identification.

The main findings from Table 23.8 are as follows.

The number of natural disasters varied positively with size of country, especially when the threshold of the largest size was exceeded. Also, higher elevation was associated with greater frequency of disasters, but not so robustly. Disaster frequency was higher in more populous countries. Frequency of disasters varied with the (lagged) frequency during 1970−1980. Or, countries more prone to disasters in 1970−1979 remained so during 1980−2004. When (lagged) deaths in 1970−1979 were treated as a measure of severity of disasters, the result implied that countries prone to severe disasters were also generally more prone to disasters in subsequent years.

Three findings were somewhat surprising: (1) occurrence of disasters was unrelated to the country's level of economic development, judged on the basis of gross domestic product (GDP) per capita (in purchasing power parity); (2) the share of coastal land was unrelated to the frequency of disasters; and (3) how democratic the political regime (measured as polity mean) was had no effect on the frequency of disasters.

Analysis of residuals of total disasters during 1980−2004 suggests that these did not bear any relationship on exposure to media—in particular, availability of newspapers per 1000 people during 1997−2000. It is surmised, therefore, that disaster reporting was not more accurate in countries better exposed to mass media than in countries with less exposure.[17]

23.3.4 **MORTALITIES**

A selection of results on the determinants of deaths from disasters is given in Table 23.9.

---

[16]For a clear and comprehensive exposition of identification conditions, see Baum (2006).

[17]Details are given in Annex 3. Note that this variable could not be incorporated in the regressions because of the smallness of the sample. Jennings (2011), however, observed that more disasters are reported in countries in which the media are independent and not subject to pressure from the government.

Table 23.8 Occurrence of Natural Disasters, 1980–2004

No. of observations = 86

F (11, 74) = 21.77

Prob. $>F$ = 0.0000

Number of Natural Disasters	Coefficient	t-value	Significance
Landlocked	− 5.673	(− 0.57)	−
Medium[a]	8.207	(0.92)	−
Large	34.079	(2.11)	**
Mean elevation (meters above sea level)	0.016	(1.56)	−
Ethnic	− 42.326	(− 1.35)	−
Log of persons/km^2	5.657	(1.78)	*
Log of GDP 1995	2.181	(0.47)	−
Log of deaths from disasters, 1970–1979	3.995	(1.73)	*
Polity mean for years 1985–1994 (range − 10 to 10)	− 1.048	(− 1.03)	−
Percentage of land within 100 km of coast or river	0.010	(0.07)	−
Number of natural disasters, 1970–1979	2.812	(5.32)	***
Constant	− 39.276	(− 1.04)	−
F-test of excluded instruments:	28.29		
F(11, 74)			
Prob. $>$ F = 0.0000			
Cragg-Donald Wald F statistic	31.95		
Kleibergen-Paap Wald rk F statistic	28.29		

*** *denotes significance at 1% level; ** at 5% level; and* * *at 10% level.*
[a] *The area dummies are as follows: small = 0–200,000 km^2 (omitted); medium = 200,000–750,000 km^2; and large = > 750,000 km^2.*

Deaths varied with size of country—highest among the largest relative to the smallest. Somewhat surprisingly, deaths were unrelated to population density. Similarly, they were unrelated to ethnic diversity that could impede collective action. Higher income levels, on the other hand, were associated with lower deaths. (Predicted) frequency of disaster had a positive effect on deaths. As (lagged) deaths—a measure of severity of disasters in 1970–1979—were related to disasters, there was an *indirect* effect of (lagged) deaths through this channel on deaths in 1980–2004. In addition, there was a *direct* effect of (lagged) deaths during this period. Degree of democracy (defined as the difference between democracy and autocracy) did not have a significant coefficient.[18] As discussed later, this flies in the face of findings from other studies that democracy makes a difference. Likewise, share of coastal land in a country did not have a significant coefficient.

---

[18](i) For details of measurement of polity, see Polity IV project, administered by the Centre for International Development and Conflict Management, University of Maryland. (ii) That freedom of the press is an important factor in averting famines in India is persuasively argued by Dreze and Sen (1989).

Table 23.9 Determinants of Mortality

No. of observations = 86
$F (11, 74) = 11.33$
Prob. $>F = 0.0000$

Number of Natural Disasters, 1980−2004	Coefficient	t-value	Significance
Predicted natural disasters, 1980−2004	0.010	(1.73)	*
Landlocked	− 0.583	(− 0.94)	−
Medium[a]	0.880	(1.88)	*
Large	2.129	(2.62)	***
Mean elevation (meters above sea level)	0.000	(1.02)	−
Ethnic	− 1.415	(− 1.16)	−
Log of persons/km^2	0.002	(0.01)	−
Log of GDP 1995	− 0.931	(− 2.9)	***
Log of deaths from disasters, 1970−1979	0.262	(3.02)	***
Polity Mean for years 1985−1994 (range −10 to 10)	0.040	(0.93)	−
Percentage of land within 100 km coast or river	0.006	(0.69)	−
Constant	12.058	(4.28)	***

*** denotes significance at the 1% level; ** at 5% level; and * at 10% level.
[a] The area dummies are as follows: small = 0−200,000 km^2 (omitted); medium = 200,000−750,000 km^2; and large = > 750,000 km^2.

23.4 DISCUSSION

As detailed simulations of disaster risk prevention and mitigation were not feasible, a broad-brush treatment based on key elasticities was adopted (Table 23.10). Two basic scenarios were considered: the first entails different assumptions on learning from past disasters and fatalities; the second assumes an increase in income levels, with greater capacity for mitigating distress and loss of human lives.[19] One aspect of learning is whether there is reduction in number of disasters relative to lagged disasters in 1970−1979, with an elasticity of 0.39. The results show that countries with more disasters in 1970−1979 were vulnerable to higher frequency of disasters in 1980−2004. This implies that whatever the prevention measures, their effectiveness was far from adequate. If lagged fatalities are looked at as an indicator of severity of disasters, the positive elasticity of disasters to

[19]Cole et al.'s (2012) analysis confirms that government responsiveness is greater when the severity of the crisis is greater. Also, voters punish incumbent politicians for crises beyond their control (e.g., a severe drought caused by monsoon failure). While voters also reward politicians for responding well to climatic events, they do not compensate them sufficiently for their "bad luck." There is thus a robust confirmation of Sen's (1998, 1999) conjecture that democracies are better at responding to more salient catastrophes. However, what undermines the plausibility of Cole et al. (2012) is its failure to account for the fact that drought relief seldom reaches the victims or, if at all, only a fraction, because of huge leakages. Besides, an analysis grounded on inter-temporal rationality of voters that allows for learning over time—for example, whether or not mandates and programs announced were implemented satisfactorily—would have been more plausible. Nevertheless, a link between democracy and fewer deaths through electoral incentives is established.

Table 23.10 Reduction in Disasters and Mortality	
First Stage Regression	**Elasticity**
Number of natural disasters, 1980–2004	
Deaths from disasters, 1970–1979	0.085
Persons/km^2	0.120
Number of natural disasters, 1970–1979	0.391
Large area	0.244
Second-stage regression	
Log of deaths from disasters, 1980–2004	
(Predicted) natural disasters, 1980–2004	0.483
Medium area	0.246
Large area	0.718
GDP 1995	−0.931
Deaths from disasters, 1970–1979	0.262 Indirect effect: 0.04[a]

[a] *The indirect effect of lagged deaths was traced through its effect on frequency of disasters and then on deaths in 1980–2004.*

this variable (0.09) suggests that more severe disasters were associated with slightly higher frequencies of disasters in the subsequent sample period. So, it may be inferred that countries experiencing more severe disasters had learned to prevent many more from occurring.[20] More populous countries were also more prone to disasters, with a positive elasticity of 0.12. Although the elasticity of disasters with respect to size of country (0.24) is largely of descriptive value, it points to the greater vulnerability of populous countries, stemming from variations in geophysical and hydrometeorological characteristics.

As regards deaths, their elasticity to (predicted) disasters was high (0.48). That is, a 5% higher frequency of disasters was associated with a 2.40% higher mortality. This indicates a high death toll from disasters. Countries that recorded higher deaths in 1970–1979 also recorded higher deaths in 1980–2004 (the elasticity being 0.26). So, countries with 5% higher (lagged) deaths recorded 1.3% higher deaths. In addition, there is an indirect effect of lagged deaths through higher frequency of natural disasters, leading to still higher deaths. However, the indirect elasticity is (relatively) small (0.04). (Absolute) elasticity of deaths to income is high (0.93). Therefore, a 5% higher GDP per capita is associated with a 4.65% reduction in deaths. Here the emphasis shifts to resources for enhancing disaster prevention and mitigation of fatalities, as well as a stronger

---

[20]A report in *Time* (Oloffson, 2009) corroborates that learning from experience saves lives: "Early in the morning on Sept. 29, an earthquake deep under the Pacific caused a massive tsunami that devastated the islands of Samoa and American Samoa, killing 111 people, ravaging villages and flattening homes. The earthquake struck at 6:48 a.m. and measured 8.3 on the Richter scale. By 7:04 a.m., an emergency alert went out from the Tsunami Warning System, a global network of sensors monitored by scientists. Less than 10 minutes later, the tsunami, with waves measuring nearly 15 ft. high, hit land. Bad as the damage was, it could have been much worse. Laura Kong, head of the International Tsunami Information Centre in Hawaii, says Independent Samoa had run a tsunami drill with planned evacuation routes in October 2007 and again last year. The preparation saved countless lives during this week's disaster."

preference for safety. While these results point to an important role of income in preventing deaths, it is plausible that the effect of income declines at higher levels.[21]

Somewhat surprising is the absence of a significant relationship between democracy and mortality. This is intriguing as good institutions (parliaments, media, and communities) are frequently associated with lower damages and deaths, since they permit public oversight. On the other hand, these institutions function differently across countries, even if they have similar legal authority and responsibility. Storm damage, for example, is more severe in Haiti than in adjoining Dominican Republic. Haiti's institutions and communities have suffered from long decades of misrule. Often, institutions are linked to democracy, but what matters more is political competition (World Bank, 2010).

23.4.1 CATASTROPHIC RISKS, INSURANCE, AND RECONSTRUCTION

An important point is that, while natural hazards cannot be controlled, they become disasters because of failures of communities, governments, and donors. In this sense, disasters are man-made. A case in point: droughts turning into famines. A general observation on greater vulnerability to natural disasters in the last three decades has been that, whatever the roles of climate change in the greater frequency and severity of natural hazards (e.g., droughts, floods) and the growth of physical assets and human settlements in areas more exposed to such hazards, their effects are compounded by government and community failures.[22] In this context, it may be emphasized that growing urbanization has compounded the problem, because even a minor event can cause enormous damage in a heavily populated area. The proportion of people in developing countries who live in cities has doubled since 1960. More than 40% of people now live in urban areas; this number is projected to rise to 55% by 2030. Nearly half of the cities are subject to extreme weather events[23] (Freeman et al., 2003; World Bank, 2010; *The Economist*, 2012a).

23.4.1.1 *Strategic considerations and priorities in disaster risk prevention and mitigation*

While developing countries bear the brunt of disasters, ironically these same countries have made fewer efforts to adapt their physical environments to mitigate the impact of such disasters and to insure themselves against disaster risks. This is partly due to a disincentive known as "Samaritan's dilemma" (i.e., nations may underinvest in protective measures since they expect foreign donors to help when such disasters strike).[24]

Within developing countries, the poor often chiefly experience the consequences of disasters due to at least two reasons: (1) they are located in areas that are more vulnerable to floods, hurricanes, and earthquakes; and (2) not only do they lose assets, but they also lack access to risk-sharing mechanisms, such as insurance.[25] It is, therefore, not surprising that disasters substantially

---

[21]See, for example, Toya and Skidmore (2007).

[22]As World Bank (2010) puts it, earthquakes, droughts, and floods are natural hazards, but the unnatural disasters are deaths and damages resulting from human acts of omission and commission.

[23]In fact, 14 of the world's 19 mega cities (with 10 million or more inhabitants) are in coastal zones, and over 70 of the world's 100 largest cities can expect a strong earthquake at least once every 50 years (Freeman et al., 2003).

[24]For an elaboration of these concerns, see Freeman et al. (2003) and World Bank (2010).

[25]For an insightful exposition of the link between poverty and risks, see Dasgupta (2007).

increase measured poverty (e.g., 50% of the increase in the head-count poverty ratio in the Philippines during the 1998 crisis was due to El Niño).

Catastrophes could be classified according to whether there are a few causal agents (e.g., the BP Deepwater Horizon oil spill) or many (e.g., depletion of the ozone layer). If there are a few causal agents, it is easier to fix responsibility (Viscusi and Zeckhauser, 2011).

Prevention is often possible and cost-effective. However, government expenditure on prevention is usually lower than on relief. While expenditure matters, how it is spent is also important. Bangladesh, for example, reduced deaths from cyclones by spending modest sums on shelters, accurate weather forecasts, warnings, and evacuation. These cost less than building large-scale embankments (World Bank, 2010).

Risk mitigation through adaptation of physical environment includes land-use planning (e.g., avoiding construction on seismic fault lines and vulnerable coastal regions and ensuring that buildings are resistant to hurricanes and earthquakes); prevention of soil erosion; and building of dams for flood control and seawalls to break storm surges. Governments could also promote farming practices so that farmers can cope better with climatic variations (e.g., use drought-resistant crops) and adapt to longer term changes.

While disaster insurance is extensive in many developed countries (in the United States, for example, 50% of direct catastrophic losses are insured), in developing countries with per capita incomes of below USD 10,000 per annum insurance coverage is less than 10%; it is only about 1% in countries with per capita incomes below USD 760.

Adverse selection is a problem in disaster insurance but not as much in other insurance markets, since many disasters can be predicted more accurately and the property at stake can be properly valued. In developing countries, however, problems arise from the thinness of insurance markets and ill-defined property rights (Freeman et al., 2003).

Two other problems are arguably more serious. One is the difficulty posed by risk spreading and the other is linked to the Samaritan's dilemma. While risk spreading in developing countries in general should not be difficult—since the losses they face are a small fraction of global resources—it often is difficult because of the segmented and shallow insurance markets.[26] The Samaritan's dilemma, on the other hand, may arise from: (1) households and firms underinvesting in insurance and undertaking adaptive measures on the presumption that governments would come to their rescue; (2) governments also underinvesting in the hope that foreign donors would bail them out; and (3) rich countries finding it difficult to scale down their *ex post* assistance in the absence of significant *ex ante* protective measures by governments in developing countries. The humanitarian urge to help when a disaster strikes is often overwhelmingly strong (Freeman et al., 2003). New financial instruments (e.g., catastrophic bonds, swaps, and weather derivatives) have been devised to deal with disaster risk but with little impact.

Governments could help correct insurance market failures through: (1) tax deductions; (2) subsidies; (3) guarantees to insurers and reinsurers; (4) hedging of such guarantees on world reinsurance and capital markets; and (5) mandatory levels of insurance. A general response to the Samaritan's dilemma is to require those at risk to undertake *ex ante* measures to reduce the harm that they will suffer if the hazard occurs. Donors, for example, may credibly commit emergency assistance to

---

[26]In the 1990s, the Caribbean countries, for example, faced insurance rate increases of between 200% and 300%, due to indemnity payments for large hurricane and earthquake losses worldwide (Freeman et al., 2003; Froot, 2001).

countries deemed to have taken disaster mitigation measures (e.g., provision for self-insurance, sea wall protection, enforcement of building guidelines in coastal and other hazard-prone areas). This would help in overcoming a basic inefficiency in disaster insurance and free up resources for other development purposes (Freeman et al., 2003).

Evidence has accumulated pointing to coordination failure turning natural catastrophes into disasters. Marris (2005), for example, reports that much of the destruction and deaths in the wake of the 2004 tsunami could have been averted. In fact, there was a chain of coordination failures.[27] Another case in point is the Orissa cyclone of 1999. A cyclone three years later (in 2002) resulted in far fewer deaths, as both official agencies and affected communities responded more quickly and in a coordinated manner (Thomalla and Schmuck, 2004).[28]

23.5 CONCLUSION

The analysis above draws attention to the higher frequency of natural disasters and deaths associated with them. Variations occur across regions and income groups, especially in deaths.

Countries prone to natural disasters in 1970–1979 continued to be so in the next decade or two. The frequency of their occurrence was higher in countries that experienced more severe disasters initially and those that were more populous. Geophysical factors (e.g., elevation and size of country) have an important role in explaining inter-country variations in the occurrence of natural disasters. Disasters, however, were unrelated to income and polity.

Income (and by implication its growth) matters a great deal in averting disaster-related deaths. Also, this assessment suggests that learning from experience, which takes diverse forms and magnitudes, has been far from adequate. (Lagged) disasters are associated with higher frequency of disasters (with a moderate elasticity), which in turn results in higher fatalities. Also, the effect of lagged deaths on subsequent deaths is moderately high, further pointing to limited learning from experience with severe disasters. While institutions matter, this analysis was not detailed enough to validate their role.

On the other hand, even moderate learning can save a large number of lives (e.g., through early warning systems and better coordination between governments and communities likely to be affected). Growth acceleration also helps avert deaths through more resources available for disaster prevention and mitigation capabilities. A combination of the two—learning from experience and more resources for disaster prevention and mitigation—would result in a massive reduction in deaths from disasters.

Disaster insurance markets remain shallow; the Samaritan's dilemma—whether a donor should bail out a country which is likely to avoid preventive measures—remains pervasive, resulting in underinvestment in disaster prevention; neglect of mainstreaming of disaster prevention and mitigation among multilateral development agencies and governments, and lack of better coordination between them; and failure to combine short-term relief with rebuilding of livelihoods and reconstruction. So an imperative for donors is to make assistance contingent on countries prone to natural catastrophe is to commit budgetary resources to prevent them on a continuing basis.

---

[27]See, for example, World Bank (2010).

[28]For an illustration of glaring coordination failures in the wake of the 2005 Kashmir earthquake, see *The Economist* (2005).

Evidence from this study indicates a growing vulnerability of developing countries to natural disasters and their grave implications on human security. Hence, a challenge for governments and development assistance is to combine growth acceleration with speedy relief and durable reduction in vulnerability. If this analysis has any validity, there are indeed grounds for optimism.

ACKNOWLEDGMENTS

This study was sponsored by T. Elhaut, Director, Asia and the Pacific Division, IFAD, Rome. His support and constructive advice were invaluable in completing it. The first draft was prepared during Gaiha's stay at Harvard Centre for Population and Development Studies, and subsequent revisions were carried out at the Department of Global Health and Population, Harvard School of Public Health. An earlier version was presented at a conference on Catastrophic Insurance, organized by NBER, Cambridge, MA, May, 2012. Discussions with David Bloom, Brian Wright, Julian Alston, Nidhi Kaicker, and editors of this book have helped enrich the study. The views expressed are, however, those of the authors' and not necessarily of the institutions to which they are affiliated.

REFERENCES

Andersen, T.J., 2005. Applications of Risk Financing Techniques to Manage Economic Exposures to Natural Hazards. Mimeo. Inter-American Development Bank, Washington, DC.

Auffret, P., 2003. High Consumption Volatility: The Impact of Natural Disasters? Mimeo. World Bank, Washington, DC.

Barton, C., Nishenko, S., 1997. Natural disasters: forecasting economic and life losses. USGS Fact Sheet, US Geological Survey, Marine and Coastal Geography Program. <http://marine.usgs.gov/fact-sheets/nat.disasters>.

Baum, C.F., 2006. An Introduction to Modern Econometrics Using Stata. Stata Press, College Station, TX.

Beech, H., 2009. Behind Asia-Pacific's unnatural disasters. Time, 19 October 2009. <http://www.time.com/time/magazine/article/0,9171,1929126,00.html#ixzz2CHhUV2FT>.

Birkmann, J. (Ed.), 2006. Measuring Vulnerability to Natural Hazards: Towards Disaster Resilient Societies. United Nations University Press, Tokyo.

Brooks, N., Adger, W.N., 2005. Country Level Risk Indicators from Outcome Data on Climate-Related Disasters: an Exploration of the Emergency Events Database. Mimeo, East Anglia, Norwich.

Cole, S., Healy, A., Werker, E., 2012. Do voters demand responsive governments? Evidence from Indian disaster relief. J. Dev. Econ. 97 (2), 167−181.

Dasgupta, P., 2007. Economics—A Very Short Introduction. Oxford University Press, Oxford.

Dreze, J., Sen, A., 1989. Hunger and Public Action. Clarendon Press, Oxford.

Freeman, P.K., Keen, M., Mani, M., 2003. Dealing with increased risk of natural disasters: challenges and options. IMF Working Paper (WP/03/197). IMF, Washington, DC.

Froot, K.A., 2001. The market for catastrophic risk: a clinical examination. J. Financ. Econ. 60 (2−3), 529−571.

Gaiha, R., Thapa, G., 2006. Natural Disasters, Vulnerability and Mortalities. Mimeo. Asia and the Pacific Division, IFAD, Rome.

IPCC, 2007. Climate change 2007: climate change impacts, adaptation and vulnerability—summary for policy makers. WGII Fourth Assessment Report, 6 April 2007. IPCC, Geneva.

Jennings, S., 2011. Time's Bitter Flood: Trends in the Number of Reported Natural Disasters. Oxfam, UK, 27 May 2011.

Kahn, M.E., 2005. The death toll from natural disasters: the role of income, geography, and institutions. Rev. Econ. Stat. 87 (2), 271–284.

Kousky, C., 2013. Informing climate adaptation: a review of the economic costs of natural disasters. Energy Econ. <http://dx.doi.org/doi:10.1016/j.eneco.2013.09.029>.

Marris, E., 2005. Inadequate warning system left Asia at the mercy of tsunami. Nature 433, 3–5.

Munich, R.E., 2011. Natural Catastrophe Year in Review, January 2011. <http://www.iii.org/assets/docs/ppt/MunichRe-010412.ppt>.

Oloffson, K., 2009. How Prepared Are Countries for a Tsunami? *Time*, 1 October 2009. <http://www.time.com/time/world/article/0,8599,1927094,00.html#ixzz2CHfuJZAc>.

Sen, A., 1998. Mortality as an indicator of economic success and failure. Econ. J. 108 (446), 1–25.

Sen, A., 1999. Development as Freedom. Alfred A. Knopf, New York, NY.

The Economist, 2005. The Kashmir earthquake, 13 October 2005.

The Economist, 2012a. The rising cost of catastrophes: how to limit the damage that natural disasters do? 14 January 2012.

The Economist, 2012b. Natural disasters. 31 March 2012.

Thomalla, F., Schmuck, H., 2004. We all knew that a cyclone was coming: disaster preparedness and cyclone of 1999 in Orissa, India. Disaster 28 (4), 373–387.

Thomas, V., Ramon, J., Albert, G., Perez, R.T., 2012. Intense Climate Disasters and Development in Asia and the Pacific. Independent Evaluation. Asian Development Bank, Manila.

Toya, H., Skidmore, M., 2007. Economic development and the impacts of natural disasters. Econ. Lett. 94/1, 20–25.

Viscusi, W.K., Zeckhauser, R.J., 2011. Addressing Catastrophic Risks: Disparate Anatomies Require Tailored Therapies. Harvard Kennedy School, Cambridge, MA, Mimeo. RWP11-045.

WDI, 2006. World Development Indicators, Supplementary Notes and Definitions. WDI, Washington, DC.

World Bank, 2006. Hazards of Nature, Risks to Development. Independent Evaluation Group. World Bank, Washington, DC.

World Bank, 2010. Natural Hazards, Unnatural Disasters: The Economics of Effective Prevention. World Bank, Washington, DC.

ANNEX 1 **435**

ANNEX 1

Table A.1.1 Definitions of Natural Hazards		
Type	**Hazard**	**Definition**
(a) Hydrometeorological	(1) Hurricanes and tropical storms	Large-scale, closed circulation system in the atmosphere with low barometric pressure and clockwise in the southern hemisphere.
	(2) Floods	Temporary inundation of normally dry land by overflowing lakes and rivers, precipitation, storm surges, tsunami, waves, mudflow, and lahar. Also caused by the failure of water-retaining structures, groundwater seepage, and water backup in sewer system.
	(3) Drought	Lack or insufficiency of rain for an extended period that causes hydrological imbalance and, consequently, water shortage, crop damage, stream flow reduction, and depletion of groundwater and soil moisture. It occurs when, for a considerable period, evaporation and transpiration (the release of underground water into the atmosphere through vegetation) exceed precipitation.
	(4) Forest fires	Uncontrolled fires whose flames can consume trees and other vegetation of more than 1.8 m in height. These often reach the proportions of a major conflagration and are sometimes begun by combustion and heat from surface and ground fires.
(b) Geophysical	(1) Earthquake	Sudden tremor of the earth's strata caused by movements of tectonic plates along fault lines in mountain ranges or mid-oceanic ranges.
	(2) Tsunami	Wave train or series of waves generated in water by an impulsive disturbance (such as earthquakes) that vertically displaces gigantic water columns. Tsunamis may reach a maximum run-up or above sea-level height of 10, 20, or even 30 m.
	(3) Slides	Downward slope movement of soil, rock, mud, or snow due to gravity. A common source of slides is prolonged torrential downpours of rain or the accumulation of heavy snow. Mass displacement of large mud, snow, or rocks can also be triggered by seismic waves.
	(4) Lahars	Mudflows caused by the melting of the ice cap by lava from a volcano or the downhill run-off of volcanic ash due to heavy rainfall.
	(5) Volcanic eruption	Process whereby molten lava, fragmented rocks, or gases are released to the earth's surface through a deep crater, vent, or fissure.
Source: *Adapted from Auffret (2003).*		

Table A.1.2 Classification by Income

For operational and analytical purposes, the World Bank's main criterion for classifying economies is gross national income (GNI) per capita. Based on its GNI per capita, economies are classified either as low income, middle income (subdivided into lower middle and upper middle), or high income. Another analytical grouping, based on geographic regions, is also used.

Income group: Economies were classified using 2004 GNI per capita, calculated using the World Bank Atlas method. The income cut-offs were: low income, USD 825 or less; lower middle income, USD 826–3255; upper middle income, USD 3256–10,065; and high income, USD 10,066 or more.

Source: *Adapted from WDI (2006).*

ANNEX 2

(a) Hydrometeorological and geophysical disasters

A disaggregated analysis of natural disasters into hydrometeorological and geophysical disasters was carried out. The empirical evidence confirms that both the frequencies and impacts of disasters differ.

- The frequencies of both hydrometeorological and geophysical disasters were considerably higher in 1995–2004, relative to 1985–1994. When these frequencies were expressed as ratio of population, the frequency of hydrometeorological disasters increased, while that of geophysical disasters remained almost unchanged.
- The vast majority of disasters during 1985–1994 were hydrometeorological, and their share rose further during the next decade.
- The share of deaths from hydrometeorological disasters rose from about 65% during 1985–1994 to about 88% in 1995–2004, while that from geophysical disasters declined sharply.
- Hydrometeorological disasters became deadlier during 1995–2004 in terms of deaths per disaster, while geophysical ones recorded a decline in their deadliness. A similar pattern is revealed by the study's preferred criterion of deaths per million population, wherein the deadliness of hydrometeorological disasters more than doubled and that of geophysical ones was halved.

In sum, both types of disasters became more frequent, but only hydrometeorological disasters became far deadlier.[29]

(b) Catastrophic mortalities

Frequently, a power law probability function is used to characterize the distribution of catastrophic events, as familiar well-behaved distributions, such as the normal distribution, are poor approximations. Catastrophes have too many extreme outliers. The distributions for such catastrophic risks are called "fat tailed," reflecting the fact that extreme outliers are much more likely than indicated by the normal (or log normal) distribution(s). Such extreme outliers cause enormous harm and account for a substantial share of expected losses from catastrophes (Viscusi and Zeckhauser, 2011).

---

[29]For an update, see Thomas et al. (2012) and Jennings (2011). Both confirm a rising trend of hydrometeorological disasters on different criteria across decades up to 2010.

Table A.2.1 Different Types of Natural Disasters and Their Death Tolls

Disaster Type	Frequency of Disasters[a]		Deaths		Deaths per Million		Deaths per Disaster		Disasters per Million	
	1985– 1994	1995– 2004	1985– 1994	1995– 2004	1985– 1994	1995– 2004	1985– 1994	1995– 2004	1985– 1994	1995– 2004
Hydrometeorological	1,111 (75.94)[b]	1,823 (80.38)	248,602 (64.60)	615,792 (88.77)	60	128	224	338	0.27	0.38
Geophysical	352 (24.06)	445 (19.62)	136,244 (35.40)	77,897 (11.23)	33	16	387	175	0.08	0.09
Total	1,463 (100.00)	2,268 (100.00)	384,846 (100.00)	693,689 (100.00)	92	144	263	306	0.35	0.47

[a] Based on the classification of Auffret (2003). Calculations based on EM-DAT. Rich OECD and non-OECD countries are excluded.
[b] Figures in parentheses show the percentage share.

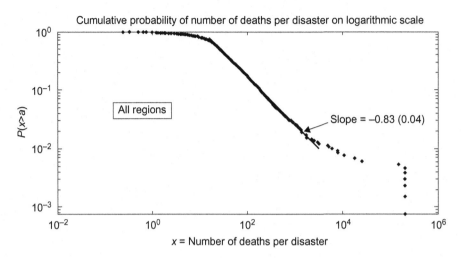

FIGURE A.2.1

Log of fatality of disasters—cumulative probabilty of deaths per disaster.

A variable x has a power law probability distribution if it can be characterized as $p(x) = cx^d$, where c is a constant and d is the scaling exponent, so that $\log p(x) = \log (c) + d \log (x)$. In the present context, x denotes fatalities per disaster over a period of time. With finite-size effects, however, it holds in a certain region of scaling $x_0 < x < x_c$. The minimum and maximum cut-off points are given by x_0 and x_c, respectively, and they define the scaling region. The idea is to see whether a substantial fraction of events lies in the universal scaling region. This may also be a good way to compare different regions. Figure A.2.1 shows that the probability of very large events/disasters is nonvanishing. A much faster decay is also seen in the last region marked by $x > x_c$. Here, x_c is a parameter that can be identified for different situations. Its higher value ensures the applicability of power law over a broader range. In general the probability density of x with finite-size scaling effects is given as $p(x) = cxa \, \psi(x)$ for $x > x_0$. Here, a is the true scaling exponent of the power law and ψ is the scaling function, which decays much faster than power law for $x > x_c$. While this function can take several forms, an exponential form, $\psi(x) \sim e - x$, is assumed.

The exponent d obtained from OLS fitting in Figure A.2.1 is the "apparent" exponent and is related to the true exponent through the scaling function. It is well established that if $\psi(x = 0) > 0$, then $a \sim d$ (almost equal). Thus assuming an exponential scaling function, the power law exponents are obtained. In some cases, there may be multiple scaling due to more than one region exhibiting power law decay. In such cases, the prediction of very large disasters is not ruled out or to be ignored.[30]

In some regions and country groups, as shown in Figures A.2.2–A.2.7, the power law applies to a narrower range of values before the curve drops sharply. This indicates the certainty of dominance of

---

[30]As pointed out by Barton and Nishenko (1997), loss of life and property due to natural disasters exhibits self-similar scaling behavior. This self-similar scaling property allows use of frequent small events to estimate the rate of occurrence of less frequent, larger events.

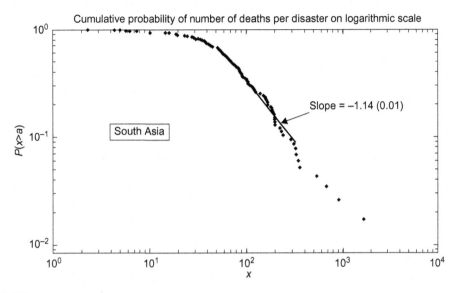

FIGURE A.2.2

Logarithm of deaths per disaster—cumulative frequency (probability) of deaths per disaster in South Asia.

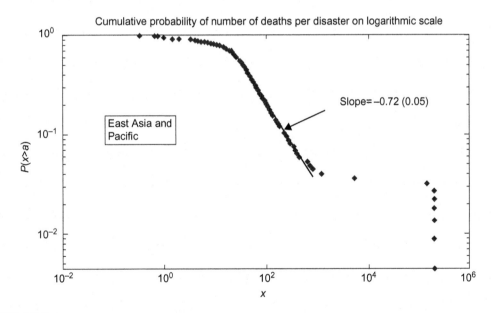

FIGURE A.2.3

Logarithm of deaths per disaster—cumulative frequency (probability) of deaths per disaster in East Asia and the Pacific.

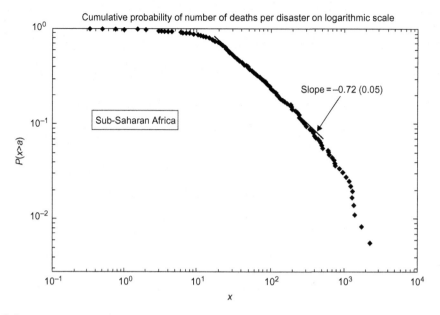

FIGURE A.2.4

Logarithm of deaths per disaster—cumulative frequency (probability) of deaths per disaster in Sub-Saharan Africa.

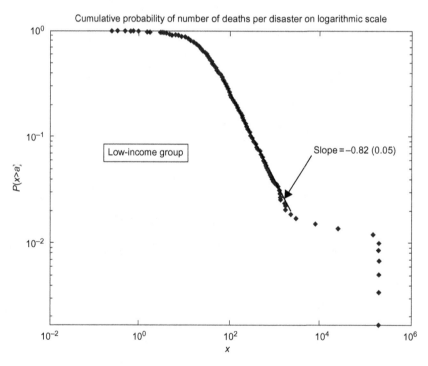

FIGURE A.2.5

Logarithm of deaths per disaster—cumulative frequency (probability) of deaths per disaster in low-income countries.

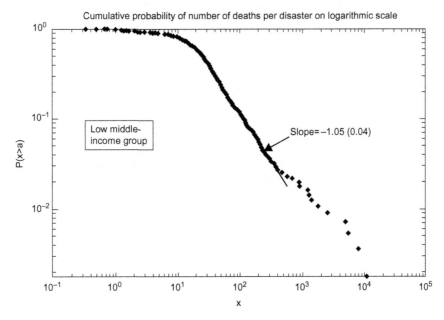

FIGURE A.2.6

Logarithm of deaths per disaster—cumulative frequency (probability) of deaths per disaster in lower middle-income countries.

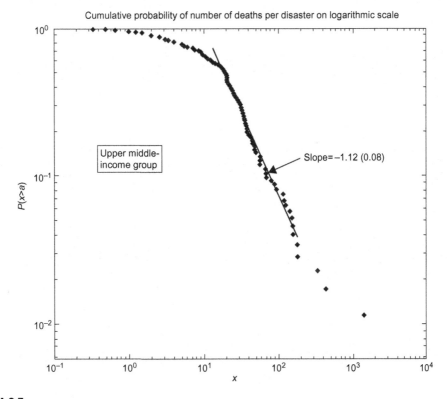

FIGURE A.2.7

Logarithm of deaths per disaster—cumulative frequency (probability) of deaths per disaster in upper middle-income countries.

power law relationship in governing the nature of deadliness. Power law over a broader range implies that the nature of deadliness or its occurrence replicates itself at multiple scales and hence the nature is complex. It describes the occurrence of large deviations from the median of the probability distribution; one of the examples where this is evident is natural disasters.

Plots of logarithms of deaths per disaster—cumulative frequency (or probability) of deaths per disaster—are given first for regional subsamples and then for groups of countries by level of income over the period 1980–2004.

ANNEX 3

RELIABILITY OF DATA ON NATURAL DISASTERS

As discussed earlier, doubts have been raised about the reliability of statistics on natural disasters and deaths associated with them. A careful scrutiny of EM-DAT and its cross-validation with other independent sources confirms that: (1) in general, the quality of data since the 1970s has been uniformly good, and (2) the mortality estimates are more reliable than those on number of people affected and economic losses.

While some doubts will persist, the results of two regressions and corresponding graphs shown in Figures A.3.1 and A.3.2 confirm that neither the frequency of natural disasters nor the deaths

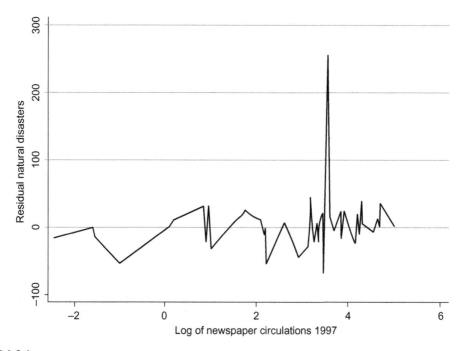

FIGURE A.3.1

Residual natural disasters and log newspaper circulation.

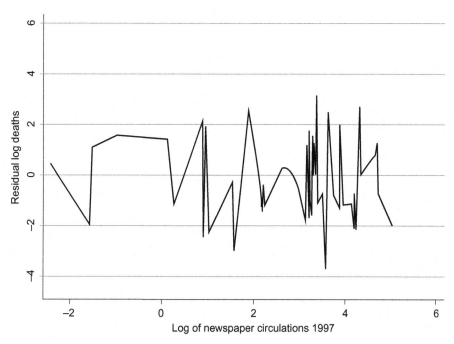

FIGURE A.3.2

Residuals of log deaths and log exposure to newspapers.

Table A.3.1 Residuals of Disasters and Log Exposure to Newspapers			
No. of observations = 55			
$F(1, 53) = 0.85$			
Prob. $>F = 0.362$			
Residual Natural Disasters, 1980–2004	**Coefficient**	**t-value**	
Log newspaper circulation 1997	1.790	0.92	–
Constant	− 6.191	− 1.00	–

associated with them vary systematically with exposure to mass media—in particular, availability of newspapers per 1000 people during 1997–2000.

Table A.3.1 shows the results of regression of the residuals from the preferred specification for occurrence of natural disasters on log of newspaper circulation during 1980–2004. As may be noted from the results, the coefficient of newspaper circulation is not significant, implying the absence of a relationship between the two variables.

A similar finding was obtained when residuals of (log) deaths were regressed on log of newspaper circulation, as shown in Table A.3.2. The robust regression results do not show any relationship between the two variables.

Table A.3.2 Residuals of (log) Deaths and Log Newspaper Circulation

No. of observations = 55
F (1, 53) = 0.63
Prob. $>F$ = 0.4298

Residual log of deaths from disasters, 1980–2004	Coefficient	t-value	
Log newspaper circulation 1997	− 0.105	− 0.80	−
Constant	0.128	0.30	−

These results are further confirmed by the two graphs. The residuals do not vary systematically with newspaper circulation.

THE GROWTH–POVERTY NEXUS: MULTIDIMENSIONAL POVERTY IN THE PHILIPPINES

24

Arsenio M. Balisacan

University of the Philippines and National Economic and Development Authority, Diliman, Philippines

CHAPTER OUTLINE

JEL: I32; O15; O53

24.1 INTRODUCTION

Poverty is increasingly recognized as a multidimensional phenomenon, yet its assessment continues to be conducted almost exclusively in terms of income (or expenditure). This practice is prevalent partly because low household incomes are casually associated with other deprivation indicators, such as low levels of literacy and life expectancy. Indeed, in recent decades, the rapid growth of household incomes in many East Asian countries has accompanied unprecedented reduction in income poverty and substantial improvement in access to human development opportunities. The same development experience, however, reveals substantial variation in welfare improvement and human development outcomes across countries, even among countries at similar income levels, as well as across space and population groups within a country (Kanbur et al., 2006; Deaton, 2010).

Moreover, the growth process has often accompanied achievements in some dimensions of household welfare but lacked progress in some other dimensions. For this reason, the Stiglitz—Sen—Fitoussi Commission recommends the simultaneous consideration of material living standards (income, consumption) and other dimensions of well being, including health, education, political voice and governance, environment, and security (Stiglitz et al., 2010). This indicates that simply raising household income (expenditure) is no longer enough to outgrow poverty in its various dimensions.

To be sure, the many faces of poverty have not escaped the lenses of the development community. The United Nations Millennium Declaration of 2000, for instance, set the framework for concerted time-bound actions at both international and national levels to achieve certain standards of human welfare and development, otherwise known as Millennium Development Goals (MDGs). The MDGs include targets for indicators associated with extreme poverty and hunger, basic education and health, and environmental sustainability.

MDG reports, whether international or national, usually present progress on each indicator singly. Indeed, no composite MDG index has been developed. The reason is plain and simple: the denominators or base populations often differ across these indicators. Total population, for example, is the base population for the income poverty indicator, while it is children for the child mortality indicator. Yet, the usefulness of such a composite index for policy design and poverty monitoring cannot be overemphasized, especially in view of the probable "interconnectedness" of the MDG indicators (i.e., progress in one goal would likely speed up progress in others).

Even more worrisome is the dearth of information on the extent of deprivations experienced *simultaneously* by the poor. The components of the Human Development Index (HDI) provide indications of basic deprivation in health, education, and living standards, but since these usually pertain to population averages for geographic areas—provinces, regions, countries—from different data sources (not from the same household survey), they fall short of informing policy discussions on what can be done to reduce abject poverty in its multiple dimensions.

Recently, Alkire and Foster (2011a) developed an empirically useful approach to measure the magnitude of multidimensional poverty. Alkire and Santos (2010a) applied the concept to assess the magnitude of abject poverty in 104 developing countries. In particular, they used a special member of the Alkire—Foster class of poverty measures, which have desirable properties useful for policy work. This measure, aptly called multidimensional poverty index (MPI), is suited to fit commonly available data, including the MDG indicators.

A common objection to aggregating the various poverty dimensions into a single number is that crucial information on the individual deprivations is lost (Ravallion, 2011). This is not so for the family of Alkire—Foster measures. The MPI, for instance, preserves the "dashboard" of dimensions of poverty—that is, the MPI can be "unpacked" to reveal the various deprivation indicators. What makes the MPI distinct and useful is that it reflects overlapping deprivations at the individual level, thereby providing a convenient analytical tool to "identify the most vulnerable people, show aspects in which they are deprived, and help to reveal the interconnections among deprivations" (Alkire and Santos, 2010b). This information is extremely useful for designing antipoverty measures and targeting scarce resources more effectively.

The 2010 HDR, as well as the Alkire—Santos paper, includes estimates for the Philippines but only for 2003. For an index such as MPI to be meaningful and useful for national policymaking and governance, especially in targeting resources and tracking the MDGs, the data would have to be as recent as possible and comparable estimates for other years and across subpopulation groups would need to be generated. Other dimensions of deprivation especially relevant to the Philippine context will also have to be incorporated in the measure. Furthermore, the link, if any, between MPI and other existing indicators of poverty and aggregate welfare, as well as the robustness of MPI comparison across space (provinces and regions), has to be established.

This study sought to systematically assess the nature, intensity, and sources of multidimensional poverty over the past two decades and across subpopulation groups in the Philippines. It found that what is generally known about the country's performance in poverty reduction in recent years, as seen in income measures of poverty, is quite different from what the lens of multidimensional poverty measures reveals. In particular, while income headcount remained largely unaffected by economic growth (albeit modest by the standards of the country's East Asian neighbors) during the past decade, multidimensional poverty did actually decline. That is, growth turned out to be beneficial to the poor who simultaneously experienced multiple deprivations. Moreover, deprivation in standard of living remains the major contributor to aggregate poverty, although there is substantial variation in the importance of various deprivations across subpopulation groups.

The chapter proceeds as follows. Sections 24.2 and 24.3 discuss the empirical approach to measuring multidimensional poverty and the data employed in the study. Section 24.4 shows the estimates of MPIs from three sets of nationally representative household survey data covering various years in the past two decades. Section 24.5 reassesses what is known about the poverty profile by subpopulation groups from the lens of multidimensional poverty. It also exploits the decomposition property of MPI to identify the sources of household deprivation. Finally, Section 24.6 provides the implications of the study for development policy and poverty research.

24.2 EMPIRICAL APPROACH

Poverty measurement involves choosing a welfare indicator, establishing a threshold level (poverty line) of this indicator, and aggregating the individual information on the poor into a summary measure of poverty. In applied work, the usual approach is to use current income (or expenditure) as a unidimensional measure summarizing a person's welfare. A person is deemed poor if the person's income is below a predetermined poverty line. The information on the poor is then combined into an aggregate measure. Numerous aggregate poverty measures have been suggested in the literature, but what has gained popularity in applied work is the so-called Foster−Greer−Thorbecke (FGT) class of poverty measures, of which the headcount is the most recognizable, owing to its simplicity. The headcount, also referred to as poverty incidence, is defined simply as the proportion of the population deemed poor.

Multidimensional poverty measurement follows generally the same track: choosing the indicators representing dimensions of deprivation, defining the deprivation thresholds (cutoffs) associated with these dimensions, and aggregating the information on individual deprivations for the population into a summary measure of poverty. While the various dimensions of poverty have long been well-articulated in the development literature, the conceptual and empirical issues in aggregating the information on multidimensional deprivations have been a fairly recent interest in poverty assessment. The past decade, in particular, has seen an explosion of the literature on approaches to multidimensional poverty assessment.[1] This study builds on this literature.

---

[1] See Alkire and Foster (2011a) and Yalonetzky (2011), and the literature cited therein.

Two practical considerations guide our choice of approach to multidimensional poverty measurement. The first is that the poverty measures inherent in the approach must be intuitive and easy to interpret and satisfy a set of desirable properties useful for policy. One such property is decomposability, which allows the aggregate index to be broken down by subpopulation group (region, type of employment, etc.) or by source of deprivation. The second consideration is that the approach should be flexible enough for application to various types of household survey data, particularly data involving a mix of ordinal (or categorical) and cardinal indicators of household welfare.

For this study, we employed a special member of the class of multidimensional poverty measures suggested by Alkire and Foster (2011a). The Alkire–Foster class of M_α poverty measures bears close affinity to what is arguably the most popular class of unidimensional poverty measures employed in the literature, the FGT P_α poverty measures, where α is a parameter reflecting society's aversion to poverty. Like the P_α poverty measures, the class of M_α poverty measures satisfies a set of desirable properties, including additive decomposability (i.e., the overall MPI is simply the weighted average of subgroup poverty indices, where the weights are population shares). Unlike the unidimensional P_α poverty measures, the M_α poverty measures can be "unpacked" to reveal the relative importance of various dimensions of deprivation to the subpopulation group. As shown below, this property proves to be very useful for policy purposes (e.g., tracking poverty and various MDGs).

The poverty measure used in this study is the "adjusted" headcount (M_0), or the proportion of the population deemed multidimensionally poor, adjusted for the average intensity of deprivation among the poor. M_0 is the counterpart of the familiar (unidimensional) headcount, P_0, or the proportion of the population deemed poor, when $\alpha = 0$.[2]

Formally, following Yalonetzky's (2011) formulation, the multidimensional headcount (H) and the average intensity (number) of deprivation among the poor (A) can be defined as

$$H(X; k, Z) \equiv \frac{1}{N} \sum_{n=1}^{N} I(c_n \geq k)$$

$$A(X; k, Z) \equiv \frac{\sum_{n=1}^{N} I(c_n \geq k) c_n}{DNH(X; k, Z)}$$

where X is a matrix of attainments, whose N rows have the information on the attainment of N individuals and whose columns represent the D attainment dimensions, $k \leq D$ is the multidimensional poverty cutoff, Z is a vector of deprivation cutoffs associated with each of the D dimensions, c_n is the weighted number of deprivations suffered by individual n, and $I(.)$ is an indicator that takes the value of 1 if the expression in the parentheses is true (otherwise it takes the value of 0).

The adjusted headcount, M_0, can then be written as

$$M_0 \equiv H(X; k, Z) A(X; k, Z) = \frac{\sum_{n=1}^{N} I(c_n \geq k) c_n}{DN}.$$

---

[2]The counterparts of poverty-gap index (P_1) and distribution-sensitive FGT index (P_2) in the class of M_α poverty measures, i.e., M_1 and M_2, respectively, are not used in this chapter since their applications require that the dimension indicators be continuous and cardinal data (e.g., income and consumption data). In contrast, M_0 is amenable to both cardinal and ordinal data. In household surveys, such as those used in this study, achievements in individual functionings are typically represented as discrete, qualitative, or ordinal data (e.g., completion of basic education, access to clean water).

Notice in the above expression that the numerator is the total number of deprivations of the multidimensionally poor, while the denominator is the maximum deprivation if all N individuals are deprived in all D dimensions. M_0 can thus also be interpreted as the actual deprivation among the poor in proportion to maximum deprivation. Furthermore, if c_n and k are normalized such that $D = 1$, then M_0 can be interpreted simply as the average of the individual poverty levels.

For instance, if 50% of households in a population are multidimensionally poor (i.e., $H_m = 0.5$) and if all households are deprived in all indicators (i.e., $A = 1$) then $MPI = H_m$. However, not all households may be equally multidimensionally poor. That is, if the average poor household is deprived of 45% of the weighted indicators, $A = 0.45$, and $MPI = 0.5 \times 0.45 = 0.225$. In effect, this population is deprived in 22.5% of total potential deprivations that the population could experience. Because of this adjustment of the headcount ratio based on intensity (A), Alkire and Foster (2011a) also call MPI the "adjusted headcount ratio."

As shown by Alkire and Foster (2011a), M_0, as well as the other members of the class of M_α poverty measures, is additively decomposable. The aggregate (population) adjusted headcount is simply a weighted sum of the subgroup headcount levels, the weights being their population shares. This property proves to be extremely useful for policy purposes and for constructing poverty profiles. For example, for a policy change that increases the functionings (in the sense of Sen) of group i and reduces those of group j, one can work out the impact of the change on each group's poverty level and then use the groups' respective population shares to estimate the new level of aggregate multidimensional poverty.

The M_0 measure is also dimensionally decomposable (Alkire and Foster, 2011a): M_0 can be shown as a weighted average of dimensional poverty values, where the weights are the predetermined dimensional weights (reflecting the relative importance attached to the dimensions). Each dimensional poverty value (censored headcount ratio) represents the proportion of the overall population deemed both poor *and* deprived in the given dimension. This also proves to be an extremely valuable property of the poverty measure. For example, for a policy change that reduces certain deprivations but not others, one can trace the impact of the policy on the dimensional poverty values and then use the dimensional weights to arrive at the overall impact of the policy change on multidimensional poverty.[3] Note, too, that both group and dimensional decompositions can be employed simultaneously to produce an even finer "resolution" of poverty impact. For example, for the same policy change involving certain dimensions of deprivation, one can work out the impact of the change on various population groups and on the overall population.

The MPI used by Alkire and Santos (2010) is a seminal cross-country application of the M_0 measure. Their results for 104 countries have found their way into the statistical annex of the 2010 Human Development Report of the United Nations Development Programme (UNDP). However, partly because the data on dimensional deprivations have to come from the same household survey and have to be defined uniformly across a large number of countries, their MPI estimates in virtually all country cases pertain to one year only and only for a small subset of deprivation indicators. While the estimates provide a useful dashboard of the character of multidimensional poverty across

---

[3]Care must be exercised in interpreting the result of such decomposition, however. As noted earlier, a policy change affecting one dimension of deprivation may have direct and (or) indirect effect on the other deprivations simultaneously experienced by the poor. For example, an improvement in access to clean water may also influence child-schooling outcomes through its impact on children's health.

developing countries, they have little to say about the *changes* in poverty within a country. The data used for the Philippines, for example, pertain to 2003, long before the global food crisis of 2007 and the subsequent global financial crisis of 2008/2009, which severely affected the poor.

24.3 HOUSEHOLD DATA AND DEPRIVATION DIMENSIONS

Of primary interest in this study are the changes in the country's performance in poverty reduction in recent years, as seen from the perspective of multiple deprivations simultaneously experienced by the poor. As such, the estimation focuses only on *nationally representative* household surveys with available unit record data and that are part of the regular household surveys of the country's statistical system. Three such surveys of the National Statistics Office (NSO) were used: (i) National Demographic and Health Survey (NDHS), which is conducted once every five years; (ii) Family Income and Expenditures Survey (FIES), conducted once every three years; and (iii) Annual Poverty Indicator System (APIS), conducted in years without FIES.[4] The sampling design of these surveys permits the generation of spatial estimates down to the regional level. It is noted that these surveys are conducted for different purposes and vary in details for the deprivation indicators of interest, hence not directly comparable even for the same variable of interest (e.g., food expenditure in APIS vs. food expenditure in FIES). For this reason, caution should be exercised in comparing values of poverty and deprivation indicators *between* data sources. It is more appropriate to focus on the changes in the values *within* data sources.

As in the construction of the HDI (UNDP, 2010) and the MPI for cross-country comparison (Alkire and Santos, 2010), this study focuses on three generally recognizable dimensions of deprivation: education, health, and standard of living. While there are other potentially measurable and policy-relevant dimensions, such as empowerment, environment, security, and participation in civil society, the binding constraint is the limitation of existing household surveys used in the study.[5] None of these surveys has been intended to measure multiple deprivations simultaneously experienced by the poor.[6] Nonetheless, the three dimensions arguably capture the most basic human functionings relevant to the Philippine context. From the perspective of consensual support, there is little disagreement that these are appropriate areas of policy concern. Moreover, parsimony dictates that focusing on the most basic forms of deprivations simplifies comparison with the conventional income measure of overall poverty.

The selection of relevant deprivation indicators associated with each dimension was guided mainly by enduring practices in policy discussions, especially in the context of the MDGs, and by available

---

[4]The years available for NDHS are 1993, 1998, 2003, and 2008. For FIES, the comparable data begin with 1985 and end with 2012, although the series in this chapter begin with 1988 because 1985 was a rather "abnormal" year—a period of highly disruptive political and economic shocks leading to an economic contraction of 7% (following a previous year's contraction of also 7%) and social unrest. For APIS, the comparable data are 1998, 2002, 2004, 2007, 2008, and 2010.

[5]The third dimension—standard of living—is actually a proxy variable for other basic needs that define human functionings, such as mobility, shelter, public amenities, and leisure.

[6]As noted above, unlike in the construction of HDI, where the relevant summary indicators are drawn from different household surveys, the construction of the M_0 index, or the MPI in UNDP's 2010 HDR, requires that all the deprivation indicators come from the same household survey and that the unit record data are accessible.

information in the household survey data used in the study. The latter consideration suggests that the set of deprivation indicators varies across the three household surveys. For example, there are more deprivation indicators linked with standard of living in both FIES and APIS than in NDHS.

For health, as in the MDGs, the two deprivation indicators are child mortality and malnutrition. In APIS and FIES, child mortality is indicated by lack of access to clean water supply and sanitation, while malnutrition is indicated by the household's difficulty in accessing basic food owing to lack of purchasing power, defined broadly to include both cash and in-kind incomes (including own-produced food). There is ample evidence in the literature pointing to a link between child mortality and access to clean water supply and sanitary facilities (Capuno and Tan, 2011; Banerjee and Duflo, 2011; Sachs, 2005). There is also no disagreement that a household whose *total income* is less than even the official *food threshold* is deemed deprived of basic food. In official MDG monitoring of income poverty, households not having enough purchasing power to meet the official food thresholds are deemed to be subsistence poor.

For education, the two complementary deprivation indicators are the years of schooling of household members and school attendance of school-age (7−16 years) children. The first indicator acts as a proxy for level of knowledge and understanding of household members. Though quite imperfect (since it may not capture well schooling quality and skills achieved by individuals), this indicator is sufficiently robust in applied work, providing a reasonably good proxy for functionings related to education. As in Alkire and Santos (2010), a household is deprived of education functionings if not one member of the household has completed basic education.[7] Similarly, a household with a school-age child not attending school is deemed deprived of educational functionings. This indicator reflects the country's MDG commitment vis-à-vis achievement of universal primary education.

As in HDI, the standard-of-living dimension is a catchall measure, reflecting access to opportunities for other human functionings not already represented in health and education. Instead of using income as the catchall measure, we used more direct, arguably sharper indicators of living standard. The basic ones are access to quality shelter, electricity, and mobility (transport); ownership of nonlabor assets, which, in an environment of highly imperfect financial markets, is an indicator of access to credit-related consumption-smoothing opportunities; and sources of incomes other than own labor employment and entrepreneurial activities. The FIES, APIS, and, to a lesser extent, the NDHS provide a relatively rich array of these and related deprivation indicators. In addition, the Census of Population and Housing (CPH), if merged with these surveys, can substantially enrich the information on household deprivation (e.g., availability of community-level indicators of living standard). However, for this study, the set of indicators was chosen in such a way that it remains parsimonious and is easily comparable over time and across subpopulation groups.

The Alkire−Foster measurement methodology identifies the poor following a two-step procedure. The first step involves setting a cutoff for each dimension and taking the weighted sum of deprivation suffered simultaneously by the individual, where the weights reflect the relative importance of each dimension in the set of poverty dimensions selected for the assessment. There is no "golden rule" to the setting of dimension weights. In practice, weight assignment is a value judgment and is thus open to arbitrary simplification. This is a weakness shared by virtually all other

---

[7]The assumption is that education confers externalities to *all* members of the household. Put differently, the effective literacy of each household member is higher if at least one household member is literate (Basu and Foster, 1998).

aggregation procedures suggested in the literature, including the convention of combining various income components into an overall income measure of welfare. This chapter follows the rule of simplicity advocated by Atkinson et al. (2002) and the convention in HDI construction: equal weights applied to each of the dimensions. This rule has intuitive appeal: "the interpretation of the set of indicators is greatly eased where the individual components have degrees of importance that, while not necessarily exactly equal, are not grossly different" (Atkinson et al., 2002, pp. 479–480 as cited in Alkire and Foster, 2011a). As noted above, one or more deprivation indicators may act as proxy for a dimension. Similarly, in cases where two or more deprivation indicators are used as proxy for a dimension, the same weighing rule is applied—that is, each deprivation indicator within a dimension is weighed equally.

The second step is setting a poverty cutoff. A household is deemed multidimensionally poor if its weighted sum of deprivations is above this cutoff. As in the setting of a poverty line for an assessment of income poverty, the determination of this cutoff is potentially controversial, partly because what constitutes a poverty norm may be influenced by current levels of living standards, the distribution of these standards across population subgroups, and other factors, including political ideology. Following the principle of consistency in setting poverty norms (Ravallion, 1994), we chose a poverty cutoff that is fixed in terms of a given level of *absolute* deprivation over time and across areas and population subgroups. The intent is to consistently rank poverty status across regions, provinces, or socioeconomic groups, as well as to monitor performance in poverty reduction over the medium term. The interest is not so much about the absolute level of poverty at any given time but the *changes* in poverty over time for various areas and population subgroups. Nonetheless, we also checked the robustness of the poverty profile to the choice of poverty cutoff.

Table 24.1 lists the dimensions and indicators, together with associated weights, included in the estimation of multidimensional poverty. Annex Tables 1.1–1.3 provide details on the definition of deprivation associated with each indicator. Again, because the deprivation indicators are not exactly comparable across the three data sources, caution needs to be exercised in comparing estimates between two data sources.

24.4 WHAT HAS BEEN HAPPENING TO POVERTY IN RECENT YEARS?

A common refrain in policy discussions is that poverty in the Philippines is high and that the economic growth in the past decade, albeit low by the standards of the country's East Asian neighbors, has largely bypassed the poor. Indeed income poverty estimates based on official assessments reveal a rather lack of response of the income poverty incidence to growth during the 2000s (World Bank, 2010; ADB, 2009; Balisacan, 2009, 2010). Alternative specifications of poverty lines, such as the "international norm" of USD 1.25 a day used by the World Bank, or the consistency-conforming poverty lines (Balisacan and Fuwa, 2004) for spatial and intertemporal poverty monitoring, report the same muted response of poverty to growth. A somewhat different picture, however, emerges when viewed from the lens of multidimensional poverty.

Table 24.2 summarizes the estimates of MPI, multidimensional headcount (*H*), and average deprivation intensity experienced by the poor (*A*), as well as the average annual rate of change of these indices for the period covered by the data. To compare these estimates with what is known about

Table 24.1 Dimensions and Indicators

	NDHS	FIES	APIS
Health			
Child mortality	x		
Water		x	x
Sanitation		x	x
Nutrition			
Food poverty	x	x	x
Education			
Years of schooling	x	x	x
Child school attendance	x		x
Potential schooling		x	
Standard of living			
Electricity	x	x	x
Shelter			
Flooring	x		
Roof		x	x
Wall		x	x
Mobility			
Access to motor vehicles	x	x	
Access to national roads		x	
Asset ownership			
Household assets	x	x	x
Transport			x
House tenure			x
Other sources of income		x	x

the profile of poverty in recent years, Figure 24.1 depicts the trends of multidimensional headcount estimates (hereafter referred to as H_m) and the official income-headcount estimates (H_y). All three data sources tend to show continued reduction in multidimensional poverty.[8] The annual rates of poverty reduction were 1.78% for FIES, 2.04% for APIS, and 2.17% for NDHS. The three data sources tend to show a deceleration of poverty reduction in the 2000s. Remarkably, the pattern of poverty is quite different when seen from the perspective of the official income headcount, which tends to show that the level of poverty was unaffected by the GDP growth since 1997. This difference is particularly evident in both FIES- and APIS-based estimates of multidimensional poverty, which show continued progress in poverty reduction in the 2000s.

Another noticeable pattern is that, in all three data series, both the proportion of the population experiencing multiple deprivations and the average intensity of deprivation (last two columns in Table 24.2) generally follow the movement of the MPI. However, the decline in the headcount is

---

[8]Note that the MPIs for before and since 2000 are not entirely comparable due to differing definitions of the "assets" variable, owing in turn to changes in the FIES questionnaire over time.

Table 24.2 MPI and Its Components: H_m and A

	Multidimensional Poverty Index (MPI)	Mutidimensional Poverty Headcount Ratio (H_m)	Average Intensity of Multidimensional Poverty (A)
NDHS			
1993	0.208	0.454	0.458
1998	0.164	0.360	0.457
2003	0.141	0.318	0.444
2008	0.137	0.306	0.449
Average annual percent change (%)	−2.257	−2.166	−0.136
FIES			
1988	0.267	0.529	0.505
1991	0.253	0.506	0.501
1994	0.234	0.471	0.496
1997	0.206	0.426	0.485
2000[a]	0.194	0.396	0.490
2003	0.189	0.386	0.490
2006	0.170	0.350	0.486
2009	0.148	0.312	0.475
2012	0.123	0.270	0.458
Average annual percent change (%)	−1.967	−1.780	−0.326
APIS			
1998	0.183	0.375	0.489
2002	0.154	0.319	0.483
2004	0.154	0.319	0.481
2007	0.140	0.301	0.467
2008	0.130	0.281	0.462
2010[b]	0.128	0.277	0.461
2011	0.124	0.275	0.451
Average annual percent change (%)	−2.475	−2.036	−0.597

[a]*Note: Before 2000 the FIES-generated MPI use a different definition of assets.*
[b]*APIS 2010 has only half of the sample size of other APIS datasets; may not be accurate below national level.*
Source: *Family Income and Expenditure Survey, Annual Poverty Indicators Survey, and National Demographic and Health Survey, various years.*

faster than that in the intensity of poverty, especially for FIES, suggesting that the reduction in MPI during the periods covered by the household surveys came largely from the reduction in the number of the poor simultaneously experiencing various deprivations.

As mentioned earlier, the MPI is dimensionally decomposable. That is, one can "unpack" the MPI to identify the relative contribution of each dimension to aggregate poverty. Table 24.3 provides the results of such decomposition for the three data series. Remarkably, the three data series

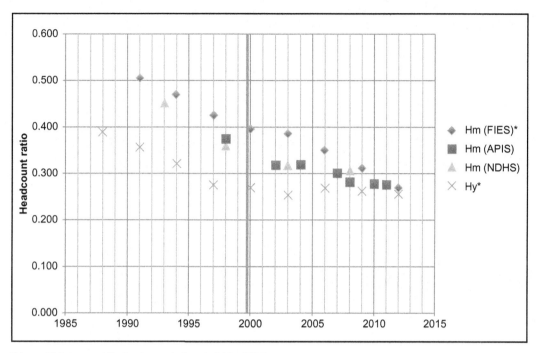

*Uses official per capita food poverty threshold for 2009.

 Note: MPI from FIES for 2000 and beyond is not strictly comparable to MPI before 2000 due to different definitions of "assets".

FIGURE 24.1

Multidimensional versus income poverty headcount ratio.

provide the same ranking of the three broad dimensions of poverty. Standard of living contributed the most to aggregate poverty, followed by health and education. The contribution of standard of living hovered around 52% for both NDHS and APIS and 43% for FIES. In the mid- to late 2000s, the contribution of health tended to decrease over time in two data series (NDHS and APIS), while that of education tended to rise, particularly in APIS. Thus it appears that while aggregate poverty, viewed from the lens of multidimensional deprivation, declined in recent years, there was much less progress in achieving universal basic education. This result is quite consistent with recent findings on the rather sorry state of the Philippine educational system (HDN, 2009), as well as on the low chances of achieving the MDG in universal primary education (NEDA-UN Country Team, 2010). However, latest data for 2011 and 2012 show that the share of education deprivation has decreased for two datasets, namely FIES and APIS. This consistent decline manifesting itself in two separate surveys may be an indication of (gradually) improving conditions on the education front. It remains to be seen if such improvements will carry on in even later data.

   Are the poverty profiles shown in Tables 24.2 and 24.3 and Figure 24.1 robust? As noted in Section 24.2, empirical measurement of multidimensional poverty is not immune to controversies (Ravallion, 2011; Alkire and Foster, 2011b; Lustig, 2011). For one, the choice of a suitable poverty

Table 24.3 Breakdown of MPI into Its Dimensions

	Dimension and Contribution to MPI (%)			Total
	Health	Education	Standard of Living	
NDHS				
1993	23.5	20.7	55.8	100.0
1998	24.3	19.6	56.1	100.0
2003	23.0	21.0	56.0	100.0
2008	23.0	24.2	52.8	100.0
FIES				
1988	38.5	21.4	40.1	100.0
1991	38.6	21.1	40.3	100.0
1994	38.0	21.7	40.3	100.0
1997	38.0	23.1	38.9	100.0
2000[a]	36.9	19.3	43.8	100.0
2003	36.3	19.9	43.8	100.0
2006	36.6	21.1	42.4	100.0
2009	36.5	21.9	41.6	100.0
2012	36.7	21.5	41.7	100.0
APIS				
1998	37.7	10.1	52.2	100.0
2002	34.0	16.3	49.6	100.0
2004	31.7	18.3	50.0	100.0
2007	32.9	18.6	48.5	100.0
2008	32.4	18.5	49.1	100.0
2010	33.1	18.6	48.3	100.0
2011	33.1	17.7	49.2	100.0

[a]Note: *Before 2000, the FIES-generated MPI use a different definition of assets.*
Source: *Family Income and Expenditure Survey, Annual Poverty Indicators Survey, and National Demographic and Health Survey, various years.*

cutoff is a judgment call. Even in the case of income poverty, deciding on the poverty line is the most controversial aspect of poverty comparison, at least in the Philippine context (Balisacan, 2003b, 2010). The issue has an important policy significance: outcomes of poverty comparison can influence policy decisions on addressing acute poverty (e.g., prioritizing resources to identified poverty groups). Thus, it is necessary to examine the robustness of poverty comparison.

Figure 24.2 summarizes the results of employing first-order dominance for the MPI.[9] A pair of nonintersecting lines representing two periods suggests that poverty comparison for these periods is

---

[9]As Alkire and Foster (2011a) show, dominance of the multidimensional headcount ensures M_0 dominance as well. Yalonetzky (2011) provides a more general dominance condition for the robustness of the multidimensional headcount to plausible values of not only the aggregate poverty cutoff but also the dimensional weights and deprivation cutoffs.

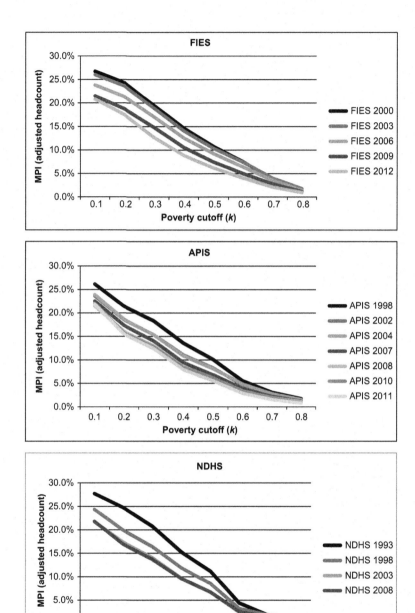

FIGURE 24.2

MPI dominance for FIES, APIS, and NDHS.

robust to all plausible poverty cutoffs. Remarkably, there are only a few pair cases where the direction of change in poverty is not robust: 2000 and 2003 for the case of FIES, 2002 and 2004 for APIS, and 2003 and 2008 for NDHS.[10] Thus, both FIES and APIS data point to an unambiguous decline in poverty between the early part and the latter part of the 2000s. For the NDHS data, a clear decline in poverty is seen when comparing the 1990s and 2000s, especially between 1998 and 2008. The change in poverty during the 2000s, i.e., between 2003 and 2008, is ambiguous. Note, however, that the comparison between the two periods (NDHS years) is problematic, since the latter year was punctuated by two major crises—the global food crisis that started in late 2007 and the global financial crisis that erupted in mid-2008. The impact of both crises on the economy and the poor was quite severe (Balisacan et al., 2010).

24.5 POVERTY PROFILE FROM THE LENS OF MPI

Much has been written about the correlates of poverty in the Philippines.[11] In fact, it has become a common practice to construct poverty profiles from the same national household surveys—particularly FIES—used in this study every time a new survey becomes available. The commonly generated profiles pertain to the incidence and distribution of poverty across geographic areas and economic sectors. For example, based on household income (or expenditure) data for recent years, Metro Manila has had the lowest headcount incidence out of the country's 17 regions, while the Autonomous Region of Muslim Mindanao, Bicol, Western Mindanao, and the Visayas, the highest. The profiles also suggest that, as in most of Asia's developing countries, poverty in the Philippines is a largely rural phenomenon. About two of every three income poor persons in the country are located in rural areas and are dependent predominantly on agricultural employment and incomes. Yet, studies also show that there are usually wide differences in income within geographic boundaries and sectors. Balisacan (2003a, 2009), for example, showed that overall income inequality at any point in time during the past two decades came mainly from differences *within* geographic boundaries and regions, not from differences in mean incomes *between* boundaries and regions.

Is the poverty profile seen from the lens of multidimensional poverty substantially different from what is already known about the distribution and relative magnitude of income poverty? Is the difference policy-relevant? In addressing these questions, we exploit the additive-decomposability property of MPI to examine not only the distribution of MPI-poor across geographic boundaries and sectors but also the composition of deprivations that differentially burden the poor. For ease of comparison, we focused on the census-augmented FIES data. This allowed a direct comparison of the MPI-based profiles with income-based poverty profiles, especially since the FIES survey instrument has remained largely unchanged since the first survey was undertaken in 1985.

Table 24.4 gives estimates of the income headcount, multidimensional headcount, and MPI, by geographic area (region and urbanity) and economic sector of employment of the household head, for 2012. A key result is that the poverty profiles are broadly similar across the three poverty measures. In both the income-headcount and multidimensional poverty measures, the concentration of

---

[10]If the "extremely" low and high values of poverty cutoffs cannot be ruled out, ambiguity also surrounds the change in NDHS-based poverty between 1993 and 1998.

[11]See, for example, Balisacan (2003a,b, 2009), Reyes (2010), ADB (2009), and World Bank (2010).

Table 24.4 Poverty Profile by Group (FIES, 2012). Note all figures in % except for population.

Subgroup	Population		Income Headcount		MPI Headcount		MPI (Adjusted Headcount)	
	Number (in millions)	Contrib.	Ratio	Contrib.	Ratio	Contrib.	Ratio	Contrib.
Sector								
Agriculture	31.01	30.9	48.40	58.24	51.11	58.57	25.73	62.71
Mining	0.61	0.6	31.14	0.74	33.20	0.75	15.28	0.73
Manufacturing	5.94	5.9	18.15	4.18	19.84	4.35	8.53	3.98
Utilities	0.48	0.5	7.62	0.14	10.37	0.19	3.94	0.15
Construction	7.62	7.6	24.71	7.31	26.76	7.54	11.25	6.74
Trade	11.08	11.0	13.20	5.68	14.71	6.02	6.11	5.32
Transportation & ICT	9.28	9.2	19.31	6.95	17.52	6.01	7.21	5.26
Finance	0.47	0.5	0.32	0.01	4.19	0.07	1.60	0.06
Services	15.77	15.7	11.16	6.83	12.61	7.35	5.35	6.64
Unemployed	18.14	18.1	14.11	9.93	13.66	9.15	5.90	8.41
	100.41	100.0		100.00		100.00		100.00
Urbanity								
Urban	44.89	44.71	12.65	22.03	14.23	23.60	6.11	21.56
Rural	55.52	55.29	36.19	77.97	37.24	76.40	17.98	78.44
	100.41	100.00		100.00		100.00		100.00
Region								
NCR	13.46	13.41	5.33	2.78	7.52	3.74	2.71	2.86
CAR	1.73	1.72	20.52	1.38	27.58	1.76	11.86	1.61
Ilocos Region	4.99	4.97	20.15	3.90	15.10	2.78	6.26	2.45
Cagayan Valley	3.44	3.43	17.62	2.35	20.08	2.55	8.64	2.34
Central Luzon	11.20	11.15	14.55	6.32	13.51	5.59	5.65	4.97
Calabarzon	14.04	13.98	12.43	6.77	13.92	7.22	5.64	6.22
Mimaropa	2.94	2.93	33.03	3.77	41.13	4.47	21.20	4.89
Bicol Region	5.77	5.74	44.15	9.88	40.48	8.63	19.03	8.62
Western Visayas	7.28	7.25	29.11	8.22	33.89	9.11	16.24	9.29
Central Visayas	7.55	7.52	35.71	10.46	33.11	9.24	15.76	9.35
Eastern Visayas	4.32	4.30	44.73	7.49	41.75	6.66	20.36	6.91
Zamboanga Peninsula	3.66	3.65	37.27	5.29	46.51	6.29	23.54	6.77
Northern Mindanao	4.67	4.65	39.38	7.13	34.82	6.01	16.58	6.08
Davao Region	4.95	4.93	29.87	5.73	32.03	5.86	15.76	6.13
Soccsksargen	4.69	4.67	43.88	7.99	43.51	7.55	22.52	8.30
ARMM	3.14	3.13	54.32	6.62	80.30	9.31	40.11	9.90
Caraga Region	2.59	2.58	38.91	3.91	33.80	3.24	16.11	3.28
	100.41	100.00		100.00		100.00		100.00

poverty is in agriculture (about 60% of the poor population), in rural areas (about 80%), and in the Central Visayas, Western Visayas, Eastern Visayas, Bicol, and Soccsksargen regions. Metro Manila, while accounting for about 13% of the country's population, contributes only about 3% to total poverty. The above results are, of course, not unexpected. In areas or sectors where the income poor are concentrated, it is likely that the same poor are also simultaneously deprived of social services, particularly health and education. Studies on the incidence of public spending on social services suggest that the benefits accrue disproportionately less to the income poor (Balisacan and Edillon, 2005; Canlas et al., 2009; World Bank, 2010). One explanation for this outcome has to do with the political economy of public provision of social services (HDN, 2009). The poor, even though numerically large, are not necessarily the more influential group in decisions concerning placements of public spending. The other explanation is that the designs of antipoverty programs are not incentive compatible—that is, the nonpoor individuals find that it is in their interest to preempt the benefits of these programs, while the poor do not (Balisacan, 2003a).

However, as also expected, because of the diversity of conditions (geography, local institutions, asset distribution, infrastructure, etc.) across the country's landscape, it is quite unlikely that multidimensional poverty measures would rank population groups in one-for-one correspondence with income poverty measures. Figure 24.3 shows the case for the country's 17 regions.

Despite regional variations in the headcount and average intensity of multidimensional poverty, our results also suggest a seeming *convergence* in these two components of MPI across all regions.

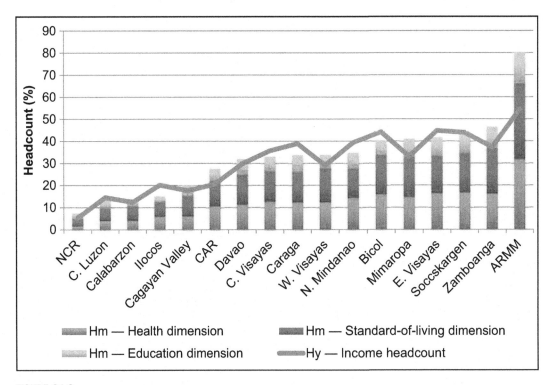

FIGURE 24.3

Multidimensional headcount (H_m) versus income headcount (H_y), by region.

Figure 24.4 shows a regional cross-section for H_m and A for the MPIs computed from FIES 2000 and FIES 2012. The size of the bubbles represents the absolute number of multidimensionally poor people. What Figure 24.4 shows is that there has been an improvement in MPI and its components across all regions (including ARMM), as depicted by a "clustering" of the region toward the graph's origin. However, ARMM stands out as the region that is getting left behind in this convergence. This only puts to light ARMM's continuing status as a sticking point in the country's thrust toward inclusive growth,

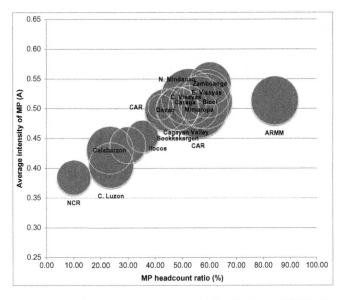

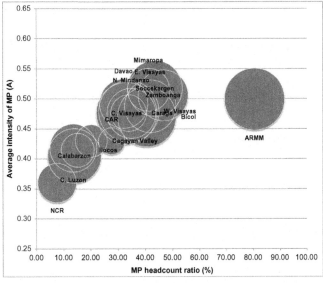

FIGURE 24.4

Intensity and headcount of multidimensional poverty by region.

and the importance of introducing reforms (e.g., through the newly established comprehensive Bangsamoro agreement) toward the alleviation of ARMM's multidimensional deprivations.

Table 24.5 provides the dimensional decomposition of the MPI for the same geographic areas and economic sectors. Although MPI varies remarkably across subpopulation groups (e.g., between Metro Manila and the Visayas regions, or between agriculture and manufacturing), there is surprisingly much less variation in the relative importance of each of the broad

Table 24.5 Poverty Profile by Group (FIES, 2012)			
Subgroup	**Dimensional decomposition of the MPI (in percent)**		
	Health	**Education**	**Standard of Living**
Sector			
Agriculture	38.38	19.93	41.69
Mining	30.76	31.18	38.06
Manufacturing	33.72	24.03	42.24
Utilities	27.88	32.07	40.04
Construction	32.86	23.93	43.21
Trade	33.96	24.74	41.30
Transportation & ICT	33.75	23.23	43.02
Finance	25.26	37.02	37.72
Services	33.40	25.49	41.11
Unemployed	35.74	23.18	41.07
Urbanity			
Urban	32.05	27.71	40.23
Rural	37.99	19.85	42.16
Region			
NCR	21.99	36.24	41.77
CAR	38.78	21.49	39.73
Ilocos Region	40.12	17.93	41.95
Cagayan Valley	30.94	25.28	43.78
Central Luzon	30.71	28.60	40.69
Calabarzon	30.83	25.64	43.53
Mimaropa	35.97	20.45	43.58
Bicol Region	39.21	17.12	43.68
Western Visayas	36.98	18.91	44.10
Central Visayas	38.48	20.90	40.61
Eastern Visayas	39.56	20.73	39.71
Zamboanga Peninsula	34.85	21.57	43.58
Northern Mindanao	41.31	21.30	37.39
Davao Region	35.61	23.26	41.13
Soccsksargen	38.41	21.00	40.60
ARMM	39.78	18.06	42.15
Caraga Region	36.95	23.70	39.35

dimensions of deprivation across these groups. This is particularly true for the relative importance of living standard, which stays within the 40−50% range of the total deprivations experienced simultaneously by the poor. It is in health and education where the geographic and sectoral differences matter. For example, it is surprising that basic education is less important as a source of multiple deprivations in agriculture than in finance and utilities. Similarly, education deprivation registers a greater importance in Metro Manila than in ARMM, Bicol, and Northern Mindanao. Note, however, that basic education services are generally of much lower quality in rural areas than in urban areas (HDN, 2010). The dimension indicators reported in the household surveys do not account for these differences.

24.6 CONCLUSION

A common refrain in policy discussions in the Philippines is that, until recently, economic growth, albeit low by the standards of the country's Southeast Asian neighbors, has largely bypassed the poor. Indeed, estimates of income poverty show that the proportion of the population deemed poor has remained largely unchanged, especially in the first decade of the current millennium, even as the economy grew at an annual average of 4.5% a year. The results of this study show that aggregate poverty, seen from the lens of multidimensional deprivation, actually declined as the economy expanded during the past decade. This finding is robust to assumptions about the poverty cutoff. From a policy perspective, the finding reinforces the view that nothing less than economic growth, even in the short term, is required to reduce poverty (broadly interpreted to include deprivations beyond income). At the same time, the diversity of both deprivation intensity and magnitude of poverty across geographic areas and sectors of the Philippine society is enormous, suggesting that, beyond growth, much needs to be done to make development more inclusive.

A multidimensional approach to poverty measurement holds promise for policy and poverty monitoring, especially given the scarcity of development finance and the government's thrust to speed up poverty reduction. A strong case, for example, can be made to prioritize poverty reduction efforts in areas or population groups with acute multiple deprivations. Getting good-quality education and health services accessible to the poor should be high in the development agenda. Investment in such services, as well as in institutions enhancing market efficiency and governance, creates favorable conditions not only for addressing other areas of human functionings (e.g., empowerment) but also for getting the country to move to a higher, sustained growth path.

However, research on multidimensional poverty and policy has to be further advanced. It would be useful, for example, to examine closely the pathways or channels by which economic growth is linked to reduction of multidimensional poverty and how the linkage is influenced by policy regimes and governance structures. The exercise is expected to deepen the understanding of what works—and what does not—in the war against acute poverty. It would be also useful to reexamine the targeting schemes employed in the government's poverty-reduction programs, including the *Pantawid Pamilyang Pilipino Program*, in light of the insights gained from the lens of multidimensional poverty.

ACKNOWLEDGMENTS

The author is grateful to Jan Carlo Punongbayan and Regina Salve Baroma for excellent research assistance; participants at the NEDA sa Makati forum, UPSE-PCED seminar, and UH Honolulu conference for useful comments; and PCED and UNDP for research support. The usual disclaimer applies.

REFERENCES

ADB [Asian Development Bank], 2009. Poverty in the Philippines: Causes, Constraints, and Opportunities. ADB, Manila.

Alkire, S., Foster, J., 2011a. Counting and multidimensional poverty measurement. J. Public Econ. 95, 476–487.

Alkire, S., Foster, J., 2011b. Understandings and misunderstandings of multidimensional poverty measurement. J. Econ. Inequal. 9 (2), 289–314.

Alkire, S., Santos, M.E., 2010a. Multidimensional Poverty Index. OPHI Research Brief. University of Oxford.

Alkire, S., Santos, M.E., 2010b. Acute Multidimensional Poverty: A New Index for Developing Countries. OPHI Working Paper No. 38, University of Oxford.

Atkinson, A.B., Cantillion, B., Marlier, E., Nolan, B., 2002. Social Indicators. The EU and Social Inclusion. Oxford University Press, Oxford.

Balisacan, A.M., 2003a. Poverty and inequality. In: Balisacan, A.M., Hill, H. (Eds.), The Philippine Economy: Development, Policies, and Challenges. Oxford University Press, New York, NY, pp. 311–341.

Balisacan, A.M., 2003b. Poverty comparison in the Philippines: is what we know about the poor robust? In: Edmonds, C.M. (Ed.), Reducing Poverty in Asia: Emerging Issues in Growth, Targeting, and Measurement. Edward Elgar, Cheltenham, UK, pp. 197–219.

Balisacan, A.M., 2009. Poverty reduction: trends, determinants, and policies. In: Canlas, D., Khan, M.E., Zhuang., J. (Eds.), Diagnosing the Philippine Economy: Toward Inclusive Growth. Anthem Press, London, pp. 261–294 (Asian Development Bank, Manila).

Balisacan, A.M., 2010. MDGs in the Philippines: setting the poverty scores right and achieving the targets. Philipp. Rev. Econ. 47 (2), 1–20.

Balisacan, A.M., Edillon, R., 2005. Poverty targeting in the Philippines. In: Weiss., J. (Ed.), Poverty Targeting in Asia. Edward Elgar, Cheltenham, UK, pp. 219–246.

Balisacan, A.M., Fuwa, N., 2004. Going beyond cross-country averages: growth, inequality, and poverty reduction in the Philippines. World Dev. 32 (11), 1891–1907.

Balisacan, A.M., Piza, S., Mapa, D., Santos, C.A., Odra, D., 2010. The Philippine economy and poverty during the global economic crisis. Philipp. Rev. Econ. 47 (1), 1–37.

Banerjee, A.V., Duflo, E., 2011. Poor Economics: A Radical Rethinking of the Way to Fight Global Poverty. Public Affairs, New York, NY.

Basu, K., Foster, J.E., 1998. On measuring literacy. Econ. J. 108 (451), 1733–1749.

Canlas, D.B., Khan, M.E., Zhuang, J. (Eds.), 2009. Diagnosing the Philippine Economy: Toward Inclusive Growth. Anthem Press, London, (Asian Development Bank, Manila).

Capuno, J.J., Tan, A.R., 2011. The Impact of Water Supply and Sanitation Facilities on Child Health in the Philippines. School of Economics, UP Diliman, Quezon City.

Deaton, A., 2010. Measuring Development: Different Data, Different Conclusions? Presentation by Angus Deaton of Princeton University in the 8th AFD/EUDC Conference: Measure for Measure, held in Paris on 1 December.

HDN [Human Development Network], 2009. Philippine Human Development Report 2008/2009. HDN, Quezon City.

HDN [Human Development Network], 2010. In search of a human face: 15 years of knowledge building for human development in the Philippines. HDN, Quezon City.

Kanbur, R., Venables, A.J., Wan, G. (Eds.), 2006. Spatial Disparities in Human Development: Perspectives from Asia. United Nations University Press, Tokyo and New York, NY.

Lustig, N., 2011. Multidimensional Indices of Achievements and Poverty: What Do We Gain and What Do We Lose? Working Paper 262. Center for Global Development, Washington, DC, Massachusetts, USA.

NEDA-UN Country Team, 2010. Philippines: Progress Report on the MDGs. NEDA, Pasig City.

Ravallion, M., 1994. Poverty Comparisons. Harwood Academic Publishers, Chur, Switzerland.

Ravallion, M., 2011. Mashup indices of development. World Bank Res. Obs. 1−32.

Reyes, C.M., 2010. The Poverty Fight: Has It Made an Impact? Philippine Institute for Development Studies, Makati City.

Sachs, J.D., 2005. The End of Poverty: Economic Possibilities for Our Time. Penguin Press, New York, NY.

Stiglitz, J.E., Sen, A., Fitoussi, J.-P., 2010. Mis-measuring Our Lives: Why GDP Doesn't Add Up. Report by the Commission on the Measurement of Economic Performance and Social Progress. New Press, New York, NY.

UNDP, 2010. Human Development Report 2010, The Real Wealth of Nations: Pathways to Human Development. Palgrave Macmillan, New York, NY.

World Bank, 2010. Philippines: Fostering More Inclusive Growth. Main Report. World Bank, Manila.

Yalonetzky, G., 2011. Conditions for the Most Robust Poverty Comparisons using the Alkire−Foster Family of Measures. OPHI Working Paper No. 44. University of Oxford.

ANNEX TABLES

Table A.1 Summary of FIES Indicators

Dimension	Indicator	Definition		Weight
Health	Child mortality			
	Sanitation	If household does not use flush toilet	**Type of Toilet:** (1) Closed pit; (2) Open Pit; (3) No toilet	1/12
	Drinking water	If household does not have access to safe drinking water	**Source of Water:** (1) Shared use, faucet, community water system; (2) Shared tube/piped well; (3) Dug well; (4) Spring, river, stream; (5) Rain; (6) Peddler	1/12
	Malnutrition			
	Food poverty	If household is food poor	Total income in 2009 prices is less than the 2009 food line	1/6
Education	Years of schooling	If no household member has completed six years of schooling		1/6
	Child potential schooling	If any school-aged child (7 to 16 years old) does not meet his/her education potential		1/6
Standard of living	Electricity	If household does not have electricity		1/15
	Shelter			
	Roof	If household's roof is composed of light/salvaged material	**Type of Roof:** (1) Light material; (2) Salvaged material; (3) Mixed but predominantly light material; (4) Mixed but predominantly salvaged material	1/30
	Wall	If household's wall is composed of light/salvaged material	**Type of Wall:** (1) Light material; (2) Salvaged material; (3) Mixed but predominantly light material; (4) Mixed but predominantly salvaged material	1/30
	Mobility			
	Ownership of vehicle	If household does not own a vehicle and is not accessible to a national highway		1/15
	Accessibility to national roads			
	Asset ownership			
	Household asset	If household does not own more than three household assets	**Household Assets:** (1) Radio; (2) Television; (3) Stereo; (4) Telephone/cellphone; (5) Refrigerator; (6) Aircon; (7) Washing machine; (8) Microcomputer/personal computer; (9) CD/DVD player; before 2000 the ff. assets were used in lieu of washing machine, CD/DVD player, and cellphone due to data availability constraints; (10) Dining set; (11) Sala set	1/15
	Other sources of income	If household's other sources of income are less than 20% of total income	**Other sources of income:** (1) Cash receipts, gifts from abroad; (2) Rentals received from non-agricultural lands/buildings; (3) Interest; (4) Pensions; (5) Dividends from investment; (6) Receipts from other sources not elsewhere classified	1/15

Table A.2 Summary of APIS Indicators

Dimension	Indicator	Definition		Weight
Health	Child mortality			
	Sanitation	If household does not use flush toilet	**Type of Toilet:** (1) Closed pit; (2) Open pit; (3) Drop/overhang; (4) No toilet	1/12
	Drinking water	If household does not have access to safe drinking water	**Source of Water:** (1) Unprotected well; (2) Developed spring; (3) Undeveloped spring; (4) River/stream; (5) Rainwater; (6) Tanker truck/peddler; (7) Others	1/12
	Malnutrition			
	Food poverty	If household is food poor	Total income in 2009 prices is less than the 2009 food line	1/6
Education	Years of schooling	If no household member has completed six years of schooling		1/6
	Child school attendance	If any school-aged child (7 to 16 years old) is out of school in years 1 to 10		1/6
	Electricity	If household does not have electricity		1/12
	Shelter			
	Roof	If household's roof is composed of light/salvaged material	**Type of Roof:** (1) Light material; (2) Salvaged material; (3) Mixed but predominantly light material; (4) Mixed but predominantly salvaged material	1/24
	Wall	If household's wall is composed of light/salvaged material	**Type of Wall:** (1) Light material; (2) Salvaged material; (3) Mixed but predominantly light material; (4) Mixed but predominantly salvaged material	1/24
	Asset ownership			
Standard of living	Household asset	If household does not own more than three household assets	**Household Assets:** (1) Radio; (2) Television; (3) Telephone; (4) Refrigerator; (5) Aircon; (6) Dining set; (7) Sala set; (8) Cellphone; (9) Gas Range; (10) Washing Machine	1/36
	Transport	If household does not own a vehicle		1/36
	House tenure	If household does not own a house and lot		1/36
	Other sources of income	If household's other sources of income are less than 20% of total income	**Other sources of income:** (1) Cash receipts, gifts from abroad; (2) Rentals received from non-agricultural lands/buildings; (3) Interest; (4) Pensions; (5) Dividends from investment; (6) Receipts from other sources not elsewhere classified	1/12

Table A.3 Summary of NDHS Indicators

Dimension	Indicator	Definition		Weight
Health	Child mortality	If any child has died in the family		1/6
	Sanitation	If household does not use flush toilet	**Type of Toilet:** (1) Closed pit; (2) Open pit; (3) Drop/overhang; (4) No toilet	1/12
	Drinking water	If household does not have access to safe drinking water	**Source of Water:** (1) Unprotected well; (2) Unprotected spring; (3) River/dam/lake; (4) Rainwater; (5) Tanker truck; (5) Cart with small tank; (6) Neighbor's tap	1/12
Education	Years of schooling	If no household member has completed six years of schooling		1/6
	Child school attendance	If any school-aged child (7 to 16 years old) is out of school in years 1 to 10		1/6
Standard of living	Electricity	If household does not have electricity		1/12
	Shelter Flooring	If the floor of shelter is dirt, sand, or dung	**Type of Floor:** (1) Earth; (2) Sand	1/12
	Mobility Ownership of vehicle	If household does not own a vehicle		1/12
	Asset ownership Household asset	If household does not own more than three household assets	**Household Assets:** (1) Radio; (2) Television; (3) Telephone; (4) Refrigerator	1/12

POVERTY REDUCTION AND THE COMPOSITION OF GROWTH IN THE MEKONG ECONOMIES

25

Peter Warr

Australian National University, Canberra, Australia

CHAPTER OUTLINE

JEL: I32; O53; F63

25.1 INTRODUCTION

Students of economic development are realizing that the quality—not just the rate—of growth is important. But what is "quality" growth? One criterion is its inclusiveness, particularly as regards the poor. What kinds of growth are most (and least) beneficial for the poor? Much of the development economics literature deals with how income distribution is affected by the rate and composition of the expansion of economic output. How do relative inequality, on the one hand, and absolute poverty, on the other, change with economic growth and how are these effects influenced by the characteristics of that growth, such as its sectoral composition? This chapter explores these issues in the context of the Mekong region.

The Mekong region is defined here as the seven entities bordering the Mekong River, consisting of five countries—Cambodia, the Lao People's Democratic Republic (Laos, for brevity), Myanmar (Burma), Thailand, and Vietnam—and two provinces of the People's Republic of China—Yunnan Province and Guangxi Zhuang Autonomous Region. For convenience, these seven entities will subsequently be referred to as the Mekong economies.

Despite substantial economic progress as a whole, most of the Mekong region remains poor relative to the rest of the world. Poverty reduction is consequently a priority for international attention. Most of the Mekong economies have experienced significant growth of economic output per person and reduction in measured poverty incidence over recent years but with varying rates both over time and across countries. Of interest in this study is the causal relationship between the rate of poverty reduction and the rate and nature of economic growth.

After describing the data on these phenomena for the Mekong economies, the study asks two statistical questions:

1. To what extent was the rate of poverty reduction that was achieved determined by the rate of growth of aggregate economic output per person?
2. Did the poverty reduction rate depend on the sectoral composition of the growth of output as well as the overall rate?

The answers to these questions are important in understanding the causes of poverty reduction within the region. Although it is now well established that economic growth is normally associated with poverty reduction (Dollar and Kraay, 2002), the manner in which the nature of the growth influences its poverty-reducing power is less well understood.

Section 25.2 reviews the aggregate output growth in each of the seven Mekong economies from 2000 to 2010. Section 25.3 performs a similar exercise for poverty reduction. The discussion makes it possible to compare the seven economies in terms of both the rate and poverty-reducing power of the economic growth. Section 25.4 looks in detail at the methods used to answer the statistical questions posed above, Section 25.5 summarizes the results and Section 25.6 concludes.

25.2 ECONOMIC GROWTH

In 2010, the levels of gross domestic product (GDP) per person among the Mekong economies varied widely, from USD 5000 in Thailand to less than USD 1000 in Cambodia, Laos, and Myanmar (Table 25.1). The Asian financial crisis of 1997–98 affected all of the Mekong economies but to

Table 25.1 Economic Output and Its Growth in the Mekong Economies

Mekong Economy	Population (million)	Population Growth Rate (2010)	GDP Per Capita (Current USD)	Average Annual Growth Rate of Real GDP Per Capita (%)	
	(2010)	(2010)	(2010)	Value	(period)
Cambodia	14.30	1.5	788.0	10.97	(2002–2010)
Guangxi	51.59	0.8	3050.0	10.77	(2000–2009)
Laos	6.23	1.7	984.2	6.87	(2003–2008)
Myanmar	59.78	1.1	702.0	3.60	(2005–2010)
Thailand	67.31	0.6	4991.5	3.59	(2000–2010)
Vietnam	86.48	1.1	1172.0	5.56	(2000–2010)
Yunnan	45.97	0.7	2326.3	8.88	(2000–2009)

GDP, gross domestic product; Laos, Lao People's Democratic Republic.
Source: Council for the Development of Cambodia, 2011. Data for Myanmar are from ADB Outlook 2006, 2010, and 2011.

widely varying degrees. Thailand was most affected, Guangxi and Yunnan the least. In the countries most affected, restoration of economic growth became a policy priority. Thailand's growth recovered but remained sluggish throughout the decade beginning in 2000, only to be battered again by the global financial crisis at the end of the decade. The level of private investment did not recover to its share of GDP before the 1997–98 financial crisis. Among the poorer countries, Cambodia, Laos, and Vietnam grew well throughout this period. In the case of Myanmar, the level of GDP per person and its growth rate are contentious. Myanmar's data in Table 25.1 are based on Asian Development Bank estimates.

25.3 POVERTY REDUCTION

Available data on poverty incidence in the seven Mekong economies are summarized in Table 25.2. Where available, the data are presented as aggregate poverty incidence and its rural and urban components.[1] The poverty lines underlying these data, which are the official national poverty lines for the Mekong economies, are held constant in real purchasing power over time. For this kind of study, there is no realistic alternative to the national poverty lines. Poverty incidence estimates using poverty lines of USD 1.25 per day and USD 2 per day are available for some but not all of the Mekong economies; at most, only a few data points are available for any one country, resulting in an insufficient number of data points in total to support statistical analysis. Many more time-series observations are available when national poverty lines are used, but in some cases, the official data have involved changes over time in the real purchasing power of the poverty line, without adjustment for the effects of these changes. This renders the poverty incidence

---

[1]Data on the decomposition of aggregate poverty incidence into its rural and urban components are available for all the Mekong economies, except Yunnan and Guangxi.

Table 25.2 Poverty Incidence and Poverty Reduction in the Mekong Economies

Mekong Economy	Level of Total Poverty Incidence		Levels of Rural and Urban Poverty Incidence			Average Annual Rate of Total Poverty Reduction	
	Value (%)	(Year)	Rural (%)	Urban (%)	(Year)	Value (%)	(Period)
Cambodia	30.1	(2007)	34.5	11.8	(2007)	1.55	(2004–2010)
Guangxi	41.7	(2007)	n.a.	n.a.	n.a.	0.43	(2001–2007)
Laos	27.6	(2008)	31.7	17.4	(2008)	1.18	(2003–2008)
Myanmar	25.6	(2010)	29.2	15.7	(2010)	1.30	(2005–2010)
Thailand	8.1	(2009)	10.4	3.0	(2009)	1.43	(2000–2009)
Vietnam	14.2	(2010)	17.4	6.9	(2010)	0.65	(2004–2010)
Yunnan	7.0	(2006)	n.a.	n.a.	n.a.	0.70	(2002–2006)

n.a., not available.
Sources: ADB Outlook, various years; United Nations, Millennium Development Goals Indicators, http://unstats.un.org/unsd/mdg/ Data.aspx; Royal Government of Cambodia, 2010; Government of Vietnam, 2010. Data on Myanmar are from UNDP, 2010, Integrated Household Living Condition Survey in Myanmar, 2009–10.

Table 25.3 Poverty Reduction Per Unit of Growth in the Mekong Economies

Mekong Economy	Poverty Reduction Per Unit of Real GDP Growth Per Capita
Cambodia	0.14
Guangxi	0.04
Laos	0.17
Myanmar	0.36
Thailand	0.40
Vietnam	0.17
Yunnan	0.08

Source: Author's calculations from the second to last columns of Tables 25.1 and 25.2.

data noncomparable over time; observations subject to this problem were omitted from the data presented below.

To what extent does economic growth actually reduce poverty? The most direct way of addressing this question is simply to relate the annual rate of poverty reduction (Table 25.2) to the rate of real economic growth per person over the same period. The results, shown in Table 25.3, indicate the average annual reduction in poverty incidence (dP) divided by the average rate of real GDP growth per person. This ratio varied widely across the seven Mekong economies. However, while these calculations are of interest, no causal relationship between poverty reduction and growth can be inferred from these simple ratios.

Table 25.4 Decomposition of Changes in Poverty Incidence, Five Mekong Countries, 2000–2010

	Cambodia	Laos	Myanmar	Thailand	Vietnam
Actual					
Total[a]	−1.467	−1.180	−1.300	−1.433	−0.650
Urban[b]	−0.324	−0.132	−0.309	−0.193	−0.086
Rural[c]	−0.916	−0.842	−0.968	−1.233	−0.440
Migration[d]	−0.227	−0.206	−0.023	−0.007	−0.123
Normalized (total = 100)					
Total[a]	100	100	100	100	100
Urban[b]	22.07	11.18	23.79	13.49	13.28
Rural[c]	62.44	71.33	74.46	86.02	67.74
Migration[d]	15.49	17.49	1.74	0.49	18.97

Notes:
Total change = urban change + rural change + migration.
[a]*Mean annual value of dP, the y-o-y change in national poverty incidence.*
[b]*Mean annual value of $a^U dP^U$, the y-o-y population share-weighted change in urban poverty incidence.*
[c]*Mean annual value of $a^R dP^R$, the y-o-y population share-weighted change in rural poverty incidence.*
[d]*Mean annual value of $(P^R - P^U) da^R$, the y-o-y migration-induced change in poverty incidence.*
Source: Author's calculations using data described in the text.

Data on total, rural, and urban poverty incidence can be used to derive a useful decomposition that indicates the contribution to reduction in total poverty incidence that is represented by poverty reduction in rural areas, urban areas, or the migration of people between the two. The quantitative relationship between total, rural, and urban poverty incidence will be reviewed first, and then the manner in which each of these measures is affected by economic growth will be discussed.

In Section A of the Appendix, Eq. (A.2) shows that the change in total poverty incidence over some period can be decomposed into three parts: (i) the change in rural poverty incidence, weighted by the rural population share; (ii) the change in urban poverty incidence, weighted by the urban population share; and (iii) the movement of people from rural to urban areas, weighted by the difference in poverty incidence between these two areas (Anand and Kanbur, 1985).

Table 25.4 shows the results of this decomposition for the Mekong economies, except for Guangxi and Yunnan, for which data on rural and urban poverty incidence were unavailable. All results shown in this table were evaluated at the mean values of the data set. For example, the mean annual change in the aggregate level of poverty incidence for Cambodia was −1.467%.

Equation (A.2) is an identity and must apply at all points in the data set. It must therefore apply at the means of the data. That is, the average annual change in total poverty incidence can be decomposed into terms capturing average poverty reduction in urban areas, average poverty reduction in rural areas, and the average movement of population between the two. The lower half of the table normalizes the decomposition by dividing all values by this mean change in aggregate poverty (− 1.467 for Cambodia, for example) and multiplying by 100. For Cambodia, reductions in rural poverty accounted for 62% of the overall reduction in poverty over the period shown in Table 25.2, reduced urban poverty accounted for 22%, and migration accounted for 15%. Migration effects were more important for Vietnam and Laos; they were least important for Thailand and Myanmar.

For these five Mekong economies, reductions in rural poverty accounted for more than 60% of the total poverty reduction that occurred; for all, except Cambodia, the proportion was more than two-thirds. Urban poverty reduction and the movement of people from rural to urban areas also played important roles, but poverty reduction within rural areas was the dominant source of aggregate poverty reduction.

The above calculations are merely descriptions of the data. The question of what caused these observed changes in poverty incidence is discussed in the following section.

25.4 THE GROWTH–POVERTY NEXUS

25.4.1 CONCEPTUAL BACKGROUND

Poverty incidence and changes in it over time depend on many factors. Economic variables are only part of the story. Among the economic variables, many issues are relevant aside from simply the overall growth rate of output. Changes in commodity prices play a role, along with tax and public expenditure policies. The policy discussion in almost all countries is heavily concerned with economic growth and its sectoral composition. Policymakers rightly wish to know the extent to which growth is inclusive. In particular, they wish to know whether or not a policy focused on growth is consistent with the goal of reducing poverty. Does economic growth cause poverty reduction? If so, to what extent? Does that growth's sectoral composition matter as far as its poverty-reducing effects are concerned?

Three methodological approaches to addressing these questions can be found in the literature. Each is problematic, but in different ways. One approach—general equilibrium modeling—has the advantage of permitting controlled experiments, changing one exogenous variable at a time and observing the effects on all endogenous variables of interest. This approach permits detailed analysis of the economic mechanisms through which output growth operates on poverty incidence (Fane and Warr, 2002). Since the distinction between exogenous and endogenous variables within the models is unambiguous, by construction, there are clear *causal* relationships between changes in exogenous variables and their effects on endogenous ones.

The objection to these models is that these causal relationships are predetermined in that they are built into the structure of the models themselves. For example, the qualitative result that economic growth reduces poverty is a direct consequence of the structural assumptions of the models. Moreover, the quantitative results are a function of the chosen values of the large numbers of behavioral parameters necessarily used within the models. Although the key parameters can be varied to show their implications for the relationships of interest, the fact remains that the true values of these parameters are largely unknown.

A second approach relies solely on household survey data to construct the distribution of real expenditures across households and to analyze the relationships among variables calculated from this distribution. For example, this approach computes the relationship between changes in the mean of the expenditure data and changes in the level of poverty incidence, calculated from the same distribution. A change in the mean of real household expenditures is taken to be a measure of (or proxy for) economic growth. These analyses have the advantage of working with internally consistent data and can provide a description of relationships among various features of the expenditure distribution.

The problem is that these statistical calculations lack a *causal* basis. They do not rest on any structural foundation that provides a basis for inferring causality. They merely describe the data. It is no more true to say that changes in the mean of the distribution of household expenditures causes changes in poverty incidence than to say the reverse. Policy application, however, requires knowledge of causal relationships. Statistical descriptions of pseudo-relationships among endogenous variables constructed from the distribution of household expenditures do not provide that. In any case, the rate of change of mean household expenditures is not the same thing as economic growth—meaning the expansion of economic output. The link between output growth and poverty incidence is the causal relationship of interest in this study. Moreover, the study wishes to ask how the sectoral composition of the output growth might influence the rate at which poverty incidence declines. That is clearly impossible with the second approach.

A third approach assembles statistical data on changes in poverty incidence and on output growth and its composition. It then regresses the former on the latter. That separate data sets are involved— the national accounts for economic growth and household survey data for poverty incidence—has advantages and disadvantages. Suppose it is found that economic growth leads to reductions in poverty incidence. It could not be said that this finding was a mere statistical artifact arising from the use of the same flawed data set to measure both variables, because quite different data sets are used to measure the two sets of variables. However, this also creates problems. The frequency of the data may not be the same. For instance, national accounts data are available at least annually, while poverty incidence data are usually available only at intervals of 2–3 years at least. The advantage of this approach is that it focuses directly on a causal relationship of strong policy interest (Ravallion and Datt, 1996). Despite its limitations, this third approach is used in the present study.

The economic development literature emphasizes the sectoral composition of growth in relation to its implications for poverty reduction, but this emphasis has been based primarily on *a priori* theorizing, rather than evidence. The obvious argument is that in most poor countries a majority of the poor live in rural areas, many of them employed in agriculture. Thus, it seems probable that growth of agriculture is more important for poverty reduction than growth of industry or services.

This conclusion does not necessarily follow, however. Rural areas are not synonymous with agriculture. Many poor people in rural areas derive incomes from petty trading and other service activities other than direct participation in agricultural production. Furthermore, sectoral growth rates may not be independent. Expansion of capacity in one sector—say, food processing—may stimulate output growth elsewhere—say, fruit production. More importantly, people are potentially mobile; given sufficient time, even poor people can presumably move to whichever sector is generating the growth. Rural poverty may therefore be reduced by urban-based growth, drawing the poor away from rural areas (Fields, 1980; Chenery and Syrquin, 1986). When sectoral interdependence and intersectoral factor mobility are considered, it is not obvious whether the sectoral composition of growth is important for poverty reduction.

Even if labors were fully and instantaneously mobile, poverty incidence could still be affected by the sectoral composition of growth. To a first order of approximation, the level of absolute poverty presumably depends on the demand for the factors of production owned by the poor, especially unskilled labor, and, to a lesser extent, agricultural land. Growth in different sectors has differential effects on the demands for these factors, depending on these sectors' factor intensities, and may have different effects on poverty, inequality, or both. Finally, the distinction rural/urban is not synonymous with agriculture/nonagriculture. Much agricultural production may occur in full or part-time farming on the fringes of urban areas and much industrial and services activity may occur in rural areas.

Only careful quantitative analysis can resolve questions of this kind, but the limited availability of data that can support statistical analysis has impeded the systematic study of poverty incidence and its determinants. Some recent studies have attempted to explore the relationships involved by analyzing cross-sectional data sets across countries, or across regions or across households for individual countries; others have attempted to assemble long-term time-series data sets on poverty incidence for individual countries. The time-series approach is generally preferable because it enables a direct study of the determinants of change in poverty at an aggregate level.

Unfortunately, the consumer expenditure surveys on which studies of poverty incidence must be based are conducted only intermittently in most developing countries. Data are thus available with intervals of several years between observations. This is true of all the Mekong economies. The data are most extensive for Thailand, but even when all national time-series observations on poverty incidence are assembled for Thailand, the number of observations is only 7, as it is for Cambodia. For Vietnam, it is 4, for Laos, 3, and for Myanmar, Guanxi, and Yunnan, 2. These numbers of observations are insufficient to sustain formal statistical analysis for any one of these economies. When all Mekong economies are pooled, the total number of observations becomes adequate. That is the approach used by the present study. The economic growth data for Myanmar are a subject of considerable debate and the empirical exercise uses Asian Development Bank estimates.

In each economy, the value of the national poverty line is held constant over time. However, since the meaning of the poverty lines is different in each of the countries, it should not be assumed that the same quantitative relationship between poverty incidence and aggregate growth would exist in all Mekong economies. In the present study, intercept dummy variables were used for six of the seven Mekong economies. It must be recognized that the use of dummy variables is an imperfect way of capturing the possible effects of different national poverty lines. The strong assumption being made is that the underlying relationship between changes in poverty incidence (dependent variable) and the growth rate (independent variable) is linear and has the same slope in all countries, differing only in the intercept terms.

Data on changes in poverty incidence between the data points indicated in Table 25.2 and similar data for the 1990s are used to construct the values of the dependent variables described below, with the calculated value on the change in poverty incidence divided by the number of years corresponding to that time interval. This gives an annual rate of change for the variable concerned. These annualized rates of change then become the variables used in the regression analysis described below.

25.4.2 POVERTY AND AGGREGATE GROWTH

For simplicity in determining the manner in which poverty incidence is affected by economic growth, the simplest possible relationship between these variables is hypothesized initially

(Ravallion and Datt, 1996), where the total number of households in poverty depends on the aggregate level of real income and population size. Section B of the Appendix derives Eq. (A.6), the implied relationship between changes in poverty and aggregate per capita growth. This equation can be used to test the statistical significance of the implied relationship between poverty reduction and aggregate per capita growth.

25.4.3 POVERTY AND SECTORAL GROWTH

Whether the sectoral composition of economic growth affects poverty reduction can be investigated as follows. The level of real GDP per person is given by the sum of value added in each sector. Section C of the Appendix derives Eq. (A.9), the implied relationship between changes in poverty and the sectoral components of aggregate per capita growth. This equation can be used to test the statistical significance of the implied relationship between poverty reduction and each sectoral component of aggregate per capita growth.

25.5 ESTIMATION RESULTS

25.5.1 POVERTY AND AGGREGATE GROWTH

Equation (A.6) was estimated as described above and the results are summarized in Table 25.5. Dummy variables were estimated for all countries, except Thailand. All country dummy variables were insignificant, except Guangxi and Yunnan. The insignificant dummy variables were all dropped and the equation was re-estimated in line with the general-to-specific approach of Hendry (1995). The estimated relationship is significant and the Ramsey RESET test suggests that it has no omitted variables. With low degrees of freedom, however, this test is relatively weak.

Table 25.5 Regression Results—Poverty and Aggregate Growth

Dependent Variable	Change in Total Poverty (dP)					
Independent Variable	Coefficient	Standard error	t-Statistic	$p > \lvert t \rvert$	95% Confidence interval	
GDP growth per capita	−0.274	0.092	−2.960	0.007	−0.467	−0.082
Intercept dummy: Guangxi	2.991	1.595	1.885	0.074	−0.316	6.299
Intercept dummy: Yunnan	2.098	1.132	1.856	0.077	−0.250	4.447
Number of observations	27					
$F(3,22)$	3.502					
Prob $>F$	0.032					
R-squared	0.322					
Adjusted R-squared	0.230					

Source: Author's calculations using data described in the text.

The Breusch-Pagan/Cook Weisberg test indicates the absence of heteroskedasticity. Nevertheless, the adjusted R-squared of 0.23 is low, suggesting that other variables may also be contributing to the behavior of the dependent variable.

25.5.2 POVERTY AND SECTORAL GROWTH

Equation (A.9) was estimated to capture the behavior of the dependent variable when the sectoral composition of growth appears on the right-hand side of the equation. The results, though suffering from the low number of observations, support the notion that growth of services in the Mekong economies is more poverty reducing than growth of either agriculture or industry. All country inter- cept dummy variables were highly insignificant, except Vietnam, and were dropped from the esti- mated equation to preserve degrees of freedom, again in line with the general-to-specific approach of Hendry (1995). The expected result that agricultural growth is the strongest contributor to pov- erty reduction was not obtained. Instead, services growth was the only GDP component that was significantly associated with poverty reduction. An F-test of the hypothesis that the coefficients on share-weighted sectoral growth rates per capita were all equal ($b_a = b_i = b_s$) was rejected at the 95% level of confidence. In short, the Greater Mekong Subregion (GMS) data indicate that growth of services is more important for poverty reduction than that of either industry or agriculture (Table 25.6).

Table 25.6 Regression Results—Poverty and Sectoral Growth

Dependent Variable	Change in Total Poverty Incidence					
Independent Variable	**Coefficient**	**Standard Error**	**t-Statistic**	**$p > \lvert t \rvert$**	**95% Confidence Interval**	
Constant	−1.214	0.770	−1.58	0.133	−2.838	0.410
Growth of agriculture GDP per capita	0.967	0.705	1.37	0.188	−0.520	2.455
Growth of industry GDP per capita	0.344	0.528	0.65	0.524	−0.770	1.457
Growth of services GDP per capita	−0.653	0.299	−2.19	0.043	−1.284	−0.023
Intercept dummy Vietnam	−1.214	0.836	−1.45	0.165	−2.977	0.549
Number of observations	27					
$F(4,22)$	1.88					
Prob $>F$	0.041					
R-squared	0.427					
Adjusted R-squared	0.292					

Source: Author's calculations using data described in the text.

25.6 **CONCLUSION**

This chapter combines time-series and cross-country data for seven Mekong economies to examine the causes of poverty reduction, particularly the role of aggregate economic growth. The results confirm that poverty reduction in the Mekong economies is strongly related to growth of real GDP per person. This finding supports earlier literature, such as Dollar and Kraay (2002), who demonstrate a similar relationship using cross-country (but not time-series) data for many countries. Moreover, it was found that the sectoral composition of this growth affects the rate of poverty reduction in so far as it affects the growth of services relative to the rest of the economy. Services growth is an important source of poverty reduction in both rural and urban areas. Structural changes and public investments that promote the growth of services are conducive to poverty reduction, given the overall rate of GDP growth. That is, expansion of aggregate output based on services contributes to the inclusiveness of the growth generated.

The contribution of that economic growth to poverty reduction rests on its contribution to the real incomes of poor people. This contribution depends, in turn, on the degree to which the growth raises the rate of return to the poor's assets. Overwhelmingly, this means unskilled labor. By raising the return to unskilled labor, economic growth can directly benefit the poor. Growth of value added in the most labor-intensive sectors may therefore be the most poverty reducing because growth of these sectors will increase the demand for unskilled labor the most. The services sector in the Mekong economies is apparently the most labor intensive, hence the most poverty reducing, even more so than agriculture.

Although industrial development is commonly assumed to be the key to overall economic development and poverty reduction, the results of this study do not confirm this assumption. Growth of services is far more important. Policies that overly emphasize growth of industry, rather than services, may impede rather than promote poverty reduction. As better data become available, this important set of policy issues needs to be explored further.

ACKNOWLEDGMENTS

The research assistance of Razib Tuhin, Ramesh Paudel, and Dung Doan and the helpful comments of an editor are gratefully acknowledged.

REFERENCES

Anand, S., Kanbur, S.M.R., 1985. Poverty under the Kuznets process. Econ. J. 95, 42–50.

Asian Development Bank, 2006, 2010, 2011. ADB Outlook. Asian Development Bank, Manila.

Chenery, H.B., Syrquin, M., 1986. Typical patterns of transformation. In: Chenery, H.B., Robinson, S., Syrquin, M. (Eds.), Industrialization and Growth. Oxford University Press, New York, NY, pp. 86–115.

Council for the Development of Cambodia, 2011. Why invest in Cambodia?. Statistical Communique on National Economic and Social Development of Guangxi; Statistical Communique on National Economic and Social Development of Yunnan.

Dollar, D., Kraay, A., 2002. Growth is good for the poor. J. Econ. Growth 7, 195–225.

Fane, G., Warr, P., 2002. How economic growth reduces poverty: a general equilibrium analysis for Indonesia. In: Shorrocks, A., Van der Hoeven, R. (Eds.), Perspectives on Growth and Poverty. United Nations University Press, Tokyo, pp. 217−234.

Fields, G.S., 1980. Poverty, Inequality and Development. Cambridge University Press, Cambridge, UK.

Government of Vietnam, 2010. Vietnam Statistical Yearbook, Hanoi.

Hendry, D.F., 1995. Dynamic Econometrics. Oxford University Press, Oxford, UK.

Ravallion, M., Datt, G., 1996. How important to India's poor is the sectoral composition of economic growth? World Bank Econ. Rev. 10, 1−25.

Royal Government of Cambodia, 2010. Cambodia Economic Statistics, Phnom Penh.

United Nations Development Program, 2010. Integrated Household Living Conditions Survey in Myanmar, 2009−10. United Nations Development Program, Yangon.

Warr, P., Wang, W., 1999. Poverty, inequality and economic growth in Taiwan. In: Ranis, G., Hu, S. (Eds.), The Political Economy of Development in Taiwan: Essays in Memory of John C. H. Fei. Edward Elgar, London, pp. 133−165.

APPENDIX

A DECOMPOSING CHANGES IN POVERTY INCIDENCE

Changes in total poverty incidence may be decomposed as follows (Warr and Wang, 1999): N, N^R, and N^U are the total, rural, and urban populations, respectively, where $N = N^R + N^U$, $\alpha^R = N^R/N$, and $\alpha^U = N^U/N$ are the rural and urban shares of the total population, respectively, where $\alpha^R + \alpha^U = 1$. The total number of people in poverty is given by $N_P = N_P^R + N_P^U$, where N_P^R and N_P^U denote the number in poverty in rural and urban areas, respectively. Total poverty incidence is given by

$$P = N_P/N = (N_P^R + N_P^U)/N = \alpha^R P^R + \alpha^U P^U \tag{A.1}$$

where $P^R = N_P^R/N^R$ denotes the proportion of the rural population that is in poverty and $P^U = N_P^U/N^U$ the corresponding incidence of poverty in urban areas.

Differentiating Eq. (A.1) totally provides a key relationship

$$dP = \alpha^R dP^R + \alpha^U dP^U + (P^R - P^U)d\alpha^R. \tag{A.2}$$

From Eq. (A.2), the change in total poverty incidence may be decomposed into three parts: (i) the change in rural poverty incidence, weighted by the rural population share; (ii) the change in urban poverty incidence, weighted by the urban population share; and (iii) the movement of people from rural to urban areas, weighted by the difference in poverty incidence between these two areas.

B POVERTY AND AGGREGATE GROWTH

Assume that the total number of households in poverty, N_P, depends on the aggregate level of real income, Y, and the size of the population, N. Thus

$$N_P = \varphi(Y, N) \tag{A.3}$$

The incidence of poverty is defined as

$$P = N_P/N = \varphi(Y, N)/N \tag{A.4}$$

Totally differentiating this equation

$$dP = (\varphi_Y Y/N)y + (\varphi_N - \varphi/N)n \tag{A.5}$$

where lowercase Roman letters represent the proportional changes of variables represented in levels by uppercase Roman letters. Thus $y = dY/Y$ and $n = dN/N$ are the growth rates of aggregate real income and of population, respectively. In the special case where the function $\varphi(.)$ is homogeneous of degree one in Y and N, Eq. (A.3) may be written as $N_P = \varphi_Y Y + \varphi_N N$ and Eq. (A.5) reduces to $dP = (\varphi_Y Y/N)(y - n)$.

In this case, we estimate relationships of the kind

$$dP = a + b(y - n) = a + b\tilde{y} \tag{A.6}$$

where $\tilde{y} = y - n$ denotes the rate of real GDP growth per person and test whether the coefficient b is significantly different from zero.

C POVERTY AND SECTORAL GROWTH

The level of real GDP per person is given by

$$\tilde{Y} = \tilde{Y}_a + \tilde{Y}_i + \tilde{Y}_s \tag{A.7}$$

where $\tilde{Y}_a$, $\tilde{Y}_i$, and $\tilde{Y}_s$ denote value added (contribution to GDP) per person in the total population, measured at constant prices, in agriculture, industry, and services, respectively. The overall real rate of growth per person can be decomposed into its sectoral components from

$$\tilde{y} = H_a\tilde{y}_a + H_i\tilde{y}_i + H_s\tilde{y}_s \tag{A.8}$$

where $H_k = Y_k/Y$, $k = (a, i, s)$, denotes the share of sector k in GDP. By estimating the equation

$$dP = a + b_aH_a\tilde{y}_a + b_iH_i\tilde{y}_i + b_sH_s\tilde{y}_s \tag{A.9}$$

and testing whether $b_a = b_i = b_s$, we may test directly whether the sectoral composition of growth affects the rate of poverty reduction.

Index

483

Printed in the United States
By Bookmasters